Advances in Experimental Medicine and Biology

Volume 1059

More information about this series at http://www.springer.com/series/5584

J. Miguel Oliveira • Sandra Pina • Rui L. Reis
Julio San Roman

Editors

Osteochondral Tissue Engineering

Challenges, Current Strategies, and Technological Advances

 Springer

Editors
J. Miguel Oliveira
3B's Research Group – Biomaterials,
Biodegradables and Biomimetics
University of Minho
Headquarters of the European Institute
of Excellence on Tissue Engineering
and Regenerative Medicine
Barco, Guimarães, Portugal

Rui L. Reis
3B's Research Group – Biomaterials,
Biodegradables and Biomimetics
University of Minho
Headquarters of the European Institute
of Excellence on Tissue Engineering
and Regenerative Medicine
Barco, Guimarães, Portugal

Sandra Pina
3B's Research Group – Biomaterials,
Biodegradables and Biomimetics
University of Minho
Headquarters of the European Institute
of Excellence on Tissue Engineering
and Regenerative Medicine
Barco, Guimarães, Portugal

Julio San Roman
Polymeric Nanomaterials and Biomaterials
Department
Institute of Polymer Science and
Technology
Spanish Council for Scientific Research
(CSIC)
Madrid, Spain

ISSN 0065-2598　　　　　ISSN 2214-8019　(electronic)
Advances in Experimental Medicine and Biology
ISBN 978-3-030-09569-7　　　　ISBN 978-3-319-76735-2　(eBook)
https://doi.org/10.1007/978-3-319-76735-2

Printed on acid-free paper

This Springer imprint is published by the registered company Springer International Publishing AG part of Springer Nature.
The registered company address is: Gewerbestrasse 11, 6330 Cham, Switzerland

Preface

In the last few years, osteochondral tissue engineering has shown an increasing development in advanced tools and technologies for damaged underlying subchondral bone and cartilage tissue repair and regeneration. Considering the limitation of articular cartilage to heal and self-repair, new therapeutic options are essential to develop approaches based on suitable strategies made of appropriate engineered biomaterials. This book overviews the most recent developments in the field of osteochondral tissue engineering. It covers the concepts and current challenges for bone and cartilage repair and regeneration, along with technological advances for osteochondral tissue. Specific topics include viscosupplementation, tissue engineering approaches, technological advances with stem cells and cell-based therapies with applications for osteochondral, bioreactors and microfluidics including multichamber bioreactors, and *in vitro* and *in vivo* mimetic models. This book presents the challenges and strategies being developed not only for bone and cartilage regeneration but also to establish the osteochondral interface formation to translate it into a clinical setting. Each chapter is prepared by world-known experts on their field, and serves as a core reference for biomedical engineering students and a wide range of established researchers and professionals working in the orthopedic field.

Barco, Guimarães, Portugal   J. Miguel Oliveira
Barco, Guimarães, Portugal   Sandra Pina
Barco, Guimarães, Portugal   Rui L. Reis
Madrid, Spain   Julio San Roman

Contents

Part I
Current Challenges in Osteochondral Repair and Regeneration: Trauma vs Disease

Chapter 1
Advances for Treatment of Knee OC Defects

Marta Ondrésik, J. Miguel Oliveira, and Rui L. Reis

Abstract Osteochondral (OC) defects are prevalent among young adults and are notorious for being unable to heal. Although they are traumatic in nature, they often develop silently. Detection of many OC defects is challenging, despite the criticality of early care. Current repair approaches face limitations and cannot provide regenerative or long-standing solution. Clinicians and researchers are working together in order to develop approaches that can regenerate the damaged tissues and protect the joint from developing osteoarthritis. The current concepts of tissue engineering and regenerative medicine, which have brought many promising applications to OC management, are overviewed herein. We will also review the types of stem cells that aim to provide sustainable cell sources overcoming the limitation of autologous chondrocyte-based applications. The various scaffolding materials that can be used as extracellular matrix mimetic and having functional properties similar to the OC unit are also discussed.

Keywords Osteochondral defects · Osteochondral tissue engineering · Regenerative medicine · Chondrocytes · Stem cell therapy · iPS cells · Scaffold design · and Bilayered scaffolds

M. Ondrésik (✉)
3B's Research Group – Biomaterials, Biodegradables and Biomimetics, University of Minho, Headquarters of the European Institute of Excellence on Tissue Engineering and Regenerative Medicine, Barco, Guimarães, Portugal

ICVS/3B's – PT Government Associate Laboratory, Braga/Guimarães, Portugal
e-mail: marta.ondresik@dep.uminho.pt

J. M. Oliveira · R. L. Reis
3B's Research Group – Biomaterials, Biodegradables and Biomimetics, University of Minho, Headquarters of the European Institute of Excellence on Tissue Engineering and Regenerative Medicine, Barco, Guimarães, Portugal

ICVS/3B's – PT Government Associate Laboratory, Braga/Guimarães, Portugal

The Discoveries Centre for Regenerative and Precision Medicine, Headquarters at University of Minho, Barco, Guimarães, Portugal

© Springer International Publishing AG, part of Springer Nature 2018
J. M. Oliveira et al. (eds.), *Osteochondral Tissue Engineering*,
Advances in Experimental Medicine and Biology 1059,
https://doi.org/10.1007/978-3-319-76735-2_1

1.1 Introduction

CC lesions are notorious for being unable to heal. They mostly affect the knee and the ankle. Usually the defects appear when the joint is used extensively, as caused by repetitive strain or direct trauma to the articulation, typical for sports activity. Genetic predisposition could be another factor responsible for OC damage [1]. Comorbidities of OC defects include joint malalignment, meniscal tear and ligamentous laxity [2]. In many cases these concomitant pathologies occur before the cartilage lesion, and known as contributors of the lesion development. Although the exact aetiology of OC defects has not yet been fully elucidated, researchers agree on two distinct phenotypes, manifested in degenerative lesions or traumatic focal defects [3, 4]. In either case, OC damages put the joint in a great risk of developing osteoarthritis; therefore, early recognition of the presence of any lesion or damage to any joint is of paramount importance [5, 6], especially because OC damages are more prevalent among adolescents and the physically more active younger population [7, 8]. Articular cartilage defects can be of many sizes and shapes and vary in depths. In most severe cases, the defects reach down to the subchondral bone. Their diagnosis is challenging, as many lesions can exist without any symptoms, and their presence is often missed in the early stages [9, 10]. Unfortunately, even the silent lesions usually progress from partial to focal defects if they are left untreated (Fig. 1.1). Having OC lesions can often cause symptoms such as discomfort, tenderness or, in more severe cases, pain, swelling of the knee and limitation in motion of the patient [11].

To establish consistency and aid communication among physicians, there were several grading scales developed, which describe the degree, severity and sometimes the location of the cartilage lesions. The most frequently used classification system was invented by Outerbridge, which was named after him, called the Outerbridge scaling system (Table 1.1) [12]. This system divides the lesions into four categories, but it does not characterize the depth of the lesion. Another widely used scaling system was developed by the International Cartilage Repair Society (ICRS), which also has four grades, complemented with a zero state marking the normal state of the cartilage. The advantage of the ICRS scaling over the Outerbridge

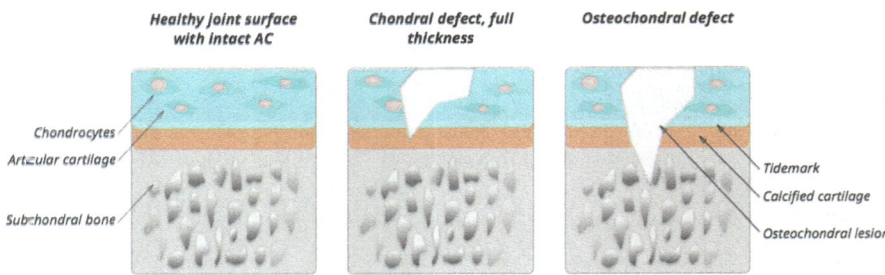

Fig. 1.1 Schematic representation of the structure of the osteochondral unit and the osteochondral defect

Table 1.1 Classification of osteochondral injuries according to the ICRS grading

ICRS grading system	
Grade 0	**Normal** Having a healthy appearance, no signs of cracks
Grade 1	**Almost normal** 1.a Soft indentation 1.b Superficial cracks and fissures
Grade 2	**Abnormal** Lesions extending down to <50% of cartilage depth
Grade 3	**Severe lesion** 3.a Cartilage defects >50% of cartilage depth 3.b Cartilage defects reach down to the calcified cartilage 3.c Cartilage defects reach down to the surface of subchondral bone but do not penetrate 3.d cartilage is bulging around the lesion site
Grade 4	**Very severe lesion** 4.a Cartilage lesion reaches down to the subchondral bone but not in the entire diameter of the lesion 4.b Cartilage lesion penetrates the subchondral bone fully and across the entire diameter of the lesion

Table 1.2 Classification of osteochondral injuries according to the Outerbridge grading system

Outerbridge system	
Grade I	Softening and swelling of the cartilage
Grade II	Fragmentation and fissuring in an area half an inch or less in diameter
Grade III	Fragmentation and fissuring, greater than 0.5 in diameter
Grade IV	Erosion of cartilage down to exposed subchondral bone

scaling is that it describes the lesions based on their extent as well as depth (Table 1.2) [4, 13, 14].

For the management of OC lesions, there are several conservative and operative techniques, which are applied according to the extent of damage, age of the patient, doctor's opinion and some other factors. These approaches have a common goal of providing pain relief, repairing the damaged tissues and improving joint functionality [15]. The conservative management of cartilage lesions includes medications to fight pain and inflammation, drilling techniques, abrasion, microfracture as well as the transplantation of OC allografts and autologous chondrocyte implantation with or without matrix [16]. Despite their continuing use in the clinics, none of the mentioned approaches brought long-term solutions so far and could not fully regenerate the OC unit. In fact, according to prospective and retrospective studies, around 60% of knee arthroplasties performed were a result of previous cartilage lesions in the joint [17]. Knee arthroplasty is the most common clinical approach used to restore the joint function. Annually there are 700,000 arthroplasties performed in the USA, and it is estimated that this number will further increase, as much as reaching 3.48 million by the year 2030 [17]. It was indicated that 5% of all knee illnesses in the general population originated from previous cartilage injuries, while 61% of the cases had OC damages, among which 19% could be specified as focal OC injury [18]. It is therefore leaving no doubt that the burden of OC injuries on today's healthcare and society is huge and its management requires the development of more effective techniques.

In this chapter, we will be briefly discussing the basics of current approaches to diagnose and treat OC defects, but our main focus will be on the future therapeutic regimens of OC lesion management with special focus on the knee joint. We will describe how the advancement of tissue engineering and regenerative medicine (TERM) brought new aspects to articular cartilage healing and will demonstrate the main components of these approaches, namely, the cell sources and matrices used in TERM.

1.2 Knee Joint

The largest and most complex joint of the body is the knee. It is a modified hinge joint, which actually consists of two articulations, namely, the patellofemoral and tibiofemoral [19, 20]. The knee is subjected to a great load while walking, jumping, running or performing other activities [21]. It provides stability via a combination of ligaments, tendons, synovial capsule and muscular components [22]. The osseous part is composed of the tibia, fibula, femur and patella. The surfaces of the joints are covered by articular cartilage, which allows frictionless movement [23, 24]. For extra cushioning, there is a pair of moon-shaped fibrocartilage tissues, called the menisci, between the femoral condyles and tibial plateau. Both the AC and the menisci have important shock-absorbing function in the joint [25].

1.2.1 Articular Cartilage

Articular cartilage is a type of hyaline cartilage which covers the joint surface providing gliding to it. It is a thin layer of tissue ranging from 3 to 7 mm of thickness at the various anatomical sites [25, 26]. The AC has a white, glassy appearance and lacks both vascularization and innervation in its healthy state. Despite its simple appearance, articular cartilage is a complex and highly specialized tissue. It consists of four layers, each of which bears with a unique structure. The main macromolecular components of the cartilage are the collagen and proteoglycan molecules, which create distinctive patterns at the different zones of the tissue and thus also define functional differences among the layers [27, 28]. Aggrecan is the most abundantly found proteoglycan molecule of the articular cartilage, while among collagen molecules, type II collagen is the most typical [29, 30]. The only cell type found in the cartilage, are the chondrocytes, which possess different properties at the different layers of the tissue. Accordingly, the (i) uppermost layer called the superficial or tangential zone has rather flattened cells localized densely. Both the cells and collagen fibrils lie parallel to the surface herein. This layer has lesser amounts of proteoglycan molecules, among which biglycan and decorin are the most representative [31, 32]. The superficial layer makes up 10–20% of the whole cartilage. The chondrocytes found in this layer also produce a protein called lubricin, which is a mucous glycoprotein. Lubricin serves as a lubricant on the cartilage surface and together with the superficial zone protein creates a film which equips the cartilage with the mentioned low-friction properties [24, 33]. This layer is also rich in fibronectin and water. The (ii) intermediate zone builds up 40–60% of the total volume of the cartilage. The cells are localized less dense, have a spheroid morphology and are larger as compared to the upper layer [34]. The extracellular matrix is abundant in proteoglycan and thick collagen fibrils. The organization of the fibrils show less structure [35, 36]. The (iii) deep or radial zone of the articular cartilage makes up 30% of the total tissue volume. This layer has the least amount of cells and water, but the most of the proteoglycan molecules and the largest collagen fibrils [37, 38]. The chondrocytes are stack in columns and have spherical shapes. Both the collagen molecules and chondrocytes lie perpendicular to the cartilage surface. The (iv) calcified cartilage is located beneath the radial zone, also called the basal layer [25, 39, 40]. It is separated from the radial zone by the tidemark, which is narrow layer, metabolically active for the calcification. The tidemark is an important interface between the soft and calcified cartilage. It also has a crucial role in damping the loads. The calcified cartilage creates a connection between the hyaline cartilage and subchondral bone [41, 42]. This layer has a sparse amount of chondrocytes, which are smaller in size and have a hypertrophic phenotype. These chondrocytes are metabolically less active and are completely embedded in the extracellular matrix. Most typical collagen in this zone is the collagen type X [43, 44]. Collagen type X is believed to have an important role in mechanical load transmission [31, 45]. Any structural change to the articular cartilage as a result of either biomechanical or biochemical changes will ultimately result in poor mechanical properties and cartilage degeneration.

1.2.2 Subchondral Bone

The subchondral bone is located underneath the calcified cartilage. Together they compose the OC unit. The AC acts in concert with the subchondral bone to absorb the mechanical load in the joint [46]. The subchondral bone consists of two anatomical entities, the subchondral plateau and the subchondral spongiosa [41]. The subchondral bone is richly perforated by veins and arteries as well as by nerves. The subchondral bone is an important exchange interface for the AC. It has been demonstrated that the AC and the subchondral bone have an intense crosstalk, especially during the onset of joint degeneration [40, 47, 48].

1.3   Limitations of Current Clinical Approaches

There are several major limitations to effective treatment of OCDs, and their diagnosis is also extremely challenging in many cases. Often subtle injuries show little or no dysfunction, which results in their delayed detection. But even if diagnosis is performed accurately, due to the low intrinsic capacity of the cartilage, conservative care is often unable to provide a long-term solution. Many of the current techniques focus on symptom management and functional improvement of the joint [7, 49]. Common treatment strategies are the use of non-steroidal anti-inflammatory drug (NSAID) agents, which are often used together with cast immobilization, or in cases where surgical intervention is required microfracture, drilling techniques and the use of allografts or autologous chondrocyte transplants are the routined applications [16].

1.3.1   Diagnosis of OC Lesions

Patients with acute joint injury usually suffer from pain, swelling and joint effusion, which appear as the consequence of the influx of inflammatory mediators. They usually are also limited in motion and have the joint inflexible in many cases or report mechanical symptoms such as clicking and locking of the knee. OC injury often goes together with the damage of the menisci and the ligaments or results in loose bodies in the joint cavity. In many cases on the other hand, OC lesions can go unnoticed, having no symptoms, and thus being challenging to diagnose [28].

Currently, the diagnosis of OC lesions relies on patient medical history, patient self-assessment, symptomatic findings, furthermore radiographic and magnetic resonance imaging, as well as arthroscopy [50]. Plain radiography is usually considered to be the gold standard as it is largely available and cost-sensitive [51, 52]. However, radiographic images may not depict the presence of small superficial cracks but can only identify more prominent lesions and loose bodies in the joint cavity. Magnetic resonance imaging (MRI) is a more precise method allowing higher resolution and thus providing more information but is also more costly [9, 53]. MRI often goes together with the application of X-ray computed tomography (CT) [54]. Both MRI and CT can use contrast agents to enhance tissue differentiation and enable better evaluation. Contrast agents are especially important for the visualization of cartilage and other soft tissue components in the joint [54, 55]. There is a large variety of contrast agents available. The use of metal chelates, i.e. gadolinium-based agents, is common for MRI application, while for CT iodine-based agents such as sodium iodide and sodium diatrizoate hydrate are the most established [56–60]. Arthroscopy is an intra-articular method used to visualize and examine the cartilage pathology. Arthroscopy is performed by orthopaedic surgeons, and during the procedure a small camera is led to the joint cavity via a small incision, which can accurately locate and classify the damage [61, 62]. Given its in situ nature, arthroscopy is a very precise technique, but consequently it is also more invasive compared to other imaging approaches. However, arthroscopy is not only used for diagnosis but also for the therapy of the OC tissue which we will introduce in the next section [61, 62].

Taken together all possibilities for diagnosis, unfortunately it is still a hurdle to detect less pronounced cartilage damages. Early diagnosis and accurate evaluation of articular lesions should be immediate, as all defects can eventually progress to focal lesions.

1.3.2   Conservative and Surgical Care

OC injuries are traumatic in nature. When patients present with OC injury without any severe symptoms, typically ICRS grades I and II (Table 1.2), they are usually advised to rest, and often cast is used to immobilize the joint. The treatment regimen also includes the use of NSAIDs. The goal is to help the potential oedema resolve and avoid the development of any necrosis [49, 63]. In grades III and IV, the injury is accompanied by pain and larger damage to the joint; therefore, usually a surgery is imperative. Current surgical concepts for the restoration of articular cartilage structure and function mostly include approaches such as arthroscopic lavage and debridement, drilling, microfracture and the implantation of OC allografts and autologous chondrocytes [64–67].

Arthroscopic lavage and debridement involves the incision of the joint and insertion of a small camera, called the arthroscope, for the evaluation of damage as mentioned before. When the location and degree of defect are assessed, it is washed thoroughly to remove all loose bodies and damaged OC tissue from the cavity [68, 69]. Microfracture and drilling are marrow stimulation techniques and are used in an attempt to initiate cartilage regeneration [70, 71]. After perforating the subchondral bone, the blood is drained into the joint covering the cartilage surface and lesions with stem cells and bioactive molecules, such as growth factors which are believed to facilitate the cartilage repair capacity. The drilling holes are in close proximity to each other, approximately 3–4 mm apart [72, 73]. The ultimate goal is to reproduce biomechanically functioning cartilage. However, this procedure mostly leads to the formation of fibrocartilage only, which has knowingly reduced capacity to bear loading. OC graft implantation is used when there are larger defects present with size around 2 cm^2 [66, 74]. Initially, autografts were harvested from the same patient's intact joint region, usually from the patellofemoral site. However, due to anatomical misalignment, this procedure often failed. Advancement was brought by the multiple use of cylindrical shape osseous grafts, a procedure called mosaicplasty, developed by Hangody and Bobić [75, 76]. Further popularity of the usage of graphs came when standardization of tissue and cell storage was established by the American Association of Tissue Bank and US Food and Drug Administration in 1998. Clear guidance on how long cells and tissues stay viable in refrigerated condition allowed commercialization and wider use of OC allografts [66].

While immunological intolerance is relatively rare, the lack of tissue incorporation and a possibility to disease transmission remain a relative complication to this procedure.

1.4   The Future of OC Repair: Advanced Therapies

The future of OC lesion repair has been taken over by the emerging approaches of TERM strategies, where the focus is on replacing and regenerating the damaged tissues, instead of restoring the function only [77, 78]. The number of TERM techniques has been substantially increasing over the past years, resulting in a wide range of procedures to tackle tissue regeneration. Given the prevalence and low chances of healing on its own, OC lesions are prime candidates for stem cell therapies and tissue engineering approaches.

1.4.1   Cell Sources

The only cell type which has the perfect capacity to regrow the native cartilage is the chondrocytes. However, continuous supply of fresh chondrocytes is hard, if not impossible, to maintain. Researchers and clinical experts are constantly looking for other more optimal cell sources and feasible strategies to repair OC lesions. Here we will list and describe the most promising cell candidates for cartilage regeneration.

1.4.1.1   Chondrocytes

Regeneration of OCDs requires the application of chondrocytes and osteoblastic cells. Autologous chondrocytes were for long the commonest choice for conservative and advanced approaches of cartilage therapy [61]. Autologous chondrocytes are the patient's own cells harvested from a nondamaged area of the joint. The derivation of biopsies from a healthy area of the cartilage can potentially initiate secondary osteoarthritis, which is a disadvantage of this technique [79]. Also, not all patients have enough healthy areas of the joint. Owing a restricted size of biopsies, the acquired cell number is limited. Therefore, to achieve sufficient amount of cells, the chondrocytes must first be expanded in vitro. This is a risk factor as the chondrocytes can lose their phenotype easily when not in their native environment [80, 81]. Once they bear with reduced chondrogenic capacity, their clinical use and therapeutic outcome are impeded [82]. Therefore, there had been various approaches developed to prevent dedifferentiation during ex vivo expansion, including the use of growth factors in the culture media, providing dynamic hydrostatic pressure to the cells while in culture and limiting the levels of oxygen in the environment [83, 84]. Another crucial factor is the amount of time of expansion; the shorter time the chondrocytes spend in culture, the better the chances are to preserve their phenotype. However, these cells have a very low proliferation rate, and consequently the shorter the time, the smaller cell number is being achieved. Ergo, donor site morbidity, cell dedifferentiation and limitations in cell number are the most critical detriment of the use of autologous chondrocytes [82].

To overcome obstacles given by the limited availability of autologous chondrocytes, non-autologous chondrocytes had also been considered and explored as a potential cell source to manage OC defects [85]. The use of allogeneic or xenogeneic chondrocytes has the advantage of acquiring larger cell populations while avoiding damaging the healthy sites of the joint and thus the development of secondary osteoarthritis. One could even obtain cells from younger populations with better chondrogenic capacity [86]. However, this technique also has drawbacks, namely, immunological intolerance and the possibility of disease transmission.

1.4.1.2 Stem Cells

Stem cells had been widely explored to overcome the limited supply of primary cells. Continuous cell sources could be provided by using off-the-shelf stem cell techniques. Most research involves the application of mesenchymal stem cells (MSCs), but embryonic stem cells (ESCs) and induced pluripotent stem cells (iPSCs) had also been considered as ideal candidates for OC regeneration (Table 1.3).

Table 1.3 Advantages and disadvantages of the various cell candidates for the regeneration of the osteochondral unit

Cell types	Benefits	Challenges
Autologous chondrocytes	Native phenotype Immunocompatibility	Insufficient cell number Prone to dedifferentiation
Allogeneic chondrocytes	Larger cell number Off-the-shelf solution	Risk of disease transmission Lack of donor availability Immune rejection
Adult stem cells (MSCs, ADSCs, SMSCs, BMSCs)	Reliable potential for differentiation Large availability Various tissue sources Easy extractability No ethical complications	Large variety of proliferative capacity, phenotype Heterogeneous cell population Differentiation problems
Embryonic stem cells (ESCs)	Multiple cell types can be produced Immortal cell source Immune-privileged cells	Risk of teratoma formation Ethical complications
Induced pluripotent stem cells (iPSCs)	Unlimited cell supply Non-invasive extraction of donor cells Large variety of cell types can be differentiated	Difficulty to achieve uniform differentiation Risk of teratoma formation

Mesenchymal stem cells (MSCs) are multipotent cells which can differentiate into osteoblast or chondroblast. They can be originated from various sources, such as the bone marrow, adipose tissue, synovium, infrapatellar fat pad, muscle, dermis, blood or the umbilical cord [87]. MSCs are defined by a minimal prerequisite established by the International Society for Cellular Therapy. This includes the expression of surface markers CD105, CD73 and CD90 and the exclusion of surface molecules CD45, CD34, CD14 or CD11b, CD79α or CD19 and HLA-DR in standard culture conditions. They also have to be plastic adherent and capable of differentiating into chondroblasts, osteoblasts and adipocytes in vitro [88].

Chondrogenic differentiation of MSCs is induced via the application of growth factors and by the manipulation of their culture environment. It is usually divided into three stages starting with (i) cell condensation and the expression of adhesion molecules, such as N-cadherin and tenascin-C, which facilitate the cell-cell interaction. This is followed by (ii) the activation of transcription mediators including bone morphogenetic proteins (BMPs), Sox-9 and FGF signalling pathways. The last step is (iii) ECM deposition and pre-chordial cell formation which is continued by full chondrocyte formation [89]. MSCs isolated from different sources may differ in their differentiation potentials [90].

Synovium-derived MSCs (SMSCs) were demonstrated to have excellent chondrogenic potential, greater than the MSCs isolated from the donor-matched bone marrow, periosteum, muscle and adipose tissue [91]. Furthermore, SMSCs are closely related and share many characteristics with the chondrocytes. AC and the synovium originate from the same pool of precursor cells and have a similar gene expression profiles, i.e. superficial zone protein, collagen type II and aggrecan are all expressed by both the chondrocytes and SMSCs [92, 93]. SMSCs also have a relatively high rate of proliferation as compared to other MSCs. Furthermore, they were shown to deposit greater amount of extracellular matrix, as compared to BMSCs. This is especially important since ECM deposition is believed to delay cell senescence and dedifferentiation [94, 95]. Additionally, SMSCs are able to form hyaline cartilage in vitro which holds extreme value for OC regeneration [96]. SMSCs are harvested via arthroscopy or from the synovial fluid. Unfortunately, the yield of cells is relatively low, having only around 14 cells per mL of synovial fluid, which have the capacity to form CFU-F colonies [97].

MSCs of other origin, such as bone marrow-derived stem cells (BMSCs) or adipose-derived stem cells (ADSCs), can also differentiate into chondrocytes or osteoblasts. Despite these cells having less chondrogenic potential, they bear with other advantages making them desirable for the OC tissue repair. In particular, ADSCs are routinely available from excised fat or lipoaspiration, which yields on average around 400,000 cells per mL, and thus provide the sufficient amount of cells easily [90]. Thorough molecular analysis had been performed to characterize these cells, and their trilineage mesodermal potential has been confirmed by several studies [98]. Different induction conditions had been identified where adipogenic, osteogenic and chondrogenic differentiation can be stimulated [99–101]. These include the presence of growth factors, hormones and mechanical stimuli, as well as 3D scaffold cultivation using scaffolding materials with different composition.

Studies using alginate scaffolds demonstrated a superficial chondrogenic capacity of chondrocytes differentiated from ADSCs as compared to normal chondrocytes after 21 days of in vitro culture in alginate scaffolds [102]. On the other hand, when using hyaluronic acid scaffolds, collagen type II expression was similar in both ADSC and chondrocyte cultures and was lower in BMSC cultures. Numerous papers had been discussing whether BMSCs or ADSCs have stronger ability to differentiate into the chondrogenic lineage. Most of them agree that BMSCs possess more chondrogenic potential. Interestingly, it was also shown that although ADSCs and BMSCs have similar surface receptor profile, they may require different induction conditions [103]. For instance, stimulating BMSCs and ADSCs by transforming growth factor-β results in different chondrogenic profile as measured by chondrogenic marker expression, such as aggrecan and collagen type II. It was pointed out that while bone morphogenetic protein-6 (BMP-6) had successfully resulted in increased expression of aggrecan by ADSCs, BMSCs required stimuli provided by TGF-β to achieve the same [104]. Unfortunately, BMSCs yield significantly less cells upon harvest as compared to ADSCs, and they are harder to obtain involving a more invasive procedure, called bone marrow aspiration. It was shown that 1 mL marrow yields between 100 and 1000 cells.

ESCs and iPSCs have also tremendous potential as a cell source for the manufacture of OC therapy. The advantage of ESCs and iPSCs over adult stem cells is that they can be expanded indefinitely without undergoing senescence. Blastocysts which are unsuitable for in vitro fertilization are used to acquire ESCs. ESCs are the inner cell mast of blastocyst. Consequently, scientific use of ESCs is ethically unaccepted in many countries despite the immense interest and potential of their application. There are four main categories of methodologies to generate chondrogenic cells from ESCs, namely, (i) targeted differentiation, (ii) differentiation of ESCs via their coculture with mature chondrocytes, (iii) embryoid body formation with stimuli of growth factors and (iv) spontaneous differentiation into MSCs and subsequent chondrogenic differentiation [105]. Besides ethical concerns, teratoma formation is another problem encountered in the application of ESCs in regenerative therapy.

iPS cells are not burdened by ethical issues and were proven to be safe and stabile when used for cartilage formation [106]. Since their first introduction in 2006, iPSCs were widely explored for various applications [107–109]. Originally, they were generated in a mouse model using fibroblasts. Four factors were manipulated, namely, two transcriptional factors, the octamer-binding transcription factors 3 and 4 (Oct3/4) and the Kruppel-like factor 4 (Klf4), and two tumour-related genes, the v-myc avian myelocytomatosis viral oncogene homolog (c-Myc) and Sox-2 [107]. Successful chondrogenic differentiation of iPSCs derived from various sources, such as fibroblast-like cells, neural stem cells and human osteoarthritic chondrocytes, was also demonstrated [110, 111]. Generation of iPSCs and subsequent differentiation into chondrogenic lineage from murine neural cells were achieved by the simultaneous overexpression of Oct4 and Klf4. Another approach, using somatic cells, had successfully generated iPSCs via a reprogramming approach using c-Myc, Klf4 and Sox-9 which were later able to differentiate into the chondrogenic lineage [112, 113]. Interestingly, even the reprogrammed cells show differences in their chondrogenic potential. Namely, iPSCs originating from chondrocytes show superior properties than iPSCs generated from neural or fibroblast-like cells [97]. iPSCs have tremendous potential for regenerative approaches. It is possible to establish allogeneic iPSC libraries, based on HLA phenotypes. These cells could be induced to differentiate into chondrogenic or osteogenic phenotype and could serve as off-the-shelf solution to heal OC lesions providing a truly advanced solution [89].

1.4.2 Biomaterials

Apart from the cell and molecular component, another pillar of the TERM approaches is the carrier. Although cells might be administered without scaffolding material, they often require support provided by a 3D structure to survive and keep their phenotype. Scaffolds need to be compatible with the cells as well as mechanically matching the environment of implantation. To design appropriate scaffolds for the OC unit, one must understand the biology of both the cartilage and the subchondral bone; therefore, it is particularly challenging. The scaffolds should integrate to their environment by being biocompatible and slowly degrade as the cells rebuild the tissue and thus also have to be biodegradable. There have been a myriad of materials proposed already originating from both natural and synthetic polymers [114]. The usage of inorganic materials, such as glasses and ceramics and metallic materials, had also been introduced. In this section we will review the advantages and disadvantages of these different materials.

1.4.2.1 Natural Polymer-Based Materials

Natural polymers are derived from the nature; therefore, they bear with high biocompatibility properties which is an advantage over the synthetic polymers. They are of animal or vegetal origin, or obtained from algae, and can be synthesized as well [115–118]. They can also aid cells in their tissue interaction, synthesis and development due to molecular domains present on their structures. Collagen polymers, for instance, contain ligands, which promote cell adhesion and matrix deposition [119, 120]. Collagen-based scaffolds showed improvement when regenerating cartilage; however, they also bear with relatively weak mechanical properties. Since both the cartilage and bone contain large amounts of collagen molecules, collagen-based scaffold seems to be an ideal candidate for OC regeneration. Current commercially available applications of collagen include Zyderm, CaReS (Arthro-Kinetics, Essingen, Germany) and Carticel (Genzyme Inc., Cambridge, UK) [121, 122]. Another favoured natural polymer is the alginate. Alginate is a linear, unbranched copolymer of L-glucuronic and D-mannuronic acid [123]. Ca-alginate porous scaffolds combined with gelatine were able to differentiate MSCs into both chondrogenic and osteogenic lineages [124]. Alginate-based 3D scaffolds were shown to have the capacity to redifferentiate chondrocytes [125]. Among the cationic polymers, chitosan is the most popular. Based on the source and method of preparation, various chitosan materials can be attained, which also bear with different properties. Successful subchondral bone and cartilage generation had been achieved by chitosan matrices [126]. Chitosan polymers have structural similarities with two major cartilage molecules, namely, hyaluronic acid and glycosaminoglycan molecules; thus, it is also prepared as a blend [127]. Silk polymers obtained from silkworm cocoon or spider silk are also widely explored for the use of TERM approaches [128]. They can be combined with synthetic polymers, which are used to improve their mechanical stability and tailorability/processability. For instance, when silk was prepared in a combination as silk fibroin/nano-CaP to obtain bilayered scaffolds, it showed superior mechanical properties compared to silk-only scaffolds [129].

Natural matrices have the advantage of being biocompatible and biodegradable and having more similarity to biological macromolecules; therefore, the native environment can recognize it and process metabolically. However, natural matrices have the danger of transition hazardous agents and have purification issues; therefore, their use on the clinics is hampered. They also bear with weaker mechanical properties as compared to synthetic polymers.

1.4.2.2   Synthetic Polymer-Based Materials

As mentioned before, the main advantage of using synthetic polymers is owing complete control over their structure and having better mechanical properties. Synthetic polymers also do not have issues with disease transmission or limitation in polymer supply. However, biocompatibility can become a major issue when using matrices fabricated from synthetic polymers. The most commonly used polymers for cartilage and bone engineering are poly(ethylene glycol) (PEG) and its derivatives poly(L-lactic acid) (PLLA) and poly(lactide-co-glycolide) (PLGA). These synthetic biodegradable polymers are often used as a blend of each other or combined with natural polymers to promote biological integration [78, 130, 131]. Despite their easy manufacture and large availability, synthetic polymers must be explored via the combination of natural polymers to enable better biocompatibility and orchestrate cellular events in vivo, e.g. cell adhesion, proliferation and matrix deposition. Accordingly, several studies demonstrated chondrogenic differentiation of PLLA in combination with polymers of natural origin, such as chitosan and silk, proving their potential for OC regeneration [132, 133]. Besides the lack of cell recognition sites, synthetic materials also tend to degrade easily once implanted promoting the development of inflammation. Thus, despite their advantages, synthetic polymer-based scaffolds are only ideal to be used as blend with natural matrices.

1.4.2.3   Other Materials (Ceramics, Glasses, Metallic Materials)

Ceramics, such as calcium phosphate ceramics and hydroxyapatite, are favoured because of their biocompatibility and osteoconductive and osteoinductive nature [85]. They were shown to promote the production of bone-like apatite when implanted, therefore being able to integrate better and induce bone regeneration [134, 135]. As it was recently summarized and demonstrated, 3D–printed Ca ceramics have tremendous potential to establish patient-specific bone grafting application; however, poor mechanical stability must be first addressed and solved [136]. Metallic materials include the application of stainless steel, titanium, titanium alloys and cobalt-based alloys. The main advantage of these materials is the structural architecture and mechanical property similar to the native grafts. Porous titanium bases were recently fabricated in combination with hydrogels for OC regeneration [137]. These biphasic scaffolds had the advantage of bearing with mechanical stability at the bone phase and having viable cell-seeded hydrogel integrated on top to support cartilage regeneration [137]. The major disadvantage of metallic materials is the lack of degeneration after implantation.

1.5   Final Remarks

There is no doubt that regeneration of OC damages is of major importance. Considering the poor healing capacity of the OC unit, so far only repair techniques had been applied, which fail to provide long-term solutions and full regeneration of the joints. Due to lack of regeneration, the danger of developing osteoarthritis is high in joints affected by OC injury. Orthopaedic surgeons and researchers are working on solutions, which can heal the damaged area by introducing cells, bioactive molecules and support matrices to the defect site. Whereas, in case of larger injuries, the application of in vitro engineered allografts could bring solutions, TERM approaches have the potential to overcome several limitations of the existing applications, such as insufficient cell number or cell differentiation, as well as can tackle the issue of lack of donors for OC grafts. TERM approaches hold great promise but also face many challenges. Finding the appropriate cell source, and establishing the environment to these cells, in which they are able to keep their phenotype or conversely differentiate into the desired phenotype, is extremely difficult, especially if we consider off-the-shelf solutions. Moreover, these cells usually need 3D support in vitro or when implanted. This is provided by scaffolding materials, which come in many shapes and forms. Finding the appropriate combination which then will facilitate tissue formation is hard. Furthermore, these materials have to be able to integrate into the native tissue and withstand the mechanical forces therein. This is especially difficult in the knee articulation, which is the largest joint in our body experiencing mechanical load over 20 MPa. Considering that the OC unit is composed of both the articular cartilage and subchondral bone, biphasic scaffolds obtained by the combination of materials and cells show promise for OC regeneration. Many routes had been proposed for material fabrication. Novel strategies using the combination of bioceramics and polymeric materials or the blend of natural and synthetic materials are considered as optimal candidates. Among the cells, MSCs including SMSCs, BMSCs and ADSCs as well iPSCs are the favoured ones due to their large availability and chondrogenic potential. The design of strategies for OC regeneration is open for further development and requires the combination of multidisciplinary approaches to reach its goal. Nevertheless, if cell and molecular biology, material science and orthopaedics are effectively brought together, chances are that we will be able to reach clinics with excellent and sustainable regenerative approaches for the OC unit.

References

1. Ding C, Cicuttini F, Scott F et al (2005) The genetic contribution and relevance of knee cartilage defects: case-control and sib-pair studies. J Rheumatol 32:1937–1942
2. Hjelle K, Solheim E, Strand T et al (2002) Articular cartilage defects in 1,000 knee arthroscopies. Arthroscopy 18:730–734. https://doi.org/10.1053/jars.2002.32839
3. Widuchowski W, Widuchowski J, Trzaska T (2007) Articular cartilage defects: study of 25,124 knee arthroscopies. Knee 14:177–182. https://doi.org/10.1016/j.knee.2007.02.001
4. Brittberg M, Lindahl A, Nilsson A et al (1994) Treatment of deep cartilage defects in the knee with autologous chondrocyte transplantation. N Engl J Med 331:889–895. https://doi.org/10.1056/NEJM199410063311401
5. Kocher MS, Tucker R, Ganley TJ, Flynn JM (2006) Management of osteochondritis dissecans of the knee. Am J Sports Med 34:1181–1191. https://doi.org/10.1177/0363546506290127
6. Espregueira-Mendes J, Pereira H, Sevivas N et al (2012) Assessment of rotatory laxity in anterior cruciate ligament-deficient knees using magnetic resonance imaging with Porto-knee testing device. Knee Surg Sports Traumatol Arthrosc 20:671–678. https://doi.org/10.1007/s00167-012-1914-9
7. Lee JE, Ryu KN, Park JS et al (2014) Osteochondral lesion of the bilateral femoral heads in a young athletic patient. Korean J Radiol 15(6):792. https://doi.org/10.3348/kjr.2014.15.6.792
8. Cahill (1995) Osteochondritis dissecans of the knee: treatment of juvenile and adult forms. J Am Acad Orthop Surg 3:237–247
9. Eckstein F, Cicuttini F, Raynauld J-P et al (2006) Magnetic resonance imaging (MRI) of articular cartilage in knee osteoarthritis (OA): morphological assessment. Osteoarthritis Cartilage 14 Suppl A:A46–A75. https://doi.org/10.1016/j.joca.2006.02.026
10. Ryd L, Brittberg M, Eriksson K et al (2015) Pre-osteoarthritis: definition and diagnosis of an elusive clinical entity. Cartilage 6:156. https://doi.org/10.1177/1947603515586048
11. Durur-Subasi I, Durur-Karakaya A, Yildirim OS (2015) Osteochondral lesions of major joints. Eurasian J Med 47:138–144. https://doi.org/10.5152/eurasianjmed.2015.50
12. Outerbridge RE (1961) The etiology of chondromalacia patellae. J Bone Jt Surg 43:752–757
13. da Cunha Cavalcanti FM, Doca D, Cohen M, Ferretti M (2012) Updating on diagnosis and treatment of chondral lesion of the knee. Rev Bras Ortop 47:12–20. https://doi.org/10.1590/S0102-36162012000100001
14. Brittberg M, Winalski CS (2003) Evaluation of cartilage injuries and repair. J Bone Joint Surg Am 85-A(Suppl 2):58–69
15. Magnussen RA, Dunn WR, Carey JL, Spindler KP (2008) Treatment of focal articular cartilage defects in the knee: a systematic review. Clin Orthop Relat Res 466:952–962. https://doi.org/10.1007/s11999-007-0097-z

16. Seo S-S, Kim C-W, Jung D-W (2011) Management of focal chondral lesion in the knee joint. Knee Surg Relat Res 23:185–196. https://doi.org/10.5792/ksrr.2011.23.4.185

17. Kurtz S, Ong K, Lau E et al (2007) Projections of primary and revision hip and knee arthroplasty in the United States from 2005 to 2030. J Bone Joint Surg Am 89:780–785. https://doi.org/10.2106/JBJS.F.00222

18. Kim J, Nakamura N, Brittberg M (2014) Techniques in cartilage repair surgery. https://doi.org/10.1007/978-3-642-41921-8

19. Last RJ (1948) Some anatomical details of the knee joint. J Bone Joint Surg Br 30B:683–688

20. Akizuki S, Mow VC, Müller F et al (1986) Tensile properties of human knee joint cartilage: I. Influence of ionic conditions, weight bearing, and fibrillation on the tensile modulus. J Orthop Res 4:379–392. https://doi.org/10.1002/jor.1100040401

21. Hasler EM, Herzog W, Wu JZ et al (1999) Articular cartilage biomechanics: theoretical models, material properties, and biosynthetic response. Crit Rev Biomed Eng 27:415–488

22. Ralphs JR, Benjamin M (1994) The joint capsule: structure, composition, ageing and disease. J Anat 184:503–509

23. Wilson R, Diseberg AF, Gordon L et al (2010) Comprehensive profiling of cartilage extracellular matrix formation and maturation using sequential extraction and label-free quantitative proteomics. Mol Cell Proteomics 9:1296–1313. https://doi.org/10.1074/mcp.M000014-MCP201

24. Fujie H, Nakamura N (2013) Frictional properties of articular cartilage-like tissues repaired with a mesenchymal stem cell-based tissue engineered construct. Conf Proc IEEE Eng Med Biol Soc. 2013:401–404

25. Thambyah A, Nather A, Goh J (2006) Mechanical properties of articular cartilage covered by the meniscus. Osteoarthr Cartil 14:580–588. https://doi.org/10.1016/j.joca.2006.01.015

26. Goldring MB (2012) Chondrogenesis, chondrocyte differentiation, and articular cartilage metabolism in health and osteoarthritis. Ther Adv Musculoskelet Dis 4:269–285. https://doi.org/10.1177/1759720X12448454

27. Kheir E, Shaw D (2009) Hyaline articular cartilage. Orthop Trauma 23:450–455. https://doi.org/10.1016/j.mporth.2009.01.003

28. Bhosale AM, Richardson JB (2008) Articular cartilage: structure, injuries and review of management. Br Med Bull 87:77–95. https://doi.org/10.1093/bmb/ldn025

29. Kiani CH, Chen LI, Wu YJ et al (2002) Structure and function of aggrecan 12:19–32

30. Li J, Anemaet W, M a D et al (2011) Knockout of ADAMTS5 does not eliminate cartilage aggrecanase activity but abrogates joint fibrosis and promotes cartilage aggrecan deposition in murine osteoarthritis models. J Orthop Res 29:516–522. https://doi.org/10.1002/jor.21215

31. Lu XL, Mow VC (2008) Biomechanics of articular cartilage and determination of material properties. Med Sci Sports Exerc 40:193–199. https://doi.org/10.1249/mss.0b013e31815cb1fc

32. Bock HC, Michaeli P, Bode C et al (2001) The small proteoglycans decorin and biglycan in human articular cartilage of late-stage osteoarthritis. Osteoarthr Cartil 9:654–663. https://doi.org/10.1053/joca.2001.0420

33. Jay GD, Tantravahi U, Britt DE et al (2001) Homology of lubricin and superficial zone protein (SZP): products of megakaryocyte stimulating factor (MSF) gene expression by human synovial fibroblasts and articular chondrocytes localized to chromosome 1q25. J Orthop Res 19:677–687. https://doi.org/10.1016/S0736-0266(00)00040-1

34. Sophia Fox AJ, Bedi A, Rodeo SA (2009) The basic science of articular cartilage: structure, composition, and function. Sports Health 1:461–468. https://doi.org/10.1177/1941738109350438

35. Eyre DR, Weis MA, Wu JJ (2006) Articular cartilage collagen: an irreplaceable framework? Eur Cells Mater 12:57–63. https://doi.org/10.22203/eCM.v012a07

36. Eyre D (2002) Collagen of articular cartilage. Arthritis Res 4:30–35

37. Bayliss MT, Venn M, Maroudas A, Ali SY (1983) Structure of proteoglycans from different layers of human articular cartilage. Biochem J 209:387–400

38. Zhytkova MA, Chizhik SA, Wierzcholski K et al (2010) Properties of cartilage on micro- and nanolevel. Adv Tribol. https://doi.org/10.1155/2010/243150
39. Redler I, Mow VC, Zimny ML, Mansell J (1975) The ultrastructure and biomechanical significance of the tidemark of articular cartilage. Clin Orthop Relat Res:357–362
40. Lories RJ, Luyten FP (2011) The bone-cartilage unit in osteoarthritis. Nat Rev Rheumatol 7:43–49. https://doi.org/10.1038/nrrheum.2010.197
41. Duncan H, Jundt J, Riddle JM et al (1987) The tibial subchondral plate. A scanning electron microscopic study. J Bone Joint Surg Am 69:1212–1220
42. Lyons TJ, Stoddart RW, McClure SF, McClure J (2005) The tidemark of the chondro-osseous junction of the normal human knee joint. J Mol Histol 36:207–215. https://doi.org/10.1007/s10735-005-3283-x
43. Eyre DR, Wu JJ (1995) Collagen structure and cartilage matrix integrity. J Rheumatol Suppl 43:82–85
44. Buckwalter JA, Mankin HJ (1998) Articular cartilage: tissue design and chondrocyte-matrix interactions. Instr Course Lect 47:477–486
45. Goldring MB, Goldring SR (2010) Articular cartilage and subchondral bone in the pathogenesis of osteoarthritis. Ann N Y Acad Sci 1192:230–237. https://doi.org/10.1111/j.1749-6632.2009.05240.x
46. Obeid EM, Adams MA, Newman JH (1994) Mechanical properties of articular cartilage in knees with unicompartmental osteoarthritis. J Bone Joint Surg Br 76:315–319
47. Pan J, Wang B, Li W et al (2012) Elevated cross-talk between subchondral bone and cartilage in osteoarthritic joints. Bone 51:212–217. https://doi.org/10.1016/j.bone.2011.11.030
48. Sharma AR, Jagga S, Lee S-S, Nam J-S (2013) Interplay between cartilage and subchondral bone contributing to pathogenesis of osteoarthritis. Int J Mol Sci 14:19805–19830. https://doi.org/10.3390/ijms141019805
49. Bentley G, Bhamra JS, Gikas PD et al (2013) Repair of osteochondral defects in joints-how to achieve success. Injury 44:S3–S10. https://doi.org/10.1016/S0020-1383(13)70003-2
50. Oei EHG, Van Tiel J, Robinson WH, Gold GE (2014) Quantitative radiologic imaging techniques for articular cartilage composition: toward early diagnosis and development of disease-modifying therapeutics for osteoarthritis. Arthritis Care Res 66:1129–1141. https://doi.org/10.1002/acr.22316
51. Wenham CYJ, Conaghan PG (2009) Imaging the painful osteoarthritic knee joint: what have we learned? Nat Clin Pract Rheumatol 5:149–158. https://doi.org/10.1038/ncprheum1023
52. Braun HJ, Gold GE (2012) Diagnosis of osteoarthritis: imaging. Bone 51:278–288. https://doi.org/10.1016/j.bone.2011.11.019
53. Van Tiel J, Reijman M, Bos PK et al (2013) Delayed gadolinium-enhanced MRI of cartilage (dGEMRIC) shows no change in cartilage structural composition after viscosupplementation in patients with early-stage knee osteoarthritis. PLoS One 8:e79785. https://doi.org/10.1371/journal.pone.0079785
54. Lusic H, Grinstaff MW (2013) X-ray-computed tomography contrast agents. Chem Rev 113:1641–1666. https://doi.org/10.1021/cr200358s
55. Shafieyan Y, Khosravi N, Moeini M, Quinn TM (2014) Diffusion of MRI and CT contrast agents in articular cartilage under static compression. Biophys J 107:485–492. https://doi.org/10.1016/j.bpj.2014.04.041
56. Lee N, Hyeon T (2012) Designed synthesis of uniformly sized iron oxide nanoparticles for efficient magnetic resonance imaging contrast agents. Chem Soc Rev 41:2575–2589. https://doi.org/10.1039/c1cs15248c
57. Zhu D, Liu F, Ma L et al (2013) Nanoparticle-based systems for T1-weighted magnetic resonance imaging contrast agents. Int J Mol Sci 14:10591–10607. https://doi.org/10.3390/ijms140510591
58. Caravan P, Ellison JJ, McMurry TJ, Lauffer RB (1999) Gadolinium(III) chelates as MRI contrast agents: structure, dynamics, and applications. Chem Rev 99:2293–2352

59. Wiewiorski M, Miska M, Kretzschmar M et al (2013) Delayed gadolinium-enhanced MRI of cartilage of the ankle joint: results after autologous matrix-induced chondrogenesis (AMIC)-aided reconstruction of osteochondral lesions of the talus. Clin Radiol 68:1031–1038. https://doi.org/10.1016/j.crad.2013.04.016

60. Freedman JD, Lusic H, Wiewiorski M et al (2015) A cationic gadolinium contrast agent for magnetic resonance imaging of cartilage. Chem Commun (Camb) 51:11166–11169. https://doi.org/10.1039/c5cc03354c

61. Filardo G, Kon E, Di Martino A et al (2012) Second-generation arthroscopic autologous chondrocyte implantation for the treatment of degenerative cartilage lesions. Knee Surg Sports Traumatol Arthrosc 20:1704–1713. https://doi.org/10.1007/s00167-011-1732-5

62. Piontek T, Ciemniewska-Gorzela K, Szulc A et al (2012) All-arthroscopic AMIC procedure for repair of cartilage defects of the knee. Knee Surgery, Sport Traumatol Arthrosc 20:922–925. https://doi.org/10.1007/s00167-011-1657-z

63. Rodrigues MT, Gomes ME, Reis RL (2011) Current strategies for osteochondral regeneration: from stem cells to pre-clinical approaches. Curr Opin Biotechnol 22:726–733. https://doi.org/10.1016/j.copbio.2011.04.006

64. Karkabi S, Rosenberg N (2015) Arthroscopic debridement with lavage and arthroscopic lavage only as the treatment of symptomatic osteoarthritic knee. Open J Clin Diagnostics 05:68–73. https://doi.org/10.4236/ojcd.2015.52013

65. Steadman JR, Rodkey WG, Briggs KK (2010) Microfracture: its history and experience of the developing surgeon. Cartilage 1:78–86. https://doi.org/10.1177/1947603510365533

66. Torrie AM, Kesler WW, Elkin J, Gallo RA (2015) Osteochondral allograft. Curr Rev Musculoskelet Med 8:413–422. https://doi.org/10.1007/s12178-015-9298-3

67. Espregueira-Mendes J, Pereira H, Sevivas N et al (2012) Osteochondral transplantation using autografts from the upper tibio-fibular joint for the treatment of knee cartilage lesions. Knee Surg Sports Traumatol Arthrosc 20:1136–1142. https://doi.org/10.1007/s00167-012-1910-0

68. Health Quality Ontario (2005) Arthroscopic lavage and debridement for osteoarthritis of the knee: an evidence-based analysis. Ont Health Technol Assess Ser 5:1–37

69. Segal NA, Buckwalter JA, Amendola A (2006) Other surgical techniques for osteoarthritis. Best Pract Res Clin Rheumatol 20:155–176. https://doi.org/10.1016/j.berh.2005.09.009

70. Chen H, Sun J, Hoemann CD et al (2009) Drilling and microfracture lead to different bone structure and necrosis during bone-marrow stimulation for cartilage repair. J Orthop Res 27:1432–1438. https://doi.org/10.1002/jor.20905

71. Steadman JR, Rodkey WG, Briggs KK, Rodrigo JJ (1999) The microfracture technic in the management of complete cartilage defects in the knee joint. Orthopade 28:26. https://doi.org/10.1007/s001320050318

72. Steadman JR, Rodkey WG, Singleton SB, Briggs KK (1997) Microfracture technique for full-thickness chondral defects: technique and clinical results. Oper Tech Orthop 7:300–304. https://doi.org/10.1016/S1048-6666(97)80033-X

73. Steadman JR, Rodkey WG, Rodrigo JJ (2001) Microfracture: surgical technique and rehabilitation to treat chondral defects. Clin Orthop Relat Res 391:S362–S369

74. Gadjanski I, Vunjak-Novakovic G (2015) Challenges in engineering osteochondral tissue grafts with hierarchical structures. Expert Opin Biol Ther 15:1583–1599. https://doi.org/10.1517/14712598.2015.1070825

75. Hangody L, Kish G, Kárpáti Z et al (1997) Arthroscopic autogenous osteochondral mosaicplasty for the treatment of femoral condylar articular defects. Knee Surgery, Sport Traumatol Arthrosc 5:262–267. https://doi.org/10.1007/s001670050061

76. Bobić V (1996) Arthroscopic osteochondral autograft transplantation in anterior cruciate ligament reconstruction: a preliminary clinical study. Knee Surg Sports Traumatol Arthrosc 3(4):262

77. Makris EA, Gomoll AH, Malizos KN et al (2014) Repair and tissue engineering techniques for articular cartilage. Nat Rev Rheumatol. https://doi.org/10.1038/nrrheum.2014.157

78. Nooeaid P, Salih V, Beier JP, Boccaccini AR (2012) Osteochondral tissue engineering: scaffolds, stem cells and applications. J Cell Mol Med 16:2247–2270. https://doi.org/10.1111/j.1582-4934.2012.01571.x
79. Foldager CB (2013) Advances in autologous chondrocyte implantation and related techniques for cartilage repair. Dan Med J 60:B4600
80. Diaz-romero J, Nesic D, Grogan SP et al (2008) Immunophenotypic changes of human articular chondrocytes during monolayer culture reflect bona fide dedifferentiation rather than amplification of progenitor. J Cell Physiol 214(1):75–83. https://doi.org/10.1002/JCP
81. Tseng A, Pomerantseva I, Cronce MJ et al (2014) Extensively expanded auricular chondrocytes form neocartilage in vivo. Cartilage 5:241–251. https://doi.org/10.1177/1947603514546740
82. Ma B, Leijten JCH, Wu L et al (2013) Gene expression profiling of dedifferentiated human articular chondrocytes in monolayer culture. Osteoarthr Cartil 21:599–603. https://doi.org/10.1016/j.joca.2013.01.014
83. Leijten JCH, Georgi N, Wu L et al (2013) Cell sources for articular cartilage repair strategies: shifting from monocultures to cocultures. Tissue Eng Part B Rev 19:31–40. https://doi.org/10.1089/ten.teb.2012.0273
84. Hubka KM, Dahlin RL, Meretoja VV et al (2014) Enhancing chondrogenic phenotype for cartilage tissue engineering: monoculture and coculture of articular chondrocytes and mesenchymal stem cells. Tissue Eng Part B Rev 20:641–654. https://doi.org/10.1089/ten.TEB.2014.0034
85. Panseri S, Russo A, Cunha C et al (2012) Osteochondral tissue engineering approaches for articular cartilage and subchondral bone regeneration. Knee Surg Sports Traumatol Arthrosc 20:1182–1191. https://doi.org/10.1007/s00167-011-1655-1
86. Hunziker EB (2002) Articular cartilage repair: basic science and clinical progress. A review of the current status and prospects. Osteoarthr Cartil 10:432–463. https://doi.org/10.1053/joca.2002.0801
87. Caplan A I, Bruder SP (2001) Mesenchymal stem cells: building blocks for molecular medicine in the 21st century. Trends Mol Med 7:259–264
88. Dominici M, Le Blanc K, Mueller I et al (2006) Minimal criteria for defining multipotent mesenchymal stromal cells. The International Society for Cellular Therapy position statement. Cytotherapy 8:315–317. https://doi.org/10.1080/14653240600855905
89. Wang M, Yuan Z, Ma N et al (2017) Advances and prospects in stem cells for cartilage regeneration. Stem Cells Int 2017:1–16. https://doi.org/10.1155/2017/4130607
90. Bourin P, Bunnell BA, Casteilla L et al (2013) Stromal cells from the adipose tissue-derived stromal vascular fraction and culture expanded adipose tissue-derived stromal/stem cells: a joint statement of the International Federation for Adipose Therapeutics and Science (IFATS) and the International Society for Cellular Therapy (ISCT). Cytotherapy 15:641–648. https://doi.org/10.1016/j.jcyt.2013.02.006
91. Sakaguchi Y, Sekiya I, Yagishita K, Muneta T (2005) Comparison of human stem cells derived from various mesenchymal tissues: superiority of synovium as a cell source. Arthritis Rheum 52:2521–2529. https://doi.org/10.1002/art.21212
92. Schumacher BL, Hughes CE, Kuettner KE et al (1999) Immunodetection and partial cDNA sequence of the proteoglycan, superficial zone protein, synthesized by cells lining synovial joints. J Orthop Res 17:110–120. https://doi.org/10.1002/jor.1100170117
93. Zhou C, Zheng H, Seol D et al (2014) Gene expression profiles reveal that chondrogenic progenitor cells and synovial cells are closely related. J Orthop Res 32:981–988. https://doi.org/10.1002/jor.22641
94. Pei M, Li JT, Shoukry M, Zhang Y (2011) A review of decellularized stem cell matrix: a novel cell expansion system for cartilage tissue engineering. Eur Cell Mater 22:333–343. discussion 343

95. Pei M, He F (2012) Extracellular matrix deposited by synovium-derived stem cells delays replicative senescent chondrocyte dedifferentiation and enhances redifferentiation. J Cell Physiol 227:2163–2174. https://doi.org/10.1002/jcp.22950

96. Ando W, Tateishi K, Katakai D et al (2008) *In vitro* generation of a scaffold-free tissue-engineered construct (TEC) derived from human synovial mesenchymal stem cells: biological and mechanical properties and further chondrogenic potential. Tissue Eng Part A 14:2041–2049. https://doi.org/10.1089/ten.tea.2008.0015

97. Vonk LA, de Windt TS, Slaper-Cortenbach ICM, Saris DBF (2015) Autologous, allogeneic, induced pluripotent stem cell or a combination stem cell therapy? Where are we headed in cartilage repair and why: a concise review. Stem Cell Res Ther 6:94. https://doi.org/10.1186/s13287-015-0086-1

98. Zuk PA, Zhu M, Ashjian P et al (2002) Human adipose tissue is a source of multipotent stem cells. Mol Biol Cell 13:4279–4295. https://doi.org/10.1091/mbc.E02-02-0105

99. Puetzer JL, Petitte JN, Loboa EG (2010) Comparative review of growth factors for induction of three-dimensional in vitro chondrogenesis in human mesenchymal stem cells isolated from bone marrow and adipose tissue. Tissue Eng Part B Rev 16:435–444. https://doi.org/10.1089/ten.TEB.2009.0705

100. Mehlhorn AT, Niemeyer P, Kaschte K et al (2007) Differential effects of BMP-2 and TGF-beta1 on chondrogenic differentiation of adipose derived stem cells. Cell Prolif 40:809–823. https://doi.org/10.1111/j.1365-2184.2007.00473.x

101. Schelbergen RF, van Dalen S, ter Huurne M et al (2014) Treatment efficacy of adipose-derived stem cells in experimental osteoarthritis is driven by high synovial activation and reflected by S100A8/A9 serum levels. Osteoarthr Cartil 22:1158–1166. https://doi.org/10.1016/j.joca.2014.05.022

102. Mardani M, Hashemibeni B, Ansar MM et al (2013) Comparison between chondrogenic markers of differentiated chondrocytes from adipose derived stem cells and articular chondrocytes in vitro. Iran J Basic Med Sci 16:763–773

103. Hildner F, Albrecht C, Gabriel C et al (2011) State of the art and future perspectives of articular cartilage regeneration : a focus on adipose-derived stem cells and platelet-derived products. J Tissue Eng Regen Med 5:36–51. https://doi.org/10.1002/term.386

104. Diekmann B, Rowland C, DP L (2010) Chondrogenesis of adult stem cells from adipose tissue and bone marrow: induction by growth factors and cartilage-derived matrix. Tissue Eng Part A 16:523–533

105. RA O (2012) Cell sources for the regeneration of articular cartilage: the past, the horizon and the future. Int J Exp Pathol 93:389–400. https://doi.org/10.1111/j.1365-2613.2012.00837.x

106. Kobayashi Y, Okada Y, Itakura G et al (2012) Pre-evaluated safe human iPSC-derived neural stem cells promote functional recovery after spinal cord injury in common marmoset without tumorigenicity. PLoS One 7:e52787. https://doi.org/10.1371/journal.pone.0052787

107. Takahashi K, Yamanaka S (2006) Induction of pluripotent stem cells from mouse embryonic and adult fibroblast cultures by defined factors. Cell 126:663–676. https://doi.org/10.1016/j.cell.2006.07.024

108. Takahashi K, Tanabe K, Ohnuki M et al (2007) Induction of pluripotent stem cells from adult human fibroblasts by defined factors. Cell 131:861–872. https://doi.org/10.1016/j.cell.2007.11.019

109. Singh VK, Kalsan M, Kumar N et al (2015) Induced pluripotent stem cells: applications in regenerative medicine, disease modeling, and drug discovery. Front Cell Dev Biol 3:2. https://doi.org/10.3389/fcell.2015.00002

110. Medvedev SP, Grigor'eva EV, Shevchenko AI et al (2011) Human induced pluripotent stem cells derived from fetal neural stem cells successfully undergo directed differentiation into cartilage. Stem Cells Dev 20:1099–1112. https://doi.org/10.1089/scd.2010.0249

111. Wei Y, Zeng W, Wan R et al (2012) Chondrogenic differentiation of induced pluripotent stem cells from osteoarthritic chondrocytes in alginate matrix. Eur Cell Mater 23:1–12

112. Hiramatsu K, Sasagawa S, Outani H et al (2011) Generation of hyaline cartilaginous tissue from mouse adult dermal fibroblast culture by defined factors. J Clin Invest 121:640–657. https://doi.org/10.1172/JCI44605

113. Outani H, Okada M, Yamashita A et al (2013) Direct induction of chondrogenic cells from human dermal fibroblast culture by defined factors. PLoS One 8:e77365. https://doi.org/10.1371/journal.pone.0077365

114. Salgado AJ, Oliveira JM, Martins A et al (2013) Tissue engineering and regenerative medicine: past, present, and future. Int Rev Neurobiol. https://doi.org/10.1016/B978-0-12-410499-0.00001-0

115. Mano JF, Silva GA, Azevedo HS et al (2007) Natural origin biodegradable systems in tissue engineering and regenerative medicine: present status and some moving trends. J R Soc Interface 4:999–1030. https://doi.org/10.1098/rsif.2007.0220

116. Kobayashi S, Fujikawa S-i, Ohmae M (2003) Enzymatic synthesis of chondroitin and its derivatives catalyzed by hyaluronidase. J. Am. Chem. Soc. 125(47):14357–14369. https://doi.org/10.1021/JA036584X

117. Silva TH, Alves A, Popa EG et al (2012) Marine algae sulfated polysaccharides for tissue engineering and drug delivery approaches. Biomatter 2:278–289. https://doi.org/10.4161/biom.22947

118. Yan LP, Oliveira JM, Oliveira AL, Reis RL (2013) Silk fibroin/nano-CaP bilayered scaffolds for osteochondral tissue engineering. Key Eng Mater 587:245–248. https://doi.org/10.4028/www.scientific.net/KEM.587.245

119. Yang C, Hillas PJ, Julio AB et al (2004) The application of recombinant human collagen in tissue engineering. BioDrugs 18:103–119

120. Rutgers M, Saris DB, Vonk LA et al (2013) Effect of collagen type I or type II on chondrogenesis by cultured human articular chondrocytes. Tissue Eng Part A 19:59–65. https://doi.org/10.1089/ten.TEA.2011.0416

121. Burdick JA, Mauck RL (2010) Biomaterials for tissue engineering applications: a review of the past and future trends. Springer Science & Business Media, Austria

122. Parenteau-Bareil R, Gauvin R, Berthod F (2010) Collagen-based biomaterials for tissue engineering applications. Materials (Basel) 3:1863–1887. https://doi.org/10.3390/ma3031863

123. Grant GT, Morris ER, Rees DA et al (1973) Biological interactions between polysaccharides and divalent cations: the egg-box model. FEBS Lett 32:195–198. https://doi.org/10.1016/0014-5793(73)80770-7

124. Petrenko YA, Ivanov RV, Petrenko AY, Lozinsky VI (2011) Coupling of gelatin to inner surfaces of pore walls in spongy alginate-based scaffolds facilitates the adhesion, growth and differentiation of human bone marrow mesenchymal stromal cells. J Mater Sci Mater Med 22:1529–1540. https://doi.org/10.1007/s10856-011-4323-6

125. Bonaventure J, Kadhom N, Cohen-Solal L et al (1994) Reexpression of cartilage-specific genes by dedifferentiated human articular chondrocytes cultured in alginate beads. Exp Cell Res 212:97–104. https://doi.org/10.1006/excr.1994.1123

126. Abarrategi A, López-Morales Y, Ramos V et al (2010) Chitosan scaffolds for osteochondral tissue regeneration. J Biomed Mater Res Part A 95A:1132–1141. https://doi.org/10.1002/jbm.a.32912

127. Deng Y, Ren J, Chen G et al (2017) Injectable in situ cross-linking chitosan-hyaluronic acid based hydrogels for abdominal tissue regeneration. Sci Rep 7:2699. https://doi.org/10.1038/s41598-017-02962-z

128. Yan L-P, Oliveira JM, Oliveira AL, Reis RL (2015) In vitro evaluation of the biological performance of macro/micro-porous silk fibroin and silk-nano calcium phosphate scaffolds. J Biomed Mater Res B Appl Biomater 103:888–898. https://doi.org/10.1002/jbm.b.33267

129. Yan L-P, Silva-Correia J, Oliveira MB et al (2015) Bilayered silk/silk-nanoCaP scaffolds for osteochondral tissue engineering: in vitro and in vivo assessment of biological performance. Acta Biomater 12:227–241. https://doi.org/10.1016/j.actbio.2014.10.021

130. Ondrésik M, Azevedo Maia FR, da Silva Morais A et al (2017) Management of knee osteo-arthritis. Current status and future trends. Biotechnol Bioeng 114:717–739. https://doi.org/10.1002/bit.26182

131. Antunes JC, Oliveira JM, Reis RL et al (2010) Novel poly(L-lactic acid)/hyaluronic acid macroporous hybrid scaffolds: characterization and assessment of cytotoxicity. J Biomed Mater Res A 94:856–869. https://doi.org/10.1002/jbm.a.32753

132. Wright LD, McKeon-Fischer KD, Cui Z et al (2014) PDLA/PLLA and PDLA/PCL nanofi-bers with a chitosan-based hydrogel in composite scaffolds for tissue engineered cartilage. J Tissue Eng Regen Med 8:946–954. https://doi.org/10.1002/term.1591

133. Liu W, Li Z, Zheng L et al (2016) Electrospun fibrous silk fibroin/poly(L-lactic acid) scaffold for cartilage tissue engineering. Tissue Eng Regen Med 13:516–526. https://doi.org/10.1007/s13770-016-9099-9

134. Miljkovic ND, Cooper GM, Marra KG (2008) Chondrogenesis, bone morphogenetic protein-4 and mesenchymal stem cells. Osteoarthr Cartil 16:1121–1130. https://doi.org/10.1016/j.joca.2008.03.003

135. Samavedi S, Whittington AR, Goldstein AS (2013) Calcium phosphate ceramics in bone tis-sue engineering: a review of properties and their influence on cell behavior. Acta Biomater 9:8037–8045. https://doi.org/10.1016/j.actbio.2013.06.014

136. Trombetta R, Inzana JA, Schwarz EM et al (2017) 3D printing of calcium phosphate ceram-ics for bone tissue engineering and drug delivery. Ann Biomed Eng 45:23–44. https://doi.org/10.1007/s10439-016-1678-3

137. Nover AB, Lee SL, Georgescu MS et al (2015) Porous titanium bases for osteochondral tis-sue engineering. Acta Biomater 27:286–293. https://doi.org/10.1016/j.actbio.2015.08.045

Chapter 2
Emerging Concepts in Treating Cartilage, Osteochondral Defects, and Osteoarthritis of the Knee and Ankle

Hélder Pereira, Ibrahim Fatih Cengiz, Carlos Vilela, Pedro L. Ripoll,
João Espregueira-Mendes, J. Miguel Oliveira, Rui L. Reis,
and C. Niek van Dijk

Abstract The management and treatment of cartilage lesions, osteochondral defects, and osteoarthritis remain a challenge in orthopedics. Moreover, these entities have different behaviors in different joints, such as the knee and the ankle, which have inherent differences in function, biology, and biomechanics. There has been a huge development on the conservative treatment (new technologies including orthobiologics) as well as on the surgical approach. Some surgical development upraises from technical improvements including advanced arthroscopic techniques but also from increased knowledge arriving from basic science research and tissue

H. Pereira (✉)
Orthopedic Department Centro Hospitalar Póvoa de Varzim, Vila do Conde, Portugal

International Centre of Sports Traumatology of the Ave, Taipas, Portugal

3B's Research Group – Biomaterials, Biodegradables and Biomimetics, University of Minho, Headquarters of the European Institute of Excellence on Tissue Engineering and Regenerative Medicine, Avepark, Parque de Ciência e Tecnologia, Zona Industrial da Gandra, Barco, Guimarães, Portugal

ICVS/3B's – PT Government Associate Laboratory, Braga, Guimarães, Portugal

Ripoll y De Prado Sports Clinic: Murcia – Madrid FIFA Medical Centre of Excellence, Madrid, Spain

I. F. Cengiz · C. Vilela
3B's Research Group – Biomaterials, Biodegradables and Biomimetics, University of Minho, Headquarters of the European Institute of Excellence on Tissue Engineering and Regenerative Medicine, Avepark, Parque de Ciência e Tecnologia, Zona Industrial da Gandra, Barco, Guimarães, Portugal

ICVS/3B's – PT Government Associate Laboratory, Braga, Guimarães, Portugal

© Springer International Publishing AG, part of Springer Nature 2018
J. M. Oliveira et al. (eds.), *Osteochondral Tissue Engineering*,
Advances in Experimental Medicine and Biology 1059,
https://doi.org/10.1007/978-3-319-76735-2_2

engineering and regenerative medicine approaches. This work addresses the state of the art concerning basic science comparing the knee and ankle as well as current options for treatment. Furthermore, the most promising research developments promising new options for the future are discussed.

Keywords Surgery · Autologous osteochondral transplantation · Bone marrow stimulation · Congruency · Alignment · Tissue engineering and regenerative medicine

P. L. Ripoll
International Centre of Sports Traumatology of the Ave, Taipas, Portugal

Ripoll y De Prado Sports Clinic: Murcia – Madrid FIFA Medical Centre of Excellence, Madrid, Spain

J. Espregueira-Mendes
3B's Research Group – Biomaterials, Biodegradables and Biomimetics, University of Minho, Headquarters of the European Institute of Excellence on Tissue Engineering and Regenerative Medicine, Avepark, Parque de Ciência e Tecnologia, Zona Industrial da Gandra, Barco, Guimarães, Portugal

ICVS/3B's – PT Government Associate Laboratory, Braga, Guimarães, Portugal

Clínica do Dragão, Espregueira-Mendes Sports Centre – FIFA Medical Centre of Excellence, Porto, Portugal

Dom Henrique Research Centre, Porto, Portugal

Orthopedic Department, University of Minho, Braga, Portugal

J. Miguel Oliveira · R. L. Reis
3B's Research Group – Biomaterials, Biodegradables and Biomimetics, University of Minho, Headquarters of the European Institute of Excellence on Tissue Engineering and Regenerative Medicine, Avepark, Parque de Ciência e Tecnologia, Zona Industrial da Gandra, Barco, Guimarães, Portugal

ICVS/3B's – PT Government Associate Laboratory, Braga, Guimarães, Portugal

The Discoveries Centre for Regenerative and Precision Medicine, Headquarters at University of Minho, Guimarães, Portugal

C. Niek van Dijk
Ripoll y De Prado Sports Clinic: Murcia – Madrid FIFA Medical Centre of Excellence, Madrid, Spain

Department of Orthopaedic Surgery and Academic Center for Evidence-Based Sports Medicine, Academic Medical Center, University of Amsterdam, Amsterdam Movement Sciences, Amsterdam, The Netherlands

Highlights
- The treatment of osteochondral defects and osteoarthritis is complex and multifactorial.
- The most commonly used surgical techniques for the treatment of osteochondral defects include microfractures, fixation, autologous or allogeneic osteochondral transplantation or mosaicplasty autologous chondrocyte implantation, and matrix-induced autologous chondrocyte implantation. So far, no method has been able to consistently achieve repair of osteochondral defects similar to the native tissue.
- Tissue engineering and regenerative medicine strategies promise new options for future treatments of cartilage and osteochondral defects.

Top 10 Suggested References
1. Dahmen J, Lambers KTA, Reilingh ML, van Bergen CJA, Stufkens SAS, Kerkhoffs Gino MMJ (2017) No superior treatment for primary osteochondral defects of the talus. Knee Surg Sports Traumatol Arthrosc. doi:https://doi.org/10.1007/s00167-017-4616-5 [Epub ahead of print]
2. Oliveira M, Reis RL (2017) Regenerative strategies for the treatment of knee joint disabilities. Studies in mechanobiology, tissue engineering and biomaterials. Springer. doi:https://doi.org/10.1007/978-3-319-44785-8
3. Ferreira C, Vuurberg G, Oliveira JM, Espregueira-Mendes J, H. P, Reis RL, Ripoll P (2016) Assessment of clinical outcome after Osteochondral Autologous Transplantation technique for the treatment of ankle lesions: a systematic review. J Isakos. doi:jisakos-2015-000020.R2
4. Andrade R, Vasta S, Papalia R, Pereira H, Oliveira JM, Reis RL, Espregueira-Mendes J (2016) Prevalence of articular cartilage lesions and surgical clinical outcomes in football (Soccer) players' knees: a systematic review. Arthroscopy 32(7):1466–1477. doi:https://doi.org/10.1016/j.arthro.2016.01.055
5. Oliveira JM, Rodrigues MT, Silva SS, Malafaya PB, Gomes ME, Viegas CA, Dias IR, Azevedo JT, Mano JF, Reis RL (2006) Novel hydroxyapatite/chitosan bilayered scaffold for osteochondral tissue-engineering applications: scaffold design and its performance when seeded with goat bone marrow stromal cells.
6. Yan LP, Silva-Correia J, Oliveira MB, Vilela C, Pereira H, Sousa RA, Mano JF, Oliveira AL, Oliveira JM, Reis RL (2015) Bilayered silk/silk-nanoCaP scaffolds for osteochondral tissue engineering: in vitro and in vivo assessment of biological performance. Acta Biomaterialia 12:227–241. doi:https://doi.org/10.1016/j.actbio.2014.10.021
7. Khan WS, Longo UG, Adesida A, Denaro V (2012) Stem cell and tissue engineering applications in orthopaedics and musculoskeletal medicine. Stem Cells Int 2012:403170. doi:https://doi.org/10.1155/2012/403170

(continued)

8. Correia SI, Silva-Correia J, Pereira H, Canadas RF, da Silva Morais A, Frias AM, Sousa RA, van Dijk CN, Espregueira-Mendes J, Reis RL, Oliveira JM (2015) Posterior talar process as a suitable cell source for treatment of cartilage and osteochondral defects of the talus. J Tissue Eng Regen Med. doi:https://doi.org/10.1002/term.2092
9. Liu M, Zeng X, Ma C, Yi H, Ali Z, Mou X, Li S, Deng Y, He N (2017) Injectable hydrogels for cartilage and bone tissue engineering. Bone Res 5:17014. doi:https://doi.org/10.1038/boneres.2017.14
10. Rai V, Dilisio MF, Dietz NE, Agrawal DK (2017) Recent strategies in cartilage repair: a systemic review of the scaffold development and tissue engineering. J Biomed Mater Res A 105(8):2343–2354. doi:https://doi.org/10.1002/jbm.a.36087

Fact Box 1 – Epidemiology of Osteoarthritis and Osteochondral Injuries of the Knee and Ankle
- Osteoarthritis is the most common joint disease with worldwide prevalence over 241 825 million people.
- The overall prevalence of full-thickness focal chondral defects in athletes has been stated as 36%.
- Osteochondral defects of the knee combined with meniscus injuries account for 3.7% of all injuries among elite football players.
- The incidence of OA in the ankle is considerably smaller than in the knee. The prevalence of symptomatic primary OA in the ankle is lower than 1% of the population.
- Moreover, ankle OA does not seem to increase with aging.
- Osteochondral defects of the talus can occur in up to 70% of acute ankle sprains and fractures.

Fact Box 2 – Osteochondral Defects (OCDs) of the Knee
- The treatment of OCDs and OA of the knee is complex and multifactorial.
- Nonoperative options include chondroprotective pharmacotherapy (glucosamines, chondroitin, diacerein, hyaluronic acid, platelet-rich plasma, and cell-based therapy), nonsteroidal anti-inflammatory medication, and physiotherapy.
- The most commonly used surgical techniques for the treatment of knee OCD lesions include microfractures, fixation, autologous (or allogenic) osteochondral transplantation (OATS) or mosaicplasty autologous chondrocyte implantation (ACI), and matrix-induced autologous chondrocyte implantation (MACI).
- So far, no method has been able to consistently achieve repair of OCD by the hyaline cartilage similar to the native.

Fact Box 3 – Osteochondral Defects (OCDs) of the Ankle
- Etiology of ankle OCD can either be traumatic and non-traumatic.
- Always consider association of ankle sprain or chronic ankle instability in the etiology of OCD.
- Fixation of a large fragment should always be attempted.
- Microfracture is still the most popular treatment.
- Similarly to that observed in the knee, no surgical treatment has proven superiority over any other.
- Tissue engineering and regenerative medicine approaches promise new options for the future.

Fact Box 4 – Tissue Engineering and Regenerative Medicine (TERM): Road for the Future
- The basic triad of TERM includes the combination of cells, scaffolds, and bioactive proteins in the healing process of any tissue.
- Orthobiologics might include conservative treatment by injection therapy including growth factors, hydrogels, cell-based therapy, or even combining gene therapy.
- Orthobiologics aim to improve symptomatic cartilage damage and also envision to delay the progressive joint degeneration.
- There has been a massive development on scaffolds assembling including nanostructure and tridimensional bioprinting.
- The road for the future seems to combine the best possible knowledge of all TERM variables aiming to achieve in the laboratory a tissue which can be matured to achieve similar features as the native and custom-made to the defect.

2.1 Introduction

Traumatic and non-traumatic etiology has been implicated in osteochondral injuries, which might or might not develop to general joint degeneration [1]. Degeneration linked to the aging process, trauma-related injuries, and deteriorating or idiopathic disorders might lead to osteochondral lesions [2]. Cartilage damage has been linked to several etiologies including some, which remain poorly understood to date. It is recognized that OA has a higher incidence in aged people [3]. However, the prevalence of articular cartilage injuries has been reported to be higher in athletes when compared to the general population [4–7]. Sports practice has been increasing worldwide. Taking as an example football (soccer), which is the most played sport worldwide, there are more than 300 million people federated and many more playing without register [8]. Any high-impact contact sport, moreover at high competitive level, might result in damage of the knee and/or structures, including articular cartilage injuries [4, 5, 9]. The large variability of the OA regarding the etiology, histological findings between individuals and groups, and response to therapies demonstrates that there is still a long way for more advanced understanding of this condition [10].

Nevertheless, cartilage injuries are often a consequence of dynamic and repetitive mechanical joint loading [11–14]. Despite the fact that cartilage is a poorly innervated and irrigated tissue when the damage reaches the subchondral bone, complaints will derive [15], including pain, swelling, catching, and locking [5, 16, 17]. Nevertheless, articular cartilage injuries may be present in asymptomatic people or even athletes. There is controversial data concerning the frequency of knee pain referred by footballers [4]. However, patellofemoral conditions are more frequent in women, while in the ankle, the lesion is mostly present at the talus [18].

If a "pure" cartilage lesion is considered, the damage occurs on the chondrocytes and articular cartilage extracellular matrix (ECM), above the subchondral plate. However, in OCDs besides cartilage injury, the subchondral bone is also involved. Many classifications have been proposed either as global OCD assessment or joint-specific scores [1]. The Outerbridge classification modified by the ICRS (International Cartilage Repair Society) is the one that most frequently used. In brief, it enrolls Grade 0, normal cartilage; Grade I, cartilage softening and swelling; Grade II, partial thickness defect not extending the subchondral bone (<1.5 cm diameter); Grade III, fissures up to the subchondral bone level (>1.5 cm diameter); and Grade IV, OCD with exposed subchondral bone. In some cases of non-traumatic etiology, usually in younger ages, in which a segment of cartilage and subchondral bone detaches from the underlying bone, a vascular or genetic etiology has been proposed, and it is referred to as osteochondritis dissecans [19]. If complete detachment of osteochondritis dissecans occurs, this might lead to intra-articular loose bodies, which further contribute to joint degeneration.

One of the major concerns related to OCDs is the secondary progression to OA [19]. However, it could never have been shown in the ankle joint that the natural history of a focal OCD is secondary OA, while most studies in the knee joint suggest it [1]. This might be related to different joint biomechanics. However, an injury to the hyaline cartilage that is related to a previous trauma is considered as a major risk factor for OA [20]. OA can be a restrictive and painful condition, which is most frequently seen in the knees, hips, ankles, and hands although it might affect any joint. Patients suffering from OA characteristically present pain, episodes of swelling, progressive deformity, and limited range of motion.

OCDs or any cartilage damage, if not adequately dealt with, may result to an earlier onset of joint degradation and osteoarthritis (OA) [21–23]. Symptomatic OCDs in any joint may lead to activity-related symptoms and require changes in lifestyle with permanent functional limitations [24–27]. Besides cartilage damage, injuries affecting the subchondral bone are frequent. However, it is still debatable whether these changes precede the biomechanical lesions of the hyaline cartilage or correspond to secondary changes.

Conservative treatments are based on the adaptation of lifestyle, anti-inflammatory or painkiller medications, supplements (e.g., glucosamine, chondroitin), and orthobiologics (hyaluronic acid, growth factors, cell therapies) [10, 28–32]. Surgical treatment ranges from arthroscopy to osteotomies, to partial or total joint replacement or fusion [10, 28, 29, 31]. Clinical history, physical examination, and imaging (standing x-rays, CT, or MRI) are mandatory for diagnosis [19, 31]. For histological assessment in specific cases or research purposes, it is possible to collect tissue and synovial fluid from joint injections of the knee or ankle without major complications [33]. This is particularly useful in rheumatologic conditions.

According to the current reports, OA has been affecting a significant number of people worldwide with a rise over time, and it represents a social and economic burden [34, 35]. Moreover, high-level sports involve high financial impact and intense social media coverage. Considering the athlete as a usually "young person" with a physically demanding profession, important factors such as age, level of completion, time into the season, and career status must be considered [4]. Therefore, dealing with OCDs and OA is a multifactorial social issue.

2.2 Epidemiology

OA is the most common joint disease [36]. It is not easy to define the global prevalence of OA, given the registered variations according to the used definition of OA for assessment, population characteristics (e.g., age, gender), geographic conditions, clinical-based or radiological-based studies, or self-reported OA [36]. One study reports the worldwide prevalence of clinical OA of 241 825 million people [34]. This number is known to be consecutively growing. The number of people with symptomatic OA has increased 71.9% between 1990 and 2013, and it is expected to keep rising in relation to the increase of life expectancy, among other factors [37, 38]. In Europe, the frequency of symptomatic knee OA is ranging between 5.4% and 29.8% [36]. According to the Framingham study and Johnston County Osteoarthritis Project, in the United States, this number was 7% and 17%, respectively [36]. When considering only a population over 45 years old, the OA prevalence ranged from 19% to 28% [36]. Flanigan et al. reported articular cartilage injuries in a cohort of 931 athletes, involving 732 men and 199 women, with a mean age of 33 years old [5]. The overall prevalence of full-thickness focal chondral defects of the knee in athletes was 36% [5]. From these, only 40% were professional athletes. The UEFA Elite Club Injury Study Group (which studies health conditions of 29 elite European football clubs) in the season of 2015/2016 reported that cartilage/meniscus injuries accounted for 3.7% of all injuries [39].

The incidence of OA in the ankle is considerably smaller than in the knee. The prevalence of symptomatic primary OA in the ankle is lower than 1% of the population [40]. Moreover, it does not seem to increase with aging [40].

Osteochondral defects of the talus can occur in up to 70% of acute ankle sprains and fractures [41]. The different incidences and prevalences of OA on both joints most probably are linked to differences in anatomy and biomechanics [10], but no definite conclusions explaining such differences are currently available.

Genetics or geographical influence might be suggested with an observed extreme variation such as the OA is present in only around 1.4% of the urban Filipinos, and increases among some rural Iranian communities up to 19.3% [42]. Moreover, gender might play a role once a high female predominance has been reported [42], suggesting some role of sex hormones in this condition. With such prevalence and the fact that there is no "cure" up to now, treatment will require continuous clinical care, institutional costs, medication, and surgeries, dictating high healthcare-related costs, besides work absence, thus representing a socioeconomic burden [43, 44]. The global impact on diminished economic productivity added to the reimbursement compensation from the impaired and sometimes the need for third-person care further dictates additional costs [43, 45, 46]. According to a recent systematic review, the social costs of OA range from 0.25% to 0.50% of a country's gross domestic product (GDP) [35]. Considering all the aforementioned, this is one of the most relevant healthcare topics and is critical to improve our effectiveness in dealing with these conditions [47].

2.3   Knee Osteochondral Defects

Normal knee hyaline cartilage has optimum biomechanical characteristics adapted to its function and adjustment capacities to the loading stresses exerted at the joint [48]. Nevertheless, when these capacities are exceeded (e.g., by high-impact loading), there is a decrease in the cartilage proteoglycans levels and an increase in the levels of degradative enzymes (e.g., metalloproteases) that ultimately lead to chondrocyte apoptosis [49, 50]. The consequence will be a loss of cartilage volume and biomechanical resistance, peak contact pressures, and ultimately cartilage defects [48]. Moreover, due to its scarce irrigation and innervation, it has very limited healing potential [25, 29, 51, 52]. Due to these biological and biomechanical conditions, cartilage repair remains a challenge in orthopedics, and so far, there is no single reliable method to achieve repair by hyaline cartilage similar to the native [31].

The first approach employed in the treatment OCDs is a conservative treatment [53]. It includes periods of rest, non-weight bearing, prevention of stiffness by an active joint mobilization, neuro-muscle and proprioceptive trainings, as well as use of medication or orthobiologics [31]. Nonoperative options include chondroprotective pharmacotherapy (glucosamines, chondroitin, diacerein, hyaluronic acid, platelet-rich plasma, and cell-based therapy), nonsteroidal anti-inflammatory medication, and physiotherapy [24, 54, 55]. Particularly in the knee joint, conservative treatment often fails, after a variable period of improvement [31, 56]. A substantial number of patients will require surgical management [31].

2.4 State of the Art in the Treatment of Osteochondral Defects of the Knee

The treatment of OCDs and OA of the knee is complex and multifactorial [57]. The goal of treatment is to provide long-lasting relief of complaints and restore function to the maximum possible [4]. The biomechanical features of the knee joint should be considered as complex.

Nowadays, there are several available surgical techniques to approach a focal OCD. The most commonly used surgical techniques for the treatment of these lesions include microfracture (Fig. 2.1), fixation, autologous osteochondral transplantation (OATS) or mosaicplasty autologous chondrocyte implantation (ACI), and matrix-induced (Fig. 2.2) autologous chondrocyte implantation (MACI) [4, 52, 58, 59]. More recently, matrix-induced autologous stem cell implantation (MASI) has been introduced given the higher mitotic rate and other biological features of these cells and constructs [60, 61].

Whenever possible, fixation of a large OCD with underlying bone (Fig. 2.3) should be attempted once it represents the most "conservative" surgical approach

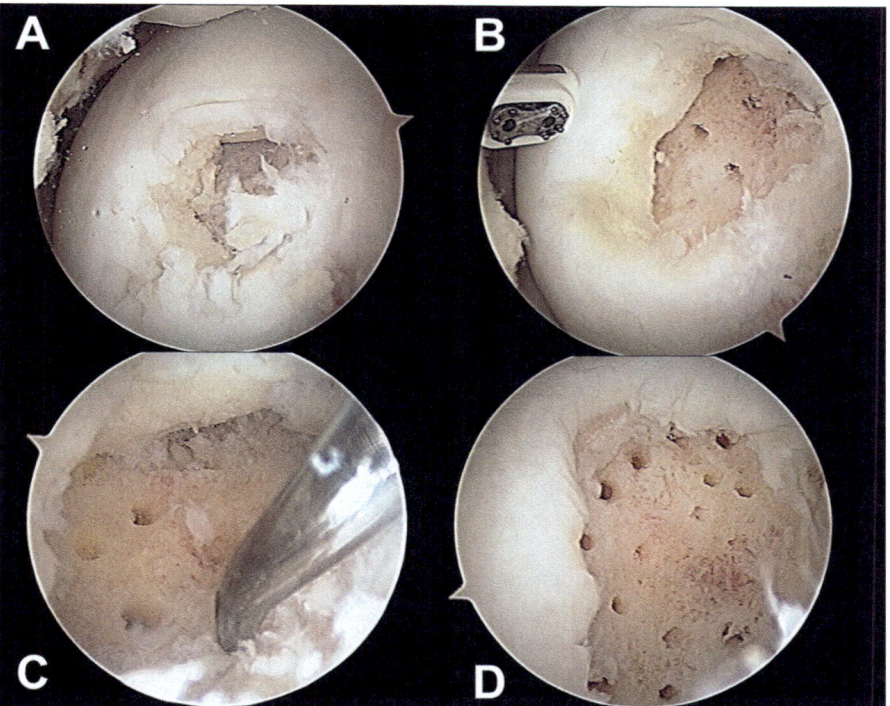

Fig. 2.1 Medial condyle grade IV osteochondral defect (**A**), debridement and microfractures with visible holes on the bone (**B–D**)

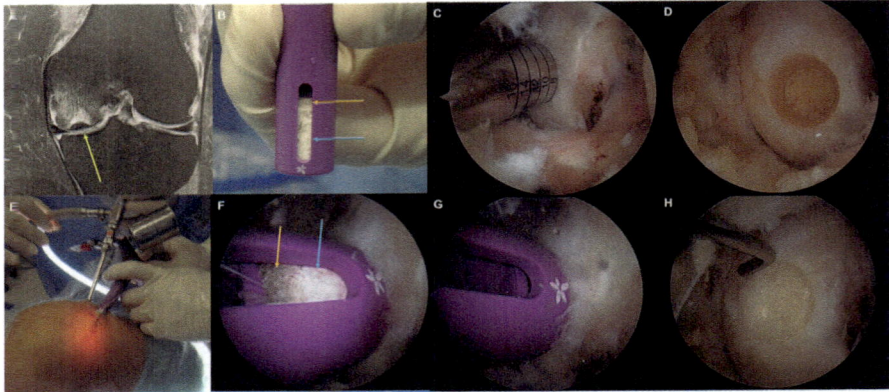

Fig. 2.2 Medial condyle unstable osteochondral defect (yellow arrow) (**A**), bilayered acellular scaffold with cartilage layer (orange arrow) and bone layer (blue arrow) (**B**), arthroscopy view with removal of the defect and preparing the receptor bone bed by means of a trephine (**C**), final arthroscopic look of the receptor zone (**D**), outside view of arthroscopic surgery (**E**), introduction of the acellular scaffold, (**H**) final aspect and palpation with a probe of the press-fit scaffold (**F, G**)

given the fact that it aims to preserve the native tissue. This is achieved by lifting the fragment (if possible keeping some partial attachment), preparing the bony beds from both sides (e.g., microfracturing) and fixation with screws or arrows [31]. Arthroscopic debridement and lavage with bone marrow stimulation such as drilling [62], microfracture (promoted by Steadman) [63], abrasion arthroplasty [64], and chondroplasty [65] are the initial surgical strategies. The rationale supporting bone marrow stimulation techniques is that by perforation of the subchondral bone, we create channels enabling the recruitment/migration of blood with growth factors and bone marrow stem cells to the defect site and formation of a stable clot, which fills the chondral defect [66–68]. Good short-term outcomes have been reported with this technique [69, 70]. Concerning histology, this treatment does not provide hyaline cartilage restoration [71, 72]. This healing process leads to fibrocartilage tissue formation, which has lower biomechanical characteristics and is more likely to break down [63, 73]. This relevant drawback is the main reason for the failure [63, 73]. Deterioration of clinical outcomes at long-term has been described, with revision surgery needed in some cases [74, 75]. Considering this fact, enhanced microfractures techniques have been recently developed with promising short-term outcomes [76–78].

More complex and anatomic strategies have been developed such as autologous or allogeneic osteochondral grafting, i.e., mosaicplasty technique [79]. Mosaicplasty is being used since 1994 when it was first performed by L. Hangody [80]. The OATS technique is used to transfer autologous (or allogeneic) bone and hyaline cartilage to the defect, providing a stable size-matched osteochondral autograft. For smaller defects, one single plug transfer to fill the defect seems to have advantages over several cylinders [79]. However, for larger defects, the mosaicplasty requires

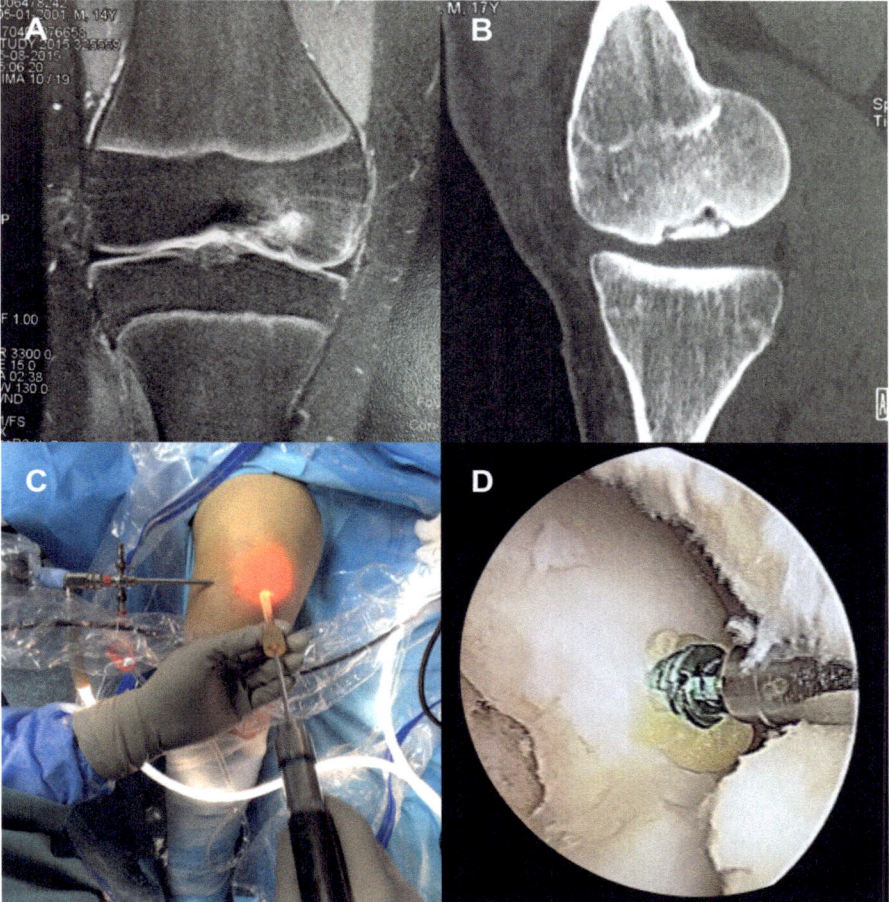

Fig. 2.3 MRI frontal view of medial condyle unstable osteochondral defect (OCD) with edema around the injury on T2 (**A**), CT lateral view assessing the underlying bone of the defect (**B**), outside view of the arthroscopic surgery (**C**), and OCD fixation with headless compression screw (**D**)

the transfer of multiple small cylinders (osteochondral plugs) to the defect [66, 67]. Nevertheless, this technique has several limitations such as restricted graft disposal and donor site-related morbidity once it creates a defect elsewhere in order to transfer tissue to the defect [81, 82]. Aiming to lower donor site morbidity, the upper tibiofemoral joint has been proposed as a potential donor site [79]. Despite its inherent risks and limitations, transplantation of osteochondral allograft is a viable option to manage larger osteochondral injuries, including those that involve an entire compartment [66, 67, 83].

The ACI approach (promoted by Mats Brittberg) is a two-stage procedure which involves harvesting of autologous chondrocytes on a first procedure, processing these in laboratory, and latterly implanting these cells in the articular cartilage defect aiming to achieve hyaline-like cartilage repair [66, 67, 84–86]. This procedure expected to accomplish higher longevity of the healed tissue improves long-term clinical and functional outcomes [86, 87]. The initial technique required a periosteal flap to cover the defect (sutured to the surrounding cartilage), and the cells were finally delivered under this coverage with a small needle. However, consistently reproducible results favoring this technique over the others have not been achieved [60, 86].

As a more advanced tissue engineering and regenerative medicine (TERM) approach, the MACI technique is an attractive alternative which involves culturing the chondrocyte cells into a tridimensional porous scaffold which is matured in the laboratory by means of bioreactors and afterward implanted into the defect [66, 67]. The MACI technique is technically less demanding and reduces surgical time, besides avoiding periosteal harvesting [88]. The reported short-to-midterm outcomes show promising results of this technique in articular cartilage injuries of the knee joint [89–91]. However, using the same principle, stem cells combined with scaffolds (MASI) have been attempted in order to improve the achieved outcome and are under development and research [60]. Some of these TERM-based approaches have been made commercially available or under commercial advertising (Table 2.1). These new techniques aim to be potential efficient options to restore OCDs; however, there is still a lack of evidence-based medicine supporting its widespread use. In the authors' opinion, it should be kept under strict research control until further conclusions can be obtained. Some of these emerging techniques include autologous matrix-induced chondrogenesis (AMIC™) [92, 93], bone marrow aspirate concentrate (BMAC) and mesenchymal stem cell-induced chondrogenesis (MCIC™) [94–96], autologous collagen-induced chondrogenesis (ACIC™) [97, 98], minced cartilage repair (DeNovo NT and CAIS) [99–101], osteochondral biomimetic scaffolds (MaioRegen®) [102–105], and hydrogels acting alone or as carriers of cells and/or proteins (BST-CarGel®) [106–109].

Correction of malalignment or unloading of an affected compartment by means of the osteotomy (Fig. 2.4) (distal femur or proximal tibia) might favor the biomechanical environment around OCD or unicompartmental OA [110, 111]. Partial or total knee replacement by means of arthroplasty or even fusion in salvage procedures is considered as the last resource [38]. Prompt diagnosis and treatment of symptomatic OCDs have enabled better clinical outcome [86, 112, 113]. Moreover, several authors advise that early treatment diminishes the risk for additional cartilage degeneration and development of secondary knee OA [9, 14, 25, 62, 86, 112]. Based on current knowledge, the treatment of OCDs relies on the defect's size, the involvement of the entire osteochondral unit, and the time from injury to repair [114]. Many algorithms for treatment have been proposed [52, 114–119].

Table 2.1 Commercial available cartilage repair systems

Product name	Main material	Trials
ACI procedures		
ChondroCelect® TiGenix, Leuven, Belgium	10,000 cells/µl suspension (Dulbecco's modified eagles medium)	First approved cell-based product in Europe
Carticel® Genzyme Biosurgery, Cambridge, MA	12 million cells suspension	First FDA-approved cell therapy product
Chondro-Gide® Geistlich Biomaterials, Wolhusen, Switzerland	Collagen	Improved clinical outcome associated to MF or as an ACI procedure
MACI® Genzyme Biosurgery, Cambridge, MA	Porcine type I/III collagen	Phase III trials Improved outcome in case series in comparison with OAT and MF
CaReS® Ars Arthro, Esslingen, Germany	Rat-tail type I collagen	Improved clinical outcomes in a multicenter study with 116 patients/ follow-up: 30 months
NeoCart® Histogenics Corporation, Waltham, MA	Bovine type I collagen Chondrocyte culture in a bioreactor	Phase III trials
Hyalograft C® Fidia Advanced Biopolymers, Abano Terme, Italy	HYAFF 11-esterified derivative of hyaluronate	Improved clinical results even when compared with MF Improved clinical outcome in case series reported in 62 patients/ follow-up: 7 years
Cartipatch® Tissue Bank of France	Agarose-alginate	Phase III trials Improved clinical outcome in case series reported in 17 patients/ follow-up: 24 months
Bioseed C® BioTissue Technologies, GmbH, Freiburg, Germany	Copolymer of PGA, PLA, and PDS – fibrin glue	Phase III trial Improved clinical outcomes in in case series reported in 52 patient/ follow-up: 4 years
BioCart II ProChon BioTech Ltd., Ness Ziona, Israel	Fibrinogen + hyaluronan	Phase II trial Improved clinical results in case series reported in 31 patients/ follow-up: 17 months
DeNovo ET® Zimmer, Warsaw, Indiana	Matrix + allogenic fetal chondrocytes	Phase III trial
Cartsystem	Sodium hyaluronate + allogeneic umbilical cord MSCs	Phase II trial
Graft		
DeNovo NT® Zimmer, Warsaw, Indiana	Matrix + allogenic chondrocytes	Good clinical outcomes in few studies reported
CAIS® Depuy-Mitek, Raynham MA	Glue + autologous morcelleied cartilage	Phase III trial

(continued)

Table 2.1 (continued)

Product name	Main material	Trials
Cell-free scaffold		
TruFit® Smith & Nephew, Andover, MA	PLGA-calcium-sulfate biopolymer bilayer porous	Suspended commercialization
BST-CarGel® Biosyntech, Quebec, Canada	Chitosan + glycerol phosphate	Phase III trial Better outcomes than MF treatment in a 5-year follow-up
CaReS-1S® Arthro-Kinetics, Esslingen, Germany	Rat-tail type I collagen	Animal trials Short case series in adults
MaioRegen® Fin-Ceramica S.p.A., Faenza, Italy	Hydroxyapatite-collagen 3D tri-layers	Few studies

Reproduced with permission of Springer [60]. Copyright, 2017 Springer International Publishing AG

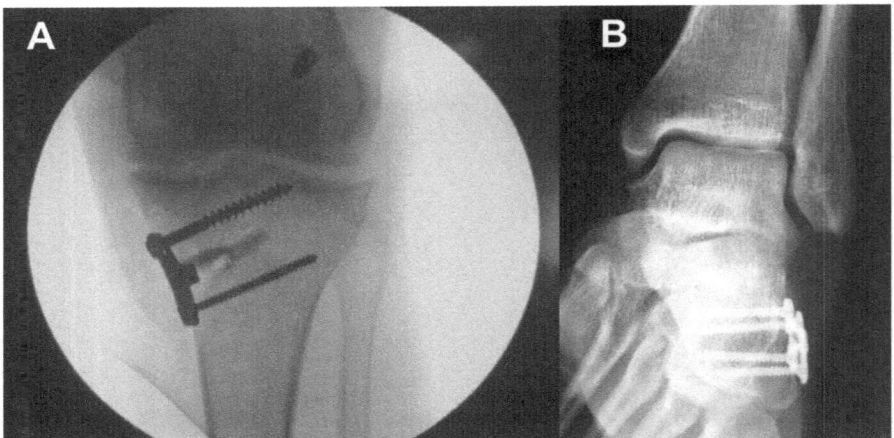

Fig. 2.4 Opening-wedge high tibial osteotomy of the knee (stereoscopy) (**A**) and calcaneal sliding osteotomy of the ankle (x-ray) (**B**)

2.5   Ankle Osteochondral Defects

An osteochondral defect (OCD) of the talus is a lesion involving the talus or distal tibia hyaline cartilage and its subchondral bone. Several classifications have been used over time, but the first comes from 1959 from Berndt and Harty [120]. The etiology of OCDs is often a single or repeated traumatic events [121]. However, ankle OCDs might also be idiopathic or non-traumatic [1, 121–123]. Similar to what happens for the knee joint, there is no single classification system, which fully addresses the topic. The anatomical grid proposed by Raikin and Elias has proven its value by making it possible to describe the location and assist in a preoperative planning [124, 125].

Shearing forces might cause superficial cartilage lesions, without damage to the underlying subchondral plate. However, after a high-impact force or repeated trauma (chronic instability), the underlying bone plate can also be damaged [126]. Ankle trauma related to an OCD frequently progresses to the formation of subchondral bone cysts. These bone cysts, surrounded by nociceptors, cause recurrent deep ankle pain leading to functional limitation. Most OCDs of the talus are found on the anterolateral or posteromedial talar dome [127]. Lateral lesions are usually narrower and oval-shaped and usually are caused by a shear mechanism. On the other hand, medial lesions usually derive from torsional impaction and axial loading, so they are frequently deeper and more cup-shaped [1, 122]. Although an OCD can have an acute onset resulting from trauma, cystic degeneration is a slower process [128]. To date, there is still not a complete understanding of the etiology or the different clinical presentation and response to treatment of ankle OCDs, despite some valid theoretical explanations [1]. While some OCDs remain asymptomatic, others present fast degradation with cyst formation and bone edema [128]. If we could predict or understand the pathogenesis of such differences, we would most likely be more efficient in dealing with this condition. The clinical presentation of a symptomatic OCD is usually deep ankle pain aggravated by effort with recurrent swelling after activity [128].

Some type of trauma is frequently accepted as the principal etiologic factor of an OCD of the talus. Trauma has been implicated in 93–98% of lateral talar defects and 61–70% of medial OCDs [129]. Etiologic factors of an OCD can be traumatic or non-traumatic [1]. Other etiologic possibilities include vascular issues and genetics [122]. Furthermore, OCDs have been found in identical twins and siblings [130] in support of the previous. Moreover, ankle OCDs are bilateral in 10% of patients [131]. Traumatic cartilage lesions of the ankle can be divided as microdamage or blunt trauma, chondral fractures (sparing the underlying bone), and osteochondral fractures [132].

Ankle sprains or chronic ankle instability is an important cause of traumatic ankle OCDs [133]. This seems to be the most frequent cause of these conditions. When the talus is inverted between the tibial plafond, medial and lateral malleoli linked by syndesmotic ligaments (the ankle mortice), the cartilage of the talus can be crushed/fractured (causing a loose body) and cause a cartilage crack or delamination, or an underlying bone bruise. Shearing forces might cause separation in the superficial layer of the cartilage [1]. OCDs might remain stable or become unstable which aggravates progression to further joint damage [1]. In testing conditions, it has been possible to reproduce lateral ankle OCD defects by intensely inverting a dorsiflexed ankle (while the foot is inverted, the lateral border of the talar dome is smashed against the fibula while the lateral ligament is ruptured). During application of excessive inverting force, the talus rotated laterally in the frontal plane within the mortise thus impacting and compressing the lateral talar margin against the articular surface of the fibula. This mechanism leads to a lateral talar OCD. A medial lesion was reproduced by plantarflexing the ankle while applying slight anterior displacement of the talus on the tibia, inversion and internal rotation of the talus on the tibia [1, 120]. Considering the previous one can assume that the treatment of ankle OCD without management of chronic ankle instability is extremely difficult and prone to failure. For this reason, one major advance in the treatment of ankle OCDs has been the concomitant arthroscopic approach of cartilage defects and lateral ligament's repair [134].

2.6   State of the Art in the Treatment of Osteochondral Defects of the Ankle

Asymptomatic incidental findings of the ankle are not infrequent, including within athletic population [135]. As aforementioned, ankle OCDs are frequently secondary to trauma, usually a consequence of ankle sprains during sports or chronic ankle instability. The available treatment options are basically similar to those on the knee. Asymptomatic OCDs can be dealt conservatively: physiotherapy, medication, orthobiologics, periods of rest, or immobilization (e.g., orthoses or walker boot) [121, 127]. However, we advise for surveillance of such injuries. Presently, there is no evidence-based or consensus in the literature concerning the superiority of any surgical treatment over another either in primary or secondary ankle OCDs [127, 136]. The final therapeutic decision relies on the patient profile and expectations as well as some characteristics of the lesion.

Preoperative planning is critical and should always include weight-bearing x-rays for alignment evaluation and global joint assessment. MRI can overestimate the size of the OCD by the presence of bone edema (usually reflects local biological activity, mostly visible in T2 sequences) surrounding the injury. The CT provides a more reliable assessment of bony defect size and volume. Additionally, CT on lateral view in plantar flexion or dorsiflexion is helpful to decide for the most advantageous anterior or posterior arthroscopic approach in a given case or even if an open approach is required (medial malleolar osteotomy for medial defects or lateral ligament detachment and afterward reinsertion for lateral defects). The arthroscopic approach is currently the preferred and most frequently used for both anterior and posterior compartments [137]. The authors advise for not using fixed distraction once this lowers the percentage of complications [138]. Moreover, as aforementioned, arthroscopy enables simultaneous treatment of concomitant pathologies (including instability) whenever required. Excision, curettage, and bone marrow stimulation techniques (ECBMS – excision of OCD fragment, curettage of subchondral bone with drilling or microfractures) aim to achieve fibrocartilaginous tissue formation which is still the less invasive surgical approach [136]. Satisfactory results with minimal aggression can be obtained depending on the patient profile and injury characteristics, and ECBMS can also be considered in bigger lesions

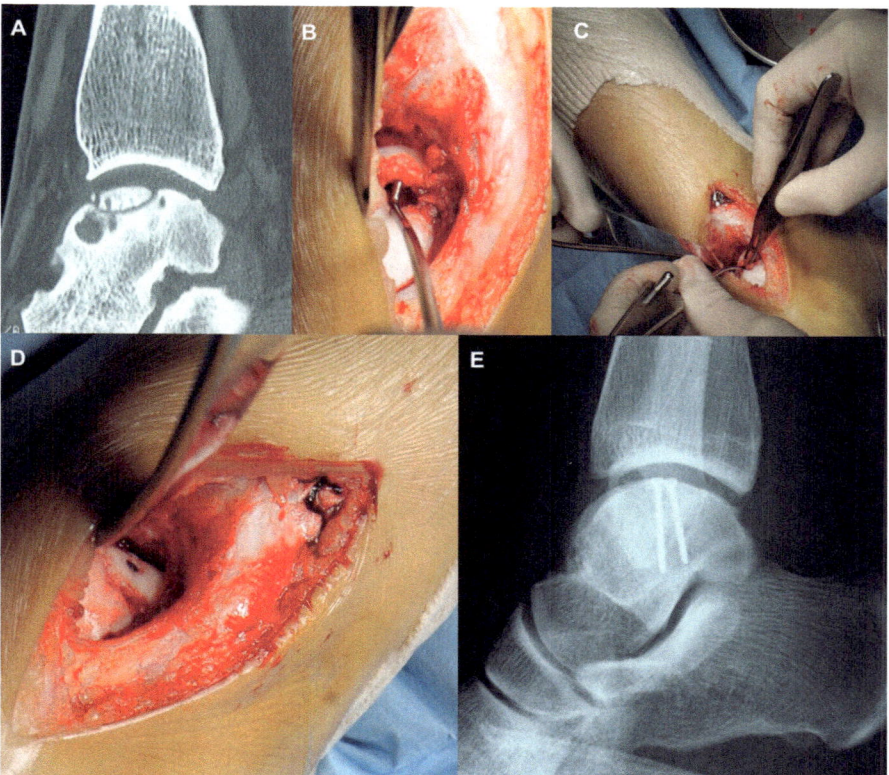

Fig. 2.5 Talar osteochondral defect with surrounding cystic lesions on CT (**A**), lifting of the defect on open surgery leaving partial attachment (**B**), filling of the defect with bone autograft after drilling (**C**), fixation of fragment with compression screw (**D**), and final x-ray look (**E**)

unable for fixation or even secondary injuries. ECBMS is considered in most cases given the outcome possibilities and lower aggression and cost. A lower percentage of good/excellent results is to be expected in larger lesions and revision surgery [136].

Preserving the native tissue by the "lift, drill, fill, and fix" surgery should be preferred whenever possible since it provides the preservation of the most of native tissue [139]. Lift the defect, drill by making microfracture or bone marrow stimulation, fill the defect with bone graft, and fix the fragment with metallic or bioabsorbable screws or pins (Fig. 2.5). This can be done fully arthroscopically in some cases or require open surgery on others. Retrograde drilling (Fig. 2.6) to decompress secondary cystic lesions linked to an OCD, and sometimes filling with bone graft, is a valid option for large cystic lesions [127]. OATS has some possible indications, but the high chance of complications must be acknowledged [82].

The osteochondral autologous transplantation surgery (OATS) technically (Fig. 2.7) is very similar to what is done in the knee joint. However, for most ankle lesions, it will require harvesting osteochondral cylinders from the knee to fill an

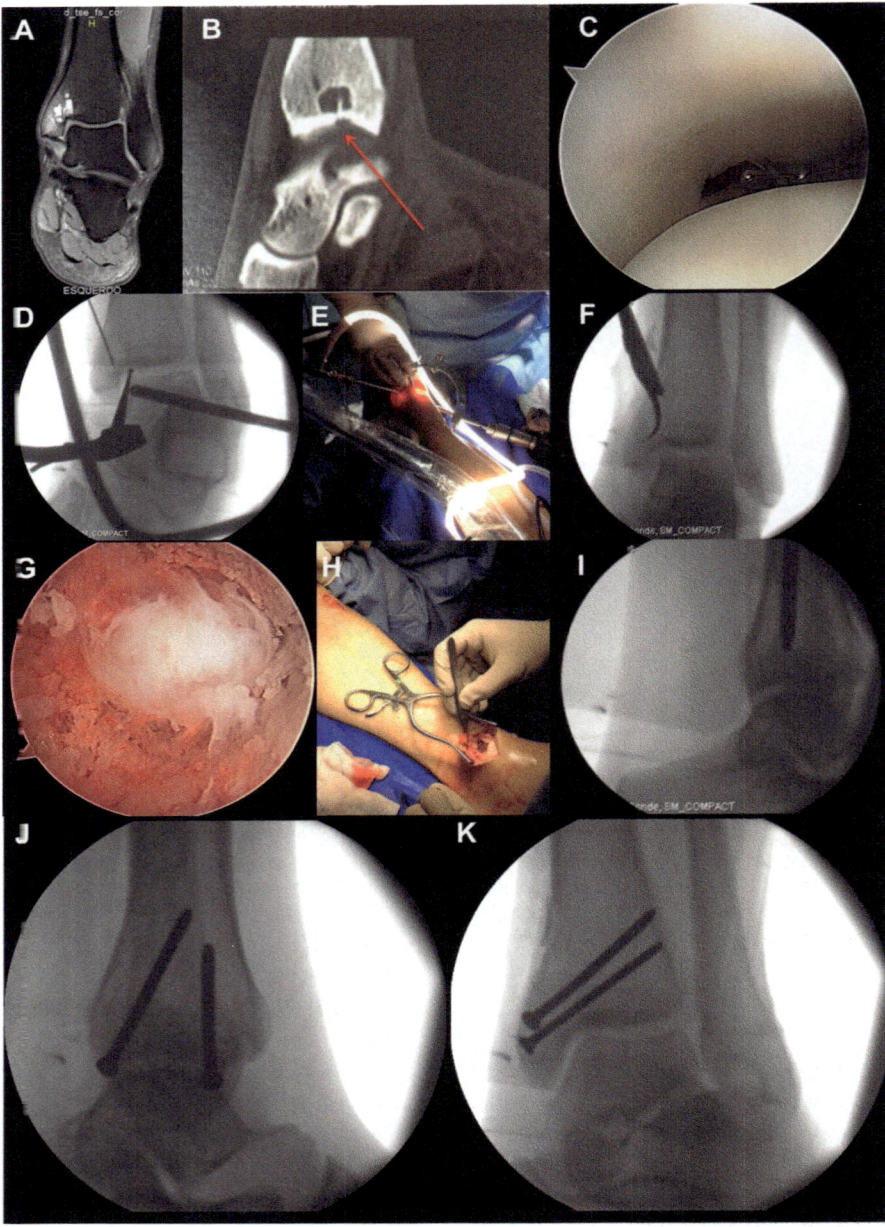

Fig. 2.6 Distal osteochondral defect (OCD) of medial malleolus – MRI view (**A**), CT view of the OCD with a small opening enabling fluid to get into the cyst (red arrow) (**B**), arthroscopic view of the cartilage small opening enabling fluid to get into the cyst (**C**), use of MicroVector guide for retrograde drilling to reach the defect under radioscopy control (**D**), outside view of the guide and drilling of a bone tunnel during arthroscopy (**E**), the arthroscope is introduced into the bone tunnel (osteoscopy) together with instruments for curettage of the cyst (**F**), inside view of the cyst from the arthroscope (**G**), bone autograft harvesting from distal tibia (**H**), the bone autograft is impacted into the defect (**I**), and two compression screws are included for extra support and compression to enhance healing (**J**, **K**)

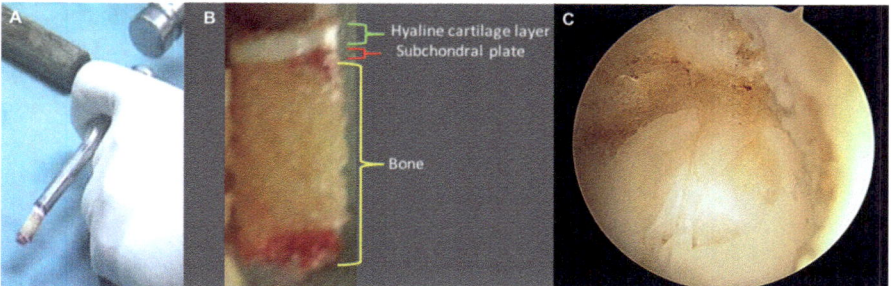

Fig. 2.7 After harvested, the osteochondral autograft is removed from the trephine used to collect it (**A**), aspect of the harvested autograft including fresh hyaline cartilage, subchondral bone and cancellous bone (**B**), arthroscopic view of a cylinder in place at 1-year follow-up (**C**)

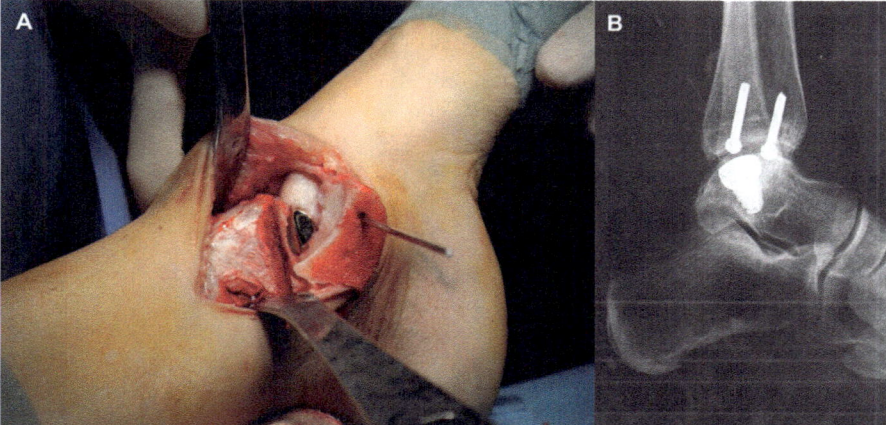

Fig. 2.8 Surgical view of the Hemicap® implant (**A**) and x-ray view of the implanted Hemicap® (**B**)

ankle defect. Although the promoters state high rate of a successful outcome, a systematic review has shown that this technique has a considerable amount of complications [82]. This must be considered by doctors and patients.

Tissue engineering and regenerative medicine (TERM) approaches promise a better and broader option for the future. However, similarly to what has been observed in the knee, cell-based therapies, scaffolds, and augmentation with hydrogels, despite very promising, so far, have not been able to provide consistently better results. Considering the former, and their higher cost, they are valid options for revision surgeries or large injuries without possibility for fixation and not amenable by any of the previous techniques, and as an approach to primary ankle OCD, we advise to keep this technology under research and controlled conditions before its extensive advertising [108, 109, 140–154]. When all biology-based surgical treatments fail, partial medial talar dome replacement by a metallic implant (Hemicap®) (Fig. 2.8) has provided positive midterm results [155]. Biomechanics remains a pillar of orthopedics. So improving the load distribution and joint alignment by means of osteotomy has proven positive effects either isolated or in combination with other procedures [28, 156]. The goal is to unload the most affected part while distributing forces to the most preserved part of the joint. Ankle fusion or ankle arthroplasty represents the last resource when dealing with very symptomatic OCDs or ankle OA [28].

2.7 Joint Anatomy, Congruency, Alignment, and Osteochondral Lesions

There are important anatomic and biomechanical differences between the knee and ankle joints, which might help to enlighten some aspects related to pathophysiology and treatment. Opposing to the ankle, the knee joint has two menisci which function as fibrocartilaginous dampers (dispersers of load), which assist in compensation on the basic incongruence of the knee joint. Menisci help to adjust the incongruity between the tibial plateau and the femoral condyles. Moreover, they increase the articulating joint surfaces, consequently reducing the load on the entire joint surface.

Another aspect is that the cartilage thickness is quite different among them. The common cartilage thickness of the talus is 1–1.7 mm, while in the knee joint, it ranges from 1 to 6 mm, depending on the location [157]. Moreover, the mechanical properties including stiffness of the talar cartilage are much more constant in the main loading area, while in the knee joint, the cartilage's properties are much more heterogeneous [158].

At higher loads, the ankle becomes a fully congruent joint [158]. The ankle has a smaller contact area than the knee in loading conditions. The contact area in the ankle at 500 N axial load is 350 mm [159–161] compared to 1120 mm^2 in the knee [162]. Therefore, it might be concluded that the total load and the load peaks in the ankle are higher than in the knee due to the smaller contact areas and the lack of damping structures. The constant hydrostatic pressure within a congruent joint like the ankle causes a permanent fluid pressure toward the subchondral plate. When the cartilage envelope of the joint is interrupted due to cartilage lesion, hydrostatic pressure might lead to secondary osteolysis and cyst formation (Fig. 2.9) [1].

The anatomical features, as well as the biomechanical differences alone, fail to explain the higher frequency of OA of the knee. Among the factors that lead to the onset and progression of OA, traumatic injuries of joint structures, as they occur in intra-articular fractures, have a critical role. A traumatic injury to the articular surface results in an immediate loss of biological features and biomechanical function [1]. A biochemical damage also occurs after trauma with loss of matrix components which might influence the risk of OA [163]. Sprains of the knee and ankle joints are among the most common injuries in sports. This can cause ligament injuries, meniscus tears (in the knee), and cartilage and bone lesions with varying degrees of severity which might be implicated in cartilage damage and OA risk.

Osteotomy is a surgery in which the bones are cut and their alignment changed with subsequent biomechanical implications in all joints. Osteotomy around the knee

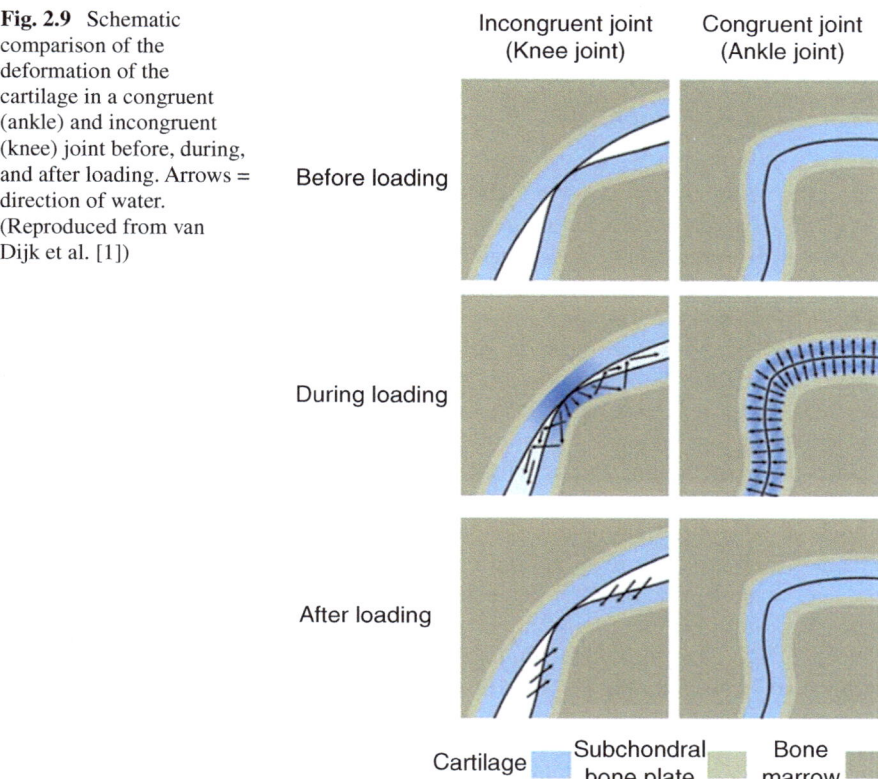

Fig. 2.9 Schematic comparison of the deformation of the cartilage in a congruent (ankle) and incongruent (knee) joint before, during, and after loading. Arrows = direction of water. (Reproduced from van Dijk et al. [1])

alters the alignment of the knee. Weight bearing will be shifted from the affected segment to a healthier part of the knee. By "unloading" the damaged cartilage, osteotomy may decrease pain, improve function, slow the joint degeneration, and possibly avoid or delay the need for (partial or) total knee replacement surgery [111].

Despite some methodological limitations on the available literature, it has been shown that valgus high tibial osteotomy reduces pain and improves function in patients with medial compartmental osteoarthritis of the knee [111]. So far, the results do not justify a conclusion on the benefit of any specific high tibial osteotomy technique for knee osteoarthritis over another [111].

Corrective ankle osteotomies enroll periarticular osteotomies of either the fibula, distal tibial metaphysis, or distal tibial metaphyseal-diaphyseal junction. Osteotomies are indicated under the presence of angular, rotational, or translational malalignment [164, 165]. Various types of realignment surgery are employed to preserve the ankle joint in cases of intermediate ankle arthritis with a partial joint space narrowing. Promising results considering pain, function, and imaging have been reported [165]. In conclusion, improvement of biomechanical environment might be helpful alone or in combination with any other "biological" treatment in either knee or ankle joints.

2.8 Current and Future Perspectives

2.8.1 Injections and Other Therapies with Growth Factors and/or Stem Cells

The orthobiologics approach, including anabolic proteins (growth factors (GFs)) [148, 166–168] and mesenchymal stem cells (MSCs) [144, 154, 167, 169–175] with or without hydrogels (e.g., hyaluronic acid, collagen, chitosan-based) [140, 150, 176–181], represents a step forward on conservative or minimally invasive therapy of both OCDs and OA.

The capacity for tissue repair is influenced by GFs, which have functions like chemotaxis, cell differentiation, proliferation, and cellular responses, which may potentially improve tissue healing (including the cartilage and bone). Therefore, the use of autologous and recombinant GFs is evolving in several fields of orthopedics. However, we need to fine-tune this technology in order to have adequate GFs acting in each tissue in proper time. As an example, platelet-rich plasma (PRP), a source of a cocktail of several autologous GFs, cannot be all things to all tissues. PRP is obtained from patient's own blood (autologous), and GFs from alpha granules of platelets become available after the platelet activation procedure. The next step will be to customize PRP for specific indications, an innovative and potentially rewarding concept [182]. The goal is to manipulate GFs and secretory proteins aiming for both cartilage and bone repair at the same time for an OCD. Many questions remain to be answered, including therapy timing (when to start therapy, how many applications, and for how long); which type of preparation, volume, or dose; and frequency of treatment [182–184]. It is difficult to compare clinical outcome PRP since there are many different methods for preparation that provide different products, for instance, regarding the GF and leukocyte concentration [168, 185].

The most widely used GFs are bone morphogenetic proteins (BMPs) and PRP [145, 167, 172]. GFs can also be genetically modified to improve its function or even use gene therapy to increase expression of a specific GF if needed for tissue healing [145, 186]. Another promising field is the use of stem cell-based therapies. Mesenchymal stem cells (MSCs) have differentiation competence for mesodermal lineages [187]. The modulation of adult MSC pathways can lead to chondro-, osteo-, and adipogenesis (chondrocytes, osteoblasts, and adipocytes, respectively) [188, 189]. The therapeutic possibilities of their use are extraordinary.

MSCs can be isolated from different tissues such as the bone marrow, skin, fat, synovia, and muscles, or from aspirates such as the bone marrow, adipose-derived [95, 169, 190–192]. MSCs allow its transplantation without provoking an immune response [193]. Depending on its source, MSCs show different performances. Bone marrow stem cells are still the most studied ones [194]. Bone marrow aspirates from the iliac crest have been used to treat chondral lesions and OCDs [195–198]. After harvesting by means of aspiration, MSCs might either be submitted to laboratory expansion within 2–3 weeks for subsequent use or the aspirate itself after concentration (centrifugation) can be immediately implanted. Moreover, in advanced TERM strategies, they might be combined with GFs, platelet-rich fibrin gel [95, 197, 199–201], fibrin glue [196] collagen gel [195, 196, 199, 200, 202] or collagen [95, 195, 196, 203] and HA [197, 199, 200, 202] scaffolds, among others [60, 204].

MSC-based treatment of focal chondral lesions and OCDs has shown promising clinical outcome in both the knee [95, 196, 197, 203, 205–208] and ankle [199, 200, 202] joints. Some reports of hyaline cartilage repair have been recently presented [209]. Moreover, a cryopreserved form of human amniotic membrane and umbilical cord (hAMUC) fetal tissues has been proposed for osteochondral injuries. These tissues have unique proteins and growth factors in the extracellular matrix and have shown to modulate inflammation, reducing adhesion and scar formation while encouraging regenerative healing [210]. Despite the very limited clinical experience, this possibility is under commercial promotion already (Amniox®). Hydrogels function by their own properties (rheological, anti-inflammatory, lubrication), but they may also function in combination of GS and/or MSCs as well as promising scaffolds which might also enable control of neovascularization process (of particular relevance concerning hyaline cartilage) [150, 177–181]

2.8.2 Tissue Engineering and Regenerative Medicine Approaches

The combination of the TERM triad (cells, scaffolds, and GFs) despite remaining a challenge is still the main goal in any tissue repair [209, 211–214]. Moreover, the possibility of one-step procedures for full OCD repair remains a major goal to fasten recovery process and avoid comorbidity and costs. Such approach has been attempted with some success [195, 199, 215, 216]. Giannini et al. [199] combined BMC and PRP gel with HA membrane or collagen powder to treat talar OCDs with positive short-term results. Moreover, histological biopsies have shown hyaline-like cartilage [199, 216].

The use of multilayered scaffolds facilitates the regeneration of the native tissue with hyaline cartilage and subchondral bone [103, 153, 204, 209, 217–221]. However, in respect for biology and the complex chain of events leading to tissue repair, enhancing scaffolds with cells and/or growth factors seems theoretically more promising in any tissue as suggested by clinical and basic science research [214, 221, 222]. The final goal of TERM [211, 223] is to develop an effective scaffold that is seeded with suitable cells and growth factors and matured in the laboratory with the use of bioreactors, and accomplishing a tissue that would be suitable for clinical implantation with similar characteristics to the native one.

Nanotechnology seems a promising field once we can use nanoparticles to deliver proteins and/or cells in different layers of a given scaffold aiming to influence the healing of different tissues according to its needs [217, 224, 225]. Moreover, it enables to label stem cells and influences their behavior in the biologic environment

[226]. Similarly, this can be used for bioactive proteins [227–229]. Besides, some authors suggest that nanoscale fibrous scaffold architecture is crucial in promoting and maintaining chondrogenic differentiation [230].

A multilayered collagen-based scaffold has been developed including the use of hydroxyapatite nanoparticles, which might enhance bone integration [103]. Another silk-based nanofibrous and nanocomposite bilayer scaffold used calcium-phosphate nanoparticles [217]. Some authors proposed bilayer scaffolds including microspheres with TGF-β for chondrogenic differentiation and BMP-2 for osteogenic differentiation [231], and several other improvements are under development [232]. Moreover, the combination of specific hydrogels or even gene therapy [233] can further enhance this process for future clinical use [153, 179, 234]. Another very promising possibility for TERM approaches is the possibility for three-dimensional (3D) bioprinting techniques which enable to fabricate injury-specific implants [235–238]. This is particularly helpful in the geometrically difficult parts of joints. 3D bioprinting can be used to produce custom-made, regenerative constructs for tissue repair [237]. 3D bioprinting techniques permit incorporation of cells and bioactive molecules during the fabrication process in order to create biologically active implants [237]. The outer shape of the construct can be made accordingly to the patient's defect based on CT and/or MRI images of the lesion. Moreover, it enables to achieve more complex zonally organized osteochondral constructs by printing with multiple bio-inks [237]. A large number of possibilities exist including hybrid printing such as thermoplastic polymers and hydrogels or incorporation of electrospun meshes in hydrogels, nanoparticles with cells, and/or bioactive molecules to optimize biomechanical and biological capacities of the construct [237].

2.9 Final Remarks

Cartilage or osteochondral defects are very frequent injuries affecting millions of people worldwide. Development of osteoarthritis (OA) is a relevant socioeconomic burden, which requires more effective possibilities for treatment. OA is more frequent in the knee than in the ankle. Most ankle OCDs are linked with the consequence of traumatic events and ankle sprains (which is one of the most frequent injuries in sports). The knee and ankle have different biological and biomechanical features, which help to understand some differences in physiopathology and response to treatment. However, a lot of further research is required in this setting. Conservative treatment remains the first option in treatment in most OCDs or OA. In this field, the development of orthobiologics (injectable hydrogels, growth factors, cell-based therapies, and so forth) has provided new options for some patients. Concerning surgical treatment, technical developments have been improving the outcome of classical approaches such as bone marrow stimulation techniques. Autologous osteochondral transplantation, despite remaining a valid option, has been linked with the significant amount of complications, which must be acknowledged. The first generation of autologous chondrocyte transplantation has not achieved the expected results. The use of acellular scaffolds has been under intense research and development. The combination and use of cells, growth factors, and cells in advanced TERM approaches promise to improve future outcome. Joint realignment by means of osteotomies is also a valid surgical tool, both in the knee and the ankle. Joint replacement offers many different possibilities including partial replacement. Results of different techniques are not the same in the knee and the ankle, which seem to be multifactorial. The road for the future will upraise most probably from TERM approaches including gene therapy, nanotechnology, and custom-made implants.

References

1. van Dijk CN, Reilingh ML, Zengerink M, van Bergen CJ (2010) Osteochondral defects in the ankle: why painful? Knee Surg Sports Traumatol Arthrosc 18(5):570–580. https://doi.org/10.1007/s00167-010-1064-x
2. Hunter DJ (2009) Risk stratification for knee osteoarthritis progression: a narrative review. Osteoarthritis Cartilage 17(11):1402–1407. https://doi.org/10.1016/j.joca.2009.04.014
3. Gelber AC, Hochberg MC, Mead LA, Wang NY, Wigley FM, Klag MJ (2000) Joint injury in young adults and risk for subsequent knee and hip osteoarthritis. Ann Intern Med 133(5):321–328
4. Andrade R, Vasta S, Papalia R, Pereira H, Oliveira JM, Reis RL, Espregueira-Mendes J (2016) Prevalence of articular cartilage lesions and surgical clinical outcomes in football (soccer) players' knees: a systematic review. Arthroscopy 32(7):1466–1477
5. Flanigan DC, Harris JD, Trinh TQ, Siston RA, Brophy RH (2010) Prevalence of chondral defects in athletes' knees: a systematic review. Med Sci Sports Exerc 42(10):1795–1801
6. Årøen A, Løken S, Heir S, Alvik E, Ekeland A, Granlund OG, Engebretsen L (2004) Articular cartilage lesions in 993 consecutive knee arthroscopies. Am J Sports Med 32(1):211–215
7. Curl WW, Krome J, Gordon ES, Rushing J, Smith BP, Poehling GG (1997) Cartilage injuries: a review of 31,516 knee arthroscopies. Arthroscopy 13(4):456–460
8. Mithoefer K, Peterson L, Saris D, Mandelbaum B, Dvorák J (2012) Special issue on articular cartilage injury in the football (soccer) player. Cartilage 3(1 suppl):4S–5S
9. Mithoefer K, Steadman RJ (2012) Microfracture in football (soccer) players: a case series of professional athletes and systematic review. Osteoarthritis Cartilage 3(1 suppl):18S–24S
10. Huch K, Kuettner KE, Dieppe P (1997) Osteoarthritis in ankle and knee joints. Semin Arthritis Rheum 26(4):667–674
11. Gomoll A, Filardo G, De Girolamo L, Esprequeira-Mendes J, Marcacci M, Rodkey W, Steadman R, Zaffagnini S, Kon E (2012) Surgical treatment for early osteoarthritis. Part I: cartilage repair procedures. Knee Surg Sports Traumatol Arthrosc 20(3):450–466
12. Krych AJ, Robertson CM, Williams RJ (2012) Return to athletic activity after osteochondral allograft transplantation in the knee. Am J Sports Med 40(5):1053–1059
13. Mithoefer K, Della Villa S (2012) Return to sports after articular cartilage repair in the football (soccer) player. Cartilage 3(1 suppl):57S–62S
14. Mithoefer K, Hambly K, Della Villa S, Silvers H, Mandelbaum BR (2009) Return to sports participation after articular cartilage repair in the knee scientific evidence. Am J Sports Med 37(1 suppl):167S–176S
15. Dvorak J, Peterson L, Junge A, Chomiak J, Graf-Baumann T (2000) Incidence of football injuries and complaints in different age groups and skill-level groups. Am J Sports Med 28(5):51–57

16. Messner K, Maletius W (1996) The long-term prognosis for severe damage to weight-bearing cartilage in the knee: a 14-year clinical and radiographic follow-up in 28 young athletes. Acta Orthop 67(2):165–168

17. Piasecki DP, Spindler KP, Warren TA, Andrish JT, Parker RD (2003) Intraarticular injuries associated with anterior cruciate ligament tear: findings at ligament reconstruction in high school and recreational athletes an analysis of sex-based differences. Am J Sports Med 31(4):601–605

18. Tandoan RN, Mann G, Verdonk R, Doral MN (2012) Sports injuries: prevention, diagnosis, treatment and rehabilitation. Springer-Verlag Berlin Heidelberg. doi:10.1007/978-3-642-15630-4

19. Bruce EJ, Hamby T, Jones DG (2005) Sports-related osteochondral injuries: clinical presentation, diagnosis, and treatment. Prim Care 32(1):253–276. https://doi.org/10.1016/j.pop.2004.11.007

20. Kuettner KE, Cole AA (2005) Cartilage degeneration in different human joints. Osteoarthritis Cartilage 13(2):93–103. https://doi.org/10.1016/j.joca.2004.11.006

21. Arendt E, Dick R (1995) Knee injury patterns among men and women in collegiate basketball and soccer NCAA data and review of literature. Am J Sports Med 23(6):694–701

22. Heijink A, Gomoll AH, Madry H, Drobnič M, Filardo G, Espregueira-Mendes J, Van Dijk CN (2012) Biomechanical considerations in the pathogenesis of osteoarthritis of the knee. Knee Surg Sports Traumatol Arthrosc 20(3):423–435

23. Vannini F, Spalding T, Andriolo L, Berruto M, Denti M, Espregueira-Mendes J, Menetrey J, Peretti G, Seil R, Filardo G (2016) Sport and early osteoarthritis: the role of sport in aetiology, progression and treatment of knee osteoarthritis. Knee Surg Sports Traumatol Arthrosc 24(6):1786–1796

24. Pánics G, Hangody LR, Baló E, Vásárhelyi G, Gál T, Hangody L (2012) Osteochondral autograft and mosaicplasty in the football (Soccer) athlete. Cartilage 3(1 suppl):25S–30S

25. Steinwachs M, Engebretsen L, Brophy R (2012) Scientific evidence base for cartilage injury and repair in the athlete. Cartilage 3(1 suppl):11S–17S

26. Drawer S, Fuller C (2001) Propensity for osteoarthritis and lower limb joint pain in retired professional soccer players. Br J Sports Med 35(6):402–408

27. Engström B, Forssblad M, Johansson C, Tornkvist H (1990) Does a major knee injury definitely sideline an elite soccer player? Am J Sports Med 18(1):101–105

28. Badekas T, Takvorian M, Souras N (2013) Treatment principles for osteochondral lesions in foot and ankle. Int Orthop 37(9):1697–1706. https://doi.org/10.1007/s00264-013-2076-1

29. Buckwalter JA (1998) Articular cartilage: injuries and potential for healing. J Orthop Sports Phys Ther 28(4):192–202

30. Cengiz IF, Oliveira JM, Ochi M, Nakamae A, Adachi N, Reis RL (2017) "Biologic" treatment for meniscal repair. In: Injuries and health problems in football. Springer, Berlin/Heidelberg, pp 679–686

31. Martel-Pelletier J, Wildi LM, Pelletier JP (2012) Future therapeutics for osteoarthritis. Bone 51(2):297–311. https://doi.org/10.1016/j.bone.2011.10.008

32. Weinraub GM (2005) Orthobiologics: a survey of materials and techniques. Clin Podiatr Med Surg 22(4):509–519, v. https://doi.org/10.1016/j.cpm.2005.08.003

33. Neustadt DH (2006) Intra-articular injections for osteoarthritis of the knee. Cleve Clin J Med 73(10):897–898. 901-894, 906-811

34. Lawrence RC, Felson DT, Helmick CG, Arnold LM, Choi H, Deyo RA, Gabriel S, Hirsch R, Hochberg MC, Hunder GG, Jordan JM, Katz JN, Kremers HM, Wolfe F, National Arthritis Data W (2008) Estimates of the prevalence of arthritis and other rheumatic conditions in the United States. Part II. Arthritis Rheum 58(1):26–35. https://doi.org/10.1002/art.23176

35. Puig-Junoy J, Ruiz Zamora A (2015) Socio-economic costs of osteoarthritis: a systematic review of cost-of-illness studies. Semin Arthritis Rheum 44(5):531–541. https://doi.org/10.1016/j.semarthrit.2014.10.012

36. Neogi T (2013) The epidemiology and impact of pain in osteoarthritis. Osteoarthritis Cartilage 21(9):1145–1153. https://doi.org/10.1016/j.joca.2013.03.018

37. Kurtz S, Ong K, Lau E, Manley M (2011) Current and projected utilization of total joint replacements. Compr Biomater 6:1–9
38. Lohmander LS (2013) Knee replacement for osteoarthritis: facts, hopes, and fears. Medicographia 35:181–188
39. Ekstrand J (2016) UEFA Elite Club Injury Study Report 2015/16.
40. Aurich M, Hofmann GO, Rolauffs B, Gras F (2014) Differences in injury pattern and prevalence of cartilage lesions in knee and ankle joints: a retrospective cohort study. Orthop Rev (Pavia) 6(4):5611. https://doi.org/10.4081/or.2014.5611
41. Hintermann B, Regazzoni P, Lampert C, Stutz G, Gachter A (2000) Arthroscopic findings in acute fractures of the ankle. J Bone Joint Surg Br 82(3):345–351
42. Haq SA, Davatchi F (2011) Osteoarthritis of the knees in the COPCORD world. Int J Rheum Dis 14(2):122–129. https://doi.org/10.1111/j.1756-185X.2011.01615.x
43. Bitton R (2009) The economic burden of osteoarthritis. Am J Manag Care 15(8 Suppl):S230–S235
44. Gupta S, Hawker GA, Laporte A, Croxford R, Coyte PC (2005) The economic burden of disabling hip and knee osteoarthritis (OA) from the perspective of individuals living with this condition. Rheumatology (Oxford) 44(12):1531–1537. https://doi.org/10.1093/rheumatology/kei049
45. Global Burden of Disease Study C (2015) Global, regional, and national incidence, prevalence, and years lived with disability for 301 acute and chronic diseases and injuries in 188 countries, 1990-2013: a systematic analysis for the global burden of disease study 2013. Lancet 386(9995):743–800. https://doi.org/10.1016/S0140-6736(15)60692-4
46. Hunter DJ, Schofield D, Callander E (2014) The individual and socioeconomic impact of osteoarthritis. Nat Rev Rheumatol 10(7):437–441. https://doi.org/10.1038/nrrheum.2014.44
47. Hiligsmann M, Cooper C, Guillemin F, Hochberg MC, Tugwell P, Arden N, Berenbaum F, Boers M, Boonen A, Branco JC, Maria-Luisa B, Bruyere O, Gasparik A, Kanis JA, Kvien TK, Martel-Pelletier J, Pelletier JP, Pinedo-Villanueva R, Pinto D, Reiter-Niesert S, Rizzoli R, Rovati LC, Severens JL, Silverman S, Reginster JY (2014) A reference case for economic evaluations in osteoarthritis: an expert consensus article from the European Society for Clinical and Economic Aspects of Osteoporosis and Osteoarthritis (ESCEO). Semin Arthritis Rheum 44(3):271–282. https://doi.org/10.1016/j.semarthrit.2014.06.005
48. Mithoefer K, Hambly K, Logerstedt D, Ricci M, Silvers H, Villa SD (2012) Current concepts for rehabilitation and return to sport after knee articular cartilage repair in the athlete. J Orthop Sports Phys Ther 42(3):254–273
49. Kiviranta I, Tammi M, Jurvelin J, Arokoski J, Säuäumäunen A-M, Helminen HJ (1992) Articular cartilage thickness and glycosaminoglycan distribution in the canine knee joint after strenuous running exercise. Clin Orthop Relat Res 283:302–308
50. Stefan Lohmander L, Roos H, Dahlberg L, Hoerrner LA, Lark MW (1994) Temporal patterns of stromelysin-1, tissue inhibitor, and proteoglycan fragments in human knee joint fluid after injury to the cruciate ligament or meniscus. J Orthop Res 12(1):21–28
51. Gomoll AH, Minas T (2014) The quality of healing: articular cartilage. Wound Repair Regen 22(S1):30–38
52. McAdams TR, Mithoefer K, Scopp JM, Mandelbaum BR (2010) Articular cartilage injury in athletes. Cartilage 1(3):165–179
53. Laskin RS (1978) Unicompartmental tibiofemoral resurfacing arthroplasty. J Bone Joint Surg Am 60(2):182–185
54. Erggelet C, Mandelbaum BR (2008) Principles of cartilage repair. Springer Science & Business Media
55. Gorsline RT, Kaeding CC (2005) The use of NSAIDs and nutritional supplements in athletes with osteoarthritis: prevalence, benefits, and consequences. Clin Sports Med 24(1):71–82
56. Tamburrino P, Castellacci E (2016) Intra-articular injections of HYADD4-G in male professional soccer players with traumatic or degenerative knee chondropathy. A pilot, prospective study. J Sports Med Phys Fitness 56(12):1534

57. Hambly K, Silvers HJ, Steinwachs M (2012) Rehabilitation after articular cartilage repair of the knee in the football (soccer) player. Cartilage 3(1 suppl):50S–56S

58. Bekkers J, de Windt TS, Brittberg M, Saris D (2012) Cartilage repair in football (soccer) athletes what evidence leads to which treatment? A critical review of the literature. Cartilage 3(1 suppl):43S–49S

59. Harris JD, Brophy RH, Siston RA, Flanigan DC (2010) Treatment of chondral defects in the athlete's knee. Arthroscopy 26(6):841–852

60. Vilela CA, Correia C, Oliveira JM, Sousa RA, Reis RL, Espregueira-Mendes J (2017) Clinical management of articular cartilage lesions. In: Oliveira M, Reis RL (eds) Regenerative strategies for the treatment of knee joint disabilities. Studies in mechanobiology, tissue engineering and biomaterials. Springer, Cham, pp 29–53

61. Pereira H, Ripoll L, Oliveira JM, Reis RL, Espregueira-Mendes J, van Dijk C (2016) A engenharia de tecidos nas lesões do desporto. Traumatologia Desportiva. LIDEL, Lisboa

62. Blevins FT, Steadman JR, Rodrigo JJ, Silliman J (1998) Treatment of articular cartilage defects in athletes: an analysis of functional outcome and lesion appearance. Orthopedics 21(7):761–767. discussion 767–768

63. Steadman JR, Rodkey WG, Rodrigo JJ (2001) Microfracture: surgical technique and rehabilitation to treat chondral defects. Clin Orthop Relat Res ((391 Suppl)):S362–S369

64. Caffey S, McPherson E, Moore B, Hedman T, Vangsness CT Jr (2005) Effects of radiofrequency energy on human articular cartilage: an analysis of 5 systems. Am J Sports Med 33(7):1035–1039. https://doi.org/10.1177/0363546504271965

65. Spahn G, Kahl E, Muckley T, Hofmann GO, Klinger HM (2008) Arthroscopic knee chondroplasty using a bipolar radiofrequency-based device compared to mechanical shaver: results of a prospective, randomized, controlled study. Knee Surg Sports Traumatol Arthrosc 16(6):565–573. https://doi.org/10.1007/s00167-008-0506-1

66. Bedi A, Feeley BT, Williams RJ (2010) Management of articular cartilage defects of the knee. J Bone Joint Surg Am 92(4):994–1009

67. Krych AJ, Gobbi A, Lattermann C, Nakamura N (2016) Articular cartilage solutions for the knee: present challenges and future direction. J ISAKOS 1:93–104. https://doi.org/10.1136/jisakos-2015-000037

68. Mithoefer K, Williams RJ, Warren RF, Potter HG, Spock CR, Jones EC, Wickiewicz TL, Marx RG (2005) The microfracture technique for the treatment of articular cartilage lesions in the knee. J Bone Joint Surg Am 87(9):1911–1920

69. Gobbi A, Nunag P, Malinowski K (2005) Treatment of full thickness chondral lesions of the knee with microfracture in a group of athletes. Knee Surg Sports Traumatol Arthrosc 13(3):213–221

70. Mithoefer K, McAdams T, Williams RJ, Kreuz PC, Mandelbaum BR (2009) Clinical efficacy of the microfracture technique for articular cartilage repair in the knee an evidence-based systematic analysis. Am J Sports Med 37(10):2053–2063

71. Upmeier H, Bruggenjurgen B, Weiler A, Flamme C, Laprell H, Willich SN (2007) Follow-up costs up to 5 years after conventional treatments in patients with cartilage lesions of the knee. Knee Surg Sports Traumatol Arthrosc 15(3):249–257. https://doi.org/10.1007/s00167-006-0182-y

72. Laupattarakasem W, Laopaiboon M, Laupattarakasem P, Sumananont C (2008) Arthroscopic debridement for knee osteoarthritis. Cochrane Database Syst Rev (1):CD005118. https://doi.org/10.1002/14651858.CD005118.pub2

73. Mankin HJ (1982) The response of articular cartilage to mechanical injury. J Bone Joint Surg Am 64(3):460–466

74. Bae DK, Song SJ, Yoon KH, Heo DB, Kim TJ (2013) Survival analysis of microfracture in the osteoarthritic knee – minimum 10-year follow-up. Arthroscopy 29(2):244–250

75. Solheim E, Hegna J, Inderhaug E, Øyen J, Harlem T, Strand T (2016) Results at 10–14 years after microfracture treatment of articular cartilage defects in the knee. Knee Surg Sports Traumatol Arthrosc 24(5):1587–1593

76. Case JM, Scopp JM (2016) Treatment of articular cartilage defects of the knee with micro-fracture and enhanced microfracture techniques. Sports Med Arthrosc 24(2):63–68
77. Koh Y-G, Kwon O-R, Kim Y-S, Choi Y-J, Tak D-H (2016) Adipose-derived mesenchymal stem cells with microfracture versus microfracture alone: 2-year follow-up of a prospective randomized trial. Arthroscopy 32(1):97–109
78. Sofu H, Kockara N, Oner A, Camurcu Y, Issın A, Sahin V (2017) Results of hyaluronic acid-based cell-free scaffold application in combination with microfracture for the treatment of osteochondral lesions of the knee: 2-year comparative study. Arthroscopy 33:209–216
79. Espregueira-Mendes J, Pereira H, Sevivas N, Varanda P, da Silva MV, Monteiro A, Oliveira JM, Reis RL (2012) Osteochondral transplantation using autografts from the upper tibio-fibular joint for the treatment of knee cartilage lesions. Knee Surg Sports Traumatol Arthrosc 20(6):1136–1142. https://doi.org/10.1007/s00167-012-1910-0
80. Hangody L, Karpati Z (1994) New possibilities in the management of severe circumscribed cartilage damage in the knee. Magy Traumatol Ortop Kezseb Plasztikai Seb 37(3):237–243
81. Andrade R, Vasta S, Pereira R, Pereira H, Papalia R, Karahan M, Oliveira JM, Reis RL, Espregueira-Mendes J (2016) Knee donor-site morbidity after mosaicplasty – a systematic review. J Exp Orthop 3(1):31. https://doi.org/10.1186/s40634-016-0066-0
82. Ferreira C, Vuurberg G, Oliveira JM, Espregueira-Mendes J, Pereira, H, Reis RL, Ripoll P (2016) Assessment of clinical outcome after osteochondral autologous transplan-tation technique for the treatment of ankle lesions: a systematic review. JISAKOS. doi:jisakos-2015-000020.R2
83. De Caro F, Bisicchia S, Amendola A, Ding L (2015) Large fresh osteochondral allografts of the knee: a systematic clinical and basic science review of the literature. Arthroscopy 31(4):757–765
84. Brittberg M, Lindahl A, Nilsson A, Ohlsson C, Isaksson O, Peterson L (1994) Treatment of deep cartilage defects in the knee with autologous chondrocyte transplantation. N Engl J Med 331(14):889–895. https://doi.org/10.1056/nejm199410063311401
85. Roberts S, McCall IW, Darby AJ, Menage J, Evans H, Harrison PE, Richardson JB (2002) Autologous chondrocyte implantation for cartilage repair: monitoring its success by magnetic resonance imaging and histology. Arthritis Res Ther 5(1):1
86. Peterson L, Vasiliadis HS, Brittberg M, Lindahl A (2010) Autologous chondrocyte implantation a long-term follow-up. Am J Sports Med 38(6):1117–1124
87. Tom Minas MDM, Arvind Von Keudell M, Bryant T, Gomoll AH (2014) The John Insall Award: a minimum 10-year outcome study of autologous chondrocyte implantation. Clin Orthop Relat Res 472(1):41
88. Bartlett W, Gooding C, Carrington R, Skinner J, Briggs T, Bentley G (2005) Autologous chondrocyte implantation at the knee using a bilayer collagen membrane with bone graft. Bone Joint J 87(3):330–332
89. Meyerkort D, Ebert JR, Ackland TR, Robertson WB, Fallon M, Zheng M, Wood DJ (2014) Matrix-induced autologous chondrocyte implantation (MACI) for chondral defects in the patellofemoral joint. Knee Surg Sports Traumatol Arthrosc 22(10):2522–2530
90. Basad E, Wissing FR, Fehrenbach P, Rickert M, Steinmeyer J, Ishaque B (2015) Matrix-induced autologous chondrocyte implantation (MACI) in the knee: clinical outcomes and challenges. Knee Surg Sports Traumatol Arthrosc 23(12):3729–3735
91. Ebert JR, Fallon M, Wood DJ, Janes GC (2017) A prospective clinical and radiological evalu-ation at 5 years after arthroscopic matrix-induced autologous chondrocyte implantation. Am J Sports Med 45(1):59
92. Gille J, Schuseil E, Wimmer J, Gellissen J, Schulz A, Behrens P (2010) Mid-term results of autologous matrix-induced chondrogenesis for treatment of focal cartilage defects in the knee. Knee Surg Sports Traumatol Arthrosc 18(11):1456–1464
93. Lee YHD, Suzer F, Thermann H (2014) Autologous matrix-induced chondrogenesis in the knee: a review. Cartilage. https://doi.org/10.1177/1947603514529445
94. Gobbi A, Karnatzikos G, Sankineani SR (2014) One-step surgery with multipotent stem cells for the treatment of large full-thickness chondral defects of the knee. Am J Sports Med 42(3):648–657

95. Gobbi A, Karnatzikos G, Scotti C, Mahajan V, Mazzucco L, Grigolo B (2011) One-step cartilage repair with bone marrow aspirate concentrated cells and collagen matrix in full-thickness knee cartilage lesions results at 2-year follow-up. Cartilage 2(3):286–299

96. Huh SW, Shetty AA, Ahmed S, Lee DH, Kim SJ (2016) Autologous bone-marrow mesenchymal cell induced chondrogenesis (MCIC). J Clin Orthop Trauma 7(3):153–156

97. Stelzeneder D, Shetty AA, Kim S-J, Trattnig S, Domayer SE, Shetty V, Bilagi P (2013) Repair tissue quality after arthroscopic autologous collagen-induced chondrogenesis (ACIC) assessed via T2* mapping. Skelet Radiol 42(12):1657–1664

98. Shetty AA, Kim SJ, Shetty V, Jang JD, Huh SW, Lee DH (2016) Autologous collagen induced chondrogenesis (ACIC: Shetty–Kim technique) – a matrix based acellular single stage arthroscopic cartilage repair technique. J Clin Orthop Trauma 7(3):164–169

99. Harris JD, Frank RM, McCormick FM, Cole BJ (2014) Minced cartilage techniques. Oper Tech Orthop 24(1):27–34

100. Farr J, Cole BJ, Sherman S, Karas V (2012) Particulated articular cartilage: CAIS and DeNovo NT. J Knee Surg 25(01):023–030

101. Farr J, Tabet SK, Margerrison E, Cole BJ (2014) Clinical, radiographic, and histological outcomes after cartilage repair with particulated juvenile articular cartilage: a 2-year prospective study. Am J Sports Med. https://doi.org/10.1177/0363546514528671

102. Kon E, Delcogliano M, Filardo G, Pressato D, Busacca M, Grigolo B, Desando G, Marcacci M (2010) A novel nano-composite multi-layered biomaterial for treatment of osteochondral lesions: technique note and an early stability pilot clinical trial. Injury 41(7):693–701

103. Kon E, Delcogliano M, Filardo G, Busacca M, Di Martino A, Marcacci M (2011) Novel nano-composite multilayered biomaterial for osteochondral regeneration a pilot clinical trial. Am J Sports Med 39(6):1180–1190

104. Brix M, Kaipel M, Kellner R, Schreiner M, Apprich S, Boszotta H, Windhager R, Domayer S, Trattnig S (2016) Successful osteoconduction but limited cartilage tissue quality following osteochondral repair by a cell-free multilayered nano-composite scaffold at the knee. Int Orthop 40(3):625–632

105. Delcogliano M, de Caro F, Scaravella E, Ziveri G, De Biase CF, Marotta D, Marenghi P, Delcogliano A (2014) Use of innovative biomimetic scaffold in the treatment for large osteochondral lesions of the knee. Knee Surg Sports Traumatol Arthrosc 22(6):1260–1269

106. Hoemann CD, Tran-Khanh N, Chevrier A, Chen G, Lascau-Coman V, Mathieu C, Changoor A, Yaroshinsky A, McCormack RG, Stanish WD (2015) Chondroinduction is the main cartilage repair response to microfracture and microfracture with BST-CarGel results as shown by ICRS-II histological scoring and a novel zonal collagen type scoring method of human clinical biopsy specimens. Am J Sports Med 43(10):2469–2480

107. Méthot S, Changoor A, Tran-Khanh N, Hoemann CD, Stanish WD, Restrepo A, Shive MS, Buschmann MD (2015) Osteochondral biopsy analysis demonstrates that BST-CarGel treatment improves structural and cellular characteristics of cartilage repair tissue compared with microfracture. Cartilage. https://doi.org/10.1177/1947603515595837

108. Shive MS, Stanish WD, McCormack R, Forriol F, Mohtadi N, Pelet S, Desnoyers J, Méthot S, Vehik K, Restrepo A (2015) BST-CarGel® treatment maintains cartilage repair superiority over microfracture at 5 years in a multicenter randomized controlled trial. Cartilage 6(2):62–72

109. Stanish WD, McCormack R, Forriol F, Mohtadi N, Pelet S, Desnoyers J, Restrepo A, Shive MS (2013) Novel scaffold-based BST-CarGel treatment results in superior cartilage repair compared with microfracture in a randomized controlled trial. J Bone Joint Surg Am 95(18):1640–1650

110. Koshino T, Wada S, Ara Y, Saito T (2003) Regeneration of degenerated articular cartilage after high tibial valgus osteotomy for medial compartmental osteoarthritis of the knee. Knee 10(3):229–236

111. Brouwer RW, Huizinga MR, Duivenvoorden T, van Raaij TM, Verhagen AP, Bierma-Zeinstra SM, Verhaar JA (2014) Osteotomy for treating knee osteoarthritis. Cochrane Database Syst Rev 12:CD004019. https://doi.org/10.1002/14651858.CD004019.pub4

112. Mithoefer K, Minas T, Peterson L, Yeon H, Micheli LJ (2005) Functional outcome of knee articular cartilage repair in adolescent athletes. Am J Sports Med 33(8):1147–1153
113. Mithoefer K, Williams RJ, Warren RF, Wickiewicz TL, Marx RG (2006) High-impact athletics after knee articular cartilage repair: a prospective evaluation of the microfracture technique. Am J Sports Med 34(9):1413–1418
114. Bekkers JE, Inklaar M, Saris DB (2009) Treatment selection in articular cartilage lesions of the knee a systematic review. Am J Sports Med 37(1 suppl):148S–155S
115. de Windt TS, Saris DB (2014) Treatment algorithm for articular cartilage repair of the knee: towards patient profiling using evidence-based tools. In: Techniques in cartilage repair surgery. Springer, Berlin/Heidelberg, pp 23–31
116. Cole BJ, Pascual-Garrido C, Grumet RC (2009) Surgical management of articular cartilage defects in the knee. J Bone Joint Surg Am 91(7):1778–1790
117. Gomoll AH, Farr J, Gillogly SD, Kercher J, Minas T (2010) Surgical management of articular cartilage defects of the knee. J Bone Joint Surg Am 92(14):2470–2490
118. Tetteh ES, Bajaj S, Ghodadra NS, Cole BJ (2012) The basic science and surgical treatment options for articular cartilage injuries of the knee. J Orthop Sports Phys Ther 42(3):243–253
119. Murray IR, Benke MT, Mandelbaum BR (2016) Management of knee articular cartilage injuries in athletes: chondroprotection, chondrofacilitation, and resurfacing. Knee Surg Sports Traumatol Arthrosc 24(5):1617–1626
120. Berndt AL, Harty M (1959) Transchondral fractures (osteochondritis dissecans) of the talus. J Bone Joint Surg Am 41-A:988–1020
121. O'Loughlin PF, Heyworth BE, Kennedy JG (2010) Current concepts in the diagnosis and treatment of osteochondral lesions of the ankle. Am J Sports Med 38(2):392–404. https://doi.org/10.1177/0363546509336336
122. Schachter AK, Chen AL, Reddy PD, Tejwani NC (2005) Osteochondral lesions of the talus. J Am Acad Orthop Surg 13(3):152–158
123. Ventura A, Terzaghi C, Legnani C, Borgo E (2013) Treatment of post-traumatic osteochondral lesions of the talus: a four-step approach. Knee Surg Sports Traumatol Arthrosc 21(6):1245–1250. https://doi.org/10.1007/s00167-012-2028-0
124. Elias I, Raikin SM, Schweitzer ME, Besser MP, Morrison WB, Zoga AC (2009) Osteochondral lesions of the distal tibial plafond: localization and morphologic characteristics with an anatomical grid. Foot Ankle Int 30(6):524–529. https://doi.org/10.3113/FAI.2009.0524
125. Elias I, Zoga AC, Morrison WB, Besser MP, Schweitzer ME, Raikin SM (2007) Osteochondral lesions of the talus: localization and morphologic data from 424 patients using a novel anatomical grid scheme. Foot Ankle Int 28(2):154–161. https://doi.org/10.3113/FAI.2007.0154
126. Vellet AD, Marks PH, Fowler PJ, Munro TG (1991) Occult posttraumatic osteochondral lesions of the knee: prevalence, classification, and short-term sequelae evaluated with MR imaging. Radiology 178(1):271–276. https://doi.org/10.1148/radiology.178.1.1984319
127. Zengerink M, Struijs PA, Tol JL, van Dijk CN (2010) Treatment of osteochondral lesions of the talus: a systematic review. Knee Surg Sports Traumatol Arthrosc 18(2):238–246. https://doi.org/10.1007/s00167-009-0942-6
128. Durr HD, Martin H, Pellengahr C, Schlemmer M, Maier M, Jansson V (2004) The cause of subchondral bone cysts in osteoarthrosis: a finite element analysis. Acta Orthop Scand 75(5):554–558. https://doi.org/10.1080/00016470410001411
129. Verhagen RA, Struijs PA, Bossuyt PM, van Dijk CN (2003) Systematic review of treatment strategies for osteochondral defects of the talar dome. Foot Ankle Clin 8(2):233–242. viii–ix
130. Woods K, Harris I (1995) Osteochondritis dissecans of the talus in identical twins. J Bone Joint Surg Br 77(2):331
131. Hermanson E, Ferkel RD (2009) Bilateral osteochondral lesions of the talus. Foot Ankle Int 30(8):723–727. https://doi.org/10.3113/FAI.2009.0723
132. Frenkel SR, DCP E (1999) Degradation and repair of articular cartilage. Front Biosci 4:671–685

133. van Dijk CN, Bossuyt PM, Marti RK (1996) Medial ankle pain after lateral ligament rupture. J Bone Joint Surg Br 78(4):562–567

134. Pereira H, Vuurberg G, Gomes N, Oliveira JM, Ripoll PL, Reis RL, Espregueira-Mendes J, van Dijk CN (2016) Arthroscopic repair of ankle instability with all-soft knotless anchors. Arthrosc Tech 5(1):e99–e107. https://doi.org/10.1016/j.eats.2015.10.010

135. Saxena A, Luhadiya A, Ewen B, Goumas C (2011) Magnetic resonance imaging and incidental findings of lateral ankle pathologic features with asymptomatic ankles. J Foot Ankle Surg 50(4):413–415. https://doi.org/10.1053/j.jfas.2011.03.011

136. Lambers KTA, Dahmen J, Reilingh ML, van Bergen CJA, Stufkens SAS, Kerkhoffs G (2017) No superior surgical treatment for secondary osteochondral defects of the talus. Knee Surg Sports Traumatol Arthrosc. https://doi.org/10.1007/s00167-017-4629-0

137. van Dijk CN, van Bergen CJ (2008) Advancements in ankle arthroscopy. J Am Acad Orthop Surg 16(11):635–646

138. Zengerink M, van Dijk CN (2012) Complications in ankle arthroscopy. Knee Surg Sports Traumatol Arthrosc 20(8):1420–1431. https://doi.org/10.1007/s00167-012-2063-x

139. Kerkhoffs GM, Reilingh ML, Gerards RM, de Leeuw PA (2016) Lift, drill, fill and fix (LDFF): a new arthroscopic treatment for talar osteochondral defects. Knee Surg Sports Traumatol Arthrosc 24(4):1265–1271. https://doi.org/10.1007/s00167-014-3057-7

140. Bhattarai N, Gunn J, Zhang M (2010) Chitosan-based hydrogels for controlled, localized drug delivery. Adv Drug Deliv Rev 62(1):83–99. https://doi.org/10.1016/j.addr.2009.07.019

141. Brittberg M (2010) Cell carriers as the next generation of cell therapy for cartilage repair: a review of the matrix-induced autologous chondrocyte implantation procedure. Am J Sports Med 38(6):1259–1271. https://doi.org/10.1177/0363546509346395

142. Candrian C, Miot S, Wolf F, Bonacina E, Dickinson S, Wirz D, Jakob M, Valderrabano V, Barbero A, Martin I (2010) Are ankle chondrocytes from damaged fragments a suitable cell source for cartilage repair? Osteoarthritis Cartilage 18(8):1067–1076. https://doi.org/10.1016/j.joca.2010.04.010

143. Caron MMJ, Emans PJ, Coolsen MME, Voss L, Surtel DAM, Cremers A, van Rhijn LW, Welting TJM (2012) Redifferentiation of dedifferentiated human articular chondrocytes: comparison of 2D and 3D cultures. Osteoarthr Cartil 20(10):1170–1178. https://doi.org/10.1016/j.joca.2012.06.016

144. Centeno CJ, Schultz JR, Cheever M, Robinson B, Freeman M, Marasco W (2010) Safety and complications reporting on the re-implantation of culture-expanded mesenchymal stem cells using autologous platelet lysate technique. Curr Stem Cell Res Ther 5(1):81–93

145. Che JH, Zhang ZR, Li GZ, Tan WH, Bai XD, Qu FJ (2010) Application of tissue-engineered cartilage with BMP-7 gene to repair knee joint cartilage injury in rabbits. Knee Surg Sports Traumatol Arthrosc 18(4):496–503. https://doi.org/10.1007/s00167-009-0962-2

146. Clar C, Cummins E, McIntyre L, Thomas S, Lamb J, Bain L, Jobanputra P, Waugh N (2005) Clinical and cost-effectiveness of autologous chondrocyte implantation for cartilage defects in knee joints: systematic review and economic evaluation. Health Technol Assess 9(47):iii–iiv. ix–x, 1–82. doi:03-65-01 [pii]

147. Dhollander AA, De Neve F, Almqvist KF, Verdonk R, Lambrecht S, Elewaut D, Verbruggen G, Verdonk PC (2011) Autologous matrix-induced chondrogenesis combined with platelet-rich plasma gel: technical description and a five pilot patients report. Knee Surg Sports Traumatol Arthrosc 19(4):536–542. https://doi.org/10.1007/s00167-010-1337-4

148. DiGiovanni CW, Lin SS, Baumhauer JF, Daniels T, Younger A, Glazebrook M, Anderson J, Anderson R, Evangelista P, Lynch SE (2013) Recombinant human platelet-derived growth factor-BB and beta-tricalcium phosphate (rhPDGF-BB/beta-TCP): an alternative to autogenous bone graft. J Bone Joint Surg Am 95(13):1184–1192. https://doi.org/10.2106/JBJS.K.01422

149. Enea D, Gwynne J, Kew S, Arumugam M, Shepherd J, Brooks R, Ghose S, Best S, Cameron R, Rushton N (2012) Collagen fibre implant for tendon and ligament biological augmentation. In vivo study in an ovine model. Knee Surg Sports Traumatol Arthrosc. https://doi.org/10.1007/s00167-012-2102-7

150. Gurkan UA, Tasoglu S, Kavaz D, Demirci U (2012) Emerging technologies for assembly of microscale hydrogels. Adv Healthc Mater 1:149–158
151. Harris JD, Siston RA, Pan X, Flanigan DC (2010) Autologous chondrocyte implantation: a systematic review. J Bone Joint Surg Am 92(12):2220–2233. https://doi.org/10.2106/JBJS.J.00049
152. Nagura I, Fujioka H, Kokubu T, Makino T, Sumi Y, Kurosaka M (2007) Repair of osteochondral defects with a new porous synthetic polymer scaffold. J Bone Joint Surg Br 89(2):258–264. https://doi.org/10.1302/0301-620X.89B2.17754
153. Pereira D, Silva-Correia J, Pereira H, Espregueira-Mendes J, Oliveira JM, Reis RL (2013) Gellan gum-based bilayered scaffolds for application in ostheochondral tissue engineering. J Tissue Eng Regen Med 6(1)
154. Vinatier C, Mrugala D, Jorgensen C, Guicheux J, Noel D (2009) Cartilage engineering: a crucial combination of cells, biomaterials and biofactors. Trends Biotechnol 27(5):307–314. https://doi.org/10.1016/j.tibtech.2009.02.005
155. Ettinger S, Stukenborg-Colsman C, Waizy H, Becher C, Yao D, Claassen L, Noll Y, Plaass C (2017) Results of HemiCAP(R) implantation as a salvage procedure for osteochondral lesions of the talus. J Foot Ankle Surg 56(4):788–792. https://doi.org/10.1053/j.jfas.2017.04.001
156. Pagenstert GI, Hintermann B, Barg A, Leumann A, Valderrabano V (2007) Realignment surgery as alternative treatment of varus and valgus ankle osteoarthritis. Clin Orthop Relat Res 462:156–168. https://doi.org/10.1097/BLO.0b013e318124a462
157. Ateshian GA, Soslowsky LJ, Mow VC (1991) Quantitation of articular surface topography and cartilage thickness in knee joints using stereophotogrammetry. J Biomech 24(8):761–776
158. Swann AC, Seedhom BB (1993) The stiffness of normal articular cartilage and the predominant acting stress levels: implications for the aetiology of osteoarthrosis. Br J Rheumatol 32(1):16–25
159. Kimizuka M, Kurosawa H, Fukubayashi T (1980) Load-bearing pattern of the ankle joint. Contact area and pressure distribution. Arch Orthop Trauma Surg 96(1):45–49
160. Beaudoin AJ, Fiore SM, Krause WR, Adelaar RS (1991) Effect of isolated talocalcaneal fusion on contact in the ankle and talonavicular joints. Foot Ankle 12(1):19–25
161. Peyron JG (1984) The epidemiology of osteoarthritis. In: Moskowitz RW, Howell DS, Goldberg VM, Mankin HJ (eds) Osteoarthritis. Diagnosis and treatment. W.B. Saunders, Philadelphia, pp S9–27
162. Ihn JC, Kim SJ, Park IH (1993) In vitro study of contact area and pressure distribution in the human knee after partial and total meniscectomy. Int Orthop 17(4):214–218
163. Rolauffs B, Muehleman C, Li J, Kurz B, Kuettner KE, Frank E, Grodzinsky AJ (2010) Vulnerability of the superficial zone of immature articular cartilage to compressive injury. Arthritis Rheum 62(10):3016–3027. https://doi.org/10.1002/art.27610
164. Roukis TS (2004) Corrective ankle osteotomies. Clin Podiatr Med Surg 21(3):353–370. vi. https://doi.org/10.1016/j.cpm.2004.03.007
165. Yi Y, Lee W (2017) Peri-talar re-alignment osteotomy for joint preservation in asymmetrical ankle osteoarthritis. EFORT Open Rev 2(7):324–331. https://doi.org/10.1302/2058-5241.2.160021
166. Gomoll AH, Madry H, Knutsen G, van Dijk N, Seil R, Brittberg M, Kon E (2010) The subchondral bone in articular cartilage repair: current problems in the surgical management. Knee Surg Sports Traumatol Arthrosc 18(4):434–447. https://doi.org/10.1007/s00167-010-1072-x
167. Chen L, Lu X, Li S, Sun Q, Li W, Song D (2012) Sustained delivery of BMP-2 and platelet-rich plasma-released growth factors contributes to osteogenesis of human adipose-derived stem cells. Orthopedics 35(9):e1402–e1409. https://doi.org/10.3928/01477447-20120822-29
168. Amable PR, Carias RB, Teixeira MV, da Cruz Pacheco I, Correa do Amaral RJ, Granjeiro JM, Borojevic R (2013) Platelet-rich plasma preparation for regenerative medicine: optimization and quantification of cytokines and growth factors. Stem Cell Res Ther 4(3):67. https://doi.org/10.1186/scrt218

169. Im G II, Shin Y-W, Lee K-B (2005) Do adipose tissue-derived mesenchymal stem cells have the same osteogenic and chondrogenic potential as bone marrow-derived cells? Osteoarthr Cartil 13(10):845–853. https://doi.org/10.1016/j.joca.2005.05.005

170. Gan Y, Dai K, Zhang P, Tang T, Zhu Z, Lu J (2008) The clinical use of enriched bone marrow stem cells combined with porous beta-tricalcium phosphate in posterior spinal fusion. Biomaterials 29(29):3973–3982. https://doi.org/10.1016/j.biomaterials.2008.06.026

171. Li M, Chen M, Han W, Fu X (2010) How far are induced pluripotent stem cells from the clinic? Ageing Res Rev 9(3):257–264. https://doi.org/10.1016/j.arr.2010.03.001

172. Lee JY, Zhou Z, Taub PJ, Ramcharan M, Li Y, Akinbiyi T, Maharam ER, Leong DJ, Laudier DM, Ruike T, Torina PJ, Zaidi M, Majeska RJ, Schaffler MB, Flatow EL, Sun HB (2011) BMP-12 treatment of adult mesenchymal stem cells in vitro augments tendon-like tissue formation and defect repair in vivo. PLoS One 6(3):e17531. https://doi.org/10.1371/journal.pone.0017531

173. Barry F, Murphy M (2013) Mesenchymal stem cells in joint disease and repair. https://doi.org/10.1038/nrrheum.2013.109

174. Lubowitz JH, Provencher MT, Poehling GG (2013) Stem cells in the knee. Arthroscopy 29(4):609–610. https://doi.org/10.1016/j.arthro.2013.01.005

175. Silva A, Sampaio R, Fernandes R, Pinto E (2014) Is there a role for adult non-cultivated bone marrow stem cells in ACL reconstruction? Knee Surg Sports Traumatol Arthrosc 22(1):66–71. https://doi.org/10.1007/s00167-012-2279-9

176. Bacelar AH, Cengiz IF, Silva-Correia J, Sousa RA, Oliveira JM, Reisa RL (2017) "Smart" hydrogels in tissue engineering and regenerative medicine applications. In: Handbook of intelligent scaffolds for tissue engineering and regenerative medicine 2:327–361. https://www.crcpress.com/Handbook-of-Intelligent-Scaffolds-for-Tissue-Engineering-and-Regenerative/Khang/p/book/9789814745123

177. Kim IL, Mauck RL, Burdick JA (2011) Hydrogel design for cartilage tissue engineering: a case study with hyaluronic acid. Biomaterials 32(34):8771–8782. https://doi.org/10.1016/j.biomaterials.2011.08.073

178. Liu M, Zeng X, Ma C, Yi H, Ali Z, Mou X, Li S, Deng Y, He N (2017) Injectable hydrogels for cartilage and bone tissue engineering. Bone Res 5:17014. https://doi.org/10.1038/boneres.2017.14

179. Pereira H, Silva-Correia J, Yan L-P, Oliveira A, Oliveira J-M, Espregueira-Mendes J, Reis R (2013) Combined application of Silk-fibroin/methacrylated gellan gum hydrogel in tissue engineering approaches for partial and/or total meniscus replacement while enabling control of neovascularization. Rev Chir Orthop Traumatol 99(8):e18-e19

180. Rao JK, Ramesh DV, Rao KP (1994) Implantable controlled delivery systems for proteins based on collagen — pHEMA hydrogels. Biomaterials 15(5):383–389. https://doi.org/10.1016/0142-9612(94)90251-8

181. Silva-Correia J, Miranda-Goncalves V, Salgado AJ, Sousa N, Oliveira JM, Reis RM, Reis RL (2012) Angiogenic potential of gellan-gum-based hydrogels for application in nucleus pulposus regeneration: in vivo study. Tissue Eng A 18(11-12):1203–1212. https://doi.org/10.1089/ten.TEA.2011.0632

182. Evans CH (2013) Advances in regenerative orthopedics. Mayo Clin Proc 88(11):1323–1339. https://doi.org/10.1016/j.mayocp.2013.04.027

183. FDA (2012) http://www.fda.gov/biologicsbloodvaccines/developmentapprovalprocess/biologicalapprovalsbyyear/ucm289012.htm

184. Foster TE, Puskas BL, Mandelbaum BR, Gerhardt MB, Rodeo SA (2009) Platelet-rich plasma: from basic science to clinical applications. Am J Sports Med 37(11):2259–2272. https://doi.org/10.1177/0363546509349921

185. Mazzocca AD, McCarthy MB, Chowaniec DM, Cote MP, Romeo AA, Bradley JP, Arciero RA, Beitzel K (2012) Platelet-rich plasma differs according to preparation method and human variability. J Bone Joint Surg Am 94(4):308–316. https://doi.org/10.2106/JBJS.K.00430

186. Cengiz IF, Silva-Correia J, Pereira H, Espregueira-Mendes J, Oliveira JM, Reis RL (2017) Advanced regenerative strategies for human knee meniscus. In: Regenerative strategies for the treatment of knee joint disabilities. Springer, pp 271–285

187. Pittenger MF, Mackay AM, Beck SC, Jaiswal RK, Douglas R, Mosca JD, Moorman MA, Simonetti DW, Craig S, Marshak DR (1999) Multilineage potential of adult human mesenchymal stem cells. Science 284(5411):143–147

188. da Silva Meirelles L, Caplan AI, Nardi NB (2008) In search of the in vivo identity of mesenchymal stem cells. Stem Cells 26(9):2287–2299. https://doi.org/10.1634/stemcells.2007-1122

189. Williams AR, Hare JM (2011) Mesenchymal stem cells: biology, pathophysiology, translational findings, and therapeutic implications for cardiac disease. Circ Res 109(8):923–940. https://doi.org/10.1161/circresaha.111.243147

190. Guilak F, Estes BT, Diekman BO, Moutos FT, Gimble JM (2010) 2010 Nicolas Andry Award: Multipotent adult stem cells from adipose tissue for musculoskeletal tissue engineering. Clin Orthop Relat Res 468(9):2530–2540. https://doi.org/10.1007/s11999-010-1410-9

191. Khan WS, Adesida AB, Tew SR, Longo UG, Hardingham TE (2012) Fat pad-derived mesenchymal stem cells as a potential source for cell-based adipose tissue repair strategies. Cell Prolif 45(2):111–120. https://doi.org/10.1111/j.1365-2184.2011.00804.x

192. Agung M, Ochi M, Yanada S, Adachi N, Izuta Y, Yamasaki T, Toda K (2006) Mobilization of bone marrow-derived mesenchymal stem cells into the injured tissues after intraarticular injection and their contribution to tissue regeneration. Knee Surg Sports Traumatol Arthrosc 14(12):1307–1314. https://doi.org/10.1007/s00167-006-0124-8

193. Meirelles Lda S, Fontes AM, Covas DT, Caplan AI (2009) Mechanisms involved in the therapeutic properties of mesenchymal stem cells. Cytokine Growth Factor Rev 20(5-6):419–427. https://doi.org/10.1016/j.cytogfr.2009.10.002

194. Bashir J, Sherman A, Lee H, Kaplan L, Hare JM (2014) Mesenchymal stem cell therapies in the treatment of musculoskeletal diseases. PM R 6(1):61–69. https://doi.org/10.1016/j.pmrj.2013.05.007

195. Kuroda R, Ishida K, Matsumoto T, Akisue T, Fujioka H, Mizuno K, Ohgushi H, Wakitani S, Kurosaka M (2007) Treatment of a full-thickness articular cartilage defect in the femoral condyle of an athlete with autologous bone-marrow stromal cells. Osteoarthritis Cartilage 15(2):226–231. https://doi.org/10.1016/j.joca.2006.08.008

196. Wakitani S, Nawata M, Tensho K, Okabe T, Machida H, Ohgushi H (2007) Repair of articular cartilage defects in the patello-femoral joint with autologous bone marrow mesenchymal cell transplantation: three case reports involving nine defects in five knees. J Tissue Eng Regen Med 1(1):74–79. https://doi.org/10.1002/term.8

197. Buda R, Vannini F, Cavallo M, Grigolo B, Cenacchi A, Giannini S (2010) Osteochondral lesions of the knee: a new one-step repair technique with bone-marrow-derived cells. J Bone Joint Surg Am 92(Suppl 2):2–11. https://doi.org/10.2106/jbjs.j.00813

198. Buda R, Vannini F, Cavallo M, Baldassarri M, Luciani D, Mazzotti A, Pungetti C, Olivieri A, Giannini S (2013) One-step arthroscopic technique for the treatment of osteochondral lesions of the knee with bone-marrow-derived cells: three years results. Musculoskelet Surg 97(2):145–151. https://doi.org/10.1007/s12306-013-0242-7

199. Giannini S, Buda R, Vannini F, Cavallo M, Grigolo B (2009) One-step bone marrow-derived cell transplantation in talar osteochondral lesions. Clin Orthop Relat Res 467(12):3307–3320. https://doi.org/10.1007/s11999-009-0885-8

200. Giannini S, Buda R, Cavallo M, Ruffilli A, Cenacchi A, Cavallo C, Vannini F (2010) Cartilage repair evolution in post-traumatic osteochondral lesions of the talus: from open field autologous chondrocyte to bone-marrow-derived cells transplantation. Injury 41(11):1196–1203. https://doi.org/10.1016/j.injury.2010.09.028

201. Haleem AM, Singergy AA, Sabry D, Atta HM, Rashed LA, Chu CR, El Shewy MT, Azzam A, Abdel Aziz MT (2010) The clinical use of human culture-expanded autologous bone marrow mesenchymal stem cells transplanted on platelet-rich fibrin glue in the treatment of artic-

ular cartilage defects: a pilot study and preliminary results. Cartilage 1(4):253–261. https://doi.org/10.1177/1947603510366027

202. Giannini S, Buda R, Battaglia M, Cavallo M, Ruffilli A, Ramponi L, Pagliazzi G, Vannini F (2013) One-step repair in talar osteochondral lesions: 4-year clinical results and t2-mapping capability in outcome prediction. Am J Sports Med 41(3):511–518. https://doi.org/10.1177/0363546512467622

203. Kasemkijwattana C, Hongeng S, Kesprayura S, Rungsinaporn V, Chaipinyo K, Chansiri K (2011) Autologous bone marrow mesenchymal stem cells implantation for cartilage defects: two cases report. J Med Assoc Thai 94(3):395–400

204. Oliveira JM, Rodrigues MT, Silva SS, Malafaya PB, Gomes ME, Viegas CA, Dias IR, Azevedo JT, Mano JF, Reis RL (2006) Novel hydroxyapatite/chitosan bilayered scaffold for osteochondral tissue-engineering applications: Scaffold design and its performance when seeded with goat bone marrow stromal cells. Biomaterials 27(36):6123–6137. https://doi.org/10.1016/j.biomaterials.2006.07.034

205. Nejadnik H, Hui JH, Feng Choong EP, Tai BC, Lee EH (2010) Autologous bone marrow-derived mesenchymal stem cells versus autologous chondrocyte implantation: an observational cohort study. Am J Sports Med 38(6):1110–1116. https://doi.org/10.1177/0363546509359067

206. Gigante A, Cecconi S, Calcagno S, Busilacchi A, Enea D (2012) Arthroscopic knee cartilage repair with covered microfracture and bone marrow concentrate. Arthrosc Tech 1(2):e175–e180. https://doi.org/10.1016/j.eats.2012.07.001

207. Enea D, Cecconi S, Calcagno S, Busilacchi A, Manzotti S, Gigante A (2015) One-step cartilage repair in the knee: collagen-covered microfracture and autologous bone marrow concentrate. A pilot study. Knee 22(1):30–35. https://doi.org/10.1016/j.knee.2014.10.003

208. Enea D, Cecconi S, Calcagno S, Busilacchi A, Manzotti S, Kaps C, Gigante A (2013) Single-stage cartilage repair in the knee with microfracture covered with a resorbable polymer-based matrix and autologous bone marrow concentrate. Knee 20(6):562–569. https://doi.org/10.1016/j.knee.2013.04.003

209. Rai V, Dilisio MF, Dietz NE, Agrawal DK (2017) Recent strategies in cartilage repair: a systemic review of the scaffold development and tissue engineering. J Biomed Mater Res A 105(8):2343–2354. https://doi.org/10.1002/jbm.a.36087

210. He H, Li W, Tseng DY, Zhang S, Chen SY, Day AJ, Tseng SC (2009) Biochemical characterization and function of complexes formed by hyaluronan and the heavy chains of inter-alpha-inhibitor (HC*HA) purified from extracts of human amniotic membrane. J Biol Chem 284(30):20136–20146. https://doi.org/10.1074/jbc.M109.021881

211. Cengiz IF, Oliveira JM, Reis RL (2014) Tissue engineering and regenerative medicine strategies for the treatment of osteochondral lesions. In: 3D multiscale physiological human. Springer, pp 25–47

212. Correia SI, Silva-Correia J, Pereira H, Canadas RF, da Silva Morais A, Frias AM, Sousa RA, van Dijk CN, Espregueira-Mendes J, Reis RL, Oliveira JM (2015) Posterior talar process as a suitable cell source for treatment of cartilage and osteochondral defects of the talus. J Tissue Eng Regen Med. https://doi.org/10.1002/term.2092

213. Khan WS, Longo UG, Adesida A, Denaro V (2012) Stem cell and tissue engineering applications in orthopaedics and musculoskeletal medicine. Stem Cells Int 2012:403170. https://doi.org/10.1155/2012/403170

214. Oliveira M, Reis RL (2017) Regenerative strategies for the treatment of knee joint disabilities. Studies in mechanobiology, tissue engineering and biomaterials. Springer. https://doi.org/10.1007/978-3-319-44785-8

215. Patrascu JM, Freymann U, Kaps C, Poenaru DV (2010) Repair of a post-traumatic cartilage defect with a cell-free polymer-based cartilage implant: a follow-up at two years by MRI and histological review. J Bone Joint Surg Br 92(8):1160–1163. https://doi.org/10.1302/0301-620x.92b8.24341

216. Siclari A, Mascaro G, Kaps C, Boux E (2014) A 5-year follow-up after cartilage repair in the knee using a platelet-rich plasma-immersed polymer-based implant. Open Orthop J 8:346–354. https://doi.org/10.2174/1874325001408010346

217. Yan LP, Silva-Correia J, Oliveira MB, Vilela C, Pereira H, Sousa RA, Mano JF, Oliveira AL, Oliveira JM, Reis RL (2015) Bilayered silk/silk-nanoCaP scaffolds for osteochondral tissue engineering: In vitro and in vivo assessment of biological performance. Acta Biomater 12:227–241. https://doi.org/10.1016/j.actbio.2014.10.021
218. Yan LP, Oliveira JM, Oliveira AL, Caridade SG, Mano JF, Reis RL (2012) Macro/microporous silk fibroin scaffolds with potential for articular cartilage and meniscus tissue engineering applications. Acta Biomater 8(1):289–301. https://doi.org/10.1016/j.actbio.2011.09.037
219. Chen Y, Bloemen V, Impens S, Moesen M, Luyten FP, Schrooten J (2011) Characterization and optimization of cell seeding in scaffolds by factorial design: quality by design approach for skeletal tissue engineering. Tissue Eng Part C Methods 17(12):1211–1221. https://doi.org/10.1089/ten.tec.2011.0092
220. Borselli C, Cezar CA, Shvartsman D, Vandenburgh HH, Mooney DJ (2011) The role of multifunctional delivery scaffold in the ability of cultured myoblasts to promote muscle regeneration. Biomaterials 32(34):8905–8914. https://doi.org/10.1016/j.biomaterials.2011.08.019
221. Christensen BB, Foldager CB, Jensen J, Jensen NC, Lind M (2016) Poor osteochondral repair by a biomimetic collagen scaffold: 1- to 3-year clinical and radiological follow-up. Knee Surg Sports Traumatol Arthrosc 24(7):2380–2387. https://doi.org/10.1007/s00167-015-3538-3
222. Pereira H, Frias AM, Oliveira JM, Espregueira-Mendes J, Reis RL (2011) Tissue engineering and regenerative medicine strategies in meniscus lesions. Arthroscopy 27(12):1706–1719. https://doi.org/10.1016/j.arthro.2011.08.283
223. Cengiz IF, Pereira H, Espregueira-Mendes J, Oliveira JM, Reis RL (2017) Treatments of meniscus lesions of the knee: current concepts and future perspectives. Regen Eng Transl Med:1–19
224. Pleshko N, Grande DA, Myers KR (2012) Nanotechnology in orthopaedics. J Am Acad Orthop Surg 20(1):60–62. https://doi.org/10.5435/JAAOS-20-01-060
225. Lu H, Lv L, Dai Y, Wu G, Zhao H, Zhang F (2013) Porous chitosan scaffolds with embedded hyaluronic acid/chitosan/plasmid-DNA nanoparticles encoding TGF-beta1 induce DNA controlled release, transfected chondrocytes, and promoted cell proliferation. PLoS One 8(7):e69950. https://doi.org/10.1371/journal.pone.0069950
226. El-Sadik AO, El-Ansary A, Sabry SM (2010) Nanoparticle-labeled stem cells: a novel therapeutic vehicle. Clin Pharm 2:9–16. https://doi.org/10.2147/CPAA.S8931
227. Santo VE, Gomes ME, Mano JF, Reis RL (2013) Controlled release strategies for bone, cartilage, and osteochondral engineering – part I: recapitulation of native tissue healing and variables for the design of delivery systems. Tissue Eng Part B Rev 19(4):308–326. https://doi.org/10.1089/ten.TEB.2012.0138
228. Lee K, Silva EA, Mooney DJ (2011) Growth factor delivery-based tissue engineering: general approaches and a review of recent developments. J R Soc Interface 8(55):153–170. https://doi.org/10.1098/rsif.2010.0223
229. Correia SI, Pereira H, Silva-Correia J, Van Dijk CN, Espregueira-Mendes J, Oliveira JM, Reis RL (2014) Current concepts: tissue engineering and regenerative medicine applications in the ankle joint. J R Soc Interface 11(92):20130784. https://doi.org/10.1098/rsif.2013.0784
230. Lim EH, Sardinha JP, Myers S (2014) Nanotechnology biomimetic cartilage regenerative scaffolds. Arch Plast Surg 41(3):231–240. https://doi.org/10.5999/aps.2014.41.3.231
231. Dormer NH, Singh M, Wang L, Berkland CJ, Detamore MS (2010) Osteochondral interface tissue engineering using macroscopic gradients of bioactive signals. Ann Biomed Eng 38(6):2167–2182. https://doi.org/10.1007/s10439-010-0028-0
232. Leping Y, Oliveira JM, Oliveira AL, Reis RL (2015) Current concepts and challenges in osteochondral tissue engineering and regenerative medicine. ACS Biomater Sci Eng. https://doi.org/10.1021/ab500038y
233. Madry H, Orth P, Cucchiarini M (2011) Gene therapy for cartilage repair. Cartilage 2(3):201–225. https://doi.org/10.1177/1947603510392914
234. Reitmaier S, Wolfram U, Ignatius A, Wilke H-J, Gloria A, Martín-Martínez JM, Silva-Correia J, Miguel Oliveira J, Luís Reis R, Schmidt H (2012) Hydrogels for nucleus replacement – facing

the biomechanical challenge. J Mech Behav Biomed Mater (0). doi:https://doi.org/10.1016/j. jmbbm.2012.05.010

235. Cengiz I, Pitikakis M, Cesario L, Parascandolo P, Vosilla L, Viano G, Oliveira J, Reis R (2016) Building the basis for patient-specific meniscal scaffolds: From human knee MRI to fabrication of 3D printed scaffolds. Bioprinting 1:1–10

236. Cengiz IF, Pereira H, Pitikakis M, Espregueira-Mendes J, Oliveira JM, Reis RL (2017) Building the basis for patient-specific meniscal scaffolds. In: Gobbi A, Espregueira-Mendes J, Lane JG, Karahan M (eds) Bio-orthopaedics: a new approach. Springer, Berlin/Heidelberg, pp 411–418. https://doi.org/10.1007/978-3-662-54181-4_32

237. Mouser VHM, Levato R, Bonassar LJ, D'Lima DD, Grande DA, Klein TJ, Saris DBF, Zenobi-Wong M, Gawlitta D, Malda J (2017) Three-dimensional bioprinting and its potential in the field of articular cartilage regeneration. Cartilage 8(4):327–340. https://doi. org/10.1177/1947603516665445

238. Oner T, Cengiz I, Pitikakis M, Cesario L, Parascandolo P, Vosilla L, Viano G, Oliveira J, Reis R, Silva-Correia J (2017) 3D segmentation of intervertebral discs: from concept to the fabrication of patient-specific scaffolds. J 3D Print Med 1(2):91–101. https://doi.org/10.2217/ 3dp-2016-0011

Chapter 3
Osteoarthritis: Trauma vs Disease

Gema Jiménez, Jesús Cobo-Molinos, Cristina Antich, and Elena López-Ruiz

Abstract Osteoarthritis (OA) is the most prevalent joint disease characterized by pain and degenerative lesions of the cartilage, subchondral bone, and other joint tissues. The causes of OA remain incompletely understood. Over the years, it has become recognized that OA is a multifactorial disease. In particular, aging and trauma are the main risk factors identified for the development of OA; however,

G. Jiménez · C. Antich
Biopathology and Regenerative Medicine Institute (IBIMER), Centre for Biomedical Research (CIBM), University of Granada, Granada, Spain

Biosanitary Research Institute of Granada (ibs.GRANADA), University Hospitals of Granada-University of Granada, Granada, Spain

Department of Human Anatomy and Embryology, Faculty of Medicine, University of Granada, Granada, Spain

Excellence Research Unit "Modelling Nature" (MNat), University of Granada, Granada, Spain
e-mail: gemajg@ugr.es; cantich@ugr.es

J. Cobo-Molinos
Department of Health Sciences, University of Jaén, Jaén, Spain
e-mail: jcobos@ujaen.es

E. López-Ruiz (✉)
Biopathology and Regenerative Medicine Institute (IBIMER), Centre for Biomedical Research (CIBM), University of Granada, Granada, Spain

Biosanitary Research Institute of Granada (ibs.GRANADA), University Hospitals of Granada-University of Granada, Granada, Spain

Department of Human Anatomy and Embryology, Faculty of Medicine, University of Granada, Granada, Spain

Excellence Research Unit "Modelling Nature" (MNat), University of Granada, Granada, Spain

Department of Health Sciences, University of Jaén, Jaén, Spain
e-mail: elop@ugr.es; elruiz@ujaen.es

© Springer International Publishing AG, part of Springer Nature 2018
J. M. Oliveira et al. (eds.), *Osteochondral Tissue Engineering*,
Advances in Experimental Medicine and Biology 1059,
https://doi.org/10.1007/978-3-319-76735-2_3

other factors such as genetic predisposition, obesity, inflammation, gender and hormones, or metabolic syndrome contribute to OA development and lead to a more severe outcome. While this disease mainly affects people older than 60 years, OA developed after joint trauma affects all range ages and has a particular impact on young individuals and people who have highest levels of physical activity such as athletes. Traumatic injury to the joint often results in joint instability or intra-articular fractures which lead to posttraumatic osteoarthritis (PTOA). In response to injury, several molecular mechanisms are activated, increasing the production and activation of different factors that contribute to the progression of OA.

In this chapter, we have focused on the interactions and contribution of the multiple factors involved in joint destruction and progression of OA. In addition, we overview the main changes and molecular mechanisms related to OA pathogenesis.

Keywords Osteoarthritis · Posttraumatic osteoarthritis · Risk factors · Joint trauma

Highlights
- OA is a multifactorial disorder and is associated with pathological changes in all joint tissues; thus, OA is considered a whole-joint disease.
- Factors that contribute to the development of OA include joint injury, obesity, aging, inflammation, genetic predisposition, gender and hormones, or metabolic syndrome. Among these factors, collective evidence indicates that aging and trauma are pivotal factors that mark OA progression.
- Comprehensive understanding of the molecular networks regulating articular cartilage homeostasis and OA pathogenesis is needed for the development of novel treatments for preventing cartilage damage and promoting repair.

3.1 Introduction

Osteoarthritis (OA) is the most prevalent joint disease characterized by pain and degenerative lesions in the cartilage and in the tissues within and surrounding the joint involved. OA has a high prevalence in the population, and it is accompanied by significant morbidity and physical disability [67]. It has been estimated that approximately 25% of the population over 18 years old is affected [22]. So far, it is also predicted that 35% of people will eventually suffer disability due to OA by 2030, and this number is expected to further expand [110]. Owing to the high incidence of this disease among population, required therapies have a substantial public health impact [87].

Any joint in the body can suffer from OA, but major joints such as the knee and hip are most commonly affected. Current OA therapies include pain management and surgical intervention for end-stage OA patients, but there are no effective therapies which can effectively prevent or reverse the progression of the disease.

Despite OA has received a lot of attention in clinical research, more studies are needed to increase our understanding of the molecular mechanism, and the etiology of this complex disease, only then the development of effective treatments will be much closer.

OA is characterized by degenerative lesions of the cartilage, but also other tissues of the joint are involved in the complex initiation and progression of the disease; thus, progression of OA also involves subchondral bone remodeling, the formation of osteophytes, the development of bone marrow lesions, and changes in the synovium, joint capsule, and ligaments (Fig. 3.1) [131].

Although OA ultimately ends to a common phenotype consisting of chronic pain, joint instability, stiffness, and loss of function, it results from a number of different etiologies. Among the multiple factors that contribute to the development and progression of OA are joint injury, obesity, aging, inflammation, and genetic risk factors [77]. Moreover, OA risk increases with the presence of other factors including gender and hormones or metabolic syndrome.

While most of the OA is idiopathic and mainly affects people older than 60 years, the risk of OA following a significant joint trauma is especially prevalent in patients at younger age and highly active individuals [6, 58]. Traumatic injury to the joint often leads to joint instability or intra-articular fractures which in the long term end in OA. The OA which is initiated after a joint injury is called posttraumatic

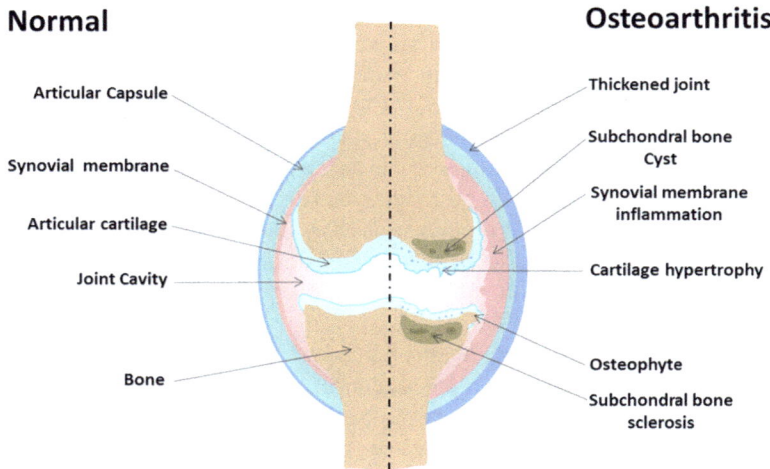

Fig. 3.1 Progression of osteoarthritis. On the left, cross section of the normal articular joint illustrates the main structural elements including the articular cartilage covering the surface of the subchondral bones and enclosed in a connective tissue capsule lined by a synovial membrane. On the right, cross section of the OA articular joint showing advanced osteoarthritic changes characterized by subchondral bone remodeling, subchondral cysts, the formation of osteophytes, cartilage hypertrophy, fissuring and fragmentation of the articular cartilage, inflammation of the synovial membrane, and joint thickening

osteoarthritis (PTOA) [81]. It is estimated that approximately 12% of all OA is a result of an injury insult, such as an articular fracture, chondral injury, or a ligament or meniscal injury [17].

Although joint trauma affects the entire joint to some degree, damage to articular cartilage is commonly the most relevant pathologic feature and primary change after joint injury and prior to joint dysfunction [16, 66]. Damage to articular cartilage leads to an imbalance of articular cartilage homeostasis which leads to the appearance of a sequence of biologic events. The chondrocytes produce and maintain the physical function of cartilage by synthesizing and degrading matrix components. In response to environmental changes, such as mechanical stress or inflammatory stimuli, the stable phenotype of the chondrocytes shift toward a catabolic phenotype increasing the production and activation of different factors that actively participate in the degeneration process [123].

In this chapter, we discuss the multiple factors involved in joint destruction, development, and progression of OA with special interest in the impact of trauma injury on OA. In addition, the identification of the molecular mechanisms related to OA pathogenesis and main changes in composition and structure of joint tissues involved are discussed.

3.2   Changes During Osteoarthritis Progress

Articular cartilage is a unique tissue composed of chondrocytes (the only cell present in cartilage) embedded in a highly hydrated extracellular matrix of collagen fibers and proteoglycans together with other non-collagenous proteins and glycoproteins present in lesser amounts [104]. Under normal conditions, chondrocytes are resting in a nonstressed steady state with low turnover conditions. In response to environmental changes such as changes in biomechanical forces, growth factors, or cytokines, chondrocytes increase its metabolic activities, and the molecular composition and organization of the extracellular matrix are altered. Factors such as age, obesity, genetic predisposition, joint instability, repetitive stress injury, or inflammation are known to disrupt the articular chondrocyte homeostasis [77].

Loss of cartilage function and quality can occur due to trauma resulting in focal or diffuse loss of cartilage or as a consequence of aging cartilage involved in the osteoarthritic process [66]. The precise molecular mechanisms of OA initiation and progression are poorly understood. However, there has been an increasing literature describing multiple growth factors and cytokines involved in the destruction of articular cartilage and subchondral bone [65]. Once the microenvironment changes, the alteration in the normal physiologic balance of cartilage tissue leads to abnormal function of chondrocytes. At early changes in cartilage tissue, chondrocytes initiate the release of oxygen free radicals which contribute to initiate progressive tissue damage [97]. Several studies have also reported the release of fibronectin fragments that induce cell damage and matrix degradation [50]. A large number of proteins and genes show altered expression in OA cartilage compared with that in cartilage from individuals without OA. For example, transforming growth factor-β (TGFβ) has been shown to be involved in OA progression. While it is known that TGFβ is expressed at low levels in mature cartilage, the expression TGFβ and signaling have been seen upregulated in OA [124]. Moreover, OA chondrocytes increase the expression of hypertrophic markers such as Runx2 and ColX [89]. Inflammatory mediators released from the synovium can also contribute to the cartilage pathology in OA [89]. Therefore, the immediate release of inflammatory cytokines such as tumor necrosis factor alpha (TNF-α), interleukin (IL)-6, and IL-1 from the synovium or from the traumatized chondrocytes themselves induces a positive feedback loop [65]. This disturbed balance also induces chondrocytes to produce matrix metalloproteinases (MMPs), aggrecanases, and other proteases, which lead to increased cartilage matrix degradation [131].The pivotal proteinase that marks OA progression is MMP-13, the major type II collagen-degrading collagenase, which is regulated by both stress and inflammatory signals [43]. In vivo studies showed that joint injuries increase the levels of metalloproteinases in synovial fluid [88]. Other catabolic enzymes such as ADAMTS5 and MMP-8 are implicated in the degradation of articular cartilage structure by cleaving the aggrecan and collagen II matrix [39].

The release of these degradative enzymes and other collagenases in joints leads to proteoglycan and collagen network breakdown which degrade the structure of articular cartilage and result in functional abnormality of chondrocytes. Consequently, chondrocytes will undergo apoptosis, and cartilage will eventually be completely lost. Numerous studies provide evidence that chondrocyte death by apoptosis is associated with the initiation and severity of articular cartilage degradation [109]. Following degradation and metabolic changes in joint tissues, the disease slowly progresses through a long clinically asymptomatic latency period to a symptomatic phase with joint pain and dysfunction. Advanced OA degeneration is associated with increased damage to cartilage and loss of type II collagen. Loss of cartilage causes friction between bones. The progressive damage of bones will cause pain and limited joint mobility. Bone remodeling and loss/degeneration of cartilage are considered central features of OA [39]. Indeed, subchondral bone has received increasing attention in OA progression, and several studies evidence that abnormalities in subchondral bone can induce joint pain and cartilage degeneration. Moreover, changes in subchondral bone could be observed even preceding cartilage lesions, and clear evidence of an association between subchondral bone mineral density and OA have been described [41]. However, the pathological changes and the role of subchondral bone in OA still require further investigation. Other pathological changes seen in advanced OA include degeneration of ligaments and menisci of the knee and hypertrophy of the joint capsule [24].

3.3   Factors that Contribute to the Development of OA

3 3.1   Joint Trauma

Patients with OA due to a joint trauma normally have a well-defined damage to the structures of the articular joint. When we talk about joint trauma, we refer mainly to those lesions that affect the articular cartilage and/or the associated subchondral bone. Numerous studies have demonstrated that joint trauma is one of the main risk factors for the development of OA. It has been demonstrated that articular fracture is associated with a loss of chondrocyte viability and increased levels of systemic biomarkers [62]. Studies have indicated that focal loss of chondrocyte viability is an initiating pathway for development of PTOA [80]. Moreover, increased intra-articular trauma severity is associated with increased acute joint trauma in a variety of joint tissues, including synovial and bone [6]. Following joint trauma, the acute symptoms include swelling of the affected joint area due to the rupture of the vessels adjacent to the joint that causes hemorrhage in the interstitial space and, consequently, the formation of a hematoma. If the injury also affects the subchondral bone, the bone marrow will be probably involved. An early consequence after the initial injury is chondrocyte death. In addition, certain enzymes are released causing an inflammatory reaction in the affected area and an accumulation of fluids out of the vessels, which result in an increased volume of the knee joint and edema [4]. Hence, acute impact joint injuries initiate a sequence of biologic events that cause the progressive joint degeneration and can lead to the development of OA [6].

The main sign of OA due to joint trauma is the pain which is predominantly related to an unfavorable biomechanical environment at the joint. Diagnosis varies according to the intensity of the trauma and the presence of soft tissue injuries or bone injures. In addition to pain, within the anamnesis, it is important to evaluate signs of inflammation that can be observed in the joint. The right diagnosis and early treatment can slow and prevent further joint damage. The nontreatment or wrong treatment may lead to joint deterioration, poor function, and compromised mobility [63]. For the diagnosis of joint trauma, it is also important to perform imaging tests such as radiography. However, radiography is very limited as it only gives indirect information of the cartilage through the image of the subchondral bone. If another imaging test is needed, computed tomography (CT) is recommended in order to study any possible loss of osteochondral fragments [13].

Acute injuries are common in young, active, and athletic individuals, but diagnosing OA in these populations has become a challenge due to their higher tolerance for pain [5]. In both professional athletes and sports fans, the most frequent cause of OA is joint overload, excessive training, or the use of incorrect techniques that cause damage to the joints. The joints that are most frequently affected are the hip and the knee as they are the ones that carry the greatest weight during the performance of the workouts; however, we should not rule out also injuries to the elbow or foot [8]. It is important to take into account that physical activity cannot be forced and that it is necessary to avoid overloads in the same area of the body.

Typically, the most common injuries that cause PTOA include chondral and osteochondral lesions, articular fracture, ligamentous lesions, or fibrocartilage lesions, among others.

3.3.1.1 Joint Injuries that Cause PTOA

Chondral and Osteochondral Lesions

The hyaline cartilage is supported by the subchondral bone. In this way, we must differentiate two types of acute lesions, hyaline cartilage lesions and lesions produced in subchondral bone [14]. However, loss of hyaline cartilage usually appears in subchondral bone lesions. The extension of the lesion depends on the intensity of the trauma and the affected structures. When a relevant injury without fracture occurs, there is an overload of the subchondral bone tissue, which causes a progressive wear of the joint [63]. Chondral injuries include abrasion, laceration (cut), or fracture. Because cartilage is an avascular and aneural tissue, the articular cartilage has a poor intrinsic healing capacity. After a chondral injury, cartilage is incapable of directly generating pain. Depending on the size and location of the chondral injury, evolution can lead to further degeneration and a great loss of cartilage surface [63].

On the other hand, when an injury to the subchondral bone occurs, the joint is filled with blood arising from the bone marrow leading to an inflammatory process. Consequently, a reparative reaction will occur due to the exposure of the joint to blood and marrow contents, and matrix will be repaired; however, a fibrocartilage tissue without the same characteristics as original articular cartilage and with reduced mechanical properties will be formed [63].

Articular Fracture

Articular fracture is a very common injury due to the high percentage of injuries caused during minor accidents. In vivo models have demonstrated that articular fracture includes physical disruption of the articular surface and underlying subchondral bone with varying degrees of severity depending on the intensity of the impact. Clinically, more complex intra-articular fractures are associated with patients with higher severity of trauma and subsequent degeneration of the articular cartilage; moreover, these injuries are closely linked with worse outcomes for the patients [17]. Articular fracture treatment includes restoration of the articular surface, correction of axial and rotational alignment of the injured limb, and surgical fixation for stabilization to allow early range of motion of the injured joint [80].

Ligamentous Lesions

Ligaments are important structures to maintain the stability of the articulation but are at risk to be injured in traumatic injuries. Knee ligamentous injuries increased injury to the patient, especially if they are unrecognized and untreated, and can lead to significant morbidity. Common diagnostic procedure to find out and confirm capsule-ligamentous lesions includes arthroscopy of the knee which is preferred compared with magnetic resonance (MR) imaging due to its ability to probe, distinguish fragile tissue from normal, and perform additional surgical procedures like removal of loose bodies [14]. Most common ligamentous injuries typically include the anterior cruciate ligament (ACL) and medial collateral ligament (MCL) [99]. Patients with these lesions and/or without a concomitant meniscus injury are at high risk for PTOA. Sports injuries are the most frequent cause of anterior cruciate ligament injuries [60].

Meniscal Lesions

Unlike hyaline cartilage, the collagen present at articular fibrocartilage is oriented variably and contains little density of proteoglycans and less presence of water. Consequently, articular fibrocartilage is less compliant and presents less capacity of regeneration. The greater number of fibrocartilage lesions often occurs at menisci of the knee, glenohumeral input, and triangular fibrocartilage of the wrist [126].

Meniscal lesions are the most common knee injuries seen in patients of all ages and especially in young or adolescent patients due to trauma. These lesions are a recognized risk factor for the development of OA. Commonly, the mechanism of injury involves a twisting injury on a semi-flexed limb through a weight-bearing knee. Overpressure on the knee due to overweight, intensive training, incorrect position of the legs such as varus or valgus, or reduced muscle strength is a typical risk factor for meniscus lesions [100]. Advances in the knowledge of meniscal anatomy, biomechanics, and function are essential to understanding meniscal pathology and treatment [2, 100]. Meniscectomy still remains as a common orthopedic procedure; however, meniscal repairs are increasingly performed over meniscectomies in young patients [1].

3.3.2   Aging

Age has been identified as one of the most important risk factors for the development of OA. The incidence and prevalence of OA increase with aging due to the combination of various risk factors, together with biological changes that lead to broke the homeostasis of the joint resulting in less capacity for healing [3, 66]. For instance, ACL injury is a cause of PTOA, and the progression of this PTOA has been seen to increase with patient age [94]. Seon et al. [99] have shown that the 54%

of the patients over 25 years at the time of ACL surgery developed OA, in front of the 26% below 25 years old [99].

The main factors affecting age-related changes include cellular senescence, reduced cell density, and altered secretory profiles [70]. The decrease in cell density is a direct consequence of cellular senescence, which is characterized by the loss of cell division capacity [111]. There are several processes that affect cell senescence such as the shortening of the telomeres [46], mitochondrial and nuclear DNA damage [15, 55], oxidative stress [68], and inflammatory process [44]. Moreover, reduced repair capacity of the cartilage increases with aging, due to the lesser capacity of chondrocytes to respond to growth factor stimulation to proliferative and anabolic process [66, 70]. For example, lower sensitivity to the stimulation with the TGFβ [57], insulin-like growth factor 1 (IGF-1) [64], and bone morphogenetic protein family (BMPs) [12, 24] leads to induced oxidative stress and prevalence of catabolic process over anabolic.

Apart from cellular changes, ECM also experiment age-associated alterations that lead to develop OA, such as the progressive calcification of cartilage that occurs before evidence OA [75]. In addition, it has been demonstrated that with aging there is a marked increase in the formation of advanced glycation end products (AGEs), which are the results of spontaneous nonenzymatic glycation of the proteins. AGE formation increase the cross-linking of collagen molecules which altered the mechanical properties of cartilage, making it more susceptible to mechanically induced damage [32, 116]. Moreover, the interaction of AGEs with cellular receptors, including the receptor for AGEs (RAGE), displayed an increment in inflammation [20] and catabolic process [108, 125].

3.3.3 Obesity

Obesity is a strong risk factor associated with development and progression of OA, especially in knee OA [36]. Moreover, the overweight increases the development of osteoarthritic processes after knee trauma [119], specifically after fixation of acetabular fractures [59, 84].

It is assumed that behind the influence of obesity on OA is mechanical overload because the joint of overweight person endures the transmission from two to five times the body weight during the course of the day, leading to wear, damage, and microtrauma [33, 74]. Apart from the increment of the mechanical loading, obesity contributes to OA through the secretion of adipose tissue-derived cytokines, called adipokines, such as a variety of interleukins (IL-1β, IL-6, IL-8, IL-15, etc.), tumor necrosis factor alpha (TNF-α), leptin, and adiponectin, among others [28]. These inflammatory factors lead to bone resorption and ECM changes through the downregulation of the synthesis of the major components of the matrix (proteoglycans and type II collagen) and the upregulation of catabolic process across the activity of MMPs and a disintegrinlike and metalloproteinase with thrombospondin type 1 motifs (ADAMTS) [28, 53].

It has been established that if obese patients with OA lose weight can reduce pain, improve the function of the joint and might reduce disease progression [23, 61]. Even, there are studies that support the use of specific strategies to weight loss in the treatment of these patients 11,121].

3.3.4 Genetic Factors

An inherited predisposition to develop OA has been established from family-based studies that supports the strong link between the genetic factors and this disease. Studies with twin have estimated a genetic influence ranging from 30% to 65% in OA, with larger influence in hands and hips, and smaller in knees [56, 73, 106]. Moreover, linkage studies with families and sibling have suggested loci linked to hip and knee OA in an area of chromosomes 2q and 11q. Related with the loci in chromosome 2q, the regions 2q 12–22 and 2q 33–35 contain genes that could be involved in the OA, like the gene for the $\alpha2$ chain of type V collagen (a major component of the bone) and fibronectin, and the receptor of IL-8 (inflammatory process) [71, 122]. In relation with chromosome 11q and the susceptibility to develop OA, it has identified a cluster related with at least seven MMP genes and a locus that is a regulator of bone mass [21]. In addition, studies in families with primary OA have detected loci on chromosomes 4, 6, and 16 [40, 72], and more recently, genome-wide association studies (GWAS) have been established loci on chromosomes 3 [76], 7 [54], 13 [31], and 19 [19] that are strongly associated with hip or knee OA susceptibility.

On the other hand, the role of specific genes involved in OA has been demonstrated by in vitro or in vivo studies. For example, in vitro gene analysis of patients and transgenic mice displayed the impact of an alteration in ECM components, (such as type II collagen and COMP), its regulators (aggrecanases), and how it contributed to the degeneration of knee joint [92, 95, 96]. In addition, development and progression of OA can be induced by alterations in signaling pathways like TGFβ/BMP [98, 131], Wnt/β-catenin [112], Indian hedgehog [18], hypoxia-inducible factor (HIF) 1α/HIF-2α [129], nuclear factor-kappaB (NF-κB) [91], and Notch [51] pathways and its downstream molecules [114] that leads to cartilage destruction. The knowledge of the genetic factors that induce or maintain osteochondral defects is a promising therapeutic strategy for novel treatments.

3.3.5 Inflammation

Over the past decade, inflammation has been established as a critical feature of OA. Many studies are opening the way to consider inflammation a key driver of OA progression after joint injury [7]. According to current research, such involvement in the pathophysiology of OA would occur through the action of inflammatory mediators [10] released by the cartilage, bone, and synovium. These mediators, such as cytokines (IL-1β, TNF-α, IL-6, etc.) and chemokines (IL-8, CCL5, CCL19, and its receptor CCR7), are produced by a variety of cell types, including macrophages, chondrocytes, and fibroblast-like synoviocytes (FLS), in response to joint trauma or chronic overuse injuries. Another source of these cytokines could be chronical inflammatory process associated to age or derived from previous injuries [7, 22, 103]. In addition to traditional mediators, it has been also found the presence of adipokines such as leptin in inflammation processes [28]. All these soluble signaling factors cause alteration of joint cell homeostasis such as pathological maturation, apoptosis, and catabolic responses by means of metalloproteinases (MMPs), prostaglandin E2 (PGE2), or nitric oxide (NO) synthesis, leading to cartilage degradation and subchondral bone remodeling. Moreover, the release of these inflammatory mediators causes alteration of joint cell homeostasis generating a loopback that would aggravate or accelerate joint degeneration [90, 93, 118].

In this way, having an inflammation either attributed to previous injuries, obesity, or age would have an increased risk to develop, aggravate, or accelerate OA progression, after tissue damage. Results from image studies using MRI and ultrasonography have evidenced a positive correlation between inflammation and the risk for structural progression of OA [37, 47, 83]. Similarly, in vivo studies have reported the association between increased serum levels of adipokines and greater cartilage loss with a higher incidence of knee joint replacement [61].

The knowledge of inflammation role in OA development and mechanisms by which it acts has provided a window of opportunities to develop disease-modifying interventions targeting inflammatory processes for the prevention and treatment of OA. In vivo and clinical studies performed so far have mainly focused on TNF and IL-1 inhibitor showing clinical symptom relief but did not achieve to stop the disease progression [69]. Then, is necessary to take into account other factors that contribute to OA development, and also the heterogeneity of the OA patients, since their phenotypes may have different pathophysiology.

3.3.6 Metabolic Syndrome

Metabolic syndrome (MetS) is a common phenotype comprised of a cluster of metabolic disorders, such as hypertension, insulin resistance, visceral obesity, and dyslipidemia, that occur together, increasing the risk of developing serious chronic disease [132]. Researchers have suggested a positive association between OA and the four central components of MetS, since epidemiological and clinical data revealed a high prevalence in patients with OA, regarding the population without OA [86, 102, 105, 115, 127]. In addition, people with MetS develop OA at an earlier age, and have more generalized pathology with higher inflammation and pain, in comparison with patients with OA in the absence of MetS [34, 86]. Thus, all these disorders have led to consider metabolic syndrome an important risk factor for OA.

During these last years, investigators are assessing how all these components that make up MetS are involved in OA. Studies have shown that all these conditions end up causing cellular damage and subsequent inflammation that, as explained in the previous section, leads to the OA development [52, 132]. Hypertension contributes to OA through subchondral ischemia that results from blood flow reduction related to narrowing of blood vessels [38]. Association between dyslipidemia and OA has also been reported, thus, the disturbance of lipid metabolism increase the risk of OA [42]. Moreover, obese people could display elevated levels of systemic oxidative stress that can be caused by insulin resistance, through hyperglycemia [29, 48, 49, 78]. This condition of local high glucose concentration can also contribute to OA by reduction of chondrocyte differentiation, therefore, decreasing the potential cartilage regeneration [30, 113]. Hence, diseases that derive from insulin resistance state such as type 2 diabetes have been robustly associated with OA in epidemiological studies [9, 29]. Concordant results from clinical studies have also reported a higher rate of knee OA progression in type 2 diabetes patients than nondiabetics on a 3-year follow-up [35]. Otherwise, visceral obesity associated to MetS, contributes directly to inflammation state due to an increase in adipokine concentration that leads to the OA development [85, 132].

So, the link of MetS to OA suggests that control or prevention of MetS conditions would modulate OA progression in humans, for example, by promoting reduction of adipose tissue in obese patients [27, 128].

3.3.7   Gender

Besides excess weight, obesity, and previous knee injury, the onset of knee OA has also been associated with female gender. Unfortunately, it represents a non-modifiable risk factor that leads to increased susceptibility and predisposition to develop OA. Numerous clinical, pathological, and epidemiological studies of OA suggest relevant difference between sexes. Women not only have higher prevalence than men, but they also have greater severity of OA [107]. In addition, the definite increase in OA in women around the time of menopause has led investigators to hypothesize that hormonal factors, in particular estrogens, may play a role in the development of OA [101]. Further support for a hormonal effect on OA comes from some, but not all, studies which have shown a higher prevalence and incidence of OA in women with hysterectomy than without it [26].

Although a lot of studies are addressing the relation between estrogens and OA, it is still not clearly defined, appearing to be concentration dependent. Despite the controversial results, the overall effect predominantly leads to inhibition of the expression and secretion of proinflammatory cytokines into the joint. It has been evidenced both in vitro and in vivo studies [90]. However, results from observational studies and clinical trials have been conflicting regarding this effect, especially about estrogen therapy [45, 79, 120].

Moreover, the disparities between sexes may be also due to the differences in the anatomical structure of joint elements, in height, weight, or just a thinner and more reduced volume of knee cartilage in women compared with that of men. Therefore, more studies are needed to evidence the role of hormones in OA and resolve these issues.

3.4 Treatment of Osteoarthritis

Currently, no available treatments are able to cure or substantially modify disease progression. In the case of joint trauma, interventions should be addressed as soon as possible to limit the degree of acute joint damage and to reduce the severity of OA [69]. The selection of the treatment depends on the intensity of the affectation. There are several methods to treat traumatic joints that include the following treatments:

Non-operative Management
- Symptomatic medical treatment: control the pain and inflammation by using cryotherapy, analgesics, and anti-inflammatories.
- Decrease early loading of injured articular surfaces after injury.
- Protect from the load to avoid detachment by shearing forces of possible fragments of cartilage detached after the trauma.
- Avoid prolonged rigid immobilization.
- Intra-articular viscosupplementation injections.
- Weight loss and exercise in obese and overweight individuals.

- **Surgical Treatment**
- Surgical management aims to reestablish the joint surface, maximizing the osteochondral biologic environment, achieve rigid fixation, and ensure early motion. The surgical techniques could be either procedures that only address cartilage repair or osteochondral procedures to treat both cartilage and subchondral bone.

- *Chondral and osteochondral defects*: For partial defects simple arthroscopic debridement with or without marrow stimulation (microfracture) is used. In the case of full-thickness defects, microfracture and autologous grafts or allografts are recommended [16]. Autologous grafts involve the extraction of healthy cartilage of the patient and its transplantation to the site of the defect. However, this technique presents several drawbacks such as the graft size limitation and donor site morbidity. In order to prevent greater damage, the cartilage is removed from areas that do not withstand heavy loads, such as the lateral margin of the femoral trochlea and notch of the knee [25].
- *Tissue engineering:* A number of promising cell sources, biocompatible tissue-engineered scaffolds, scaffoldless techniques, biological factors, and mechanical stimuli are currently being investigated in the field of articular cartilage tissue

engineering, which aims to repair, regenerate, and/or improve injured or diseased articular cartilage functionality [130]. For example, autologous chondrocyte implantation (ACI) is the first generation of cell transplantation techniques for cartilage repair and is used widely for patients who have cartilage lesions between 1 cm^2 and 12 cm^2 or had previously failed restoration treatments of the knee such as microfracture surgeries [82].

- *Total and partial joint replacements:* For severe joint injuries as well as for advanced OA, articular cartilage cannot be recovered by any of the above-discussed treatments. In these cases, the damaged osteochondral tissue is partially or totally removed, and total or partial joint replacements are performed to help patients restore normal function. However, joint replacement therapies are not recommended in younger patients due to relatively short life spans of current implants, and revision surgery offers less favorable outcomes [117].

3.5   Conclusions

OA is a complex process without a full understood etiology. Changes observer during progression of OA not only target cartilage tissue, also affect to subchondral bone and synovial tissue. There is a crosstalk between cartilage and bone cells in the course of the disease that play a major role in the joint homeostasis.

Among the risk factors that result in structural and functional failure of joints are joint trauma, obesity, aging, inflammation, genetic predisposition, gender and hormones, or metabolic syndrome. Despite cartilage senescence could be considered part of "normal" chronological age, and represent an important individual risk factor for the development of OA, all risk factors of OA are inter-related, not inter-dependent. In addition to aging, another pivotal factor that marks OA is damage due to trauma.

The majority of individuals with a significant traumatic joint injury develop PTOA. In the young patient, the pathogenesis of knee OA is predominantly related to joint trauma and an unfavorable biomechanical environment at the joint. Once the damage occurs, a sequence of events is initiated at the joint tissues and leads to progressive articular surface damage.

Future treatments must take into account all the specific characteristic of individuals and clinically relevant factors associated like severity of joint injury. Therefore, a multi-varied therapy which includes the knowledge of the OA risk factors of a specific patient could be used to make the clinical diagnosis.

Therapies focused on joint injuries with a clear trauma origin should address the earliest symptoms such as inflammation, stiffness, joint dysfunction, or pain. This is especially important in the young individual since changes in these patients could still be reversible, and therefore, early treatment could prevent further progression of the disease. A need also exists for therapies that stimulate intrinsic repair of the damage tissue and inhibiting catabolic pathways that lead to chondrocyte death and matrix loss.

Acknowledgments This work was supported by Fundación Progreso y Salud (Junta de Andalucía, project number PIN-0379-2016) and by the Ministerio de Economía, Industria y Competitividad (FEDER funds, project RTC-2016-5451-1). G.J. acknowledges the Junta de Andalucía for providing a postdoctoral fellowship. C.A. acknowledges the predoctoral fellowship from the Spanish Ministry of Education, Culture and Sports (BOE-A-2014-13539). Also, E.L-R. acknowledges the MINECO for providing a postdoctoral fellowship through the project RTC-2016-5451-1.

References

1. Abrams GD et al (2013) Trends in meniscus repair and meniscectomy in the United States, 2005-2011. Am J Sports Med 41(10):2333–2339. https://doi.org/10.1177/0363546513495641
2. Ahmad J, Maltenfort M (2017) Arthroscopic treatment of osteochondral lesions of the talus with allograft cartilage matrix. Foot Ankle Int 38(8):855–862. https://doi.org/10.1177/1071100717709571
3. Aigner T, Richter W (2012) OA in 2011: Age-related OA? A concept emerging from infancy? Nat Rev Rheumatol 8(2):70. https://doi.org/10.1038/nrrheum.2011.206
4. Altman R et al (1991) The American College of Rheumatology criteria for the classification and reporting of osteoarthritis of the hip. Arthritis Rheum 34(5):505–514. Available at: http://www.ncbi.nlm.nih.gov/pubmed/2025304 (Accessed: 12 July 2017)
5. Amoako AO, Pujalte GGA (2014) Osteoarthritis in young, active, and athletic individuals. Clinical medicine insights. Arthritis Musculoskeletal Dis 7:27–32. https://doi.org/10.4137/CMAMD.S14386
6. Anderson DD et al (2011a) Post-traumatic osteoarthritis: improved understanding and opportunities for early intervention. J Orthop Res 29(6):802–809. https://doi.org/10.1002/jor.21359
7. Badalà F, Nouri-mahdavi K, Raoof DA (2008) The role of synovitis in osteoarthritis pathogenesis. Bone 144(5):724–732. https://doi.org/10.1038/jid.2014.371
8. Bauer KL, Polousky JD (2017) Management of Osteochondritis Dissecans Lesions of the knee, elbow and ankle. Clin Sports Med 36(3):469–487. https://doi.org/10.1016/j.csm.2017.02.005
9. Berenbaum F (2012) Diabetes-induced osteoarthritis: from a new paradigm to a new phenotype. Postgrad Med J 88(1038):240–242. https://doi.org/10.1136/pgmj.2010.146399rep
10. Berenbaum F (2013) Osteoarthritis as an inflammatory disease (osteoarthritis is not osteoarthrosis!). Osteoarthr Cartil 21(1):16–21. https://doi.org/10.1016/j.joca.2012.11.012
11. Bliddal H, Christensen R (2006) The management of osteoarthritis in the obese patient: practical considerations and guidelines for therapy. Obes Rev 7(4):323–331. https://doi.org/10.1111/j.1467-789X.2006.00252.x
12. Bobacz K et al (2003) Expression of bone morphogenetic protein 6 in healthy and osteoarthritic human articular chondrocytes and stimulation of matrix synthesis in vitro. Arthritis Rheum 48(9):2501–2508. https://doi.org/10.1002/art.11248
13. Boesen M et al (2017) Osteoarthritis year in review 2016: imaging. Osteoarthr Cartil 25(2):216–226. https://doi.org/10.1016/j.joca.2016.12.009
14. Borgohain B et al (2014) Risks of concomitant trauma to the knee in lower limb long bone shaft fractures: a retrospective analysis from a prospective study population. Adv Biomed Res 3(1):49. https://doi.org/10.4103/2277-9175.125764
15. Botter SM et al (2011) Analysis of osteoarthritis in a mouse model of the progeroid human DNA repair syndrome trichothiodystrophy. Age 33(3):247–260. https://doi.org/10.1007/s11357-010-9175-3

16. Brittberg M et al (2016a) Cartilage repair in the degenerative ageing knee. Acta Orthop 87(sup363):26–38. https://doi.org/10.1080/17453674.2016.1265877
17. Brown TD et al (2006) Posttraumatic osteoarthritis: a first estimate of incidence, prevalence, and burden of disease. J Orthop Trauma 20(10):739–744. https://doi.org/10.1097/01.bot.0000246468.80635.ef
18. Buckland J (2010) Osteoarthritis: blocking hedgehog signaling might have therapeutic potential in OA. Nat Rev Rheumatol 6(2):61–61. https://doi.org/10.1038/nrrheum.2009.270
19. Castaño Betancourt MC et al (2012) Genome-wide association and functional studies identify the DOT1L gene to be involved in cartilage thickness and hip osteoarthritis. Proc Nat Acad Sci USA 109(21):8218–8223. https://doi.org/10.1073/pnas.1119899109
20. Cecil DL et al (2005) Inflammation-induced chondrocyte hypertrophy is driven by receptor for advanced glycation end products. J Immunol (Baltimore, Md: 1950) 175(12):8296–8302. Available at: http://www.ncbi.nlm.nih.gov/pubmed/16339570 (Accessed: 25 June 2017)
21. Chapman K et al (1999) Osteoarthritis-susceptibility locus on chromosome 11q, detected by linkage. Am J Hum Genet 65:167–174. Available at: https://www.ncbi.nlm.nih.gov/pmc/articles/PMC1378087/pdf/10364529.pdf (Accessed: 28 June 2017)
22. Chen D et al (2017a) Osteoarthritis: toward a comprehensive understanding of pathological mechanism. Bone Res 5(august 2016):16044. https://doi.org/10.1038/boneres.2016.44
23. Christensen R et al (2007) Effect of weight reduction in obese patients diagnosed with knee osteoarthritis: a systematic review and meta-analysis. Ann Rheumatic Dis BMJ Pub Group 66(4):433–439. https://doi.org/10.1136/ard.2006.065904
24. Chubinskaya S et al (2002) Age-related changes in cartilage endogenous osteogenic protein-1 (OP-1). Biochim Biophys Acta (BBA) - Mol Basis Dis 1588(2):126–134. https://doi.org/10.1016/S0925-4439(02)00158-8
25. Chubinskaya S et al (2015) Articular Cartilage Injury and Potential Remedies. J Orthopaed Trauma 29(Suppl 12):S47–S52. https://doi.org/10.1097/BOT.0000000000000462
26. Cicuttini FM, Spector T, Baker J (1997) Risk factors for osteoarthritis in the tibiofemoral and patellofemoral joints of the knee. J Rheumatol 24(6):1164–1167. Available at: http://www.ncbi.nlm.nih.gov/pubmed/9195526 (Accessed: 16 July 2017)
27. Clockaerts S et al (2012) Statin use is associated with reduced incidence and progression of knee osteoarthritis in the Rotterdam study. Ann Rheum Dis 71(5):642–647. https://doi.org/10.1136/annrheumdis-2011-200092
28. Conde J et al (2011a) Adipokines and osteoarthritis: novel molecules involved in the pathogenesis and progression of disease. Arthritis 2011(2090–1992 (electronic)):203901. https://doi.org/10.1155/2011/203901
29. Courties A, Sellam J (2016) Osteoarthritis and type 2 diabetes mellitus: what are the links? Diabetes Res Clin Pract 122:198–206. https://doi.org/10.1016/j.diabres.2016.10.021
30. Cramer C et al (2010) Persistent high glucose concentrations Alter the regenerative potential of mesenchymal stem cells. Stem Cells Dev 19(12):1875–1884. https://doi.org/10.1089/scd.2010.0009
31. Day-Williams A et al (2011) A variant in MCF2L is associated with osteoarthritis. Am J Hum Genet 89(3):446–450. https://doi.org/10.1016/j.ajhg.2011.08.001
32. DeGroot J et al (2004) Accumulation of advanced glycation end products as a molecular mechanism for aging as a risk factor in osteoarthritis. Arthritis Rheum 50(4):1207–1215. https://doi.org/10.1002/art.20170
33. Ding C et al (2012) Body fat is associated with increased and lean mass with decreased knee cartilage loss in older adults: a prospective cohort study. Int J Obes 37:822–827. https://doi.org/10.1038/ijo.2012.136
34. Engstrm G et al (2008) 325 C-reactive protein metabolic syndrome and incidence of severe hip and knee osteoarthritis. A population-based cohort study. Osteoarthr Cartil 16:S143–S144. https://doi.org/10.1016/S1063-4584(08)60369-6
35. Eymard F et al (2015) Diabetes is a risk factor for knee osteoarthritis progression. Osteoarthr Cartil 23(6):851–859. https://doi.org/10.1016/j.joca.2015.01.013

36. Felson DT et al (2000) Osteoarthritis: new insights. Part 1: the disease and its risk factors. Ann Intern Med 133(8):635. https://doi.org/10.7326/0003-4819-133-8-200010170-00016
37. Felson DT et al (2016) Synovitis and the risk of knee osteoarthritis: the most study HHS public access. Osteoarthr Cartil 24(3):458–464. https://doi.org/10.1016/j.joca.2015.09.013
38. Findlay DM (2007) Vascular pathology and osteoarthritis. Rheumatology 46(12):1763–1768. https://doi.org/10.1093/rheumatology/kem191
39. Findlay DM, Kuliwaba JS (2016) Bone-cartilage crosstalk: a conversation for understanding osteoarthritis. Bone Res 4:16028. https://doi.org/10.1038/boneres.2016.28
40. Forster T et al (2004) Finer linkage mapping of primary osteoarthritis susceptibility loci on chromosomes 4 and 16 in families with affected women. Arthri Rheumat 50(1):98–102. https://doi.org/10.1002/art.11427
41. Funck-Brentano T, Cohen-Solal M (2015) Subchondral bone and osteoarthritis. Curr Opin Rheumatol 27(4):420–426. https://doi.org/10.1097/BOR.0000000000000181
42. Gkretsi V, Simopoulou T, Tsezou A (2011) Lipid metabolism and osteoarthritis: lessons from atherosclerosis. Prog Lipid Res 50(2):133–140. https://doi.org/10.1016/j.plipres.2010.11.001
43. Goldring MB et al. (2011) Roles of inflammatory and anabolic cytokines in cartilage metabolism: signals and multiple effectors converge upon MMP-13 regulation in osteoarthritis. Eur Cells Mat 21: 202–20. Available at: http://www.ncbi.nlm.nih.gov/pubmed/21351054 (Accessed: 11 July 2017)
44. Greene MA, Loeser RF (2015) Aging-related inflammation in osteoarthritis. Osteoarthr Cartil 23(11):1966–1971. https://doi.org/10.1016/j.joca.2015.01.008
45. Hannan MT et al (1990) Estrogen use and radiographic osteoarthritis of the knee in women. Arthritis Rheum 33(4):525–532. https://doi.org/10.1002/art.1780330410
46. Harbo M et al (2012) The distribution pattern of critically short telomeres in human osteoarthritic knees. Arthrit Res Therapy 14(1):R12. https://doi.org/10.1186/ar3687
47. Hasson CJ, Caldwell GE, Van Emmerik REA (2009) NIH public access. Mot Control 27(4):590–609. https://doi.org/10.1016/j.humov.2008.02.015.Changes
48. Henrotin YE, Bruckner P, Pujol JPL (2003) The role of reactive oxygen species in homeostasis and degradation of cartilage. Osteoarthr Cartil 11(10):747–755. https://doi.org/10.1016/S1063-4584(03)00150-X
49. Hiraiwa H et al (2011) Inflammatory effect of advanced glycation end products on human meniscal cells from osteoarthritic knees. Inflamm Res 60(11):1039–1048. https://doi.org/10.1007/s00011-011-0365-y
50. Homandberg GA (2001) Cartilage damage by matrix degradation products: fibronectin fragments. Clin Orthopaed Relat Res (391 Suppl): S100–7. Available at: http://www.ncbi.nlm.nih.gov/pubmed/11603694 (Accessed: 16 July 2017)
51. Hosaka Y et al (2013) Notch signaling in chondrocytes modulates endochondral ossification and osteoarthritis development. Proc Natl Acad Sci U S A 110(5):1875–1880. https://doi.org/10.1073/pnas.1207458110
52. Hotamisligil GS (2006) Inflammation and metabolic disorders. Nature 444(7121):860–867. https://doi.org/10.1038/nature05485
53. Kapoor M et al (2011) Role of proinflammatory cytokines in the pathophysiology of osteoarthritis. Nat Rev Rheumatol 7(1):33–42. https://doi.org/10.1038/nrrheum.2010.196
54. Kerkhof HJM et al (2010) A genome-wide association study identifies a locus on chromosome 7q22 to influence susceptibility for osteoarthritis. Arthritis Rheum 62(2):NA. https://doi.org/10.1002/art.27184
55. Kim J et al (2010) Mitochondrial DNA damage is involved in apoptosis caused by proinflammatory cytokines in human OA chondrocytes. Osteoarthr Cartil 18(3):424–432. https://doi.org/10.1016/j.joca.2009.09.008
56. Kirk KM et al (2002) The validity and heritability of self-report osteoarthritis in an Australian older twin sample. Twin Res 5(2):98–106. https://doi.org/10.1375/1369052022965
57. van der Kraan PM, Blaney Davidson EN, van den Berg WB (2010) A role for age-related changes in TGF? Signaling in aberrant chondrocyte differentiation and osteoarthritis. Arthritis Res Therapy 12(1):201. https://doi.org/10.1186/ar2896

58. Kuijt M-TK et al (2012) Knee and ankle osteoarthritis in former elite soccer players: a systematic review of the recent literature. J Sci Med Sport 15(6):480–487. https://doi. org/10.1016/j.jsams.2012.02.008

59. Lawyer TJ et al (2014) Prevalence of post-traumatic osteoarthritis in morbidly obese patients after acetabular fracture fixation. J Long-Term Eff Med Implants 24(2–3):225–231. Available at: http://www.ncbi.nlm.nih.gov/pubmed/25272222 (Accessed: 1 July 2017)

60. Lee HH, Chu CR (2012) Clinical and basic science of cartilage injury and arthritis in the Football (Soccer) Athlete. Cartilage 3(1 Suppl):63S–68S. https://doi.org/10.1177/1947603511426882

61. Lee R, Kean WF (2012) Obesity and knee osteoarthritis. InflammoPharmacology 20(2):53–58. https://doi.org/10.1007/s10787-011-0118-0

62. Lewis JS et al (2011) Acute joint pathology and synovial inflammation is associated with increased intra-articular fracture severity in the mouse knee. Osteoarthr Cartil 19(7):864–873. https://doi.org/10.1016/j.joca.2011.04.011

63. Li H et al (2017) Treatment of talus osteochondral defects in chronic lateral unstable ankles: small-sized lateral chondral lesions had good clinical outcomes. Knee Surg Sports Traumatol Arthrosc. https://doi.org/10.1007/s00167-017-4591-x

64. Loeser RF et al (2000) Reduction in the chondrocyte response to insulin?Like growth factor 1 in aging and osteoarthritis: studies in a non?Human primate model of naturally occurring disease. Arthritis Rheum 43(9):2110–2120. https://doi.org/ 10.1002/1529-0131(200009)43:9<2110::AID-ANR23>3.0.CO;2-U

65. Loeser RF (2006) Molecular mechanisms of cartilage destruction: mechanics, inflammatory mediators, and aging collide. Arthritis Rheum 54(5):1357–1360. https://doi.org/10.1002/ art.21813

66. Loeser RF (2011) Aging and osteoarthritis. Current Opinion Rheum 23(5):492–496. https:// doi.org/10.1097/BOR.0b013e3283494005

67. Loeser RF et al (2012) Osteoarthritis: a disease of the joint as an organ. Arthritis Rheum 64(6):1697–1707. https://doi.org/10.1002/art.34453

68. Loeser RF et al (2014) Aging and oxidative stress reduce the response of human articular chondrocytes to insulin-like growth factor 1 and osteogenic protein 1. Arthritis Rheum 66(8):2201–2209. https://doi.org/10.1002/art.38641

69. Lotz MK (2010a) New developments in osteoarthritis. Posttraumatic osteoarthritis: pathogenesis and pharmacological treatment options. Arthritis Res Therapy 12(3):211. https://doi.org/10.1186/ar3046

70. Lotz M, Loeser RF (2012) Effects of aging on articular cartilage homeostasis. Bone 51(2):241–248. https://doi.org/10.1016/j.bone.2012.03.023

71. Loughlin J et al (2000) Linkage analysis of chromosome 2q in osteoarthritis. Rheumatology 39:377–381. Available at: https://oup.silverchair-cdn.com/oup/backfile/Content_public/ Journal/rheumatology/39/4/10.1093_rheumatology_39.4.377/3/390377.pdf?Expires=1498 762747&Signature=TLcBIZF13-QNxl0d-uQ0aLbOke6TiUBveOt~oJDDuLFsU7fy5cxleS-SelEKgm-7N-GMLPbnGeh17Bkc3rssk3G4OxIVM8Heg0 (Accessed: 28 June 2017)

72. Loughlin J et al (2002) Finer linkage mapping of a primary hip osteoarthritis susceptibility locus on chromosome 6. Eur J Hum Genet 10(9):562–568. https://doi.org/10.1038/ sj.ejhg.5200848

73. MacGregor AJ et al (2000) The genetic contribution to radiographic hip osteoarthritis in women: results of a classic twin study. Arthritis Rheum 43(11):2410–2416. https://doi. org/10.1002/1529-0131(200011)43:11<2410::AID-ANR6>3.0.CO;2-E

74. Maquet PG, Pelzer GA (1977) Evolution of the maximum stress in osteo-arthritis of the knee. J Biomech 10(2):107–117. Available at: http://www.ncbi.nlm.nih.gov/pubmed/858709 (Accessed: 26 June 2017)

75. Mitsuyama H et al (2007) Calcification of human articular knee cartilage is primarily an effect of aging rather than osteoarthritis. Osteoarthr Cartil 15(5):559–565. https://doi.org/10.1016/j. joca.2006.10.017

76. Miyamoto Y et al (2008) Common variants in DVWA on chromosome 3p24.3 are associated with susceptibility to knee osteoarthritis. Nat Genet 40(8):994–998. https://doi.org/10.1038/ng.176
77. Moskowitz RW (2009) The burden of osteoarthritis: clinical and quality-of-life issues. Am J Manag Care 15(8 Suppl):S223–S229. Available at: http://www.ncbi.nlm.nih.gov/pubmed/19817508 (Accessed: 10 July 2017)
78. Nah SS et al (2008) Effects of advanced glycation end products on the expression of COX-2, PGE2 and NO in human osteoarthritic chondrocytes. Rheumatology 47(4):425–431. https://doi.org/10.1093/rheumatology/kem376
79. Nevitt MC et al. (1996) Association of estrogen replacement therapy with the risk of osteoarthritis of the hip in elderly white women. Study of Osteoporotic Fractures Research Group. Arch Intern Med 156(18):2073-80. Avalaible at: https://www.ncbi.nlm.nih.gov/pubmed/8862099
80. Olson SA et al (2015) Therapeutic opportunities to prevent post-traumatic arthritis: lessons from the natural history of arthritis after articular fracture. J Orthop Res 33(9):1266–1277. https://doi.org/10.1002/jor.22940
81. Onur TS et al (2014) Joint instability and cartilage compression in a mouse model of post-traumatic osteoarthritis. J Orthopaed Res: Off Pub Orthopaed Res Soc 32(2):318–323. https://doi.org/10.1002/jor.22509
82. Pascual-Garrido C, McNickle AG, Cole BJ (2009) Surgical treatment options for osteochondritis dissecans of the knee. Sports Health 1(4):326–334. https://doi.org/10.1177/1941738109334216
83. Pelletier JP et al (2008) A new non-invasive method to assess synovitis severity in relation to symptoms and cartilage volume loss in knee osteoarthritis patients using MRI. Osteoarthr Cartil 16(SUPPL. 3):8–13. https://doi.org/10.1016/S1063-4584(08)60004-7
84. Porter SE et al (2008) Complications of acetabular fracture surgery in morbidly obese patients. J Orthop Trauma 22(9):589–594. https://doi.org/10.1097/BOT.0b013e318188d6c3
85. Pottie P et al (2006) Obesity and osteoarthritis: more complex than predicted! Ann Rheum Dis 65(11):1403–1405. https://doi.org/10.1136/ard.2006.061994
86. Puenpatom RA, Victor TW (2009) Increased prevalence of metabolic syndrome in individuals with osteoarthritis: an analysis of NHANES III data. Postgrad Med 121(6):9–20. https://doi.org/10.3810/pgm.2009.11.2073
87. Puig-Junoy J, Ruiz Zamora A (2015) Socio-economic costs of osteoarthritis: a systematic review of cost-of-illness studies. Semin Arthritis Rheum 44(5):531–541. https://doi.org/10.1016/j.semarthrit.2014.10.012
88. Qi C, Changlin H, Zefeng H (2007) Matrix metalloproteinases and inhibitor in knee synovial fluid as cartilage biomarkers in rabbits: the effect of high-intensity jumping exercise. J Surg Res 140(1):149–157. https://doi.org/10.1016/j.jss.2006.12.556
89. Reynard LN, Loughlin J (2013) The genetics and functional analysis of primary osteoarthritis susceptibility. Expert Rev Mol Med 15:e2. https://doi.org/10.1017/erm.2013.4
90. Richette P et al (2007) Oestrogens inhibit interleukin 1beta-mediated nitric oxide synthase expression in articular chondrocytes through nuclear factor-kappa B impairment. Ann Rheum Dis 66(3):345–350. https://doi.org/10.1136/ard.2006.059550
91. Rigoglou S, Papavassiliou AG (2013) The NF-κB signalling pathway in osteoarthritis. Int J Biochem Cell Biol 45(11):2580–2584. https://doi.org/10.1016/j.biocel.2013.08.018
92. Rodriguez-Lopez J et al (2008) Genetic variation including nonsynonymous polymorphisms of a major aggrecanase, ADAMTS-5, in susceptibility to osteoarthritis. Arthritis Rheum 58(2):435–441. https://doi.org/10.1002/art.23201
93. Roos EM, Arden NK (2015) Strategies for the prevention of knee osteoarthritis. Nat Rev Rheumatol 12(2):92–101. https://doi.org/10.1038/nrrheum.2015.135
94. Roos H et al (1995) Osteoarthritis of the knee after injury to the anterior cruciate ligament or meniscus: the influence of time and age. Osteoarthr Cartil 3(4):261–267. Available at: http://www.ncbi.nlm.nih.gov/pubmed/8689461 (Accessed: 1 July 2017)

95. Säämänen A-M et al (2000) Osteoarthritis-like lesions in transgenic mice harboring a small deletion mutation in type II collagen gene. Osteoarthr Cartil 8(4):248–257. https://doi.org/10.1053/joca.2000.0298

96. Salminen H et al (2000) Up-regulation of cartilage oligomeric matrix protein at the onset of articular cartilage degeneration in a transgenic mouse model of osteoarthritis. Arthritis Rheum 43(8):1742–1748. https://doi.org/10.1002/1529-0131(200008)43:8<1742::AID-ANR10>3.0.CO;2-U

97. Sauter E et al (2012) Cytoskeletal dissolution blocks oxidant release and cell death in injured cartilage. J Orthop Res 30(4):593–598. https://doi.org/10.1002/jor.21552

98. Schmal H et al (2012) Expression of BMP-receptor type 1A correlates with progress of osteoarthritis in human knee joints with focal cartilage lesions. Cytotherapy 14(7):868–876. https://doi.org/10.3109/14653249.2012.681039

99. Seon JK, Song EK, Park SJ (2006) Osteoarthritis after anterior cruciate ligament reconstruction using a patellar tendon autograft. Int Orthop 30(2): 94–8. https://doi.org/10.1007/s00264-005-0036-0

100. Sigurdsson U et al. (2016) Delayed gadolinium-enhanced MRI of meniscus (dGEMRIM) and cartilage (dGEMRIC) in healthy knees and in knees with different stages of meniscus pathology. BMC Musculoskeletal Dis 17(1): 406. https://doi.org/10.1186/s12891-016-1244-z

101. Silman AJ, Newman J (1996) Obstetric and gynaecological factors in susceptibility to peripheral joint osteoarthritis. Ann Rheum Dis 55(9):671–673. Available at: http://ard.bmj.com/content/55/9/671.full.pdf

102. Singh G et al (2002) Prevalence of cardiovascular disease risk factors among US adults with self reported osteoarthritis. Am J Manag Care 8(15):383–391

103. Sokolove J, Lepus CM (2013) Role of inflammation in the pathogenesis of osteoarthritis: latest findings and interpretations. Therapeutic Advances in Musculoskeletal Disease 5(2):77–94. https://doi.org/10.1177/1759720X12467868

104. Sophia Fox AJ, Bedi A, Rodeo SA (2009) The basic science of articular cartilage: structure, composition, and function. Sports health 1(6):461–468. https://doi.org/10.1177/1941738109350438

105. Sowers M et al (2009) Knee osteoarthritis in obese women with cardiometabolic clustering. Arthritis Care Res 61(10):1328–1336. https://doi.org/10.1002/art.24739

105. Spector TD et al (1996) Genetic influences on osteoarthritis in women: a twin study. BMJ 312(7036):940–943. Available at: http://www.ncbi.nlm.nih.gov/pubmed/8616305 (Accessed: 27 June 2017)

107. Srikanth VK et al (2005) A meta-analysis of sex differences prevalence, incidence and severity of osteoarthritis. Osteoarthr Cartil 13(9):769–781. https://doi.org/10.1016/j.joca.2005.04.014

108. Steenvoorden MMC et al (2006) Activation of receptor for advanced glycation end products in osteoarthritis leads to increased stimulation of chondrocytes and synoviocytes. Arthritis Rheum 54(1):253–263. https://doi.org/10.1002/art.21523

109. Thomas CM et al (2011) Chondrocyte death by apoptosis is associated with the initiation and severity of articular cartilage degradation. Int J Rheum Dis 14(2):191–198. https://doi.org/10.1111/j.1756-185X.2010.01578.x

110. Thomas E, Peat G, Croft P (2014) Defining and mapping the person with osteoarthritis for population studies and public health. Rheumatology (Oxford) 53(2):338–345. https://doi.org/10.1093/rheumatology/ket346

111. Toh WS et al (2016) Cellular senescence in aging and osteoarthritis. Acta Orthopaed 87(sup363):6–14. https://doi.org/10.1080/17453674.2016.1235087

112. Tornero-Esteban P et al (2015) Altered expression of Wnt signaling pathway components in osteogenesis of mesenchymal stem cells in osteoarthritis patients. PLoS One 10(9):e0137170. https://doi.org/10.1371/journal.pone.0137170

113. Tsai TL, Manner PA, Li WJ (2013) Regulation of mesenchymal stem cell chondrogenesis by glucose through protein kinase C/transforming growth factor signaling. Osteoarthr Cartil 21(2):368–376. https://doi.org/10.1016/j.joca.2012.11.001

114. Valdes AM et al (2010) Genetic variation in the SMAD3 gene is associated with hip and knee osteoarthritis. Arthritis Rheum 62(8):2347–2352. https://doi.org/10.1002/art.27530
115. Velasquez MT, Katz JD (2010) Osteoarthritis: another component of metabolic syndrome? Metab Syndr Relat Disord 8(4):295–305. https://doi.org/10.1089/met.2009.0110
116. Verzijl N et al (2003) AGEing and osteoarthritis: a different perspective. Curr Opin Rheumatol 15(5):616–622. Available at: http://www.ncbi.nlm.nih.gov/pubmed/12960490 (Accessed: 25 June 2017)
117. Wainwright C et al (2011) Age at hip or knee joint replacement surgery predicts likelihood of revision surgery. Bone Joint J 93–B(10):1411–1415. https://doi.org/10.1302/0301-620X.93B10.27100
118. Wei Y, Bai L (2016) Recent advances in the understanding of molecular mechanisms of cartilage degeneration, synovitis and subchondral bone changes in osteoarthritis. Connect Tissue Res 57(4):245–261. https://doi.org/10.1080/03008207.2016.1177036
119. Whittaker JL et al (2015) Outcomes associated with early post-traumatic osteoarthritis and other negative health consequences 3?10 years following knee joint injury in youth sport. Osteoarthr Cartil 23(7):1122–1129. https://doi.org/10.1016/j.joca.2015.02.021
120. Wluka AE, Cicuttini FM, Spector TD (2000) Menopause, oestrogens and arthritis. Maturitas 35(3):183–199. https://doi.org/10.1016/S0378-5122(00)00118-3
121. Wluka AE, Lombard CB, Cicuttini FM (2012) Tackling obesity in knee osteoarthritis. Nat Rev Rheumatol 9(4):225–235. https://doi.org/10.1038/nrrheum.2012.224
122. Wright GD et al (1996) Association of two loci on chromosome 2q with nodal osteoarthritis. Ann Rheum Dis 55(5):317–319. Available at: http://www.ncbi.nlm.nih.gov/pubmed/8660106 (Accessed: 28 June 2017)
123. Xia B et al (2014) Osteoarthritis pathogenesis: a review of molecular mechanisms. Calcif Tissue Int 95(6):495–505. https://doi.org/10.1007/s00223-014-9917-9
124. Xu L et al (2014) Induction of high temperature requirement A1, a serine protease, by TGF-beta1 in articular chondrocytes of mouse models of OA. Histol Histopathol 29(5):609–618. https://doi.org/10.14670/HH-29.10.609
125. Yammani RR et al (2006) Increase in production of matrix metalloproteinase 13 by human articular chondrocytes due to stimulation with S100A4: role of the receptor for advanced glycation end products. Arthritis Rheum 54(9):2901–2911. https://doi.org/10.1002/art.22042
126. Yoon KH, Park KH (2014) Meniscal repair. Knee Surg Relat Res 26(2):68–76. https://doi.org/10.5792/ksrr.2014.26.2.68
127. Yoshimura N et al (2012) 'Accumulation of metabolic risk factors such as overweight, hypertension, dyslipidaemia, and impaired glucose tolerance raises the risk of occurrence and progression of knee osteoarthritis: a 3-year follow-up of the ROAD study. Osteoarthr Cartil 20(11):1217–1226. https://doi.org/10.1016/j.joca.2012.06.006
128. Yudoh K, Karasawa R (2010) Statin prevents chondrocyte aging and degeneration of articular cartilage in osteoarthritis (OA). Aging 2(12):990–998. https://doi.org/10.18632/aging.100213
129. Zhang F-J, Luo W, Lei G-H (2015) Role of HIF-1? And HIF-2? In osteoarthritis. Joint Bone Spine 82(3):144–147. https://doi.org/10.1016/j.jbspin.2014.10.003
130. Zhang L, Hu J, Athanasiou KA (2009) The role of tissue engineering in articular cartilage repair and regeneration. Crit Rev Biomed Eng 37(1–2):1–57. Available at: http://www.ncbi.nlm.nih.gov/pubmed/20201770 (Accessed: 16 July 2017)
131. Zhao W et al (2016) Cartilage degeneration and excessive subchondral bone formation in spontaneous osteoarthritis involves altered TGF-β signaling. Journal of orthopaedic research : official publication of the Orthopaedic Research Society 34(5):763–770. https://doi.org/10.1002/jor.23079
132. Zhuo Q et al (2012) Metabolic syndrome meets osteoarthritis. Nature Rev Rheum 8(12):729–737. https://doi.org/10.1038/nrrheum.2012.135

Chapter 4
Surgical Treatment Paradigms of Ankle Lateral Instability, Osteochondral Defects and Impingement

Hélder Pereira, Gwendolyn Vuurberg, Pietro Spennacchio, Jorge Batista, Pieter D'Hooghe, Kenneth Hunt, and Niek Van Dijk

Abstract Ankle sprain is amongst the most frequent musculoskeletal injuries, particularly during sports activities. Chronic ankle instability (CAI) resulting from an ankle sprain might have severe long-lasting consequences on the ankle joint.

H. Pereira (✉)
3B's Research Group – Biomaterials, Biodegradables and Biomimetics, University of Minho, Headquarters of the European Institute of Excellence on Tissue Engineering and Regenerative Medicine, Avepark, Parque de Ciência e Tecnologia, Zona Industrial da Gandra, Barco, Guimarães, Portugal

ICVS/3B's – PT Government Associated Laboratory, Braga/Guimarães, Portugal

Orthopedic Department of Póvoa de Varzim – Vila do Conde Hospital Centre, Póvoa de Varzim, Portugal

Ripoll y De Prado Sports Clinic: Murcia-Madrid FIFA Medical Centre of Excellence, Murcia/Madrid, Spain

International Centre of Sports Traumatology of the Ave, Taipas, Portugal

G. Vuurberg
Department of Orthopaedic Surgery, Academic Medical Centre, Amsterdam Movement Sciences, University of Amsterdam, Amsterdam, The Netherlands

Academic Center for Evidence Based Sports Medicine, Amsterdam, The Netherlands

Amsterdam Collaboration for Health and Safety in Sports, Amsterdam, The Netherlands

P. Spennacchio
Clinique du Sport, Centre Hospitalier Luxembourg, Luxembourg, Luxembourg

J. Batista
Clinical Department Club Atletico Boca Juniores and CAJB – Centro Artroscopico, Jorge batista SA, Buenos Aires, Argentina

P. D'Hooghe
Aspetar Orthopaedic and Sports Medicine Hospital, Department of Orthopaedic Surgery, Aspire Zone, Doha, Qatar

K. Hunt
Department of Orthopaedic Surgery, University of Colorado School of Medicine, Aurora, CO, USA

© Springer International Publishing AG, part of Springer Nature 2018   85
J. M. Oliveira et al. (eds.), *Osteochondral Tissue Engineering*,
Advances in Experimental Medicine and Biology 1059,
https://doi.org/10.1007/978-3-319-76735-2_4

Despite the fact that most patients will respond favourably to appropriate conservative treatment, around 20% will develop symptomatic CAI with sense of giving away and recurrent sprains leading to functional impairment. "Classical" surgical repair by Brostrom-like surgery in one of its many modifications has achieved good results over the years. Recently, major advances in surgical techniques have enabled arthroscopic repair of ankle instability with favourable outcome while also enabling the treatment of other concomitant lesions: loose bodies, osteochondral defects (OCDs) or ankle impingement. Moreover, when the tissue remnant does not permit a repair technique, anatomic reconstruction by means of using a free graft has been developed. In many cases, OCDs occur as a consequence of CAI. However, traumatic and non-traumatic aetiologies have been described. There is no evidence favouring any surgical treatment over another concerning OCDs. Considering lower cost and limited aggression, microfracture is still the most frequent surgical approach. Herein, the authors describe their algorithm in the treatment of these conditions. Similarly, anterior or posterior impingement might be linked with CAI. These are clinical syndromes based on clinical diagnosis which are currently managed arthroscopically upon failure of conservative treatment.

Keywords Ankle impingement syndromes · Ankle sprain · Chronic lateral ankle instability · Osteochondral defects

Top 10 Evidence-Based References
1. Doherty C, Bleakley C, Delahunt E, Holden S (2017) Treatment and prevention of acute and recurrent ankle sprain: an overview of systematic reviews with meta-analysis. Br J Sports Med 51:113–125
2. Hunt KJ, Fuld RS, Sutphin BS, Pereira H, D'Hooghe P (2017) Return to sport following lateral ankle ligament repair is under-reported: a systematic review. J ISAKOS Joint Disorders Orthop Sports Med EPub ahed of Print. https://doi.org/10.1136/jisakos-2016-000064
3. Hélder Pereira, Gwen Vuurberg,f Nuno Gomes, Joaquim Miguel Oliveira, Pedro L. Ripoll, Rui Luís Reis, João Espregueira-Mendes, C. Niek van Dijk (2016) Arthroscopic repair of ankle instability with all-soft knotless anchors. Arthrosc Tech 5(1):e99–e107. Published online 2016 Feb 1. https://doi.org/10.1016/j.eats.2015.10.010

N. Van Dijk
Ripoll y De Prado Sports Clinic: Murcia-Madrid FIFA Medical Centre of Excellence, Murcia/Madrid, Spain

Department of Orthopaedic Surgery, Academic Medical Centre, Amsterdam Movement Sciences, University of Amsterdam, Amsterdam, The Netherlands

Academic Center for Evidence Based Sports Medicine, Amsterdam, The Netherlands

Amsterdam Collaboration for Health and Safety in Sports, Amsterdam, The Netherlands

4. Guillo S, Takao M, Calder J, Karlson J, Michels F, Bauer T, Ankle Instability Group (2016) Arthroscopic anatomical reconstruction of the lateral ankle ligaments. Knee Surg Sports Traumatol Arthrosc 24(4):998–1002. https://doi.org/10.1007/s00167-015-3789-z. Epub 2015 Sep 25. PMID: 26408309

5. Michels F, Pereira H, Calder J, Matricali G, Glazebrook M, Guillo S, Karlsson J, ESSKA-AFAS Ankle Instability Group (2017) Searching for consensus in the approach to patients with chronic lateral ankle instability: ask the expert. Knee Surg Sports Traumatol Arthrosc. https://doi.org/10.1007/s00167-017-4556-0. [Epub ahead of print]

6. Jari Dahmen, Kaj TA Lambers, Mikel L Reilingh, Christiaan JA van Bergen, Sjoerd AS Stufkens, Gino MMJ Kerkhoffs (2017) No superior treatment for primary osteochondral defects of the talus. Knee Surg Sports Traumatol Arthrosc. https://doi.org/10.1007/s00167-017-4616-5. [Epub ahead of print]

7. Zengerink M, Struijs PAA, Tol JL et al (2010) Treatment of osteochondral lesions of the talus: a systematic review. Knee Surg Sports Traumatol Arthrosc 18:238. https://doi.org/10.1007/s00167-009-0942-6

8. VIDEO: Athletic Injuries of the Ankle - Anterior ankle impingement ESSKA Academy. Pereira H. Jul 14, 2014; 58936 Topic: Ankle: https://academy.esska.org/esska/2014/dvds/58936/helder.pereira.athletic.injuries.of.the.ankle.-.anterior.ankle.impingement.html?f=topic=7059*media=1

9. van Dijk CN1, van Bergen C (2008) Advancements in ankle arthroscopy J Am Acad Orthop Surg 16(11):635–46.

10. Ross KA, Murawski CD, Smyth NA, Zwiers R, Wiegerinck JI, van Bergen CJ, Dijk CN, Kennedy JG (2017) Current concepts review: arthroscopic treatment of anterior ankle impingement. Foot Ankle Surg 23(1):1–8.

Fact Box 1 – Ankle Injury Epidemiology

- Inversion ankle sprain most frequent mechanisms: often during landing on the lateral border of the foot, or if the foot gets locked on the ground, while the body continues to turn.
- Isolated lesions of the ATFL occur in 65% of all injuries, while combined rupture of the ATFL and CFL occurs in approximately 20%.
- Despite adequate conservative treatment, approximately 20% of patients develop chronic lateral ankle instability.
- Nonanatomic reconstruction techniques significantly change ankle and subtalar biomechanics
- Upon failure of conservative treatment, anatomic repair or reconstruction techniques have achieved high percentage of good results.

Fact Box 2 – Surgical Options for Treatment of Lateral Ankle Instability
- The so-called anatomic techniques include isolated repair of ATFL remnant and combined ATFL and CFL repair, with or without Gould augmentation by pants-over-vest reinforcement with inferior extensor retinaculum.
- If the remnant tissues are considered as irreparable, or in revision surgeries, anatomic reconstruction by using a tendon graft (e.g. gracilis tendon) either open, percutaneous or arthroscopic has produced favourable outcome.
- Arthroscopic surgical techniques are under development with promising results (at least similar to open techniques while enabling treatment of comorbidities), but more studies are required, particularly in in high-level athletes.

Fact Box 3 – Most Frequent Risk Factors for Surgical Treatment of Lateral Ankle Instability
- Stiffness <5% (reduced ROM >5°)
- Re-rupture
- Nerve damage
- Complications with skin closure
- Risk factors for worst surgical outcome:
- Patients with hyperlaxity
- Very long-standing ligamentous injury (over 10 years)
- Previous surgery for ankle ligament repair

Fact Box 4 – Osteochondral Defects (OCDs) of the Ankle
- Traumatic and non-traumatic aetiologies have been described.
- Ankle sprain or chronic ankle instability might be implicated in the aetiology of OCD.
- Fixation of a large fragment shall be performed whenever possible.
- Microfracture is still the most popular treatment once it has favourable results, low aggression and low cost.
- Moreover, no surgical treatment has proven superiority over any other in this field so far.
- Tissue engineering and regenerative medicine approaches promise new options for the future.

Fact Box 5 – Ankle Anterior and Posterior Impingement
- Both are based on clinical diagnosis while imaging might be helpful in preoperative planning.
- Arthroscopic approach of bony or soft tissue impingement is the rule upon failure of conservative treatment.
- Both are treated in outpatient clinic with immediate range of motion and weightbearing. Full return to activity is usually achieved between 4 and 6 weeks.
- It is very important to start active dorsiflexion-plantarflexion exercises from day one to avoid stiffness.

4.1 Introduction

Ankle lateral instability is a very frequent injury which might cause functional limitations in both athletes and in the general population. It has been stated that ankle sprain is one of the most frequent injuries during sports activity; however, criteria for return to activity are under-reported [1]. The rapid direction and step's changes in addition to landings from falls, collisions and jumps present players with high injury risk during sports. These manoeuvres, which are key elements of the sport at the top level, produce high loads to the hindfoot, frequently exceeding the mechanical resistance of the ankle joint [2, 3].

An inversion ankle sprain is the most frequent cause of acute ankle injury in sports [4]. This typically occurs after a jump, when landing on the lateral border of the foot, but might also occur if the foot gets locked on the ground, while the body continues to turn.

This sudden increase in inversion and internal rotation forces, combined with either dorsi- or plantarflexion, produces sufficient strains to rupture the ankle lateral ligaments, causes concomitant osteochondral lesions or aggravates anterior or posterior joint impingement [5, 6]. The anterior talofibular ligament (ATFL) is injured first; then with increased inversion and rotation, the calcaneofibular ligament (CFL) is also torn (Fig. 4.1) [7]. In about 65% of cases, an isolated lesion of the ATFL will occur, while combined ruptures of the ATFL and CFL happen in around 20% [8, 9]. The posterior talofibular ligament (PTFL) is rarely injured during inversion sprain [10, 11]. In approximately 10–15% of all inversion injuries, there is a total rupture of the lateral ankle ligaments [12]. Moreover, 50% of these cases have concomitant other injuries in the joint (medial ligament injuries, syndesmotic injuries, loose bodies, osteochondral defects (OCDs)) [13].

If not treated adequately and in due time, these injuries will lead to chronic ankle instability (CAI) and might have severe consequences such as osteochondral defects, ankle impingement, synovitis and post-traumatic ankle arthrosis (given the recurrence of ankle sprains) [14–17]. Furthermore, patients with CAI have altered joint kinematics which in turn lead to an increased chance on recurrent ankle sprains

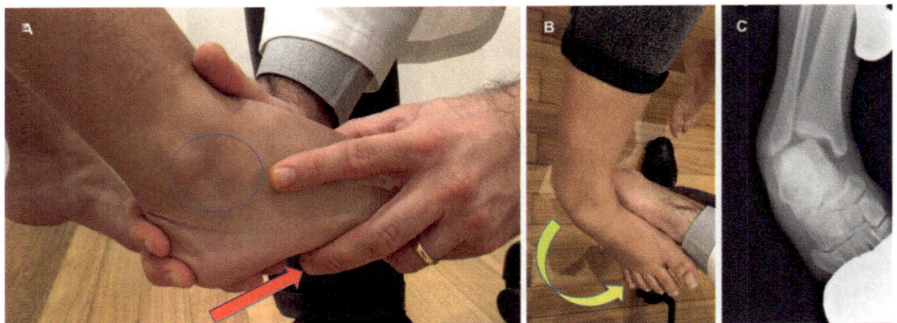

Fig. 4.1 (**A**) Anterior drawer test in which the surgeon induces anterior translation force (red arrow). The anterior dislocation of the talus makes visible a sulcus sign (blue circle). (**B**) Tilt test in which a rotational force (yellow arrow) is induced suggesting calcaneofibular ligament injury. (**C**) Varus stress X-ray reproducing the tilt test and demonstrating impingement of the talus within the ankle mortise

[18]. These persistent "microtraumatisms" will increase the possibility for osteochondral injuries as well as anterior or posterior impingement. In case of failure of conservative treatment, patients who suffer from recurrent ankle sprains can be effectively treated by means of surgical stabilization [19–21]. With the objective to minimize surgical aggression and enable immediate treatment of comorbidities, arthroscopic techniques have been developed and optimized, providing at least similar outcome as open techniques [20].

In order to preserve joint kinematics and optimize clinical results, present surgical techniques aim to restore the "normal" anatomy [22]. Use of peroneal tendons as used in the past is therefore not advised unless this is considered to be the last option [22]. The two most popular techniques include anatomic repair and anatomic reconstruction [22]. A third technique, receiving less attention in current literature, is capsular shrinkage [23]. By use of radiofrequency, the joint capsule is heated which induces shrinkage of collagenous structures aiming to tighten the ATFL (without any foreign or allogeneic material such as suture anchors or tendon grafts) [24]. Despite overall good results, de Vries et al. [23] reported the technique to be unable to modify objective ankle joint laxity.

4.2   From Ankle Sprain to Chronic Lateral Ankle Instability

Although the natural history of ankle sprains is not completely understood, the inherent stability of the ankle mortice and its congruency might contribute to the fact that complete but isolated ATFL ruptures have good prognosis. Most patients are successfully treated with functional treatment [25]. In some selected cases, especially in elite athletes, it has been proposed that early surgery can be considered as a first-line treatment to achieve a faster return to play [26, 27].

If no ligament rupture occurs, functional rehabilitation treatment will enable to resume activities in few days/weeks. Pain is used as a guide for patients and doctors. Ruptured lateral ankle ligaments usually require a period of rigid/semi-rigid immobilization followed by soft brace protection or taping (taping has some risk of skin irritation) [28].

Despite adequate conservative treatment, around 20–30% of patients will develop CAI with persistent symptoms (fear of reinjury limiting activity, sense of giving away, and recurrent sprains) [26, 29, 30]. Standardized and reproducible criteria for reporting return to play for athletes are scarce in literature, and there are no objective guidelines to assist us in this determinant decision [1].

CAI derives from several functional and mechanical factors [7, 31, 32]. These include lower-leg proprioceptive deficits, disturbance of normal reflexes and (peroneal) muscle weakness which are relevant contributors to the persistence of the symptoms [31]. Subsequently, a thorough rehabilitation programme that emphasizes proprioceptive, neuromuscular control and balance training must always be followed. Available data report success rates up to 80% after functional rehabilitation programmes [10, 26].

4.2.1 Principles of Surgical Treatment of Lateral Ankle Instability

Surgery is indicated to restore functional stability upon failure of conservative treatment [25, 30]. The surgical options to treat CAI range from anatomic repair to nonanatomic reconstructions.

Currently, there is insufficient evidence to support any specific superior surgical intervention in the treatment of chronic ankle instability [29, 33].

Nevertheless, nonanatomic reconstruction, as the classic Evans, Watson-Jones or Chrisman-Snook procedures, has been shown to significantly alter the normal biomechanics of the ankle complex, particularly the subtalar joint [8, 9, 34, 35]. Given these concerns [33], and the favourable outcome of anatomic techniques, the former are currently the first line of surgical treatment [36–38].

Anatomic open repair was first described in 1966 by Bröstrom et al. [39]. This technique respects the original anatomy by tightening the torn ATFL and CFL to the distal fibula (Fig. 4.2). Two modifications were introduced over time by Gould et al. [40] that advises to suture the inferior retinaculum extensorum (RE) over the proximal ATFL end to augment the repair, and the modification by Karlsson et al. [41] advises to shorten the ligaments were often not disrupted but elongated.

The functional outcomes of these techniques in its many modifications have been excellent, with success rates reported as high as 87–95% [12, 40, 41]. Retrospective case series of arthroscopic repair techniques have shown successful postoperative

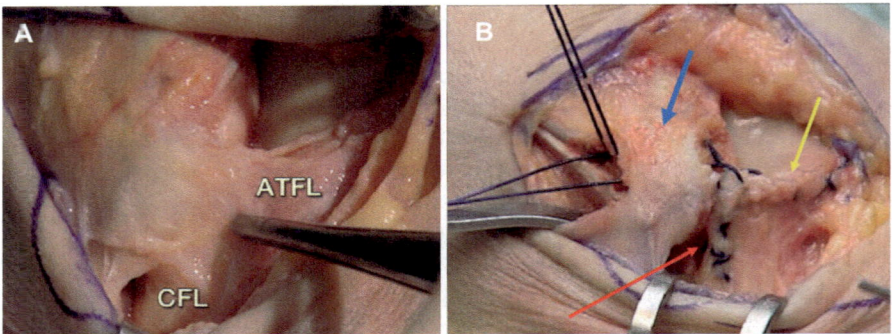

Fig. 4.2 (**A**) Open surgery where anterior tibiofibular (ATFL) and calcaneofibular ligament (CFL) are visible. (**B**) Open Brostrom repair with repair of the anterior tibiofibular (yellow arrow, ATFL) and calcaneofibular ligament (red arrow, CFL) is reattached to the fibula (blue arrow)

results with a high rate of self-reported satisfaction (94.5%), with low rate of complications (0.5–3%) [42–45].

The rehabilitation protocol after anatomic repair of the lateral ligament follows the functional treatment for acute ligament rupture, with a lower-leg cast for 1 or 2 weeks, followed by 2–4 weeks in a functional brace [37]. To encourage earlier return to play, range of movement exercises and protected loading are recommended after 2 weeks as tolerated. Inversion and rotational exercises should be limited during the first 4–6 weeks. Return to sport is usually possible between 10 and 12 weeks; dynamic postural control tests are considered valuable functional assessment tools to progress in return to full activities [4, 33, 46].

4.2.2   Recent Advances in Surgery for Ankle Instability

All the anatomic repair techniques depend on the quality of the ligaments' remnant in order to achieve an effective repair [36]. Karlsson et al. determined risk factors for worst outcome: hyperlaxity, long-standing injuries and previous surgical treatment [41].

When the tissue remnant is considered inadequate for repair, then anatomic reconstruction using a free tendon graft (autograft or allograft), usually the gracilis tendon, has been proposed with favourable outcome [47, 48]. Available clinical data suggest that these anatomic free graft-based reconstructions, either by arthroscopic, percutaneous or open techniques [49], enable favourable outcome in properly selected cases: inadequate remnant or as a salvage/revision procedure [47, 48, 50, 51].

Graft-based reconstructions may lead to increased stiffness once the graft is much stronger than the native tissue [47]. Usually a more aggressive rehabilitation is possible, depending on the intraoperative achieved tension and graft fixation [52].

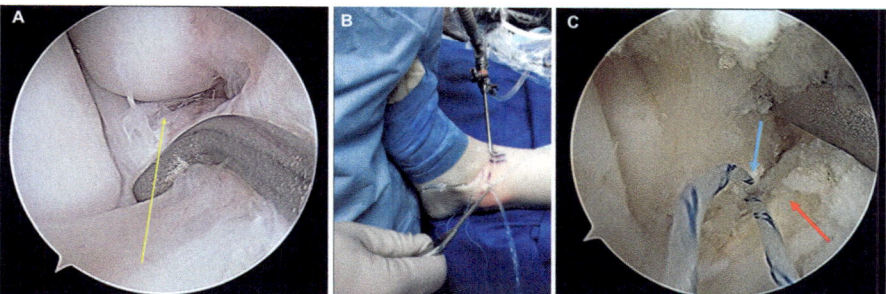

Fig. 4.3 (**A**) Arthroscopic view of the ATFL remnant detached from the fibula (yellow arrow); (**B**) outside view of arthroscopic ATFL repair; (**C**) arthroscopic view of reattachment of the ATFL remnant (red arrow) to the fibula and knot tying (light blue arrow)

Song et al. recently showed a midterm better ankle joint function in patients who received an ATFL reconstruction, compared with the Broström procedure [51], but this finding requires further research with larger series and uniform selection criteria followed by randomized studies.

The current trend is on the pursuit of minimally invasive arthroscopic techniques (Fig. 4.3). Based on the favourable outcome of open ligament repair, several authors have described repair techniques aiming to replicate what has been learnt with open surgery and achieve similar repairs with arthroscopic anchor-based approaches [36–38, 42, 49, 53]. This might lower the surgical morbidity and shorten the time of recovery [54, 55]. Arthroscopy also enables the treatment of concomitant intra-articular lesions in addition to ankle stabilization [14, 56]. Considering the aforementioned retrospective series, comparative studies for open and arthroscopic anatomic lateral ligament repair have shown similar clinical and biomechanical outcome [20, 54, 55].

4.3   Ankle Osteochondral Defects

An osteochondral defect (OCD) of the talus is a lesion involving the talar articular cartilage and its subchondral bone. Several classifications have been used over time, but the first comes from 1959 from Berndt and Harty [57]. OCDs are usually caused by a single or multiple traumatic events, but non-traumatic, idiopathic OCDs of the ankle have been described [58–61]. No classification fully addresses the problem, but the anatomic grid proposed by Raikin and Elias has proven to be useful both in the talus and the tibial plafond [62, 63]. The defect initially may involve only superficial cartilage damage caused by shearing stresses, without damage to the underlying subchondral bone, but a bony injury after a high-impact force also can cause a defect [64]. Ankle trauma associated with an OCD often develops leading to the formation of subchondral bone cysts. These cysts are related with persistent deep ankle pain thereby causing functional impairment. Most OCDs of the talus are found on the anterolateral or posteromedial talar dome [65]. Lateral lesions are usually shallow oval shaped, and a shear mechanism has been proposed to be more frequently implicated. This opposes to medial lesions which are usually deeper, and cup shaped, suggesting a mechanism of torsional impaction and axial loading [58, 60]. Despite several theories and basic science studies concerning OCDs of the talus, its aetiology and pathogenesis are still not fully comprehended.

An OCD might have an acute onset. However, the process leading to subchondral cyst formation requires some time, and it's a slower process [66]. The reason why some OCDs remain asymptomatic is still unclear, while others with apparently similar features cause pain on weightbearing (aggravated by effort), show persistent bone oedema on magnetic resonance imaging and ultimately lead to a subchondral cyst. Understanding this process would be critical in order to prevent progressive joint damage [66].

A traumatic event is commonly accepted as the most important aetiologic factor of an OCD of the talus. For lateral talar defects, trauma has been implicated in 93–98% and for medial defects in 61–70% [67]. OCD aetiology can be divided in non-traumatic and traumatic defects [60]. Vascular aetiology, ischemia, subsequent necrosis, and genetics have been accepted as aetiologic factors [53]. Moreover, OCDs have been found in identical twins and siblings [68]. OCDs are bilateral in 10% of patients [69]. Traumatic cartilage lesions include three categories: microdamage or blunt trauma, chondral fractures and osteochondral fractures [70].

Ankle sprains have a predominant role in the aetiology of traumatic OCDs, once these are probably the most frequent traumatic events leading to these injuries [13]. When a talus twists inside its "bony mortice" during an ankle sprain, the cartilage covering of the talus can be damaged by direct impactions causing a real OCD, bone bruise, cartilage crack or delamination. Shearing forces might cause separation in superficial layer of the cartilage [60]. Loose bodies can be created (and cause even more cartilage damage), or OCDs might remain partially stable in its position (Fig. 4.4). The lesions can either heal and remain asymptomatic, or progress to deep ankle pain on weightbearing and form subchondral bone cysts. IBerndt and Harty were able to reproduce lateral ankle OCDs under laboratory conditions by intensely inverting a dorsiflexed ankle. As the foot was inverted, the lateral border of the talar dome was compacted against the face of the fibula, and when the lateral ligament ruptured it lead to cartilage avulsion. During application of excessive inverting force, the talus rotated laterally in the frontal plane within the mortise, thus impacting and compressing the lateral talar margin against the articular surface of the fibula. With this mechanism, a portion of the talar margin was sheared off from the main body of the talus, causing a lateral OCD. A medial lesion was reproduced by plantarflexing the ankle in combination with slight anterior displacement of the talus on the tibia and inversion and internal rotation of the talus on the tibia [57, 60].

For this reason, one can assume a tight connection between most ankle OCDs and CAI which is the topic for reflection in the herein presented paper.

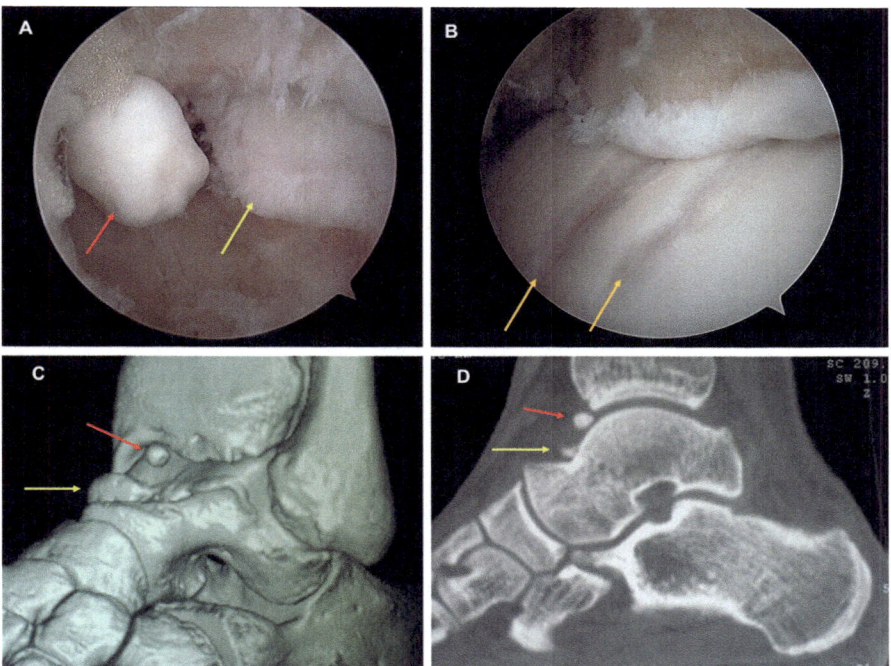

Fig. 4.4 (**A, B**) Arthroscopic view of loose body (red arrow) and talar spur (yellow arrow), causing osteochondral ridge defects (orange arrow on the talar dome). (**C, D**) CT view of the loose body (red arrow) and talar spur (yellow arrow)

4.3.1 Principles of Surgical Treatment of Osteochondral Defects

Asymptomatic incidental findings of the ankle are not infrequent, including within athletic population [71].

Asymptomatic and/or low symptomatic OCDs can usually be treated conservatively, even if kept under clinical and/or image surveillance. Conservative treatment includes orthobiologics, physiotherapy, periods of rest or immobilization (e.g. Walker Boot) [59, 65].

Regarding the symptomatic ankle OCDs, several approaches are possible depending on the characteristics of the lesion and patient profile. There is no current consensus in literature of clear superiority of any surgical treatment over another either in primary or secondary ankle OCDs [65, 72, 73].

Preoperative planning is of paramount relevance, and it should always include X-rays for alignment assessment and global evaluation. The computed tomography (CT) is a critical method since it provides a relatively more reliable assessment of bone defects, which can be overestimated by the MRI oedema around the defect

Table 4.1 Practical algorithm for surgical treatment of osteochondral defects of the ankle

Type of osteochondral defect	Treatment option
Asymptomatic/low-symptomatic lesions	Conservative: periods of rest/walker boot
Symptomatic lesions ≤ 15 mm	Excision, curettage and bone marrow stimulation (ECBMS)
Symptomatic lesions ≥ 15 mm	Fixation* / OATS Consider ECBMS
Large talar cystic lesion	Retrograde drilling ± bone transplant Consider ECBMS
Secondary lesions	OATS/ACI/Hemicap ®/Osteotomy Consider ECBMS

ECBMS excision, curettage and bone marrow stimulation, *OATS* osteochondral autologous transplantation surgery, *ACI* autologous chondrocyte implantation (last generation); Hemicap®, metallic implant for partial replacement of the medial talar dome. ECBMS is considered in most cases given the outcome possibilities and lower aggression and cost. Lower percentage of good/excellent results is to be expected in larger lesions and revision surgery

mainly in T2 sequences. However, the presence of such oedema in T2 suggests activity around the lesion. Moreover, CT lateral view in plantar flexion or dorsiflexion is helpful to determine if it's possibly an anterior or posterior arthroscopic approach or if an open approach is required (medial malleolar osteotomy for medial defects or lateral ligament detachment and afterwards reinsertion for lateral defects). Arthroscopic approach is currently the preferred and most frequently used for both anterior and posterior compartments [74]. Moreover, when no fixed distraction is used, the percentage of complications is extremely low [75].

Given the lack of evidence of any superior treatment, the author's approach favours to prefer the less aggressive options. More aggressive, thus more prone to complications or higher cost procedures are considered for secondary or revision surgeries (Table 4.1).

Whenever possible, an ankle OCD which is possible to fix in place (with sufficient size and preferably with some underlying bone) will constitute our first option (Fig. 4.5). Either open or arthroscopic, the "lift, drill, fill, fix" technique should always be considered once it is the one who preserves the most of the native tissue and hyaline cartilage [76] (lift, the defect; drill, by making microfracture or bone marrow stimulation; fill, the defect with bone graft; and fix, the fragment with metallic or bioabsorbable screws or pins).

In OCDs smaller than 15 mm, excision, curettage and bone marrow stimulation, usually by microfractures (Fig. 4.6), aims to stimulate the underlying subchondral bone bringing "blood" containing growth factors (GFs) and mesenchymal stem cells (MSCs) which will promote fibrocartilage coverage of the defect and provide around 85% of successful outcome at a 5-year follow-up [77]. Given the satisfactory results with minimal aggression, depending on the patient profile and injury characteristics, this approach can also be considered in bigger lesions unable for fixation or secondary injuries [65, 72, 73].

Large cystic lesions, including tibial OCDs, can be addressed by retrograde drilling to lower the pressure within the cyst and filling with bone graft when possible or required.

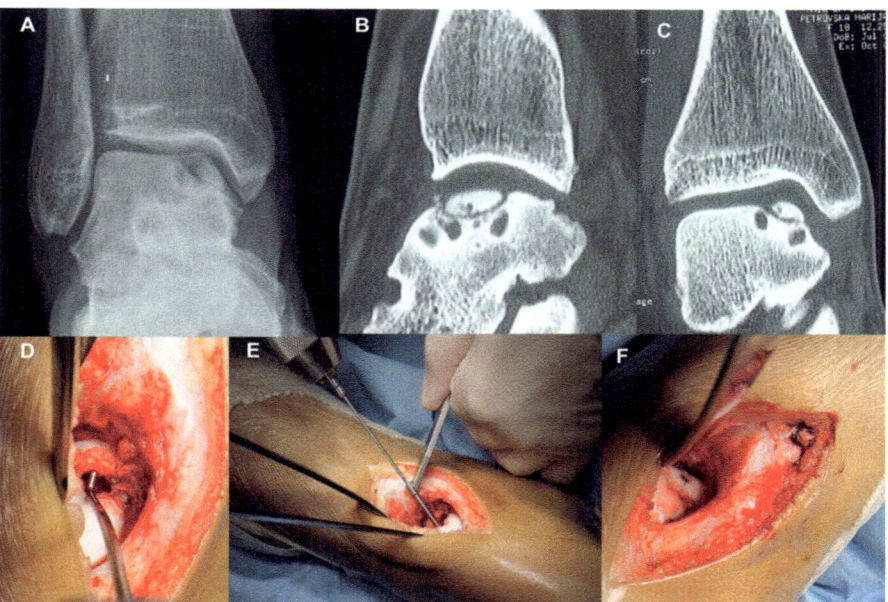

Fig. 4.5 (**A**) X-ray with visible medial OCD on the talar dome. (**B, C**) CT confirms OCD with underlying bone and cystic lesions around it. (**D**) After medial malleolus osteotomy, the OCD is lifted, submitted to bone marrow stimulation and filled with bone autograft. (**E**) The fragment is fixed. (**F**) Final view with fixation with compression screw

The osteochondral autologous transplantation surgery (OATS) consists in harvesting osteochondral cylinders from the knee to fill an ankle defect. Despite a high rate of successful outcome stated by the promoters, a systematic review has shown that this technique is linked to a high percentage of complication [78]. So, in our algorithm it remains a salvage procedure for large OCDs or secondary lesions (after failure of previous surgeries).

Cell-based therapies, scaffolds and augmentation with hydrogels, despite being quite promising, have not been able to consistently present superiority to the previously described techniques on the clinical setting. For this reason, and considering their high cost, they remain options for revision surgeries or large injuries without possibility for fixation and not amenable by any of the previous techniques [79–95]. However, we strongly believe in advanced tissue engineering and regenerative medicine approaches for the future.

When all biologic surgical treatments fail, a novel metallic implant designed for secondary defects of the medial talar dome (Hemicap®) has provided favourable outcome [96].

Finally, realignment by means of the osteotomy (calcaneal sliding (Fig. 4.7) or supramalleolar) is a powerful tool to provide a more favourable biomechanical environment for OCD healing by unloading the affected site [97, 98].

As a last resource, ankle fusion or ankle arthroplasty in very selective cases might be the end line treatment [98].

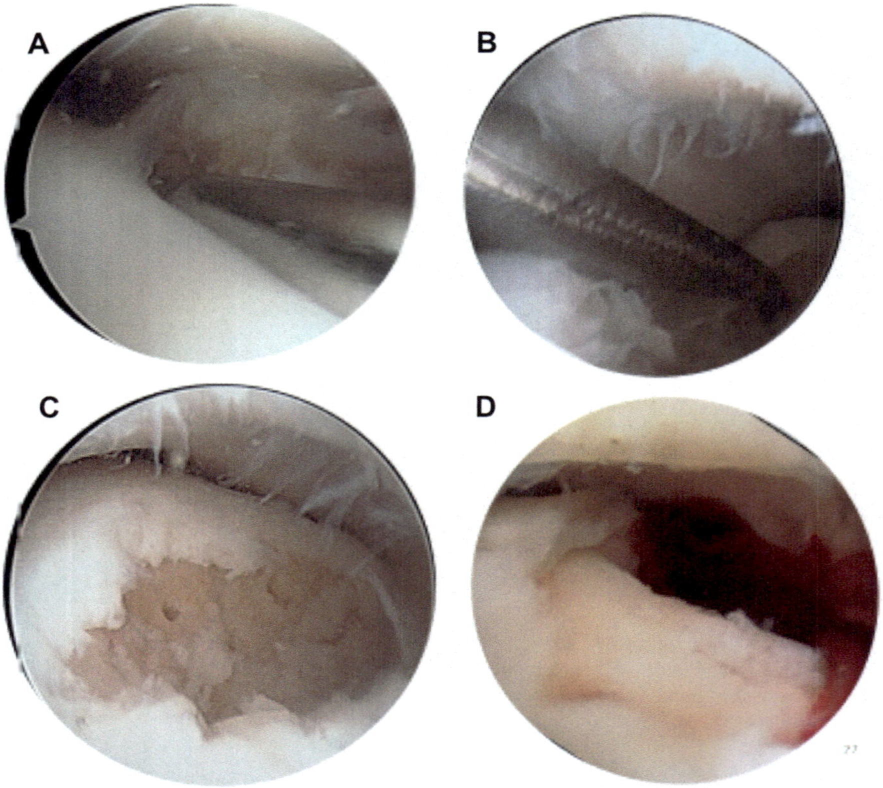

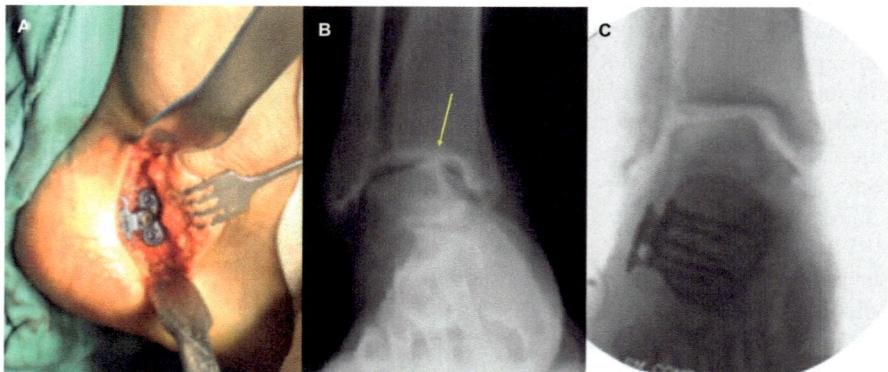

Fig. 4.6 (**A**) Ankle OCD arthroscopic view and removal of unstable fragment; (**B**) microfracture probe; (**C**) final look after microfractures; (**D**) blood coming from the microfracture holes after relieving the tourniquet

Fig. 4.7 (**A**) Surgical procedure of calcaneal sliding osteotomy; (**B**) preoperative X-ray demonstrating severe varus with impingement of the talus on the tibial plafond; (**C**) final position achieved with improved alignment enabling better load distribution

4.4   Ankle Impingement Syndromes

Repetitive microtrauma to the anterior aspect of the ankle joint might lead to bony spur formation ultimately causing anterior impingement syndrome [99]. This microtrauma might be linked to CAI or repetitive direct impact force (e.g. kicking a ball) [100]. About one third of patients with CAI will experience pain related to ankle impingement. Injury of the anterior-inferior talofibular ligament might lead to the development of a "meniscoid lesion" which might cause soft tissue anterolateral impingement [99]. Impingement is considered as a syndrome, meaning that it is basically a clinical diagnosis in which the key sentence is superficial recognizable pain on palpation. Patients complain of persistent pain in walking, aggravated by climbing stairs (dorsiflexion or local pressure might cause entrapment of soft tissue/synovitis between two hard surfaces). Anterior or anteromedial impingement is usually caused by osteophytes, which are not enthesophytes (Fig. 4.8). They do not result traction once they are included in the limits of the capsule [101]. X-ray (including the AMIC view –anteromedial oblique view) [102] or CT (less frequently MRI) can be useful for preoperative planning and identification of concomitant loose bodies or painful broken osteophytes.

Posterior impingement syndrome concerns a mechanical conflict due to hyper-plantarflexion [103]. It can be either acute (*os trigonum* or Stieda process fracture or dislocation) after trauma [104] or chronic, caused by repetitive microtrauma (which might also be linked to CAI) (Fig. 4.9). Chronic cases can be linked to hypertrophic os trigonum or posterior talar process as well as related fractures or soft tissue impingement (e.g. cysts). It is often observed in footballers, cyclers, swimmers, acro-gymnasts and ballet dancers [105, 106]. It is also a syndrome, where posterior impingement test is most helpful and imaging is used for preoperative planning in most cases [107]. Upon failure of conservative treatment (physiotherapy, injections, shoe wear), surgical treatment is recommended.

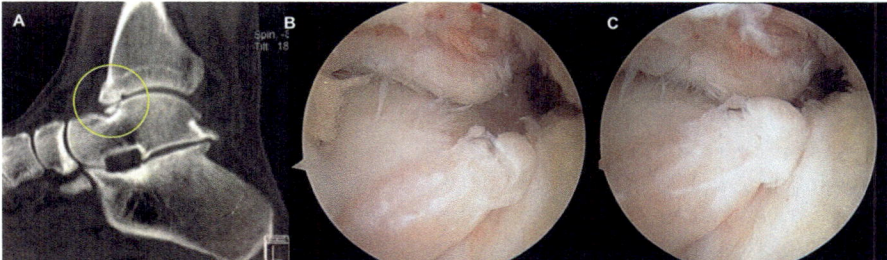

Fig. 4.8 (**A**) CT demonstrating anterior impingement (yellow circle); (**B, C**) arthroscopic view in neutral position and dorsiflexion where bony impingement is confirmed

Fig. 4.9 CT 3D view of plantarflexion ankle with posterior impingement with os trigonum (yellow arrow)

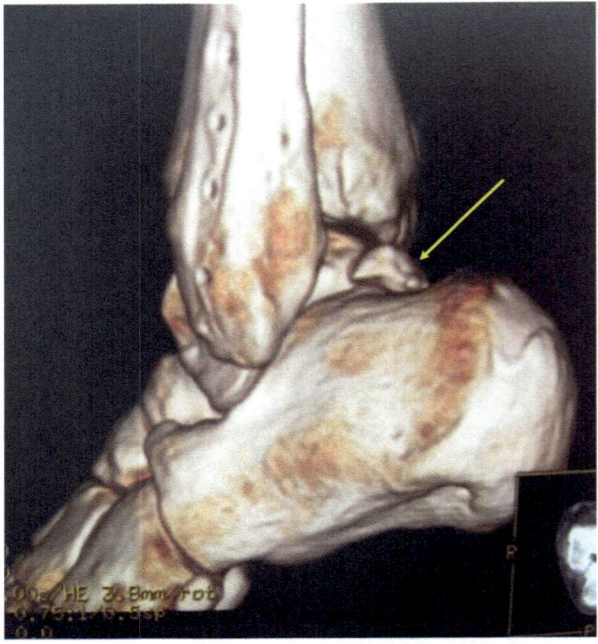

4.4.1 Principles of Surgical Treatment of Anterior Impingement

The treatment of anterior, anteromedial, anterolateral, bony or soft tissue ankle impingement is nowadays achieved mainly by arthroscopic approach. The medial portal is created in dorsiflexion, medial to the crossing line between the anterior tibialis tendon and the joint line [108]. This way the cartilage surface is protected under the tibial plafond, and the working space is "opened". The lateral portal is performed under transillumination and again in dorsiflexion to avoid nerve damage (the superficial peroneal nerve moves posteriorly). The tibial osteophyte shall be removed from superior to inferior and the talar osteophyte from distal to proximal to fully control the bone morphology [101, 107]. It is recommended to minimize aggression which will ultimately lead to a faster recovery and avoid secondary instability due to loss of bony contact (if too much bone is removed) [101, 107]. This is an outpatient procedure, and the patient can weight bear from day 1 if tolerated. It is very important to start active dorsiflexion-plantarflexion exercises from day 1 to avoid stiffness. Stiches are removed at 2 weeks, and full return to activity is possible within 4–6 weeks. Satisfactory results have been published around 85–90% at a 5-year follow-up, and around 80% remain asymptomatic at an 8-year follow-up [101, 107].

4.4.2 Principles of Surgical Treatment of Posterior Impingement

The two-portal endoscopic approach for the hindfoot described by Van Dijk et al. created a revolution in the treatment of these conditions [109], either bony or soft tissue impingement. It lowered dramatically the surgical aggression as it is an outpatient procedure, and the patient can weight bear from day 1 if tolerated. Once more, it is very important to start active dorsiflexion-plantarflexion exercises from day 1. Stiches are removed at 2 weeks, and full return to activity is possible within 4–6 weeks for isolated procedures [103]. The flexor hallucis longus tendon is used as a medial landmark to define a safe working area to avoid the medial neurovascular bundle.

The knowledge of anatomy is fundamental, and the step-by-step technique has been described elsewhere [110]. Effort shall be made to remove the *os trigonum* in one piece to avoid living small loose bodies behind.

4.5 Final Remarks

- The majority of inversion ankle sprains are effectively managed with functional conservative treatment, even in the case of ligament rupture.
- There is increasing evidence on the effectiveness of arthroscopic approach for CAI treatment. So far, the reported outcomes are at least equivalent to open techniques. However, more high-level studies are still needed.
- When repair of the remnant tissue is no longer possible, anatomic reconstruction by using a free graft (auto- or allograft) has provided good results and is also suitable for revision cases. Moreover, replication of the anatomy may facilitate to overcome the limitations of previous nonanatomic techniques.
- Osteochondral defects can have traumatic and non-traumatic aetiology. CAI is a major cause of traumatic OCDs.
- Fixation of an OCD should be performed whenever possible. Besides this, the most frequent surgical treatment remains bone marrow stimulation (e.g. microfractures). This relies on the high percentage of satisfactory results and lower aggression, as well as the fact that no surgical procedure has, so far, demonstrated consistent advantage over the former.
- Tissue engineering and regenerative medicine promises to provide new more effective options for the future.
- Anterior and posterior impingement syndromes are based on clinical diagnosis while imaging is helpful in preoperative planning.
- Aetiology can be traumatic with the contribution of repeated microtrauma connected to CAI.

- Arthroscopic/endoscopic approaches for both these entities enable high percentage of good results with minimal complications and fast return to activity.
- CAI, as herein described, is a major entity which can cause further damage through time in the ankle joint. Effective and timely treatment will avoid further joint damage.

References

1. Hunt KJ, Fuld RS, Sutphin BS, Pereira H, D'Hooghe P (2017) Return to sport following lateral ankle ligament repair is under-reported: a systematic review. J ISAKOS Joint Disorders Orthop Sports Med. EPub ahed of Print. https://doi.org/10.1136/jisakos-2016-000064
2. Lindner M, Kotschwar A, Zsoldos RR, Groesel M, Peham C (2012) The jump shot - a biomechanical analysis focused on lateral ankle ligaments. J Biomech 45(1):202–206. https://doi.org/10.1016/j.jbiomech.2011.09.012
3. Kristianslund E, Krosshaug T (2013) Comparison of drop jumps and sport-specific sidestep cutting: implications for anterior cruciate ligament injury risk screening. Am J Sports Med 41(3):684–688. https://doi.org/10.1177/0363546512472043
4. D'Hooge P, Giza E, Longo G Torn ankle ligaments in elite handball: does a player require surgery? Aspetar Sports Medicine Journal N.A. 186–194. http://www.aspetar.com/journal/upload/PDF/2014410115142.pdf
5. Fong DT, Ha SC, Mok KM, Chan CW, Chan KM (2012) Kinematics analysis of ankle inversion ligamentous sprain injuries in sports: five cases from televised tennis competitions. Am J Sports Med 40(11):2627–2632. https://doi.org/10.1177/0363546512458259
6. Kristianslund E, Bahr R, Krosshaug T (2011) Kinematics and kinetics of an accidental lateral ankle sprain. J Biomech 44(14):2576–2578. https://doi.org/10.1016/j.jbiomech.2011.07.014
7. Brostroem L (1964) SPRAINED ANKLES. I. ANATOMIC LESIONS IN RECENT SPRAINS. Acta Chir Scand 128:483–495
8. Pereira H, D'Hooghe P, Anderson N, Fuld R, Kelley JZ, Baldini TH, Isakos Laf Committee, Hunt K (2017) Ankle and subtalar joint kinematics following lateral ligament repair-implications for early surgical treatment. Arthroscopy J Arthrosc Relat Surg 33(10):e101. https://doi.org/10.1016/j.arthro.2017.08.106
9. Hunt K, D'Hooghe P, Kelley JZ, Anderson N, Fuld R, Baldini TH, Isakos Laf Committee, Pereira H (2017) The role of calcaneofibular ligament (CFL) injury in ankle instability: implications for surgical management. Arthroscopy J Arthrosc Relat Surg 33(10, Supplement):e51–e52 (10). https://doi.org/10.1016/j.arthro.2017.08.024
10. Baumhauer JF, O'Brien T (2002) Surgical considerations in the treatment of ankle instability. J Athl Train 37(4):458–462
11. Lynch SA (2002) Assessment of the injured ankle in the athlete. J Athl Train 37(4):406–412
12. Brostroem L (1966) Sprained ankles. V. Treatment and prognosis in recent ligament ruptures. Acta Chir Scand 132(5):537–550
13. van Dijk CN, Bossuyt PM, Marti RK (1996) Medial ankle pain after lateral ligament rupture. J Bone Joint Surg 78(4):562–567
14. Ferkel RD, Chams RN (2007) Chronic lateral instability: arthroscopic findings and long-term results. Foot Ankle Int 28(1):24–31. https://doi.org/10.3113/fai.2007.0005
15. Lee J, Hamilton G, Ford L (2011) Associated intra-articular ankle pathologies in patients with chronic lateral ankle instability: arthroscopic findings at the time of lateral ankle reconstruction. Foot Ankle Spec 4(5):284–289
16. Liszka H, Depukat P, Gadek A (2016) Intra-articular pathologies associated with chronic ankle instability. Folia Med Cracov 56(2):95–100

17. Harrington KD (1979) Degenerative arthritis of the ankle secondary to long-standing lateral ligament instability. J Bone Joint Surg Am 61(3):354–361
18. Kobayashi T, Gamada K (2014) Lateral ankle sprain and chronic ankle instability: a critical review. Foot Ankle Spec 7(4):298–326. https://doi.org/10.1177/1938640014539813
19. Guillo S, Takao M, Calder J, Karlson J, Michels F, Bauer T (2015) Arthroscopic anatomical reconstruction of the lateral ankle ligaments. Knee Surg Sports Traumatol Arthrosc Off J ESSKA. https://doi.org/10.1007/s00167-015-3789-z
20. Drakos MC, Behrens SB, Paller D, Murphy C, DiGiovanni CW (2014) Biomechanical comparison of an open vs arthroscopic approach for lateral ankle instability. Foot Ankle Int 35(8):809–815. https://doi.org/10.1177/1071100714535765
21. Struijs PA, Kerkhoffs GM (2010) Ankle sprain. BMJ Clin Evid 2010:1115
22. Ferran NA, Oliva F, Maffulli N (2009) Ankle instability. Sports Med Arthrosc Rev 17(2):139–145. https://doi.org/10.1097/JSA.0b013e3181a3d790
23. de Vries JS, Krips R, Blankevoort L, Fievez AW, van Dijk CN (2008) Arthroscopic capsular shrinkage for chronic ankle instability with thermal radiofrequency: prospective multicenter trial. Orthopedics 31(7):655
24. Maiotti M, Massoni C, Tarantino U (2005) The use of arthroscopic thermal shrinkage to treat chronic lateral ankle instability in young athletes. Arthroscopy J Arthrosc Relat Surg Off Publ Arthrosc Assoc North Am Int Arthrosc Assoc 21(6):751–757. https://doi.org/10.1016/j.arthro.2005.03.016
25. Michels F, Pereira H, Calder J, Matricali G, Glazebrook M, Guillo S, Karlsson J, Group E-AAI, Acevedo J, Batista J, Bauer T, Calder J, Carreira D, Choi W, Corte-Real N, Glazebrook M, Ghorbani A, Giza E, Guillo S, Hunt K, Karlsson J, Kong SW, Lee JW, Michels F, Molloy A, Mangone P, Matsui K, Nery C, Ozeki S, Pearce C, Pereira H, Perera A, Pijnenburg B, Raduan F, Stone J, Takao M, Tourne Y, Vega J (2017) Searching for consensus in the approach to patients with chronic lateral ankle instability: ask the expert. Knee Surg Sports Traumatol Arthrosc Off J ESSKA. https://doi.org/10.1007/s00167-017-4556-0
26. Kerkhoffs GM, van den Bekerom M, Elders LA, van Beek PA, Hullegie WA, Bloemers GM, de Heus EM, Loogman MC, Rosenbrand KC, Kuipers T, Hoogstraten JW, Dekker R, Ten Duis HJ, van Dijk CN, van Tulder MW, van der Wees PJ, de Bie RA (2012) Diagnosis, treatment and prevention of ankle sprains: an evidence-based clinical guideline. Br J Sports Med 46(12):854–860. https://doi.org/10.1136/bjsports-2011-090490
27. White WJ, McCollum GA, Calder JD (2016) Return to sport following acute lateral ligament repair of the ankle in professional athletes. Knee Surg Sports Traumatol Arthrosc Off J ESSKA 24(4):1124–1129. https://doi.org/10.1007/s00167-015-3815-1
28. Kemler E, van de Port I, Schmikli S, Huisstede B, Hoes A, Backx F (2015) Effects of soft bracing or taping on a lateral ankle sprain: a non-randomised controlled trial evaluating recurrence rates and residual symptoms at one year. J Foot Ankle Res 8:13. https://doi.org/10.1186/s13047-015-0069-6
29. de Vries JS, Krips R, Sierevelt IN, Blankevoort L (2006) Interventions for treating chronic ankle instability. Cochrane Database Syst Rev 4:CD004124. https://doi.org/10.1002/14651858.CD004124.pub2
30. Doherty C, Bleakley C, Delahunt E, Holden S (2017) Treatment and prevention of acute and recurrent ankle sprain: an overview of systematic reviews with meta-analysis. Br J Sports Med 51(2):113–125. https://doi.org/10.1136/bjsports-2016-096178
31. Hertel J (2002) Functional Anatomy, Pathomechanics, and Pathophysiology of Lateral Ankle Instability. J Athl Train 37(4):364–375
32. McKay GD, Goldie PA, Payne WR, Oakes BW (2001) Ankle injuries in basketball: injury rate and risk factors. Br J Sports Med 35(2):103–108
33. de Vries JS, Krips R, Sierevelt IN, Blankevoort L, van Dijk CN (2011) Interventions for treating chronic ankle instability. Cochrane Database Syst Rev 8:CD004124. https://doi.org/10.1002/14651858.CD004124.pub3

34. Karlsson J, Bergsten T, Lansinger O, Peterson L (1989) Surgical treatment of chronic lateral instability of the ankle joint. A new procedure. Am J Sports Med 17(2):268–273.; discussion 273-264. https://doi.org/10.1177/036354658901700220

35. Vuurberg G, Veen OC, Pereira H, Blankevoort L, van Dijk CN (2017) Tenodesis reconstruction in patients with chronic lateral ankle instability is associated with a high risk of complications compared with anatomic repair and reconstruction: a systematic review and meta-analysis. J ISAKOS Joint Disorders Orthop Sports Med. https://doi.org/10.1136/jisakos-2016-000121

36. Vega J, Golano P, Pellegrino A, Rabat E, Pena F (2013) All-inside arthroscopic lateral collateral ligament repair for ankle instability with a knotless suture anchor technique. Foot Ankle Int 34(12):1701–1709. https://doi.org/10.1177/1071100713502322

37. Pereira H, Vuurberg G, Gomes N, Oliveira JM, Ripoll PL, Reis RL, Espregueira-Mendes J, Niek van Dijk C (2016) Arthroscopic repair of ankle instability with all-soft knotless anchors. Arthrosc Tech 5(1):e99–e107. https://doi.org/10.1016/j.eats.2015.10.010

38. Takao M, Matsui K, Stone JW, Glazebrook MA, Kennedy JG, Guillo S, Calder JD, Karlsson J, Ankle Instability G (2016) Arthroscopic anterior talofibular ligament repair for lateral instability of the ankle. Knee Surg Sports Traumatol Arthrosc Off J ESSKA 24(4):1003–1006. https://doi.org/10.1007/s00167-015-3638-0

39. Brostrom L (1966) Sprained ankles. VI. Surgical treatment of "chronic" ligament ruptures. Acta Chir Scand 132(5):551–565

40. Gould N, Seligson D, Gassman J (1980) Early and late repair of lateral ligament of the ankle. Foot Ankle 1(2):84–89

41. Karlsson J, Bergsten T, Lansinger O, Peterson L (1988) Reconstruction of the lateral ligaments of the ankle for chronic lateral instability. J Bone Joint Surg Am 70(4):581–588

42. Matsui K, Burgesson B, Takao M, Stone J, Guillo S, Glazebrook M, Group EAAI (2016) Minimally invasive surgical treatment for chronic ankle instability: a systematic review. Knee Surg Sports Traumatol Arthrosc Off J ESSKA 24(4):1040–1048. https://doi.org/10.1007/s00167-016-4041-1

43. Acevedo JI, Mangone P (2015) Arthroscopic Brostrom technique. Foot Ankle Int 36(4):465–473. https://doi.org/10.1177/1071100715576107

44. Lui TH (2016) Modified arthroscopic brostrom procedure with bone tunnels. Arthrosc Tech 5(4):e775–e780. https://doi.org/10.1016/j.eats.2016.03.003

45. Prissel MA, Roukis TS (2014) All-inside, anatomical lateral ankle stabilization for revision and complex primary lateral ankle stabilization: a technique guide. Foot Ankle Spec 7(6):484–491. https://doi.org/10.1177/1938640014548418

46. Pearce CJ, Tourne Y, Zellers J, Terrier R, Toschi P, Silbernagel KG, Group E-AAI (2016) Rehabilitation after anatomical ankle ligament repair or reconstruction. Knee Surg Sports Traumatol Arthrosc Off J ESSKA 24(4):1130–1139. https://doi.org/10.1007/s00167-016-4051-z

47. Takao M, Oae K, Uchio Y, Ochi M, Yamamoto H (2005) Anatomical reconstruction of the lateral ligaments of the ankle with a gracilis autograft: a new technique using an interference fit anchoring system. Am J Sports Med 33(6):814–823. https://doi.org/10.1177/0363546504272688

48. Coughlin MJ, Schenck RC Jr, Grebing BR, Treme G (2004) Comprehensive reconstruction of the lateral ankle for chronic instability using a free gracilis graft. Foot Ankle Int 25(4):231–241. https://doi.org/10.1177/107110070402500407

49. Guillo S, Takao M, Calder J, Karlson J, Michels F, Bauer T, Ankle Instability G (2016) Arthroscopic anatomical reconstruction of the lateral ankle ligaments. Knee Surg Sports Traumatol Arthrosc Off J ESSKA 24(4):998–1002. https://doi.org/10.1007/s00167-015-3789-z

50. Okuda R, Kinoshita M, Morikawa J, Jotoku T, Abe M (1999) Reconstruction for chronic lateral ankle instability using the palmaris longus tendon: is reconstruction of

the calcaneofibular ligament necessary? Foot Ankle Int 20(11):714–720. https://doi. org/10.1177/107110079902001107

51. Song B, Li C, Chen N, Chen Z, Zhang Y, Zhou Y, Li W (2017) All-arthroscopic anatomical reconstruction of anterior talofibular ligament using semitendinosus autografts. Int Orthop 41(5):975–982. https://doi.org/10.1007/s00264-017-3410-9

52. Glazebrook M, Stone J, Matsui K, Guillo S, Takao M, Group EAAI (2016) Percutaneous ankle reconstruction of lateral ligaments (Perc-Anti RoLL). Foot Ankle Int 37(6):659–664. https://doi.org/10.1177/1071100716633648

53. Corte-Real NM, Moreira RM (2009) Arthroscopic repair of chronic lateral ankle instability. Foot Ankle Int 30(3):213–217. https://doi.org/10.3113/FAI.2009.0213

54. Yeo ED, Lee KT, Sung IH, Lee SG, Lee YK (2016) Comparison of all-inside arthroscopic and open techniques for the modified brostrom procedure for ankle instability. Foot Ankle Int 37(10):1037–1045. https://doi.org/10.1177/1071100716666508

55. Matsui K, Takao M, Miyamoto W, Matsushita T (2016) Early recovery after arthroscopic repair compared to open repair of the anterior talofibular ligament for lateral instability of the ankle. Arch Orthop Trauma Surg 136(1):93–100. https://doi.org/10.1007/s00402-015-2342-3

56. Hintermann B, Boss A, Schafer D (2002) Arthroscopic findings in patients with chronic ankle instability. Am J Sports Med 30(3):402–409. https://doi.org/10.1177/03635465020300031601

57. Berndt AL, Harty M (1959) Transchondral fractures (osteochondritis dissecans) of the talus. J Bone Joint Surg Am 41-A:988–1020

58. Schachter AK, Chen AL, Reddy PD, Tejwani NC (2005) Osteochondral lesions of the talus. J Am Acad Orthop Surg 13(3):152–158

59. O'Loughlin PF, Heyworth BE, Kennedy JG (2010) Current concepts in the diagnosis and treatment of osteochondral lesions of the ankle. Am J Sports Med 38(2):392–404. https://doi. org/10.1177/0363546509336336

60. van Dijk CN, Reilingh ML, Zengerink M, van Bergen CJ (2010) Osteochondral defects in the ankle: why painful? Knee Surgery Sports Traumatol Arthrosc Off J ESSKA 18(5):570–580. https://doi.org/10.1007/s00167-010-1064-x

61. Ventura A, Terzaghi C, Legnani C, Borgo E (2013) Treatment of post-traumatic osteochondral lesions of the talus: a four-step approach. Knee Surg Sports Traumatol Arthrosc Off J ESSKA 21(6):1245–1250. https://doi.org/10.1007/s00167-012-2028-0

62. Elias I, Raikin SM, Schweitzer ME, Besser MP, Morrison WB, Zoga AC (2009) Osteochondral lesions of the distal tibial plafond: localization and morphologic characteristics with an anatomical grid. Foot Ankle Int 30(6):524–529. https://doi.org/10.3113/FAI.2009.0524

63. Elias I, Zoga AC, Morrison WB, Besser MP, Schweitzer ME, Raikin SM (2007) Osteochondral lesions of the talus: localization and morphologic data from 424 patients using a novel anatomical grid scheme. Foot Ankle Int 28(2):154–161. https://doi.org/10.3113/FAI.2007.0154

64. Vellet AD, Marks PH, Fowler PJ, Munro TG (1991) Occult posttraumatic osteochondral lesions of the knee: prevalence, classification, and short-term sequelae evaluated with MR imaging. Radiology 178(1):271–276. https://doi.org/10.1148/radiology.178.1.1984319

65. Zengerink M, Struijs PA, Tol JL, van Dijk CN (2010) Treatment of osteochondral lesions of the talus: a systematic review. Knee Surg Sports Traumatol Arthrosc Off J ESSKA 18(2):238–246. https://doi.org/10.1007/s00167-009-0942-6

66. Durr HD, Martin H, Pellengahr C, Schlemmer M, Maier M, Jansson V (2004) The cause of subchondral bone cysts in osteoarthrosis: a finite element analysis. Acta Orthop Scand 75(5):554–558. https://doi.org/10.1080/00016470410001411

67. Verhagen RA, Struijs PA, Bossuyt PM, van Dijk CN (2003) Systematic review of treatment strategies for osteochondral defects of the talar dome. Foot Ankle Clin 8(2):233–242. viii-ix

68. Woods K, Harris I (1995) Osteochondritis dissecans of the talus in identical twins. J Bone Joint Surg 77(2):331

69. Hermanson E, Ferkel RD (2009) Bilateral osteochondral lesions of the talus. Foot Ankle Int 30(8):723–727. https://doi.org/10.3113/FAI.2009.0723

70. Frenkel SR, DCP E (1999) Degradation and repair of articular cartilage. Front Biosci 4:671–685
71. Saxena A, Luhadiya A, Ewen B, Goumas C (2011) Magnetic resonance imaging and incidental findings of lateral ankle pathologic features with asymptomatic ankles. J Foot Ankle Surg 50(4):413–415. https://doi.org/10.1053/j.jfas.2011.03.011
72. Dahmen J, Lambers KTA, Reilingh ML, van Bergen CJA, Stufkens SAS, Kerkhoffs G (2017) No superior treatment for primary osteochondral defects of the talus. Knee Surg Sports Traumatol Arthrosc Off J ESSKA. https://doi.org/10.1007/s00167-017-4616-5
73. Lambers KTA, Dahmen J, Reilingh ML, van CJA B, Stufkens SAS, Kerkhoffs G (2017) No superior surgical treatment for secondary osteochondral defects of the talus. Knee Surg Sports Traumatol Arthrosc Off J ESSKA. https://doi.org/10.1007/s00167-017-4629-0
74. van Dijk CN, van Bergen CJ (2008) Advancements in ankle arthroscopy. J Am Acad Orthop Surg 16(11):635–646
75. Zengerink M, van Dijk CN (2012) Complications in ankle arthroscopy. Knee Surg Sports Traumatol Arthrosc Off J ESSKA 20(8):1420–1431. https://doi.org/10.1007/s00167-012-2063-x
76. Kerkhoffs GM, Reilingh ML, Gerards RM, de Leeuw PA (2016) Lift, drill, fill and fix (LDFF): a new arthroscopic treatment for talar osteochondral defects. Knee Surg Sports Traumatol Arthrosc Off J ESSKA 24(4):1265–1271. https://doi.org/10.1007/s00167-014-3057-7
77. Doral MN, Bilge O, Batmaz G, Donmez G, Turhan E, Demirel M, Atay OA, Uzumcugil A, Atesok K, Kaya D (2012) Treatment of osteochondral lesions of the talus with microfracture technique and postoperative hyaluronan injection. Knee Surg Sports Traumatol Arthrosc Off J ESSKA 20(7):1398–1403. https://doi.org/10.1007/s00167-011-1856-7
78. Ferreira C, Vuurberg G, Oliveira JM, Espregueira-Mendes J, Pereira H, Reis RL, Ripoll P (2016) Assessment of clinical outcome after osteochondral autologous transplantation technique for the treatment of ankle lesions: a systematic review. J Isakos. doi:jisakos-2015-000020.R2
79. Bhattarai N, Gunn J, Zhang M (2010) Chitosan-based hydrogels for controlled, localized drug delivery. Adv Drug Deliv Rev 62(1):83–99. https://doi.org/10.1016/j.addr.2009.07.019
80. Brittberg M (2010) Cell carriers as the next generation of cell therapy for cartilage repair: a review of the matrix-induced autologous chondrocyte implantation procedure. Am J Sports Med 38(6):1259–1271. https://doi.org/10.1177/0363546509346395
81. Candrian C, Miot S, Wolf F, Bonacina E, Dickinson S, Wirz D, Jakob M, Valderrabano V, Barbero A, Martin I (2010) Are ankle chondrocytes from damaged fragments a suitable cell source for cartilage repair? Osteoarthr Cartil/OARS Osteoarthr Res Soc 18(8):1067–1076. https://doi.org/10.1016/j.joca.2010.04.010
82. Caron MMJ, Emans PJ, Coolsen MME, Voss L, Surtel DAM, Cremers A, van Rhijn LW, Welting TJM (2012) Redifferentiation of dedifferentiated human articular chondrocytes: comparison of 2D and 3D cultures. Osteoarthr Cartil 20(10):1170–1178. https://doi.org/10.1016/j.joca.2012.06.016
83. Centeno CJ, Schultz JR, Cheever M, Robinson B, Freeman M, Marasco W (2010) Safety and complications reporting on the re-implantation of culture-expanded mesenchymal stem cells using autologous platelet lysate technique. Curr Stem Cell Res Ther 5(1):81–93
84. Che JH, Zhang ZR, Li GZ, Tan WH, Bai XD, Qu FJ (2010) Application of tissue-engineered cartilage with BMP-7 gene to repair knee joint cartilage injury in rabbits. Knee Surg Sports Traumatol Arthrosc Off J ESSKA 18(4):496–503. https://doi.org/10.1007/s00167-009-0962-2
85. Clar C, Cummins E, McIntyre L, Thomas S, Lamb J, Bain L, Jobanputra P, Waugh N (2005) Clinical and cost-effectiveness of autologous chondrocyte implantation for cartilage defects in knee joints: systematic review and economic evaluation. Health Technol Assess 9 (47):iii-iv, ix-x, 1–82. doi:03-65-01 [pii]
86. Dhollander AA, De Neve F, Almqvist KF, Verdonk R, Lambrecht S, Elewaut D, Verbruggen G, Verdonk PC (2011) Autologous matrix-induced chondrogenesis combined with platelet-rich

plasma gel: technical description and a five pilot patients report. Knee Surg Sports Traumatol Arthrosc Off J ESSKA 19(4):536–542. https://doi.org/10.1007/s00167-010-1337-4

87. DiGiovanni CW, Lin SS, Baumhauer JF, Daniels T, Younger A, Glazebrook M, Anderson J, Anderson R, Evangelista P, Lynch SE (2013) Recombinant human platelet-derived growth factor-BB and beta-tricalcium phosphate (rhPDGF-BB/beta-TCP): an alternative to autogenous bone graft. J Bone Joint Surg Am 95(13):1184–1192. https://doi.org/10.2106/JBJS.K.01422

88. Enea D, Gwynne J, Kew S, Arumugam M, Shepherd J, Brooks R, Ghose S, Best S, Cameron R, Rushton N (2012) Collagen fibre implant for tendon and ligament biological augmentation. In vivo study in an ovine model. Knee Surg Sports Traumatol Arthrosc Off J ESSKA. https://doi.org/10.1007/s00167-012-2102-7

89. Gurkan UA, Tasoglu S, Kavaz D, Demirci U (2012) Emerging technologies for assembly of microscale hydrogels. Adv Healthc Mater 1:149–158

90. Harris JD, Siston RA, Pan X, Flanigan DC (2010) Autologous chondrocyte implantation: a systematic review. J Bone Joint Surg Am 92(12):2220–2233. https://doi.org/10.2106/JBJS.J.00049

91. Nagura I, Fujioka H, Kokubu T, Makino T, Sumi Y, Kurosaka M (2007) Repair of osteochondral defects with a new porous synthetic polymer scaffold. J Bone Joint Surg Br 89(2):258–264. https://doi.org/10.1302/0301-620X.89B2.17754

92. Pereira D, Silva-Correia J, Pereira H, Espregueira-Mendes J, Oliveira JM, Reis RL (2012) Gellan gum-based bilayered scaffolds for application in osteochondral tissue engineering. J Tissue Eng Regen Med 6(1):185

93. Shive MS, Stanish WD, McCormack R, Forriol F, Mohtadi N, Pelet S, Desnoyers J, Methot S, Vehik K, Restrepo A (2015) BST-CarGel(R) treatment maintains cartilage repair superiority over microfracture at 5 years in a multicenter randomized controlled trial. Cartilage 6(2):62–72. https://doi.org/10.1177/1947603514562064

94. Stanish WD, McCormack R, Forriol F, Mohtadi N, Pelet S, Desnoyers J, Restrepo A, Shive MS (2013) Novel scaffold-based BST-CarGel treatment results in superior cartilage repair compared with microfracture in a randomized controlled trial. J Bone Joint Surg Am 95(18):1640–1650. https://doi.org/10.2106/JBJS.L.01345

95. Vinatier C, Mrugala D, Jorgensen C, Guicheux J, Noel D (2009) Cartilage engineering: a crucial combination of cells, biomaterials and biofactors. Trends Biotechnol 27(5):307–314. https://doi.org/10.1016/j.tibtech.2009.02.005

96. Ettinger S, Stukenborg-Colsman C, Waizy H, Becher C, Yao D, Claassen L, Noll Y, Plaass C (2017) Results of HemiCAP(R) implantation as a salvage procedure for osteochondral lesions of the talus. J Foot Ankle Surg 56(4):788–792. https://doi.org/10.1053/j.jfas.2017.04.001

97. Pagenstert GI, Hintermann B, Barg A, Leumann A, Valderrabano V (2007) Realignment surgery as alternative treatment of varus and valgus ankle osteoarthritis. Clin Orthop Relat Res 462:156–168. https://doi.org/10.1097/BLO.0b013e318124a462

98. Badekas T, Takvorian M, Souras N (2013) Treatment principles for osteochondral lesions in foot and ankle. Int Orthop 37(9):1697–1706. https://doi.org/10.1007/s00264-013-2076-1

99. Ross KA, Murawski CD, Smyth NA, Zwiers R, Wiegerinck JI, van Bergen CJ, Dijk CN, Kennedy JG (2017) Current concepts review: arthroscopic treatment of anterior ankle impingement. Foot Ankle Surg 23(1):1–8. https://doi.org/10.1016/j.fas.2016.01.005

100. Tol JL, Slim E, van Soest AJ, van Dijk CN (2002) The relationship of the kicking action in soccer and anterior ankle impingement syndrome. A biomechanical analysis. Am J Sports Med 30(1):45–50. https://doi.org/10.1177/03635465020300012101

101. Tol JL, Verheyen CP, van Dijk CN (2001) Arthroscopic treatment of anterior impingement in the ankle. J Bone Joint Surg 83(1):9–13

102. van Dijk CN, Wessel RN, Tol JL, Maas M (2002) Oblique radiograph for the detection of bone spurs in anterior ankle impingement. Skelet Radiol 31(4):214–221. https://doi.org/10.1007/s00256-002-0477-0

103. Miyamoto W, Miki S, Kawano H, Takao M (2017) Surgical outcome of posterior ankle impingement syndrome with concomitant ankle disorders treated simultaneously in

patient engaged in athletic activity. J Orthop Sci 22(3):463–467. https://doi.org/10.1016/j.jos.2016.12.017

104. Kose O, Okan AN, Durakbasa MO, Emrem K, Islam NC (2006) Fracture of the os trigonum: a case report. J Orthop Surg (Hong Kong) 14(3):354–356. https://doi.org/10.1177/230949900601400326

105. Walls RJ, Ross KA, Fraser EJ, Hodgkins CW, Smyth NA, Egan CJ, Calder J, Kennedy JG (2016) Football injuries of the ankle: a review of injury mechanisms, diagnosis and management. World J Orthop 7(1):8–19. https://doi.org/10.5312/wjo.v7.i1.8

106. Lees A, Asai T, Andersen TB, Nunome H, Sterzing T (2010) The biomechanics of kicking in soccer: a review. J Sports Sci 28(8):805–817. https://doi.org/10.1080/02640414.2010.481305

107. Niek van Dijk C (2006) Anterior and posterior ankle impingement. Foot Ankle Clin 11(3):663–683. https://doi.org/10.1016/j.fcl.2006.06.003

108. de Leeuw PA, Golano P, Clavero JA, van Dijk CN (2010) Anterior ankle arthroscopy, distraction or dorsiflexion? Knee Surg Sports Traumatol Arthrosc Off J ESSKA 18(5):594–600. https://doi.org/10.1007/s00167-010-1089-1

109. van Dijk CN, Scholten PE, Krips R (2000) A 2-portal endoscopic approach for diagnosis and treatment of posterior ankle pathology. Arthroscopy J Arthrosc Relat Surg Off Publ Arthrosc Assoc North Am Int Arthrosc Assoc 16(8):871–876. https://doi.org/10.1053/jars.2000.19430

110. Haverkamp D, Bech N, de Leeuw P, d'Hooghe P, Kynsburg A, Calder J, Ogut T, Batista J, Pereira H (2016) Posterior compartment of the ankle joint: a focus on arthroscopic treatment (ICL 17). In: Becker R, Kerkhoffs GMMJ, Gelber PE, Denti M, Seil R (eds) ESSKA instructional course lecture book. Springer, Berlin/Heidelberg. https://doi.org/10.1007/978-3-662-49114-0_15

Part II
Viscosupplementation

Chapter 5
Clinical Management in Early OA

Rita Grazina, Renato Andrade, Ricardo Bastos, Daniela Costa,
Rogério Pereira, José Marinhas, António Maestro,
and João Espregueira-Mendes

Abstract Knee osteoarthritis affects an important percentage of the population throughout their life. Several factors seem to be related to the development of knee osteoarthritis including genetic predisposition, gender, age, meniscal deficiency, lower limb malalignments, joint instability, cartilage defects, and increasing sports participation. The latter has contributed to a higher prevalence of early onset of knee osteoarthritis at younger ages with this active population demanding more consistent and durable outcomes. The diagnosis is complex and the common signs and symptoms are often cloaked at these early stages. Classification systems have been developed and are based on the presence of knee pain and radiographic

R. Grazina
Orthopaedic Surgery at Centro Hospitalar de Vila Nova de Gaia/Espinho E.P.E,
Vila Nova de Gaia, Portugal

R. Andrade
Clínica do Dragão, Espregueira-Mendes Sports Centre - FIFA Medical Centre of Excellence,
Porto, Portugal

Dom Henrique Research Centre, Porto, Portugal

Faculty of Sports, University of Porto, Porto, Portugal

R. Bastos
Clínica do Dragão, Espregueira-Mendes Sports Centre - FIFA Medical Centre of Excellence,
Porto, Portugal

Dom Henrique Research Centre, Porto, Portugal

Fluminense Federal University, Niteroi/Rio de Janeiro, Brazil

D. Costa
SMIC Dragão - Serviço Médico de Imagem Computorizada, Porto, Portugal

R. Pereira
Clínica do Dragão, Espregueira-Mendes Sports Centre - FIFA Medical Centre of Excellence,
Porto, PortugalDom Henrique Research Centre, Porto, Portugal

Faculty of Sports, University of Porto, Porto, Portugal

Faculty of Health Sciences, University Fernando Pessoa, Porto, Portugal

© Springer International Publishing AG, part of Springer Nature 2018   111
J. M. Oliveira et al. (eds.), *Osteochondral Tissue Engineering*,
Advances in Experimental Medicine and Biology 1059,
https://doi.org/10.1007/978-3-319-76735-2_5

findings coupled with magnetic resonance or arthroscopic evidence of early joint degeneration. Nonsurgical treatment is often the first-line option and is mainly based on daily life adaptations, weight loss, and exercise, with pharmacological agents having only a symptomatic role. Surgical treatment shows positive results in relieving the joint symptomatology, increasing the knee function and delaying the development to further degenerative stages. Biologic therapies are an emerging field showing early promising results; however, further high-level research is required.

Keywords Early osteoarthritis · Knee · Cartilage

Highlights
- Diagnosis of early knee OA relies in three main criteria including knee pain, radiographic image showing osteophytes, and MRI or arthroscopic findings of cartilage injury.
- Recommended conservative measures include weight loss (0.25%/week), aquatic/land exercises (strength and aerobic training), and pharmacological agents including oral NSAIDs and opioids.
- Lower limb malalignments and meniscal deficiency should be corrected to avoid overloading the joint and prevent further damage.

J. Marinhas
Orthopaedic Surgery at Centro Hospitalar de Vila Nova de Gaia/Espinho E.P.E,
Vila Nova de Gaia, Portugal

Clínica do Dragão, Espregueira-Mendes Sports Centre - FIFA Medical Centre of Excellence,
Porto, Portugal

Dom Henrique Research Centre, Porto, Portugal

A. Maestro
Real Sporting de Gijón SAD, Gijón, Spain

FREMAP Mutua de Accidentes, Gijón, Spain

J. Espregueira-Mendes (✉)
Clínica do Dragão, Espregueira-Mendes Sports Centre - FIFA Medical Centre of Excellence,
Porto, Portugal

Dom Henrique Research Centre, Porto, Portugal

Orthopaedics Department of Minho University, Minho, Portugal

3B's Research Group–Biomaterials, Biodegradables and Biomimetics, University of Minho,
Headquarters of the European Institute of Excellence on Tissue Engineering and Regenerative
Medicine, Guimarães, Portugal

ICVS/3B's–PT Government Associate Laboratory, Braga/Guimarães, Portugal
e-mail: espregueira@dhresearchcentre.com

5.1   Introduction

Osteoarthritis (OA) is characterized by several articular dysfunctions with consequent anatomical changes in joint structures. OARSI (Osteoarthritis Research Society International) defines osteoarthritis as "a disorder involving movable joints characterized by cell stress and extracellular matrix degradation initiated by micro- and macro-injury that activates maladaptive repair responses including pro-inflammatory pathways of innate immunity." This in turn manifests initially as abnormal joint tissue metabolism and subsequently by anatomic and physiologic derangements. Contrary to what is believed, the disease is not exclusively related to cartilage deterioration. The physiopathology is complex and all anatomical components presented in the articular space are deteriorated. Besides the cartilage, subchondral bone, menisci, ligaments, synovial membrane, and synovial fluid are involved [1]. Process of tissue destruction, abnormal metabolism, and reparative attempt leads to a deleterious vicious cycle which interruption is challenging. The final consequence is an inflammatory, painful, and deformed joint with stiffness and limited range of motion that leads to an important functional disability. In fact, the disease is one of the most common disabilities in adults.

Osteoarthritis affects 33.6% of the population over 65 years old and, surprisingly, 13.9% of the population over 25 years old [2]. Worldwide estimates are that 9.6% of men and 18.0% of women aged over 60 years have symptomatic osteoarthritis, 80% of those will have limitations in movement and 25% cannot perform their major daily activities of life [3]. As considered a degenerative disorder with tissue deterioration, it is comprehensible that the prevalence increases with age and some predictions must be considered for the comprehension of this health problem [4]:

- By 2050, global life expectancy at birth is projected to increase by almost 8 years, climbing from 68.6 years in 2015 to 76.2 years in 2050.
- The global population of the "oldest old" – people aged 80 and older – is expected to more than triple between 2015 and 2050, growing from 126.5 million to 446.6 million. The oldest old population in some Asian and Latin American countries is predicted to quadruple by 2050.

It is projected a relative increase on 10% over two decades (by year 2032) in the occurrence of "any OA" that leads to healthcare consultation considering the future age and sex structure of the population as well as the prevalence of overweight and obesity. By 2032, over 26,000 new OA cases per 1000,000 population aged 45 or older will have consulted healthcare. To a large extent, it will be the primary care physicians who will face the increased workload, but a crisis in supply of total joint replacement surgery is also anticipated [5].

Basic science and new technologies are being developed aiming the drastic change of OA natural history and inevitable progression. Available statistics are based in current concepts and knowledge about the disease, and predictions can be modified according to new treatment options.

5.2 Physiopathology

5.2.1 OA Physiopathology

Articular cartilage damage is claimed to be the initial event in the development of osteoarthritis. This specialized tissue must be under controlled conditions of homeostasis in order to complete its load-bearing functions. Its properties are mostly due to the organization of the extracellular matrix, mainly composed by collagen (responsible for the volume and shape) and proteoglycan (responsible for the elasticity). After receiving the mechanical loads, the extracellular matrix dissipates it, and only then it is transmitted to the chondrocytes [6].

Physiological loads are crucial for the organization of the articular cartilage, allowing the homeostasis between anabolic and catabolic reactions. OA is believed to happen from an imbalance in these reactions [6]. At first, there is an increase in anabolic activities with greater production of type II collagen and proteoglycan in response to any derangement in the normal physiology of the knee that causes loss of glycosaminoglycan. Additionally, this will lead to changes in the cartilage resistance. In the early phase of OA, there is an increase in catabolic reactions along with an increase of inflammatory mediators, proteinases, and stress phase reactants promoted by the fragmented type II collagen which will ultimately culminate in cartilage loss [7].

5.2.2 Relation to Genetic Factors and Gender

OA is most commonly defined as idiopathic when no factor is known for its occurrence [8]. Many studies have been carried in order to understand the role of genetics in the development of OA. Knee OA is not inherited according to the Mendelian pattern, suggesting a much more complex genetic pattern [9]. Still, the sibling recurrence risk for knee OA is 2.08–2.31 for radiographic knee OA, with a risk of 2.8–4.8 for total knee replacement [10]. The increased risk of knee OA seems to persist even after adjusting for environmental risk factors [11]. Many genes have been claimed to be related to the disease, namely, those coding for structural proteins, such as collagen type II [12–14]. A deeper knowledge of the involved genes in OA might help defining treatment targets in order to prevent late-stage OA [8].

The prevalence of OA also varies with genders, with women being more affected [15]. In fact, articular cartilage is hormone sensitive and specific genes coding for estrogen receptors increase the risk of OA when in homozygota [16]. Anatomical differences also exist between genders and might play a role in the development of the disease. In this sense, women tend to have a thinner layer of articular cartilage [17], Q angle around 3° higher than that of men [18], and narrower distal femur [19], and there is also a difference in the femur and tibial plate sizes [20, 21].

5.2.3 Relation to Meniscal Tears

Arthroscopic partial meniscectomy is one of the most commonly performed orthopedic interventions around the world [22, 23]. Incidence of meniscal tears is around 2 per 1000 patients/year in the Netherlands and are responsible for about 25,000 hospital admissions every year in England and Wales [24, 25]. Arthroscopic partial meniscectomy is commonly applied due to its short-term relief of symptoms [26]. However, it is crucial to preserve as much as meniscal tissue as possible since a disturbance on the close relation between the menisci and the articular cartilage may contribute to accelerate the process of osteoarthritis development [27]. In fact, to account to the lack of meniscal tissue coverage at the central region of the tibia, the subchondral bone plate is thicker to support the role of the meniscal structures in shock absorption and load transmission [28].

Partial meniscectomies induce less radiographic apparent OA than total meniscectomy, with worst outcomes in patients with lateral meniscectomy [29]. Moreover, it is well documented that OA develops faster after a meniscal tear or after partial meniscectomy [6]. In young, active patients without cartilage damage, partial meniscectomy and meniscus replacement have shown beneficial effects in long-term studies [27].

Partial meniscectomy is indicated in patients not responding to pharmacological and physical therapies or patients showing symptoms such as blocking or locking due to instable meniscal lesions [27]. Still, it is crucial to restrict the amount of meniscal resection to the unstable meniscal tissue, without removing the degenerated but stable tissue. This will prevent the eventual progression of OA caused by decreased joint contact area and increased peak contact pressure [30].

5.2.4 Relation to Axis Deviation

Lower limb axis deviations are responsible for the asymmetric load distribution across the knee joint [31]. A systematic review on this topic concluded that there is still little evidence establishing a causative effect between axis deviation and OA [32]. Brouwer et al. concluded in their study that both varus and valgus malalignments are related to incident OA [33] and considered as independent risk factor for progressive OA [32].

5.2.5 Relation to Joint Instabilities

Knee joint instability has been considered to be a main contributing factor for knee OA. In this sense, joint stability is "the ability of a joint to maintain an appropriate functional position throughout its range of motion" [34]. In an unstable joint, the load-bearing area is moved to the periphery, overloading this region of the articular cartilage [35]. Patient-reported knee instability referred as "feeling of giving way, shifting or buckling of the knee during daily activities" has systematically been reported in patients with knee OA [36–38].

The anterior cruciate ligament (ACL) rupture is the most common ligament injury in the knee [39–41]. A systematic review showed that knee radiologic changes are fivefold more common after an ACL injury when compared to the contralateral knee, with a relative risk of 3.89 for the development of knee OA [42]. Within this line, a significant long-term prevalence of knee OA after ACL reconstruction up to 80% has been reported [43, 44]. Nevertheless, those patients treated with ACL reconstruction seem to have a lower risk of knee OA than those treated conservatively (relative risk 3.62 vs 4.98) [42].

5.2.6 Relation to Cartilage Defects

Articular cartilage defects are a well-known pathological entity that have been hypothesized to be closely related to early OA [45]. These can be divided according to their depth, into partial-thickness and full-thickness lesions [35].

Focal cartilage defects are common, being found in approximately 20% of knee arthroscopies [46, 47] and in 40% of magnetic resonance imaging (MRI) studies in healthy individuals [48]. This prevalence is considerably higher when accounting for recreational and professional athletes [49–52]. However, there is insufficient evidence on when and how these lesions evolve into OA [35]. Cicuttini et al. [45] demonstrated that patients with tibiofemoral cartilage defects showed significantly larger reductions in both tibial and femoral cartilage volume after 2 years of follow-up, with medial compartment defects leading to a greater loss of articular cartilage. Even in asymptomatic healthy individuals with non-full-thickness medial tibiofemoral cartilage defects, without the radiographic knee OA, the presence of knee cartilage defects is most likely going to lead to further loss of knee cartilage [45, 53]. However, longitudinal studies in symptomatic patients have provided conflicting findings [54–57].

5.2.7 Relation to Sports

Sports practice has long been associated with the premature occurrence of OA [58]. In fact, there are many studies reporting early OA in athletes that practice football but also running, dance, tennis, and other high-impact sports [59–61]. Still, a causative association between running and knee OA has not been found and, in fact, there is no association between running and knee OA diagnosis and it has even been suggested a protective effect against surgery [62–64]. In this sense, either a more sedentary lifestyle or long exposure to high-volume and/or high-intensity running are both associated with knee OA [65]. Hence, incidence of knee OA should rather be more closely related to higher volume/intensities of exercise and to the sport-specific movement patterns.

The most affected athletes are those that play at elite levels and those practicing pivoting sports with rapid acceleration/deceleration and high-impact loads on the joint [66, 67]. Within this line, football players seem to develop OA 4–5 years earlier than the general population, at a higher rate (5–12 times more frequent), with incidences varying between 16% and 80% [59, 60] The pathophysiological mechanism under the relationship between sports and OA is the magnitude of loads transmitted during sports activities. For instance, standing from a chair produces a load 3.3-fold the body weight to the knee. However, sports activities can produce even higher loads as, for instance, jumping, which applies a 20-fold load the body weight, often overloading the knee [58].

5.3 Diagnosis

The diagnosis of early knee OA is complex and multifactorial. In earlier phases, the common signs and symptoms may still be cloaked and appear infrequently. Moreover, the radiographic evaluation to establish the OA is more challenging and arthroscopy procedures are seldom used for solely diagnostic purposes. In these earlier phases, the use of MRI is desirable as it is capable of detecting the whole spectrum of pathological joint tissue changes. In this sense, and according to the criteria defined by Luyten and colleagues [68] and further adapted by Madry et al. [69], early knee OA is present when three main criteria are fulfilled:

1. Presence of knee pain (at least two episodes of pain for more than 10 days in the last year)
2. Standard radiographs Kellgren and Lawrence grades 0–2 (standing weight-bearing position with knees in approximately 20° of flexion and the feet in 5° of external rotation; the radiographs should be done bilaterally from a posteroanterior view in the frontal plane)
3. At least one of the proposed functional criteria:

    (a) MRI findings demonstrating articular cartilage degeneration and/or menis-cal degeneration and/or subchondral bone marrow lesions, with at least two of the following findings:

        (i) Cartilage morphology WORMS 3–6
        (ii) Cartilage BLOKS grades 2 and 3
        (iii) Meniscus BLOKS grades 3 and 4
        (iv) Bone marrow lesions WORMS 2 and 3

Fig. 5.1 Arthroscopic view of the knee joint showing softening and fibrillation of the tibial plateau cartilage with early OA

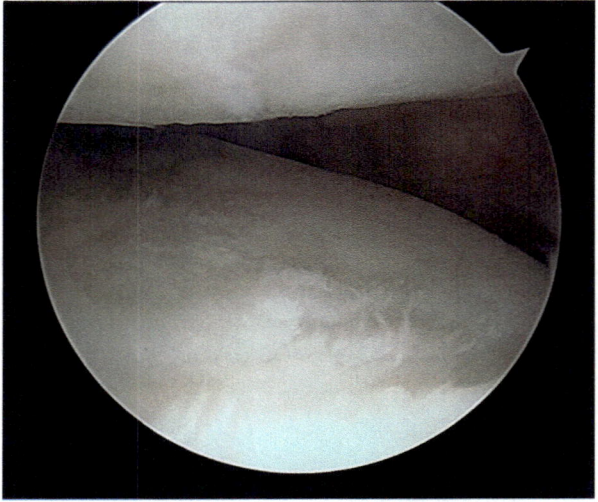

(b) Arthroscopic findings of cartilage lesions (ICRS grades 1–4 in at least two compartments or grades 2–4 in one compartment with surrounding softening and swelling) as shown in Fig. 5.1

Specific classification systems for early knee OA are still under development, and for diagnosis the abovementioned criteria have been used.

5.3.1 Clinical Evaluation

Knee pain is the main symptom of knee OA and should be present for at least 10 days and at least twice in the past, in order to suggest a diagnose of early knee OA [68, 69]. The pattern of pain is also associated with the severity of OA and usually divided into two stages. The first, happening in early phases of the pathology, is mainly associated with local tissue response and peripheral nociceptor activation [70]. It is mostly associated with activity [71], especially when climbing up or going down the stairs [72]. The second type of pain is often present in later phases of the disease, consisting of a persistent unpredictable pain [71]. This is mostly associated with central neural processing, as hyperexcitability of the central nervous system often occurs in these patents due to chronic nociceptor stimulation [70, 73].

In addition to pain, most patients also report swelling and stiffness. Knee swelling is most often present after sports participation and stiffness after a period of inactivity (usually in the morning or in the late evening). Knee stiffness is often resolved after a few minutes of exercise. Limitations on the knee range of motion, muscle weakness, decreased knee proprioception, and increased knee laxity are also often present in these patients and are associated with functional activity restrictions. Evaluation of joint space narrowing and crepitus is also important to be included in the clinical examination [69].

5.3.2   *Radiography*

When the clinical history and examination are suggestive of knee OA, imaging studies should be obtained in order to confirm the diagnosis and assess the extent of the damage. In this sense, radiographs are the gold standard for the diagnosis of OA due to their relative low cost and represent a simple and easily available exam [74].

The commonly used study series in the evaluation of knee OA are similar to the one of other joints, consisting on two radiographs obtained in perpendicular planes to each other [75]. Anterior–posterior (AP) view should be weight-bearing to allow adequate visualization of the joint space (Fig. 5.2). This is mostly important when the joint narrowing is still incipient, as the standing position allows a greater apposition of the articular surfaces of the femur and tibia [76]. According to Ahlback [77], abnormal joint narrowing occurs when, in the standing position, the joint space is 3 mm or less or is less than 50% of the other healthy knee. An asymmetric joint space is usually the initial change observed in radiographs [75]. If one of the knee compartments has a thinner articular cartilage, the joint will narrow on weight-bearing. On the other hand, the other side of the joint might open 1–2 mm due to weight transfer from one compartment to the other [76]. As the disease progresses, other typical changes occur, including the formation of cysts, subchondral sclerosis, and osteophytes [75].

When solely an AP or PA view is obtained, about 4–7% of OA cases can be missed. The addition of a lateral or a skyline view can increase diagnostic sensibility

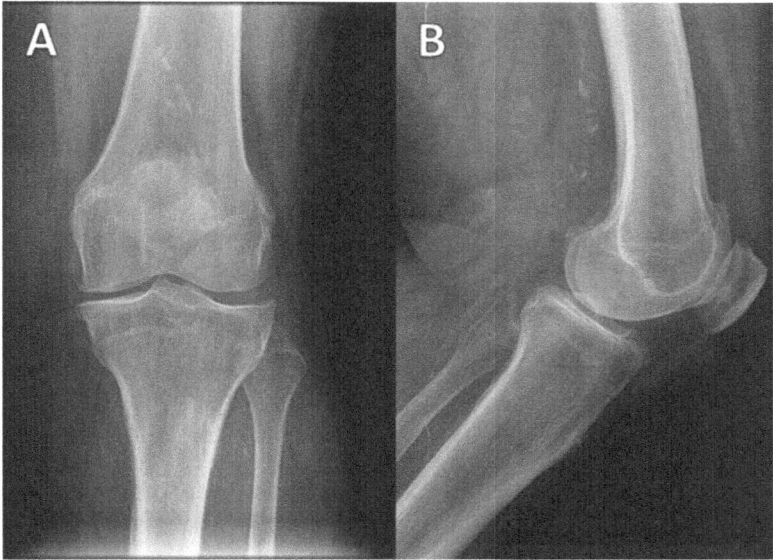

Fig. 5.2 Weight-bearing radiography of the knee showing signs of joint degeneration suggesting early knee OA. (**A**) AP view showing joint line narrowing in the medial compartment and presence of osteophyte. (**B**) Lateral view showing patellofemoral joint degeneration

(from 73% to 96–97% – when OA is defined as the presence of any osteophyte). The former might be easier to obtain for the simplicity of its acquisition, sparing the need for a highly trained technician [78].

The gold standard radiographic classification is the Kellgren and Lawrence grading, subgrouping the arthritic changes into five groups [68, 69]:

- *Grade 0:* no changes
- *Grade 1:* possible narrowing of the joint space and possible osteophytic process
- *Grade 2:* osteophytes and possible narrowing of the joint space
- *Grade 3:* osteophytes, narrowing of the joint space, some degree of sclerosis and possible bone deformity
- *Grade 4:* large osteophytes, marked narrowing of the joint line, severe sclerosis and bone deformity.

Nevertheless, it has to be taken into account that this classification lacks precision, especially in the evaluation of earlier stages of OA. Still, Kellgren–Lawrence grades 0–2 (osteophytes only) are suggestive of initial stages of knee OA [69]. Additionally, AP long-axis views are centered at the knee and include the full length of the limb. They are also acquired in the standing position. These are adequate for the evaluation of axis alignment as well as leg length discrepancy [79], whose main purpose is to obtain an adequate preoperative planning for either osteotomies or total knee replacements [80]. Hence, this view might be of interest only upon surgical planning and postoperative follow-up.

5.3.3 MRI

The MRI due to its capability of detecting the whole spectrum of pathological joint tissue changes is the preferable imaging procedure to detect early knee OA. MRI is a noninvasive imaging study that has the advantage of allowing the evaluation of soft tissue with high contrast, thus providing information on the status of the articular cartilage, menisci, and ligaments [81]. On the other hand, knee radiographs allow the detection of joint space narrowing, which is an indirect measurement of the amount of articular cartilage. However, it is known that changes happen at the level of the cartilage long before any manifestation is detected on the radiographs. In this sense, the first radiographic abnormalities are only visible when more than 10% of the cartilage volume is lost [82]. In this sense, cartilage loss usually starts around the 40 years old with normal radiological findings, with about 5% of the cartilage volume is lost each year in patients with established OA [83].

MRI identifies with more feasibility patients with tricompartmental OA, whose radiographs or CT scans were misleading, showing only bicompartmental OA [81]. However, the question that is raised is whether there is a favorable benefit/cost relation concerning the evaluation of these patients using MRI [84]. As a matter of fact, when considering early OA of the knee, especially if considering high-demand individuals, MRI can provide crucial information, namely, a better knowledge of the stage of the disease and the diagnosis of soft-tissue injuries closely related to the development of OA (such as meniscal tears or ligament rupture). Moreover, it allows the diagnosis of early stages of OA without radiographic repercussion [81] and may play an important role in classification systems [68].

5.4   Treatment in Early OA

5.4.1   Objectives

The objectives of the available treatment options are to decrease pain and swelling, improve knee function, allow the return to sports or working activities, and delay the disease progression.

5.4.2   Daily Life Changes

- *Weight Management*

The American College of Rheumatology and the OA Research Society International recommend overweight patients with knee OA to implement strategies to lose weight [85, 86]. In this sense, the weight loss might reduce pain and physical disability in these patients [87], and a weight loss of 0.25% per week should be achieved [85].

- *Exercise*

The American College of Rheumatology recommends patients with symptomatic knee OA to comply with exercise programs [86]. Aquatic exercise seems to have benefits in both function and quality of life, although improvement in pain is only discrete [88, 89]. In this sense, land-based exercises seem to achieve higher improvements in pain and physical function [85], providing short-term benefits that are often sustained for 2–6 months after treatment cessation and comparable with the estimates reported for nonsteroidal anti-inflammatory drugs (NSAIDs) [90]. Strength training has also been associated with an improvement in pain and physical function [91]. In this sense, isometric, isotonic, and isokinetic strengthening and aerobic activities (e.g., walking, cycling, or jogging) have similar outcomes, and the choice will depend upon the specific and individual patient's characteristics [92]. Within this line, education of the patient plays a fundamental role to increase the patient's adherence, compliance, and long-term maintenance of regular exercise implying a social commitment and continuous improvement of self-management skills [93–95].

In summary, the patient should adequate the exercise to his physical condition. Deconditioned patients should consider initiating an aquatic exercise program and progress to land-based exercises and strength training as the aerobic conditioning improves [86].

Several randomized controlled trials have showed the efficacy of muscle strengthening exercises with or without weight-bearing and aerobic exercises for pain relief in people with knee osteoarthritis [96–99]. In this sense, it has been recommended that the therapeutic exercise programs focus on improving aerobic capacity, quadriceps muscle strength, and lower extremity performance, carried out three times a week under supervision [97]. Additionally, since pain and psychological distress often lead to avoidance of activities and consequently to muscle weakness and activity restrictions [100, 101], education of the patient should also be included and associated to the other treatments being applied aiming the enhancement of the outcomes achieved [102–104].

5.4.3 Pharmacological Interventions

- *NSAIDs and Tramadol*

For patients with symptomatic OA, the American Academy of Orthopaedic Surgeons (AAOS) recommends the use of NSAIDs or tramadol. On the other hand, the use of acetaminophen, opioids, and pain patches is not recommended, as the effect is not shown to be superior to placebo [105]. Among the numerous pharmaceutical agents, acetaminophen is usually recommended as a first-line therapy for the management of OA pain; however, when it is ineffective, the NSAIDs are the second choice [106, 107]. In a network meta-analysis, comparing three different classes of pharmacological interventions, including oral NSAIDs (diclofenac, naproxen, piroxicam), potent opioids (hydromorphone, oxycodone), or less potent opioids (tramadol) for knee OA, found that the three approaches provided similar outcomes in terms of pain reduction and functional improvements [108].

- *Glucosamine and Chondroitin*

The OA Research Society International (OARSI) does not recommend the use of either chondroitin or glucosamine for disease modification, due to conflicting results reported. Moreover, their use for symptom relief are still uncertain [85].

- *Viscosupplementation*

Hyaluronic acid is a component of human cartilage and synovial fluid [109], with a normal concentration in the adult knee of 2.5–4.0 mg/mL [110]. It has an important role in shock absorption and lubrication [111]. In patients with OA, a depolymerization of hyaluronic acid occurs, and it is cleared faster [112, 113], leading to a loss of the normal viscoelasticity of the synovial fluid [109]. Intra-articular injections of hyaluronic acid have the main goal of lubricating the joint and restoring viscoelasticity [109]. Additional supposed effects are analgesia, anti-inflammatory effect, and eventual chondroprotection [110]. Nevertheless, scientific evidence is not clear regarding the benefits of hyaluronic acid. In fact, studies have contradictory results [109], which led the OARSI to provide an uncertain recommendation for the use of hyaluronic acid in patients with knee OA [85]. The AAOS, on the other hand, recommends against the use of hyaluronic acid for symptomatic knee OA, as scientific literature that shows beneficial effects did not achieve the minimum clinically important improvement thresholds [105]. Before the recommendation of its use, the true benefits of intra-articular injections of hyaluronic acid in early stages of knee OA need to be assessed through well-designed randomized controlled trials.

- *Intra-articular Corticosteroids*

The OARSI considers the use of intra-articular corticosteroids adequate for pain relief in patients with OA [85]. On the other hand, the AAOS does not recommend for or is against its use, as studies testing intra-articular corticosteroids display methodological limitations and conflicting results [105].

The effects of corticosteroid injections usually last up to 4 weeks. Due to the deleterious effects of large doses and its continuous use, their use should respect the empirical 3-month rule (time lapse from each injection) [114]. Additionally, their use should be limited in patients presenting early OA with prevalent synovitis after the failure of common nonsurgical therapies [106].

5.4.4 Surgical Treatments

- *Arthroscopic Procedures*

Arthroscopic procedures for treating knee OA include joint lavage and debridement which, theoretically, remove debris and inflammatory cytokines from inside the joint [115, 116]. Additionally, loose bodies can also be removed [117]. Despite the theoretical advantages and its wide use, the role of arthroscopy in knee OA is still controversial [117, 118]. In fact, systematic reviews and meta-analyses have shown that joint lavage alone or in combination with steroid injection do not lead to clinically relevant benefits in patients with knee OA and there is no evidence that arthroscopic debridement has benefit for OA [119–121].

Adequate patient selection is crucial for the success of these interventions. Those that mostly benefit from this surgery are middle-aged patients with early OA or those with meniscal tears and/or cartilage flaps which did not responded to conservative measures [117]. Arthroscopic debridement provides only short-term symptoms relief and should not be used routinely [117, 122].

- *Cartilage Repair and Reconstruction Techniques*

Techniques for cartilage repair have been proposed due the knowledge of the lack of repair capacity of the articular cartilage [123]. Different techniques have been described including drilling and microfractures, autologous chondrocyte implantation, and cell-free implant or even replacement techniques, such as autologous or allographic osteochondral transplantation (either single plug or mosaicplasty) [117, 124, 125]. However, these cartilage procedures have limited indications, as those patients that might benefit from them do not actually have OA but small-sized cartilage lesions that may evolve to generalized OA. It must be taken into account that the technical success will be limited by other factors promoting OA, including limb malalignment, ligamentous instability, or patella maltracking, which must be addressed properly to delay the development of the OA [117].

In summary, although cartilage repair and reconstruction techniques have shown good to excellent outcomes in focal articular cartilage injuries of the knee, their role or early knee OA is still limited as they do not address other key aspects of early OA, such as degenerative and inflammatory joint environment or unfavorable genetic factors [124].

- *Meniscal Replacement*

Meniscal replacement using cell-free scaffolds following partial meniscectomy have been proposed as a viable option for symptomatic OA, aiming the improvement of symptoms while potentially delaying the progression of joint degeneration. While it is optimal for earlier stages of knee OA, their use is not recommended for grade 3–4 OA [27].

Meniscal allograft replacement is indicated in patients with a previous total meniscectomy presenting localized pain in the meniscus-deficient compartment (Fig. 5.3). Additionally, the surgeon must certify that the patient has a stable knee joint, no malalignment, and the articular cartilage with only minor evidence of degenerative changes (under grade 3 according to the ICRS score) [27]. However, there is still little evidence showing that meniscal allograft replacement reduces the progression of osteoarthritis and that the allographic meniscal tissue is as effective as the native meniscus [126].

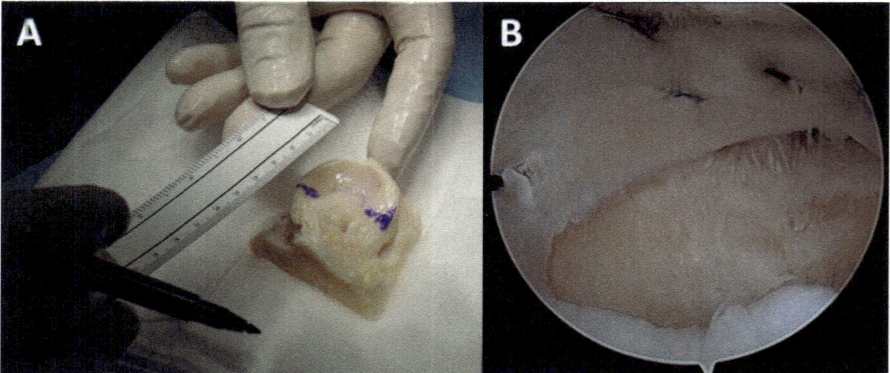

Fig. 5.3 Meniscal allograft transplantation. (**A**) Measuring the meniscal allograft. (**B**) All-inside meniscal suture

These findings highlight the important role of the menisci in the load transmission and protection of the underlying articular cartilage [29]. Hence, to prevent and delay the onset or development of knee OA, meniscal injuries should be treated whenever indicated, and meniscal tissue should be preserved as much as possible.

- *Load-Modifying Procedures*

  (a) Osteotomies

Osteotomies have been performed since the nineteenth century [127] and are indicated for unicompartmental OA with an associated varus or valgus deformity [117]. The osteotomy can be either femoral or tibial and open or close-wedged. This surgical technique is highly differentiated and more demanding than a total knee replacement [117]. The objective of the procedure is to change the mechanical axis of the lower limb, in order to transfer load from the more damaged compartment to the healthy one, slowing joint degeneration [128].

As for other surgical procedures, the most important to obtain a good clinical result is a good patient selection. The patient should be younger than 60–65 and should not have symptomatic femoropatellar OA, and the knee should be stable. However, if an ACL tear exists, a reconstruction can be made at the same time as the osteotomy [129]. Though, these interventions should be avoided in patients that are overweighed or have chondrocalcinosis [117].

Joint unloading through high tibial osteotomy (HTO) has been traditionally applied as conservative intervention to delay the OA progression. Better outcomes are reported when the HTO is applied in earlier OA stages [130]. A Cochrane systematic review showed that HTO is effective in reducing pain and improving knee function in patients with medial knee OA; however, there is still lack of comparative studies against other surgical options or nonoperative treatment to assess if HTO is more effective than other treatments for knee OA [131]. Nonetheless, HTO has showed positive outcomes [132] with a survival rate of 70–97.6% at 10 years [133]. The survival rate is increased when the HTO is performed in association with a cartilage procedure, but it is well known that in both approaches, the outcomes deteriorate overtime [134, 135].

(b) Knee Distraction

This surgical procedure uses external fixators to make distraction over the knee joint [118]. It was initially used for the treatment of malalignment and joint contracture [136]. The objective is the reduction of mechanical stress, allowing for cartilage repair [118]. Many devices have been developed for this purpose showing promising early results [137–139, 136]. Interestingly, MRI studies showed that denuded bone area diminished in size after knee distraction, suggesting a mechanism of cartilage repair [136]. Patients also seem to have symptomatology benefits from this intervention [118].

- *Joint Replacement*

Knee replacement is not routinely indicated in patients with early knee OA, and if performed, the achieved results are worse [140, 141]. Total or unicompartmental knee arthroplasty is more indicated for end-stage joint OA.

5.4.5 Biological Treatments: Injections

- *PRP (Platelet-Rich Plasma)*

Platelets contain alpha-granules comprising a variety of growth factors which can change these events of OA [142]. In fact, PRP decreases catabolic reactions, promoting anabolism and chondral remodeling [143]. Additionally, the inflammatory response is downregulated [144, 145], which might justify the pain improvement seen in these patients [143].

There is a trend toward the study of PRP in OA, with most studies showing promising results in the improvement of pain and knee scores [146–150]. Nevertheless, many controversies still exist about the treatment with PRP, namely, the adequate dosage schedule. Literature diverges in this matter, with authors varying not only the number of injections but also the optimal timing, frequency, and dose–response, preparation methods and reporting of injection procedures, platelet concentration and growth factor levels, confounders (including degree of injury, comorbidities, use of anti-inflammatory agents), and variability in control groups (e.g., anti-inflammatory drugs or saline). According to Dhillon et al. [143], single injections achieve results as good as those for two injections, but these results are not consistent as Görmelli et al. [151] showed that knee improvement was better with three injections.

The addition of PRP to articular cartilage surgical procedures through PRP-augmented scaffolds have shown promising beneficial results in the articular cartilage repair process in animals and humans based on macroscopic, histologic, and biochemical analysis and based on clinical outcome scores, respectively [152].

PRP seems to be a potential tool in the treatment of symptomatic OA; however, guidelines for its use are still scarce. As a matter of fact, the AAOS guidelines for the treatment of knee OA state that the use of PRP is inconclusive, since the existing studies do not provide a comparative analysis of clinical effectiveness [105].

- *Stem Cells*

Mesenchymal stem cells have the ability to proliferate, regenerate, and differentiate into more specialized cells [153, 154]. They have been widely used for their additional anti-inflammatory properties, along with their ability to stimulate cell repair and proliferation [154]. Studies have been conducted in order to test the safety and benefits of stem cell injection in knee OA.

Clinical trials seem to lead to positive and promising results in patients with knee OA [155–159]. Nevertheless, in the absence of high-level evidence of its effectiveness, the use of stem cell therapy for knee OA is still not recommended [160].

5.5 Conclusions

Knee OA is a common disease that affects an important percentage of the population throughout their life. Nowadays, with an increased trend for sports practice, OA occurs at younger ages, and people are less amenable to change their lifestyles, stimulating the development of novel and more effective approaches. Classification systems adequate for early OA are still lacking, and treatment is still insufficient for the requirements of an active population. Nonsurgical treatment is often the first-line option and is mainly based on daily life adaptations, weight loss, and exercise, with pharmacological agents having only a symptomatic role. Surgical treatment shows consistent results and however is more invasive and more prone to complications. The treatment of early OA is an emerging field due to the high incidence of the pathology and high requirements of the affected individuals.

References

1. Brandt K, Radin E, Dieppe P, Van De Putte L (2006) Yet more evidence that osteoarthritis is not a cartilage disease. Ann Rheum Dis 65(10):1261–1264
2. Lawrence RC, Felson DT, Helmick CG, Arnold LM, Choi H, Deyo RA, Gabriel S, Hirsch R, Hochberg MC, Hunder GG (2008) Estimates of the prevalence of arthritis and other rheumatic conditions in the United States: part II. Arthritis Rheumatol 58(1):26–35
3. World Health Organization. Chronic diseases and health promotion. http://www.who.int/chp/topics/rheumatic/en/. Accessed 18/10/2017
4. He W, Goodkind D, Kowal P (2016) Census Bureau, international population reports. P95/16–1, An Aging World: 2015. US Government Publishing Office, Washington, DC
5. Fehring TK, Odum SM, Troyer JL, Iorio R, Kurtz SM, Lau EC (2010) Joint replacement access in 2016: a supply side crisis. J Arthroplast 25(8):1175–1181
6. Madry H, Luyten FP, Facchini A (2012) Biological aspects of early osteoarthritis. Knee Surg Sports Traumatol Arthrosc 20(3):407–422
7. Favero M, Ramonda R, Goldring MB, Goldring SR, Punzi L (2015) Early knee osteoarthritis. RMD Open 1(Suppl 1):e000062
8. Clement N (2013) Is osteoarthritis of the knee hereditary? A Review of the Literature. Hereditary Genet 1:2161–1041
9. Fernández-Moreno M, Rego I, Carreira-Garcia V, Blanco FJ (2008) Genetics in osteoarthritis. Curr Genomics 9(8):542–547
10. Valdes AM, Spector TD (2011) Genetic epidemiology of hip and knee osteoarthritis. Nat Rev Rheumatol 7(1):23–32
11. Neame R, Muir K, Doherty S, Doherty M (2004) Genetic risk of knee osteoarthritis: a sibling study. Ann Rheum Dis 63(9):1022–1027
12. Cicuttini FM, Spector TD (1996) Genetics of osteoarthritis. Ann Rheum Dis 55(9):665–667
13. Mustafa Z, Chapman K, Irven C, Carr A, Clipsham K, Chitnavis J, Sinsheimer J, Bloomfield V, McCartney M, Cox O (2000) Linkage analysis of candidate genes as susceptibility loci for osteoarthritis—suggestive linkage of COL9A1 to female hip osteoarthritis. Rheumatology (Oxford) 39(3):299–306
14. Zhai G, Rivadeneira F, Houwing-Duistermaat J, Meulenbelt I, Bijkerk C, Hofman A, van Meurs J, Uitterlinden A, Pols H, Slagboom P (2004) Insulin-like growth factor I gene promoter polymorphism, collagen type II α1 (COL2A1) gene, and the prevalence of radiographic osteoarthritis: the Rotterdam Study. Ann Rheum Dis 63(5):544–548
15. Peyron J, Altman R (1984) The epidemiology of osteoarthritis. In: Osteoarthritis diagnosis and treatment. WB Saunders, Philadelphia, pp 9–27
16. O'Connor MI (2006) Osteoarthritis of the hip and knee: sex and gender differences. Orthop Clin North Am 37(4):559–568
17. Faber S, Eckstein F, Lukasz S, Mühlbauer R, Hohe J, Englmeier K-H, Reiser M (2001) Gender differences in knee joint cartilage thickness, volume and articular surface areas: assessment with quantitative three-dimensional MR imaging. Skelet Radiol 30(3):144–150

18. Hsu RW, Himeno S, Coventry MB, Chao EY (1990) Normal axial alignment of the lower extremity and load-bearing distribution at the knee. Clin Orthop Relat Res 255:215–227

19. Poilvache PL, Insall JN, Scuderi GR, Font-Rodriguez DE (1996) Rotational landmarks and sizing of the distal femur in total knee arthroplasty. Clin Orthop Relat Res 331:35–46

20. Hitt K, Shurman JR, Greene K, McCarthy J, Moskal J, Hoeman T, Mont MA (2003) Anthropometric measurements of the human knee: correlation to the sizing of current knee arthroplasty systems. J Bone Joint Surg Am 85(suppl 4):115–122

21. Fernandes MS, Pereira R, Andrade R, Vasta S, Pereira H, Pinheiro JP, Espregueira-Mendes J (2017) Is the femoral lateral condyle's bone morphology the trochlea of the ACL? Knee Surg Sports Traumatol Arthrosc 25(1):207–214

22. Abrams GD, Frank RM, Gupta AK, Harris JD, McCormick FM, Cole BJ (2013) Trends in meniscus repair and meniscectomy in the United States, 2005-2011. Am J Sports Med 41(10):2333–2339

23. Mitchell J, Graham W, Best TM, Collins C, Currie DW, Comstock RD, Flanigan DC (2016) Epidemiology of meniscal injuries in US high school athletes between 2007 and 2013. Knee Surg Sports Traumatol Arthrosc 24(3):715–722

24. Baker P, Coggon D, Reading I, Barrett D, McLaren M, Cooper C (2002) Sports injury, occupational physical activity, joint laxity, and meniscal damage. J Rheumatol 29(3):557–563

25. Draijer L, Belo J, Berg H, Geijer R, Goudswaard A (2010) Summary of the practice guideline Traumatic knee problems'(first revision) from the Dutch College of General Practitioners. Ned Tijdschr Geneeskd 154:A2225–A2225

25. Paxton ES, Stock MV, Brophy RH (2011) Meniscal repair versus partial meniscectomy: a systematic review comparing reoperation rates and clinical outcomes. Arthroscopy 27(9):1275–1288

27. Verdonk R, Madry H, Shabshin N, Dirisamer F, Peretti GM, Pujol N, Spalding T, Verdonk P, Seil R, Condello V (2016) The role of meniscal tissue in joint protection in early osteoarthritis. Knee Surg Sports Traumatol Arthrosc 24(6):1763–1774

28. Ziegler R, Goebel L, Seidel R, Cucchiarini M, Pape D, Madry H (2015) Effect of open wedge high tibial osteotomy on the lateral tibiofemoral compartment in sheep. Part III: analysis of the microstructure of the subchondral bone and correlations with the articular cartilage and meniscus. Knee Surg Sports Traumatol Arthrosc 23(9):2704–2714

29. Englund M, Lohmander L (2004) Risk factors for symptomatic knee osteoarthritis fifteen to twenty-two years after meniscectomy. Arthritis Rheumatol 50(9):2811–2819

30. Peretti GM, Gill TJ, Xu J-W, Randolph MA, Morse KR, Zaleske DJ (2004) Cell-based therapy for meniscal repair. Am J Sports Med 32(1):146–158

31. Gomoll AH, Angele P, Condello V, Madonna V, Madry H, Randelli P, Shabshin N, Verdonk P, Verdonk R (2016) Load distribution in early osteoarthritis. Knee Surg Sports Traumatol Arthrosc 24(6):1815–1825

32. Tanamas S, Hanna FS, Cicuttini FM, Wluka AE, Berry P, Urquhart DM (2009) Does knee malalignment increase the risk of development and progression of knee osteoarthritis? A systematic review. Arthritis Care Res (Hoboken) 61(4):459–467

33. Brouwer G, Van Tol A, Bergink A, Belo J, Bernsen R, Reijman M, Pols H, Bierma-Zeinstra S (2007) Association between valgus and varus alignment and the development and progression of radiographic osteoarthritis of the knee. Arthritis Rheumatol 56(4):1204–1211

34. Burstein A, Wright T (1994) Joint stability. In: Burstein A, Wright T (eds) Fundamentals of orthopedic biomechanics, 1st edn. Williams & Wilkins, Baltimore, pp 63–93

35. Heijink A, Gomoll AH, Madry H, Drobnič M, Filardo G, Espregueira-Mendes J, Van Dijk CN (2012) Biomechanical considerations in the pathogenesis of osteoarthritis of the knee. Knee Surg Sports Traumatol Arthrosc 20(3):423–435

36. Fitzgerald GK, Piva SR, Irrgang JJ (2004) Reports of joint instability in knee osteoarthritis: its prevalence and relationship to physical function. Arthritis Care Res (Hoboken) 51(6):941–946

37. van der Esch M, Knoop J, van der Leeden M, Voorneman R, Gerritsen M, Reiding D, Romviel S, Knol DL, Lems WF, Dekker J (2012) Self-reported knee instability and activity

limitations in patients with knee osteoarthritis: results of the Amsterdam osteoarthritis cohort. Clin Rheumatol 31(10):1505–1510

38. Knoop J, Van Der Leeden M, Van Der Esch M, Thorstensson CA, Gerritsen M, Voorneman RE, Lems WF, Roorda LD, Dekker J, Steultjens MP (2012) Association of lower muscle strength with self-reported knee instability in osteoarthritis of the knee: results from the Amsterdam Osteoarthritis Cohort. Arthritis Care Res (Hoboken) 64(1):38–45

39. Clayton RA, Court-Brown CM (2008) The epidemiology of musculoskeletal tendinous and ligamentous injuries. Injury 39(12):1338–1344

40. Majewski M, Susanne H, Klaus S (2006) Epidemiology of athletic knee injuries: a 10-year study. Knee 13(3):184–188

41. Griffin LY, Albohm MJ, Arendt EA, Bahr R, Beynnon BD, Demaio M, Dick RW, Engebretsen L, Garrett WE Jr, Hannafin JA, Hewett TE, Huston LJ, Ireland ML, Johnson RJ, Lephart S, Mandelbaum BR, Mann BJ, Marks PH, Marshall SW, Myklebust G, Noyes FR, Powers C, Shields C Jr, Shultz SJ, Silvers H, Slauterbeck J, Taylor DC, Teitz CC, Wojtys EM, Yu B (2006) Understanding and preventing noncontact anterior cruciate ligament injuries: a review of the Hunt Valley II meeting, January 2005. Am J Sports Med 34(9):1512–1532

42. Ajuied A, Wong F, Smith C, Norris M, Earnshaw P, Back D, Davies A (2014) Anterior cruciate ligament injury and radiologic progression of knee osteoarthritis: a systematic review and meta-analysis. Am J Sports Med 42(9):2242–2252

43. Risberg MA, Oiestad BE, Gunderson R, Aune AK, Engebretsen L, Culvenor A, Holm I (2016) Changes in knee osteoarthritis, symptoms, and function after anterior cruciate ligament reconstruction: a 20-year prospective follow-up study. Am J Sports Med 44(5):1215–1224

44. Øiestad BE, Holm I, Aune AK, Gunderson R, Myklebust G, Engebretsen L, Fosdahl MA, Risberg MA (2010) Knee function and prevalence of knee osteoarthritis after anterior cruciate ligament reconstruction: a prospective study with 10 to 15 years of follow-up. Am J Sports Med 38(11):2201–2210

45. Cicuttini F, Ding C, Wluka A, Davis S, Ebeling PR, Jones G (2005) Association of cartilage defects with loss of knee cartilage in healthy, middle-age adults: a prospective study. Arthritis Rheumatol 52(7):2033–2039

46. Curl WW, Krome J, Gordon ES, Rushing J, Smith BP, Poehling GG (1997) Cartilage injuries: a review of 31,516 knee arthroscopies. Arthroscopy 13(4):456–460

47. Hjelle K, Solheim E, Strand T, Muri R, Brittberg M (2002) Articular cartilage defects in 1,000 knee arthroscopies. Arthroscopy 18(7):730–734

48. Ding C, Garnero P, Cicuttini F, Scott F, Cooley H, Jones G (2005) Knee cartilage defects: association with early radiographic osteoarthritis, decreased cartilage volume, increased joint surface area and type II collagen breakdown. Osteoarthr Cartil 13(3):198–205

49. Mithoefer K, Hambly K, Della Villa S, Silvers H, Mandelbaum BR (2009) Return to sports participation after articular cartilage repair in the knee scientific evidence. Am J Sports Med 37(1 suppl):167S–176S

50. Krych AJ, Pareek A, King AH, Johnson NR, Stuart MJ, Williams RJ (2016) Return to sport after the surgical management of articular cartilage lesions in the knee: a meta-analysis. Knee Surg Sports Traumatol Arthrosc 25(10):3186–3196

51. Campbell AB, Pineda M, Harris JD, Flanigan DC (2016) Return to sport after articular cartilage repair in athletes' knees: a systematic review. Arthroscopy 32(4):651–668. e651

52. Andrade R, Vasta S, Papalia R, Pereira H, Oliveira JM, Reis RL, Espregueira-Mendes J (2016) Prevalence of articular cartilage lesions and surgical clinical outcomes in football (soccer) players' knees: a systematic review. Arthroscopy 32(7):1466–1477

53. Wang Y, Ding C, Wluka A, Davis S, Ebeling P, Jones G, Cicuttini F (2005) Factors affecting progression of knee cartilage defects in normal subjects over 2 years. Rheumatology (Oxford) 45(1):79–84

54. Davies-Tuck M, Wluka A, Wang Y, Teichtahl A, Jones G, Ding C, Cicuttini F (2008) The natural history of cartilage defects in people with knee osteoarthritis. Osteoarthr Cartil 16(3):337–342

55. Ding C, Cicuttini F, Scott F, Cooley H, Boon C, Jones G (2006) Natural history of knee cartilage defects and factors affecting change. Arch Intern Med 166(6):651–658

56. Hunter DJ, Zhang Y, Niu J, Goggins J, Amin S, LaValley MP, Guermazi A, Genant H, Gale D, Felson DT (2006) Increase in bone marrow lesions associated with cartilage loss: a longitudinal magnetic resonance imaging study of knee osteoarthritis. Arthritis Rheumatol 54(5):1529–1535

57. Amin S, LaValley MP, Guermazi A, Grigoryan M, Hunter DJ, Clancy M, Niu J, Gale DR, Felson DT (2005) The relationship between cartilage loss on magnetic resonance imaging and radiographic progression in men and women with knee osteoarthritis. Arthritis Rheumatol 52(10):3152–3159

58. Vannini F, Spalding T, Andriolo L, Berruto M, Denti M, Espregueira-Mendes J, Menetrey J, Peretti G, Seil R, Filardo G (2016) Sport and early osteoarthritis: the role of sport in aetiology, progression and treatment of knee osteoarthritis. Knee Surg Sports Traumatol Arthrosc 24(6):1786–1796

59. Krajnc Z, Vogrin M, Rečnik G, Crnjac A, Drobnič M, Antolič V (2010) Increased risk of knee injuries and osteoarthritis in the non-dominant leg of former professional football players. Wien Klin Wochenschr 122:40–43

60. Kuijt M-TK, Inklaar H, Gouttebarge V, Frings-Dresen MH (2012) Knee and ankle osteoarthritis in former elite soccer players: a systematic review of the recent literature. J Sci Med Sport 15(6):480–487

61. Spector TD, Harris PA, Hart DJ, Cicuttini FM, Nandra D, Etherington J, Wolman RL, Doyle DV (1996) Risk of osteoarthritis associated with long-term weight-bearing sports: a radiologic survey of the hips and knees in female ex-athletes and population controls. Arthritis Rheumatol 39(6):988–995

62. Timmins KA, Leech RD, Batt ME, Edwards KL (2017) Running and knee osteoarthritis: a systematic review and meta-analysis. Am J Sports Med 45(6):1447–1457

63. Miller RH (2017) Joint loading in runners does not initiate knee osteoarthritis. Exerc Sport Sci Rev 45(2):87–95

64. Bastick AN, Belo JN, Runhaar J, Bierma-Zeinstra SM (2015) What are the prognostic factors for radiographic progression of knee osteoarthritis? A meta-analysis. Clin Orthop Relat Res 473(9):2969–2989

65. Alentorn-Geli E, Samuelsson K, Musahl V, Green CL, Bhandari M, Karlsson J (2017) The Association of recreational and competitive running with hip and knee osteoarthritis: a systematic review and meta-analysis. J Orthop Sports Phys Ther 47(6):373–390

66. Kujala UM, Kaprio J, Sarno S (1994) Osteoarthritis of weight bearing joints of lower limbs in former elite male athletes. BMJ 308(6923):231–234

67. Saxon L, Finch C, Bass S (1999) Sports participation, sports injuries and osteoarthritis. Sports Med 28(2):123–135

68. Luyten FP, Denti M, Filardo G, Kon E, Engebretsen L (2012) Definition and classification of early osteoarthritis of the knee. Knee Surg Sports Traumatol Arthrosc 20(3):401–406

69. Madry H, Kon E, Condello V, Peretti GM, Steinwachs M, Seil R, Berruto M, Engebretsen L, Filardo G, Angele P (2016) Early osteoarthritis of the knee. Knee Surg Sports Traumatol Arthrosc 24(6):1753–1762

70. Malfait A-M, Schnitzer TJ (2013) Towards a mechanism-based approach to pain management in osteoarthritis. Nat Rev Rheumatol 9(11):654–664

71. Neogi T (2013) The epidemiology and impact of pain in osteoarthritis. Osteoarthr Cartil 21(9):1145–1153

72. Hensor E, Dube B, Kingsbury SR, Tennant A, Conaghan PG (2015) Toward a clinical definition of early osteoarthritis: onset of patient-reported knee pain begins on stairs. Data from the osteoarthritis initiative. Arthritis Care Res (Hoboken) 67(1):40–47

73. Finan PH, Buenaver LF, Bounds SC, Hussain S, Park RJ, Haque UJ, Campbell CM, Haythornthwaite JA, Edwards RR, Smith MT (2013) Discordance between pain and radiographic severity in knee osteoarthritis: findings from quantitative sensory testing of central sensitization. Arthritis Rheumatol 65(2):363–372

74. Buckland-Wright J (1994) Quantitative radiography of osteoarthritis. Ann Rheum Dis 53(4):268
75. Swagerty D, Hellinger D (2001) Radiographic assessment of osteoarthritis. Am Fam Physician 64(2):279–288
76. Leach RE, Gregg T, Siber FJ (1970) Weight-bearing radiography in osteoarthritis of the knee. Radiology 97(2):265–268
77. Ahlbäck S (1968) Osteoarthrosis of the knee. A radiographic investigation. Acta Radiol Diagn (Stockh) Suppl 277:7–72
78. Chaisson C, Gale D, Gale E, Kazis L, Skinner K, Felson D (2000) Detecting radiographic knee osteoarthritis: what combination of views is optimal? Rheumatology (Oxford) 39(11):1218–1221
79. Sabharwal S, Zhao C (2008) Assessment of lower limb alignment: supine fluoroscopy compared with a standing full-length radiograph. J Bone Joint Surg Am 90(1):43–51
80. Matos LF, Giordano M, Cardoso GN, Farias RB (2015) Comparative radiographic analysis on the anatomical axis in knee osteoarthritis cases: inter and intraobserver evaluation. Rev Bras Ortop 50(3):283–289
81. Chan WP, Lang P, Stevens MP, Sack K, Majumdar S, Stoller DW, Basch C, Genant HK (1991) Osteoarthritis of the knee: comparison of radiography, CT, and MR imaging to assess extent and severity. AJR Am J Roentgenol 157(4):799–806
82. Burgkart R, Glaser C, Hinterwimmer S, Hudelmaier M, Englmeier KH, Reiser M, Eckstein F (2003) Feasibility of T and Z scores from magnetic resonance imaging data for quantification of cartilage loss in osteoarthritis. Arthritis Rheumatol 48(10):2829–2835
83. Cicuttini FM, Jones G, Forbes A, Wluka AE (2004) Rate of cartilage loss at two years predicts subsequent total knee arthroplasty: a prospective study. Ann Rheum Dis 63(9):1124–1127
84. Menashe L, Hirko K, Losina E, Kloppenburg M, Zhang W, Li L, Hunter DJ (2012) The diagnostic performance of MRI in osteoarthritis: a systematic review and meta-analysis. Osteoarthr Cartil 20(1):13–21
85. McAlindon TE, Bannuru RR, Sullivan M, Arden N, Berenbaum F, Bierma-Zeinstra S, Hawker G, Henrotin Y, Hunter D, Kawaguchi H (2014) OARSI guidelines for the non-surgical management of knee osteoarthritis. Osteoarthr Cartil 22(3):363–388
86. Hochberg MC, Altman RD, April KT, Benkhalti M, Guyatt G, McGowan J, Towheed T, Welch V, Wells G, Tugwell P (2012) American College of Rheumatology 2012 recommendations for the use of nonpharmacologic and pharmacologic therapies in osteoarthritis of the hand, hip, and knee. Arthritis Care Res (Hoboken) 64(4):465–474
87. Christensen R, Bartels EM, Astrup A, Bliddal H (2007) Effect of weight reduction in obese patients diagnosed with knee osteoarthritis: a systematic review and meta-analysis. Ann Rheum Dis 66(4):433–439
88. Bartels EM, Juhl CB, Christensen R, Hagen KB, Danneskiold-Samsøe B, Dagfinrud H, Lund H (2016) Aquatic exercise for the treatment of knee and hip osteoarthritis. Cochrane Libr. https://doi.org/10.1002/14651858.CD005523.pub3
89. Lu M, Su Y, Zhang Y, Zhang Z, Wang W, He Z, Liu F, Li Y, Liu C, Wang Y (2015) Effectiveness of aquatic exercise for treatment of knee osteoarthritis. Z Rheumatol 74(6):543–552
90. Fransen M, McConnell S, Harmer AR, Van der Esch M, Simic M, Bennell KL (2015) Exercise for osteoarthritis of the knee. Cochrane Libr. https://doi.org/10.1002/14651858.CD004376.pub3
91. Jansen MJ, Viechtbauer W, Lenssen AF, Hendriks EJ, de Bie RA (2011) Strength training alone, exercise therapy alone, and exercise therapy with passive manual mobilisation each reduce pain and disability in people with knee osteoarthritis: a systematic review. Physiotherapy 57(1):11–20
92. Kon E, Filardo G, Drobnic M, Madry H, Jelic M, van Dijk N, Della Villa S (2012) Non-surgical management of early knee osteoarthritis. Knee Surg Sports Traumatol Arthrosc 20(3):436–449

93. Mazieres B, Thevenon A, Coudeyre E, Chevalier X, Revel M, Rannou F (2008) Adherence to, and results of, physical therapy programs in patients with hip or knee osteoarthritis. Development of French clinical practice guidelines. Joint Bone Spine 75(5):589–596

94. Bennell KL, Buchbinder R, Hinman RS (2015) Physical therapies in the management of osteoarthritis: current state of the evidence. Curr Opin Rheumatol 27(3):304–311

95. Gay C, Chabaud A, Guilley E, Coudeyre E (2016) Educating patients about the benefits of physical activity and exercise for their hip and knee osteoarthritis. Systematic literature review. Ann Phys Rehabil Med 59(3):174–183

96. Tanaka R, Ozawa J, Kito N, Moriyama H (2013) Efficacy of strengthening or aerobic exercise on pain relief in people with knee osteoarthritis: a systematic review and meta-analysis of randomized controlled trials. Clin Rehabil 27(12):1059–1071

97. Juhl C, Christensen R, Roos EM, Zhang W, Lund H (2014) Impact of exercise type and dose on pain and disability in knee osteoarthritis: a systematic review and meta-regression analysis of randomized controlled trials. Arthritis Rheumatol 66(3):622–636

98. Brosseau L, MacLeay L, Robinson V, Wells G, Tugwell P (2003) Intensity of exercise for the treatment of osteoarthritis. Cochrane Libr. https://doi.org/10.1002/14651858.CD004259

99. Focht BC (2006) Effectiveness of exercise interventions in reducing pain symptoms among older adults with knee osteoarthritis: a review. J Aging Phys Act 14(2):212–235

100. Holla JF, Sanchez-Ramirez DC, van der Leeden M, Ket JC, Roorda LD, Lems WF, Steultjens MP, Dekker J (2014) The avoidance model in knee and hip osteoarthritis: a systematic review of the evidence. J Behav Med 37(6):1226–1241

101. Hart HF, Collins NJ, Ackland DC, Crossley KM (2015) Is impaired knee confidence related to worse kinesiophobia, symptoms, and physical function in people with knee osteoarthritis after anterior cruciate ligament reconstruction? J Sci Med Sport 18(5):512–517

102. Cruz-Almeida Y, King CD, Goodin BR, Sibille KT, Glover TL, Riley JL, Sotolongo A, Herbert MS, Schmidt J, Fessler BJ (2013) Psychological profiles and pain characteristics of older adults with knee osteoarthritis. Arthritis Care Res (Hoboken) 65(11):1786–1794

103. Fitzgerald GK, White DK, Piva SR (2012) Associations for change in physical and psychological factors and treatment response following exercise in knee osteoarthritis: an exploratory study. Arthritis Care Res (Hoboken) 64(11):1673–1680

104. Beckwée D, Vaes P, Cnudde M, Swinnen E, Bautmans I (2013) Osteoarthritis of the knee: why does exercise work? A qualitative study of the literature. Ageing Res Rev 12(1):226–236

105. AAOS (2013) Treatment of osteoarthritis of the knee. Evidence-based guideline, 2nd edn. American Academy of Orthopaedic Surgeons

106. Filardo G, Kon E, Longo UG, Madry H, Marchettini P, Marmotti A, Van Assche D, Zanon G, Peretti GM (2016) Non-surgical treatments for the management of early osteoarthritis. Knee Surg Sports Traumatol Arthrosc 24(6):1775–1785

107. Zhang W, Jones A, Doherty M (2004) Does paracetamol (acetaminophen) reduce the pain of osteoarthritis?: a meta-analysis of randomised controlled trials. Ann Rheum Dis 63(8):901–907

108. Smith SR, Deshpande BR, Collins JE, Katz JN, Losina E (2016) Comparative pain reduction of oral non-steroidal anti-inflammatory drugs and opioids for knee osteoarthritis: systematic analytic review. Osteoarthr Cartil 24(6):962–972

109. Hunter DJ (2015) Viscosupplementation for osteoarthritis of the knee. N Engl J Med 372(11):1040–1047

110. Strauss EJ, Hart JA, Miller MD, Altman RD, Rosen JE (2009) Hyaluronic acid viscosupplementation and osteoarthritis. Am J Sports Med 37(8):1636–1644

111. Conrozier T, Chevalier X (2008) Long-term experience with hylan GF-20 in the treatment of knee osteoarthritis. Expert Opin Pharmacother 9(10):1797–1804

112. Balazs EA, Watson D, Duff IF, Roseman S (1967) Hyaluronic acid in synovial fluid. I. Molecular parameters of hyaluronic acid in normal and arthritic human fluids. Arthritis Rheumatol 10(4):357–376

113. Balazs EA, Denlinger JL (1993) Viscosupplementation: a new concept in the treatment of osteoarthritis. J Rheumatol Suppl 39:3–9

114. Law TY, Nguyen C, Frank RM, Rosas S, McCormick F (2015) Current concepts on the use of corticosteroid injections for knee osteoarthritis. Phys Sportsmed 43(3):269–273
115. Chang RW, Falconer J, David Stulberg S, Arnold WJ, Manheim LM, Dyer AR (1993) A randomized, controlled trial of arthroscopic surgery versus closed-needle joint lavage for patients with osteoarthritis of the knee. Arthritis Rheumatol 36(3):289–296
116. Ogilvie-Harris D, Fitsialos D (1991) Arthroscopic management of the degenerative knee. Arthroscopy 7(2):151–157
117. Rönn K, Reischl N, Gautier E, Jacobi M (2011) Current surgical treatment of knee osteoarthritis. Arthritis. https://doi.org/10.1155/2011/454873
118. Palmer JS, Monk AP, Hopewell S, Bayliss LE, Jackson W, Beard DJ, Price AJ (2016) Surgical interventions for early structural knee osteoarthritis. Cochrane Libr. https://doi.org/10.1002/14651858.CD012128
119. Avouac J, Vicaut E, Bardin T, Richette P (2009) Efficacy of joint lavage in knee osteoarthritis: meta-analysis of randomized controlled studies. Rheumatology (Oxford) 49(2):334–340
120. Laupattarakasem W, Laopaiboon M, Laupattarakasem P, Sumananont C (2008) Arthroscopic debridement for knee osteoarthritis. Cochrane Libr. https://doi.org/10.1002/14651858.CD005118.pub2
121. Reichenbach S, Rutjes AW, Nüesch E, Trelle S, Jüni P (2010) Joint lavage for osteoarthritis of the knee. Cochrane Libr. https://doi.org/10.1002/14651858.CD007320.pub2
122. Moseley JB, O'malley K, Petersen NJ, Menke TJ, Brody BA, Kuykendall DH, Hollingsworth JC, Ashton CM, Wray NP (2002) A controlled trial of arthroscopic surgery for osteoarthritis of the knee. N Engl J Med 347(2):81–88
123. Widuchowski W, Lukasik P, Kwiatkowski G, Faltus R, Szyluk K, Widuchowski J, Koczy B (2008) Isolated full thickness chondral injuries. Prevalence and outcome of treatment. A retrospective study of 5233 knee arthroscopies. Acta Chir Orthop Traumatol Cechoslov 75(5):382–386
124. Angele P, Niemeyer P, Steinwachs M, Filardo G, Gomoll AH, Kon E, Zellner J, Madry H (2016) Chondral and osteochondral operative treatment in early osteoarthritis. Knee Surg Sports Traumatol Arthrosc 24(6):1743–1752
125. Gomoll A, Filardo G, De Girolamo L, Esprequeira-Mendes J, Marcacci M, Rodkey W, Steadman R, Zaffagnini S, Kon E (2012) Surgical treatment for early osteoarthritis. Part I: cartilage repair procedures. Knee Surg Sports Traumatol Arthrosc 20(3):450–466
126. Smith NA, Parkinson B, Hutchinson CE, Costa ML, Spalding T (2016) Is meniscal allograft transplantation chondroprotective? A systematic review of radiological outcomes. Knee Surg Sports Traumatol Arthrosc 24(9):2923–2935
127. Smith J, Wilson A, Thomas N (2013) Osteotomy around the knee: evolution, principles and results. Knee Surg Sports Traumatol Arthrosc 21(1):3–22
128. Amis AA (2013) Biomechanics of high tibial osteotomy. Knee Surg Sports Traumatol Arthrosc 21(1):197–205
129. Portner O (2014) High tibial valgus osteotomy: closing, opening or combined? Patellar height as a determining factor. Clin Orthop Relat Res 472(11):3432–3440
130. Bonasia DE, Dettoni F, Sito G, Blonna D, Marmotti A, Bruzzone M, Castoldi F, Rossi R (2014) Medial opening wedge high tibial osteotomy for medial compartment overload/arthritis in the varus knee: prognostic factors. Am J Sports Med 42(3):690–698
131. Brouwer RW, Huizinga MR, Duivenvoorden T, van Raaij TM, Verhagen AP, Bierma-Zeinstra S, Verhaar JA (2014) Osteotomy for treating knee osteoarthritis. Cochrane Libr. https://doi.org/10.1002/14651858.CD004019.pub3
132. Sun H, Zhou L, Li F, Duan J (2017) Comparison between closing-wedge and opening-wedge high tibial osteotomy in patients with medial knee osteoarthritis: a systematic review and meta-analysis. J Knee Surg 30(02):158–165
133. Rossi R, Bonasia DE, Amendola A (2011) The role of high tibial osteotomy in the varus knee. J Am Acad Orthop Surg 19(10):590–599
134. Harris JD, McNeilan R, Siston RA, Flanigan DC (2013) Survival and clinical outcome of isolated high tibial osteotomy and combined biological knee reconstruction. Knee 20(3):154–161

135. Bastos Filho R, Magnussen RA, Duthon V, Demey G, Servien E, Granjeiro JM, Neyret P (2013) Total knee arthroplasty after high tibial osteotomy: a comparison of opening and closing wedge osteotomy. Int Orthop 37(3):427–431
136. van der Merwe W (2016) New and evolving surgical techniques. In: Parker D (ed) Management of knee osteoarthritis in the younger, active patient, Springer, pp 149–155
137. Kajiwara R, Ishida O, Kawasaki K, Adachi N, Yasunaga Y, Ochi M (2005) Effective repair of a fresh osteochondral defect in the rabbit knee joint by articulated joint distraction following subchondral drilling. J Orthop Res 23(4):909–915
138. Kamei G, Ochi M, Okuhara A, Fujimiya M, Deie M, Adachi N, Nakamae A, Nakasa T, Ohkawa S, Takazawa K (2013) A new distraction arthroplasty device using magnetic force; a cadaveric study. Clin Biomech (Bristol, Avon) 28(4):423–428
139. Gabriel S, Clifford A, Maloney W, O'connell M, Tornetta Iii P (2012) Unloading the OA knee with a novel implant system. J Appl Biomech 29(6):647–654
140. Dowsey M, Nikpour M, Dieppe P, Choong P (2012) Associations between pre-operative radiographic changes and outcomes after total knee joint replacement for osteoarthritis. Osteoarthr Cartil 20(10):1095–1102
141. Niinimäki TT, Murray DW, Partanen J, Pajala A, Leppilahti JI (2011) Unicompartmental knee arthroplasties implanted for osteoarthritis with partial loss of joint space have high re-operation rates. Knee 18(6):432–435
142. do Amaral RJFC, da Silva NP, Haddad NF, Lopes LS, Ferreira FD, Bastos Filho R, Cappelletti PA, de Mello W, Cordeiro-Spinetti E, Balduino A (2016) Platelet-rich plasma obtained with different anticoagulants and their effect on platelet numbers and mesenchymal stromal cells behavior in vitro. Stem Cells Int. https://doi.org/10.1155/2016/7414036
143. Dhillon MS, Patel S, John R (2017) PRP in OA knee–update, current confusions and future options. SICOT-J 3:27. https://doi.org/10.1051/sicotj/2017004
144. Pereira RC, Scaranari M, Benelli R, Strada P, Reis RL, Cancedda R, Gentili C (2013) Dual effect of platelet lysate on human articular cartilage: a maintenance of chondrogenic potential and a transient proinflammatory activity followed by an inflammation resolution. Tissue Eng Part A 19(11–12):1476–1488
145. Wu C-C, Chen W-H, Zao B, Lai P-L, Lin T-C, Lo H-Y, Shieh Y-H, Wu C-H, Deng W-P (2011) Regenerative potentials of platelet-rich plasma enhanced by collagen in retrieving pro-inflammatory cytokine-inhibited chondrogenesis. Biomaterials 32(25):5847–5854
146. Laver L, Marom N, Dnyanesh L, Mei-Dan O, Espregueira-Mendes J, Gobbi A (2016) PRP for degenerative cartilage disease: a systematic review of clinical studies. Cartilage. 1947603516670709
147. Meheux CJ, McCulloch PC, Lintner DM, Varner KE, Harris JD (2016) Efficacy of intra-articular platelet-rich plasma injections in knee osteoarthritis: a systematic review. Arthroscopy 32(3):495–505
148. Campbell KA, Saltzman BM, Mascarenhas R, Khair MM, Verma NN, Bach BR, Cole BJ (2015) Does intra-articular platelet-rich plasma injection provide clinically superior outcomes compared with other therapies in the treatment of knee osteoarthritis? A systematic review of overlapping meta-analyses. Arthroscopy 31(11):2213–2221
149. Khoshbin A, Leroux T, Wasserstein D, Marks P, Theodoropoulos J, Ogilvie-Harris D, Gandhi R, Takhar K, Lum G, Chahal J (2013) The efficacy of platelet-rich plasma in the treatment of symptomatic knee osteoarthritis: a systematic review with quantitative synthesis. Arthroscopy 29(12):2037–2048
150. Anitua E, Sánchez M, Aguirre JJ, Prado R, Padilla S, Orive G (2014) Efficacy and safety of plasma rich in growth factors intra-articular infiltrations in the treatment of knee osteoarthritis. Arthroscopy 30(8):1006–1017
151. Görmeli G, Görmeli CA, Ataoglu B, Çolak C, Aslantürk O, Ertem K (2017) Multiple PRP injections are more effective than single injections and hyaluronic acid in knees with early osteoarthritis: a randomized, double-blind, placebo-controlled trial. Knee Surg Sports Traumatol Arthrosc 25(3):958–965

152. Sermer C, Devitt B, Chahal J, Kandel R, Theodoropoulos J (2015) The addition of platelet-rich plasma to scaffolds used for cartilage repair: a review of human and animal studies. Arthroscopy 31(8):1607–1625
153. Bastos Filho R, Lermontov S, Borojevic R, Schott PC, Gameiro VS, Granjeiro JM (2012) Cell therapy of pseudarthrosis. Acta Ortop Bras 20(5):270–273
154. Weinberg ME, Kaplan DJ, Pham H, Goodwin D, Dold A, Chiu E, Jazrawi LM (2017) Injectable biological treatments for osteoarthritis of the knee. JBJS Rev 5(4):e2
155. Koh Y-G, Choi Y-J (2012) Infrapatellar fat pad-derived mesenchymal stem cell therapy for knee osteoarthritis. Knee 19(6):902–907
156. Koh Y-G, Kwon O-R, Kim Y-S, Choi Y-J (2014) Comparative outcomes of open-wedge high tibial osteotomy with platelet-rich plasma alone or in combination with mesenchymal stem cell treatment: a prospective study. Arthroscopy 30(11):1453–1460
157. Saw K-Y, Anz A, Jee CS-Y, Merican S, Ng RC-S, Roohi SA, Ragavanaidu K (2013) Articular cartilage regeneration with autologous peripheral blood stem cells versus hyaluronic acid: a randomized controlled trial. Arthroscopy 29(4):684–694
158. Vega A, Martín-Ferrero MA, Del Canto F, Alberca M, García V, Munar A, Orozco L, Soler R, Fuertes JJ, Huguet M (2015) Treatment of knee osteoarthritis with allogeneic bone marrow mesenchymal stem cells: a randomized controlled trial. Transplantation 99(8):1681–1690
159. Wong KL, Lee KBL, Tai BC, Law P, Lee EH, Hui JH (2013) Injectable cultured bone marrow–derived mesenchymal stem cells in varus knees with cartilage defects undergoing high tibial osteotomy: a prospective, randomized controlled clinical trial with 2 Years' follow-up. Arthroscopy 29(12):2020–2028
160. Pas HI, Winters M, Haisma HJ, Koenis MJ, Tol JL, Moen MH (2017) Stem cell injections in knee osteoarthritis: a systematic review of the literature. Br J Sports Med 51:1125–1133

Chapter 6
Hyaluronic Acid

Hélder Pereira, Duarte Andre Sousa, António Cunha, Renato Andrade, J. Espregueira-Mendes, J. Miguel Oliveira, and Rui L. Reis

Abstract In recent times, the field of tissue engineering and regenerative medicine (TERM) has considerably increased the extent of therapeutic strategies for clinical application in orthopedics. However, TERM approaches have its rules and requirements, in the respect of the biologic response of each tissue and bioactive agents which need to be considered, respected, and subject of ongoing studies. Different medical devices/products have been prematurely available on the market and used in clinics with limited success. However, other therapeutics, when used in a serious and evidence-based approach, have achieved considerable success, considering the respect for solid expectations from doctors and patients (when properly informed).

H. Pereira (✉)
3B's Research Group – Biomaterials, Biodegradables and Biomimetics, University of Minho, Headquarters of the European Institute of Excellence on Tissue Engineering and Regenerative Medicine, Barco, Guimarães, Portugal

ICVS/3B's – PT Government Associated Laboratory, Braga, Portugal

Orthopedic Department of Póvoa de Varzim – Vila do Conde Hospital Centre, Póvoa de Varzim, Portugal

Ripoll y De Prado Sports Clinic: Murcia-Madrid FIFA Medical Centre of Excellence, Murcia/Madrid, Spain

International Centre of Sports Traumatology of the Ave, Taipas, Portugal
e-mail: helderduartepereira@gmail.com

D. A. Sousa
Orthopedic Department of Póvoa de Varzim – Vila do Conde Hospital Centre, Póvoa de Varzim, Portugal

A. Cunha
International Centre of Sports Traumatology of the Ave, Taipas, Portugal

R. Andrade
Clínica do Dragão, Espregueira-Mendes Sports Centre – FIFA Medical Centre of Excellence, Porto, Portugal

© Springer International Publishing AG, part of Springer Nature 2018   137
J. M. Oliveira et al. (eds.), *Osteochondral Tissue Engineering*,
Advances in Experimental Medicine and Biology 1059,
https://doi.org/10.1007/978-3-319-76735-2_6

Orthobiologics has appeared as a recent technological trend in orthopedics. This includes the improvement or regeneration of different musculoskeletal tissues by means of using biomaterials (e.g., hyaluronic acid), stem cells, and growth factors (e.g., platelet-rich plasma). The potential symbiotic relationship between biologic therapies and surgery makes these strategies suitable to be used in one single intervention.

However, herein, the recent clinical studies using hyaluronic acid (HA) in the treatment of orthopedic conditions will mainly be overviewed (e.g., osteochondral lesions, tendinopathies). The possibilities to combine different orthobiologic agents as TERM clinical strategies for treatment of orthopedic problems will also be briefly discussed.

Keywords Osteochondral lesions · Tendinopathies · Orthobiologics · Hyaluronic acid · Stem cells · Platelet-rich plasma · Growth factors · Tissue engineering and regenerative medicine

J. Espregueira-Mendes
3B's Research Group – Biomaterials, Biodegradables and Biomimetics, University of Minho, Headquarters of the European Institute of Excellence on Tissue Engineering and Regenerative Medicine, Barco, Guimarães, Portugal

ICVS/3B's – PT Government Associated Laboratory, Braga, Portugal

Clínica do Dragão, Espregueira-Mendes Sports Centre – FIFA Medical Centre of Excellence, Porto, Portugal

Dom Henrique Research Centre, Porto, Portugal

Orthopedic Department, University of Minho, Braga, Portugal

J. M. Oliveira
3B's Research Group – Biomaterials, Biodegradables and Biomimetics, University of Minho, Headquarters of the European Institute of Excellence on Tissue Engineering and Regenerative Medicine, Barco, Guimarães, Portugal

ICVS/3B's – PT Government Associated Laboratory, Braga, Portugal

Rioll y De Prado Sports Clinic: Murcia-Madrid FIFA Medical Centre of Excellence, Murcia/Madrid, Spain

The Discoveries Centre for Regenerative and Precision Medicine, Headquarters at University of Minho, Guimarães, Portugal

R. L. Reis
3B's Research Group – Biomaterials, Biodegradables and Biomimetics, University of Minho, Headquarters of the European Institute of Excellence on Tissue Engineering and Regenerative Medicine, Barco, Guimarães, Portugal

ICVS/3B's – PT Government Associated Laboratory, Braga, Portugal

The Discoveries Centre for Regenerative and Precision Medicine, Headquarters at University of Minho, Guimarães, Portugal

Top 10 Evidence-Based References

1. Gigante A, Callegari L (2011) The role of intra-articular hyaluronan (Sinovial) in the treatment of osteoarthritis. Rheumatol Int 31(4):427–444. https://doi.org/10.1007/s00296-010-1660-6
2. Strauss EJ, Hart JA, Miller MD, Altman RD, Rosen JE (2009) Hyaluronic acid viscosupplementation and osteoarthritis: current uses and future directions. Am J Sports Med 37(8):1636–1644. https://doi.org/10.1177/0363546508326984
3. Xing D, Wang B, Liu Q, Ke Y, Xu Y, Li Z, Lin J (2016) Intra-articular hyaluronic acid in treating knee osteoarthritis: a PRISMA-compliant systematic review of overlapping meta-analysis. Sci Rep 6:32790. https://doi.org/10.1038/srep32790
4. Bellamy N, Campbell J, Robinson V, Gee T, Bourne R, Wells G (2006) Viscosupplementation for the treatment of osteoarthritis of the knee. Cochrane Database Syst Rev (2):CD005321. https://doi.org/10.1002/14651858.CD005321.pub2
5. Witteveen AG, Hofstad CJ, Kerkhoffs GM (2015) Hyaluronic acid and other conservative treatment options for osteoarthritis of the ankle. Cochrane Database Syst Rev (10):CD010643. https://doi.org/10.1002/14651858.CD010643.pub2
6. Colen S, Geervliet P, Haverkamp D, Van Den Bekerom MP (2014) Intra-articular infiltration therapy for patients with glenohumeral osteoarthritis: a systematic review of the literature. Int J Shoulder Surg 8(4):114–121. https://doi.org/10.4103/0973-6042.145252
7. Trellu S, Dadoun S, Berenbaum F, Fautrel B, Gossec L (2015) Intra-articular injections in thumb osteoarthritis: a systematic review and meta-analysis of randomized controlled trials. Joint Bone Spine 82(5):315–319. https://doi.org/10.1016/j.jbspin.2015.02.002
8. Lynen N, De Vroey T, Spiegel I, Van Ongeval F, Hendrickx NJ, Stassijns G (2017) Comparison of peritendinous hyaluronan injections versus extracorporeal shock wave therapy in the treatment of painful achilles' tendinopathy: a randomized clinical efficacy and safety study. Arch Phys Med Rehabil 98(1):64–71. https://doi.org/10.1016/j.apmr.2016.08.470
9. Kumai T, Muneta T, Tsuchiya A, Shiraishi M, Ishizaki Y, Sugimoto K, Samoto N, Isomoto S, Tanaka Y, Takakura Y (2014) The short-term effect after a single injection of high-molecular-weight hyaluronic acid in patients with enthesopathies (lateral epicondylitis, patellar tendinopathy, insertional Achilles tendinopathy, and plantar fasciitis): a preliminary study. J Orthop Sci 19 (4):603–611. https://doi.org/10.1007/s00776-014-0579-2

10. Tsaryk R, Gloria A, Russo T, Anspach L, De Santis R, Ghanaati S, Unger RE, Ambrosio L, Kirkpatrick CJ (2015) Collagen-low molecular weight hyaluronic acid semi-interpenetrating network loaded with gelatin microspheres for cell and growth factor delivery for nucleus pulposus regeneration. Acta Biomater 20:10–21. https://doi.org/10.1016/j.actbio.2015.03.041

Fact Box 1 – What Is Hyaluronic Acid

- Hyaluronic acid (HA) is a high molecular weight biopolysaccharide, discovered in 1934.
- HA is a naturally occurring biopolymer, found in most connective tissues, and is particularly concentrated in synovial fluid, the vitreous fluid of the eye, umbilical cords, and chicken combs.
- It is a high viscoelastic fluid capable to reproduce and repair the rheological proprieties of the synovial fluid.
- Besides its rheological proprieties, it acts as a shock absorber and as a lubricant and has anti-angiogenic, anti-inflammatory, and analgesic properties as well as immunosuppressive capacities.
- As a hydrogel or tridimensional scaffold, it can be used as protein or cell carrier.

Fact Box 2 – Clinical Experience with Hyaluronic Acid in Orthopedics

- Hyaluronic acid (HA) has been used mostly in the treatment of cartilage and osteoarthritis.
- It has been used in several joints (knee, ankle, shoulder, hip, first carpometacarpal, etc.).
- It has shown transient benefits in pain relief and improved range of motion (minimum 6 months).
- It has rare adverse effects (mainly self-limited pain and swelling (spontaneously solved within 48 h)).
- It has shown promising results in the treatment of tendinopathies including enthesopathies.
- Intra-tendon injections might have deleterious effects.

> **Fact Box 3 – Hyaluronic Acid and Tissue Engineering and Regenerative Medicine**
>
> - Tissue engineering and regenerative medicine (TERM) aims for more advanced approaches of tissue regeneration and disease control (combining scaffolds, cells, growth factors, prolotherapy, nanotechnology, bioreactors, gene therapy, etc.).
> - Advanced TERM approaches using hyaluronic acid (HA) have been tested including combination of HA with growth factors, cells, nanotechnology, and advanced scaffolds.
> - Dealing with cartilage (including hyaline or fibrocartilage) and tendon regeneration represents two of the most challenging tissues in the field of clinical orthopedics, and the road for the future for sure will comprise advanced TERM approach.
> - The achieved knowledge from the clinical experience with HA must be used as a launching platform for future basic science studies. Furthermore, higher-quality clinical studies are required for more accurate conclusions.

6.1 Introduction

The treatment of osteochondral (OC) defects and/or osteoarthritis in different joints remains a challenge, and the quest for the optimal conservative treatment continues. This represents an important socioeconomic burden given the high prevalence of these diseases and possibilities to cause functional impairment. In epidemiology, half of the world's population aged 65 years or older has osteoarthritis (OA), which is the most prevalent disorder of articulating joints in humans [1]. The estimated social cost of OA might range between 0.25% and 0.50% of a country's gross domestic product [2]. Degeneration related to the aging process, trauma-related injuries, and degenerative or idiopathic disorders can lead to OC lesions [3]. A distinction should be made between the chondral lesion in which the damage occurs on chondrocytes and articular cartilage extracellular matrix (ECM) opposing to OC defects in which, besides cartilage damage, the subchondral bone tissue is also affected [4]. The hyaline cartilage layer has highly flexible and supportive characteristics. The unique features of the cartilage tissue limit its regenerative capacity due to the lack of vascularization and innervation [5]. Similarly, cartilage has very limited remodeling possibilities given the low number of cells and low metabolic activity of chondrocytes (mainly the mature ones) which produces less extracellular matrix (ECM) [6, 7].

Damage to articular cartilage preceded by joint trauma is a major risk factor leading to progression of OA [8]. However, OA has a higher incidence in aged people and it is strongly correlated with natural aging process [9]. The huge variability of the OA tissues between individuals is proven to be one of the most important factors affecting the understanding of the disease and the variability in response to therapeutics. The current treatments are based on the adaptation of lifestyle and on the use of anti-inflammatory and painkiller drugs or clinically solved with the substitution by an artificial implant.

Another very important field within orthopedics is related to tendinopathies within its many presentation forms which affect a high number of patients [10, 11]. Tendon is also a tissue with low mitotic activity and with limited self-repair capacity [12].

Currently, there has been growing popularity of non-operative therapies, which are able to induce the body's self-repair and recovery from injuries or simply improve the symptoms without relevant secondary effects. These therapies include hyaluronic acid injections which are included in a new therapeutic field named orthobiologics [13]. Despite hyaluronic acid, orthobiologics also enrolls the clinical use of growth factors (e.g., platelet-rich plasma, bone morphogenetic proteins) or mesenchymal stem cells [14–21].

Viscosupplementation, which concerns to hyaluronic acid injections, was the first generation of orthobiologics. The treatment was firstly done in 1997 to relieve patients from pain symptoms of OA [1]. The outcome was satisfactory, and when compared to oral drug administration (NSAID), viscosupplementation was able to diminish patient's pain.

Herein only the clinical use and potential future applications of hyaluronic acid will be presented. Most studies related to orthobiologics have several methodological limitations. Despite the growing evidence supporting some therapeutic strategies, more high-level studies with uniform outcome measures are required in order to assess the present and prepare the future.

6.2   What Is Hyaluronic Acid?

Hyaluronic acid (HA) injection (Fig. 6.1) at the injury site has been used as a conservative method to treat OC lesions, osteoarthritis, and tendinopathies [1, 14, 16, 22–28]. In brief, HA is a high molecular weight biopolysaccharide, discovered

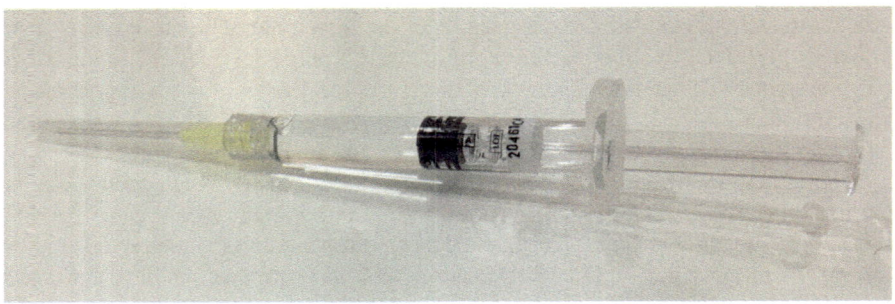

Fig. 6.1 Commercial formulation of hyaluronic acid injection ready for clinical use under strict aseptic conditions

in 1934, by Karl Meyer and John Palmer in the vitreous of bovine eyes [29]. HA is a naturally occurring biopolymer, which has important biological functions in bacteria and higher animals including humans. It is found in most connective tissues and is particularly concentrated in synovial fluid, the vitreous fluid of the eye, umbilical cords, and chicken combs [29]. It is naturally synthesized by a class of integral membrane proteins called hyaluronan synthases and degraded by a family of enzymes called hyaluronidases. The first medical HA application in humans was in the vitreous substitution/replacement during eye surgery, in the late 1950s [22, 29]. Viscosupplementation (VS) came into clinical use in Japan and Italy in 1987 and in Canada in 1992 but was adopted in Europe and the USA in the second half of the 1990s [29].

VS can be described as the intra-articular administration of a high viscoelastic fluid into the synovial joint to reproduce and repair the rheological proprieties of the synovial fluid. VS can enhance the vital joint lubrication and shock absorption ability, essential functions for mobility improvements, and pain relief. All market available products for VS are based on hyaluronic acid (HA), a high molecular weight (105–107 Da) and unbranched glycosaminoglycan that can be found in the extracellular matrix of human tissue.

Despite HA being described as safe and effective, several of its functions within the body remain unknown [30, 31]. HA has shown chondroprotective effects in vivo and in vitro [29]. Combined with its rheological proprieties, this helps to explain the beneficial long-term effects on articular cartilage. HA also reduces pain-associated nerve impulses and sensitivity. HA is a free, non-sulfated, and negatively charged glycosaminoglycan (GAG) capable of interacting with receptors and ECM proteins [32].

HA can be derived from different sources such as rooster combs, bacterial production, or either animal or human sources [33]. Its properties (e.g., rheological properties) depend on its source. Nevertheless, HA solutions always present high viscosity. It is a water-soluble polymer and has specific enzymatic degradation. There are two forms of HA, based in the chain length: low and high molecular weight ($\leq 2 \times 10^6$ Da and $2 \times 10^6 - \geq 4 \times 10^6$ Da, respectively). Structural and biological functions of HA vary among these different presentations as suggested by Stern et al. [34, 35]. The interactions between tissues and HA occur through hyaladherins. Functions such as cell communication, motility, and morphogenesis occur due to interactions between hyaladherins and tissue receptors, mainly CD44 and RHAMM at the cell surface [29]. The hyaluronic acid high molecular weight (HMWHA) molecule plays a structural role by being able to bind 10 to 10,000 time its weight in water [34, 36, 37]. Thus, osmotically active in a completely hydrated state, it is able to fill the space acting as a shock absorber and also as a lubricant. From a biological point of view, the HMW chains are anti-angiogenic and anti-inflammatory and possess immunosuppressive capacities [38, 39]. Wide ranges of studies have reported a decrease in the inflammatory response and apoptosis through the downregulation of many factors responsible for ECM. These results suggest that HMWHA impairs the phenomena of phagocytosis, macrophage activation, and inflammatory cytokines production. This can also explain its possible beneficial effects in tendinopathies.

However, HMWHA chains can break down into low molecular weight chains (LMWHA), which are found to have a pro-inflammatory effect [26]. These fragments have been shown to secrete inflammatory cytokines and stimulate angiogenesis and tissue remodeling after activation of endogenous signaling pathways. They can promote the activation and maturation of dendritic cells and release of pro-inflammatory cytokines [35, 39, 40]. Molecular changes in the ECM of damaged joints alter the composition and structure of natural HA. Along with molecule secretion and tissue remodeling, the development of pathologies also occurs [34]. So, a lot remains to understand in order to improve efficacy of this therapeutic agent.

6.3 Most Frequent Isolated Injection Therapy in Different Joints and Tendons

HA injection, also known as viscosupplementation (VS), constitutes a conservative treatment to improve the biomechanical function of the joint and/or tendons mainly due to HA physicochemical characteristics (i.e., hydrogel state) [41, 42]. HA is a gel-like constituent injected in the joint or tendon sheath [1, 43]. In some cases, to ensure safety or effectiveness of HA delivery, its application can be guided by ultrasound or X-ray fluoroscopy [44]. It is thought that HA acts as lubricant displaying a cushion effect. However, the biological mechanism behind this role in cartilage/tendon repair still remains a debate in medical community [45].

6.3.1 Intra-Articular Application

Positive effects of HA intra-articular injections have been reported in the conservative treatment of symptomatic osteoarthritis [46, 47]. It can be used as alternative to surgical approach mainly in the early phases of osteoarthritis or in patients without medical conditions to undergo more aggressive surgical treatment (e.g., replacement arthroplasty) [1, 27, 31, 48–51]. Viscosupplementation can be performed as augmentation after surgical intervention to treat osteochondral lesions [52]. The Food and Drug Administration (FDA) regulates the use of HA for intra-articular injection. Several forms of HA have received FDA approval for clinical use [53], despite the biological mechanism of HA is not fully explained as well as the differences between the several HA presentations. There is still debate and controversy in literature concerning the best approach for each joint and grade of disease [54].

HA has a half-life of less than 24 h after intra-articular injection [29]. Short half-life, due to the rapid breakdown and reabsorption of HA, might represent a limitation in the intra-articular injections. For this reason, sustained-release approaches are under development. It has been already suggested in clinical trials that combination of treatments is shown to be more effective than HA alone [55–60].

The highest clinical experience available is related to knee (Fig. 6.2) and ankle (Fig. 6.3) use of HA [1, 31]. Protocols range from one single shot treatment to treatments requiring three to five injections, depending on the formulation of HA

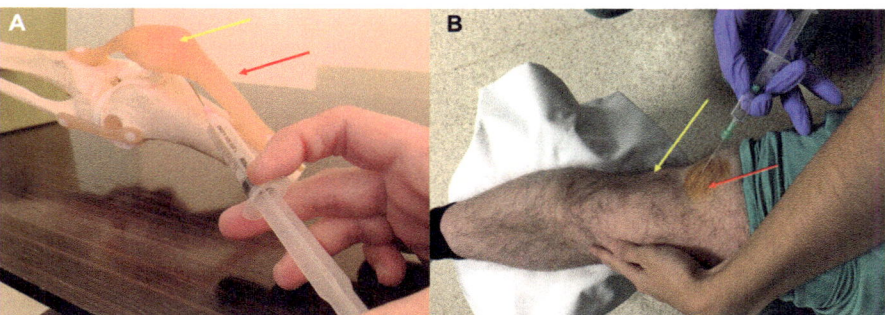

Fig. 6.2 A schematic representation of the knee injection of hyaluronic acid. (**A**) Model where the patella is visible (yellow arrow) which must be relaxed; the needle should be introduced in the lateral suprapatellar pouch (red arrow) on an oblique orientation toward the joint. Usually, an experienced operator is capable to sense the exact moment when he trespasses the joint capsule with the needle. After this, the gel can be introduced safely. (**B**) Clinical representation where the patella (yellow arrow) and the suprapatellar pouch (red arrow) are visible. The operator slightly moves the patella laterally to relax the suprapatellar pouch

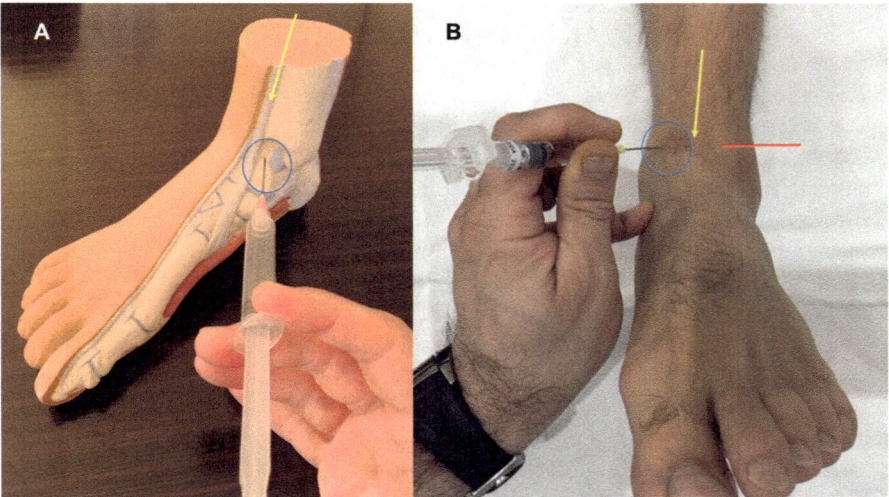

Fig. 6.3 A schematic representation of ankle injection of hyaluronic acid (**A**). The anterior tibial tendon (ATT) is visible (yellow arrow). The injection is given with the ankle in dorsiflexion, which dislocates the ATT centrally, in the soft spot where you can feel the joint line (blue circle). Again, an experienced operator is capable to sense the exact moment when he trespasses the joint capsule with the needle. After this, the gel can be introduced safely. (**B**) Clinical representation of the ankle injection. Yellow arrow represents the ATT, red line represents the joint line, and the little finger palpates the tip of the medial malleolus (usually the joint line will be 1.5–2 cm above)

and the condition of the patient [26]. The possible adverse reactions are rare (1–2% of cases) and consist usually in transient pain and/or swelling, which usually resolve spontaneously in 48 h [26]. No serious adverse effects have been reported.

A recent meta-analysis of all high-quality previously published meta-analysis on the subject of HA use in the knee joint by Xing et al. showed that HA provided a moderate but real benefit for patients with knee OA [1]. This study, a systematic review of overlapping meta-analyses, that investigated efficacy and safety of HA, in treating knee OA, concluded that currently, the best evidence suggested that HA is an effective intervention in treating knee OA without increased risk of adverse events. It concluded that US-approved HA is safe and efficacious through an average of 26 weeks in treating symptomatic knee OA.

Therefore, the evidence supports the use of the HA in the treating knee OA. Further studies with effect size statistic are still required to qualify the clinical efficacy [1].

Concerning the ankle, a recent Cochrane review by Witteveen et al. identified six randomized controlled studies on the topic and concluded that it remains unclear which patients (age, grade of ankle OA) benefit the most from HA injections and which dosage schedule should be used [31]. Considering the best available evidence, HA can be conditionally recommended if patients have an inadequate response to simple analgesics. Moreover, the authors highlight the study limitations due to methodological issues of the published studies [31]. Some authors highlight the difference in congruency when comparing the ankle (congruent) joint to the knee (incongruent) joint [4]. If that would be the case, theoretically, the rheological properties of HA should be more prone to work on the ankle when compared to the knee.

More recently, several studies have stated the benefits of HA application in the shoulder. The shoulder is the most "unstable" joint of the body given its wide range of motion (including circumferential motion). On a recent study enrolling 41 patients suffering from chronic shoulder pain with limitation of motion due to glenohumeral joint OA, the authors concluded that HA may be a safe and effective treatment option for pain and stiffness and that the effects of the injections are still present for up to 6 months after the treatment [51]. However, a systematic review on the topic failed to provide definitive indications once more for methodological limitations of available literature [61]. However, HA was shown to provide consecutively better outcomes when compared to placebo.

Hip OA has also been approached, either in severe cases or in young patients (in which there is advantage to delay arthroplasty replacement) [27, 44, 49]. In the hip, several authors advise for the benefits of ultrasound-assisted injection (Fig. 6.4) (particularly in more severe OA which makes it technically much more difficult to do blindly) [44]. The preliminary results have been positive in diminishing pain and improving mobility for a minimum period of 6 months.

HA has also been tried in smaller joints like the first carpometacarpal joint of the thumb [50]. From a recent meta-analysis, it seems that the most symptomatic patients are the most prone for improvement and, when compared to corticosteroid, the positive results stand for a longer period [62].

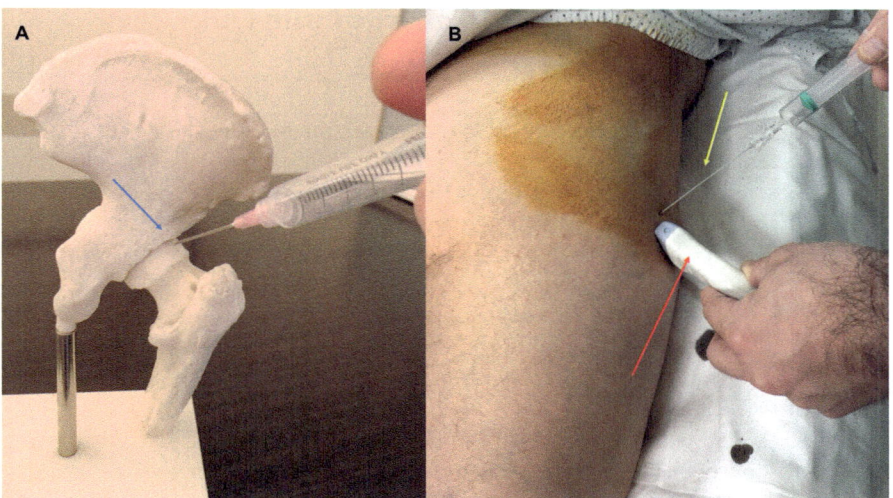

Fig. 6.4 A schematic representation of injection of hyaluronic acid in the hip joint (blue line) (**A**). (**B**) Clinical representation of hip joint injection assisted by ultrasound (red arrow), with progressive introduction of the needle (yellow arrow)

6.3.2   *Clinical Use in Tendinopathies*

Tendinopathies, in its many presentations, represent a current challenge in orthopedics, and in many cases, surgical options provide poor outcome [10, 11, 63]. As aforementioned, the tendon has reduced self-repair capacity, poor vascularity, and low mitotic activity [12, 63].

Promising results have been achieved in the rotator cuff with HA injection treatment. A multicenter, randomized, controlled trial assessing peritendinous HA outcome in patients with supraspinatus tendinopathy reported improved outcome and faster rehabilitation with lower number of physiotherapy sessions [25]. Also in non-calcific rotator cuff tendinopathies, when comparing HA to extracorporeal shockwaves therapy (ESWT), injections of HA provided faster clinical improvement compared to ESWT, which might result in more gradual improvement over time [64].

Some promising results have also been achieved in a preliminary study on enthesopathies (lateral epicondylitis, patellar tendinopathy, insertional Achilles tendinopathy, and plantar fasciitis) [65]. However, further research is required.

Concerning the Achilles tendon, a recent randomized study comparing peritendinous HA injections with standard ESWT has shown better outcome with HA [43]. However, one anima-based study suggests that intra-tendon HA injection might have deleterious effect and should be avoided [66]. Once more, more evidence is required.

6.4   Can Hyaluronic Acid Be Combined with Other Orthobiologic Therapies?

Not only basic science research but also some preliminary clinical data suggest that HA can be combined with other orthobiologics (growth factors, cells, biomaterials, hydrogels/scaffolds) but further suggest that there might be several advantages in doing it [67–69]. However, doing it properly and adequately, the rules of TERM must be followed which promise to change the paradigm of medicine and will, most likely, take years of intense work.

Moreover, HA can also be used as augmentation of a surgical procedure as proposed by Doral et al. combining microfractures and HA in the treatment of OC defects [56]. Other authors used MSCs and HA aiming to improve surgical results of OC lesions [55, 57, 60]. Synergistic anabolic actions of HA and PRP have been demonstrated [70, 71]. Similarly, a biocompatible carrier-forming HA-based microgel (PlnD1-HA) in order to preserve BMP2 activities has also been tested in vitro [72] and in vivo [73]. Collagen-low molecular weight hyaluronic acid semi-interpenetrating network loaded with gelatin microspheres for cell and growth factor delivery for nucleus pulposus regeneration has also been proposed [74].

6.5   Future Perspectives

HA can be used in isolation, taking advantage of its inherent properties. However, the road for the future will include, from one side, optimization of HA itself (better understanding of adequate formulation, dosage, number of treatments according to the pathology, and the patients' profile). But from another much more ambitious perspective, what we have learnt from this biopolysaccharide and its possibilities and clinical results should be used to develop advanced TERM strategies combining growth factors, cells, scaffolds, bioreactors, nanotechnology, etc. aiming for the ultimate full repair of the tissues and control of injuries/diseases. Some initial steps have been given but a long road needs to be traveled with this goal.

6.6   Final Remarks

Hyaluronic acid is a safe alternative for conservative treatment of several orthopedic conditions.

Intra-articular joint injections (with or without image-assisted application) have proven to be a safe procedure. Rare and mostly self-limited adverse reactions have been reported (mainly transitory pain and swelling).

It has shown fair midterm results in symptomatic control of osteoarthritis in different joints and in different grades of the disease. The highest clinical experience reported comes from the knee and ankle joint.

Promising results have been achieved in the treatment of tendinopathies with peri-tendon injections. Some authors advise for possible deleterious effects from intra-tendon injections.

More high-quality studies are required before further clinical conclusions at this point.

In the future, TERM approaches promise to improve current results, mainly by combining several factors with advanced strategies, specifically designed for each clinical condition.

References

1. Xing D, Wang B, Liu Q, Ke Y, Xu Y, Li Z, Lin J (2016) Intra-articular hyaluronic acid in treating knee osteoarthritis: a PRISMA-compliant systematic review of overlapping meta-analysis. Sci Rep 6:32790. https://doi.org/10.1038/srep32790
2. Puig-Junoy J, Ruiz Zamora A (2015) Socio-economic costs of osteoarthritis: a systematic review of cost-of-illness studies. Semin Arthritis Rheum 44(5):531–541. https://doi.org/10.1016/j.semarthrit.2014.10.012
3. Hunter DJ (2009) Risk stratification for knee osteoarthritis progression: a narrative review. Osteoarthr Cartil/OARS Osteoarthr Res Soc 17(11):1402–1407. https://doi.org/10.1016/j.joca.2009.04.014
4. van Dijk CN, Reilingh ML, Zengerink M, van Bergen CJ (2010) Osteochondral defects in the ankle: why painful? Knee Surg Sports Traumatol Arthrosc Off J ESSKA 18(5):570–580. https://doi.org/10.1007/s00167-010-1064-x
5. Temenoff JS, Mikos AG (2000) Review: tissue engineering for regeneration of articular cartilage. Biomaterials 21(5):431–440
6. Hunziker EB (2000) Articular cartilage repair: problems and perspectives. Biorheology 37(1-2):163–164
7. Pacifici M, Koyama E, Iwamoto M, Gentili C (2000) Development of articular cartilage: what do we know about it and how may it occur? Connect Tissue Res 41(3):175–184
8. Kuettner KE, Cole AA (2005) Cartilage degeneration in different human joints. Osteoarthr Cartil/OARS Osteoarthr Res Soc 13(2):93–103. https://doi.org/10.1016/j.joca.2004.11.006
9. Gelber AC, Hochberg MC, Mead LA, Wang NY, Wigley FM, Klag MJ (2000) Joint injury in young adults and risk for subsequent knee and hip osteoarthritis. Ann Intern Med 133(5):321–328
10. de Vos RJ, van PLJ V, Moen MH, Weir A, Tol JL, Maffulli N (2010) Autologous growth factor injections in chronic tendinopathy: a systematic review. Br Med Bull 95(1):63–77. https://doi.org/10.1093/bmb/ldq006
11. van Sterkenburg MN, van Dijk CN (2011) Injection treatment for chronic midportion Achilles tendinopathy: do we need that many alternatives? Knee Surg Sports Traumatol Arthrosc Off J ESSKA 19(4):513–515. https://doi.org/10.1007/s00167-011-1415-2
12. Snedeker JG, Foolen J (2017) Tendon injury and repair - a perspective on the basic mechanisms of tendon disease and future clinical therapy. Acta Biomater 63:18–36. https://doi.org/10.1016/j.actbio.2017.08.032
13. Weinraub GM (2005) Orthobiologics: a survey of materials and techniques. Clin Podiatr Med Surg 22(4):509–519., v. https://doi.org/10.1016/j.cpm.2005.08.003

14. Correia SI, Pereira H, Silva-Correia J, Van Dijk CN, Espregueira-Mendes J, Oliveira JM, Reis RL (2014) Current concepts: tissue engineering and regenerative medicine applications in the ankle joint. J R Soc Interface R Soc 11(92):20130784. https://doi.org/10.1098/rsif.2013.0784

15. de Mos M, van der Windt AE, Jahr H, van Schie HT, Weinans H, Verhaar JA, van Osch GJ (2008) Can platelet-rich plasma enhance tendon repair? A cell culture study. Am J Sports Med 36(6):1171–1178. https://doi.org/10.1177/0363546508314430

16. DeChellis DM, Cortazzo MH (2011) Regenerative medicine in the field of pain medicine: prolotherapy, platelet-rich plasma therapy, and stem cell therapy—theory and evidence. Tech Reg Anesth Pain Manag 15:74–80

17. Evans CH (2013) Platelet-rich plasma a la carte: commentary on an article by Satoshi Terada, MD, et al.: "use of an antifibrotic agent improves the effect of platelet-rich plasma on muscle healing after injury". J Bone Joint Surg Am 95(11):e801–e802. https://doi.org/10.2106/JBJS.M.00485

18. Luyten FP, Vanlauwe J (2012) Tissue engineering approaches for osteoarthritis. Bone 51(2):289–296. https://doi.org/10.1016/j.bone.2011.10.007

19. Martel-Pelletier J, Wildi LM, Pelletier JP (2012) Future therapeutics for osteoarthritis. Bone 51(2):297–311. https://doi.org/10.1016/j.bone.2011.10.008

20. Qi Y, Feng G, Yan W (2012) Mesenchymal stem cell-based treatment for cartilage defects in osteoarthritis. Mol Biol Rep 39(5):5683–5689. https://doi.org/10.1007/s11033-011-1376-z

21. Nordsletten L (2006) Recent developments in the use of bone morphogenetic protein in orthopaedic trauma surgery. Curr Med Res Opin 22(s1):S13–S17. https://doi.org/10.1185/030079906X80585

22. Myers KR (2013) Trends in biological joint resurfacing. Bone Joint Res 2(9):193–199. https://doi.org/10.1302/2046-3758.29.2000189

23. Pereira H, Ripoll L, Oliveira JM, Reis RL, Espregueira-Mendes J, van Dijk C (2016) A Engenharia de tecidos nas lesões do Desporto. Traumatologia Desportiva. LIDEL, Lisboa

24. Di Giacomo G, De Gasperis N (2015) The role of hyaluronic acid in patients affected by glenohumeral osteoarthritis. J Biol Regul Homeost Agents 29(4):945–951

25. Flores C, Balius R, Alvarez G, Buil MA, Varela L, Cano C, Casariego J (2017) Efficacy and tolerability of Peritendinous hyaluronic acid in patients with supraspinatus Tendinopathy: a multicenter, randomized, controlled trial. Sports Med Open 3(1):22. https://doi.org/10.1186/s40798-017-0089-9

26. Gigante A, Callegari L (2011) The role of intra-articular hyaluronan (Sinovial) in the treatment of osteoarthritis. Rheumatol Int 31(4):427–444. https://doi.org/10.1007/s00296-010-1660-6

27. Van Den Bekerom MP, Mylle G, Rys B, Mulier M (2006) Viscosupplementation in symptomatic severe hip osteoarthritis: a review of the literature and report on 60 patients. Acta Orthop Belg 72(5):560–568

28. Zengerink M, Struijs PA, Tol JL, van Dijk CN (2010) Treatment of osteochondral lesions of the talus: a systematic review. Knee Surg Sports Traumatol Arthrosc Off J ESSKA 18(2):238–246. https://doi.org/10.1007/s00167-009-0942-6

29. Necas J, Bartosikova L, Brauner P, Kolar J (2008) Hyaluronic acid (hyaluronan): a review. Veterinarni Medicina 53(8):397–411

30. Collins MN, Birkinshaw C (2013) Hyaluronic acid based scaffolds for tissue engineering—a review. Carbohydr Polym 92(2):1262–1279. https://doi.org/10.1016/j.carbpol.2012.10.028

31. Witteveen AG, Hofstad CJ, Kerkhoffs GM (2015) Hyaluronic acid and other conservative treatment options for osteoarthritis of the ankle. Cochrane Database Syst Rev 10:CD010643. https://doi.org/10.1002/14651858.CD010643.pub2

32. Bonnet F, Dunham DG, Hardingham TE (1979) Structure and interactions of cartilage proteoglycan binding region and link protein. Biochem J 228:77–85

33. McArthur BA, Dy CJ, Fabricant PD, Valle AG (2012) Long term safety, efficacy, and patient acceptability of hyaluronic acid injection in patients with painful osteoarthritis of the knee. Patient Prefer Adherence 6:905–910. https://doi.org/10.2147/ppa.s27783

34. Stern R, Asari AA, Sugahara KN (2006) Hyaluronan fragments: an information-rich system. Eur J Cell Biol 85(8):699–715. https://doi.org/10.1016/j.ejcb.2006.05.009
35. Stern R, Kogan G, Jedrzejas MJ, Šoltés L (2007) The many ways to cleave hyaluronan. Biotechnol Adv 25(6):537–557. https://doi.org/10.1016/j.biotechadv.2007.07.001
36. Toole BP (2001) Hyaluronan in morphogenesis. Semin Cell Dev Biol 12(2):79–87. https://doi.org/10.1006/scdb.2000.0244
37. Strauss EJ, Hart JA, Miller MD, Altman RD, Rosen JE (2009) Hyaluronic acid Viscosupplementation and osteoarthritis: current uses and future directions. Am J Sports Med 37(8):1636–1644. https://doi.org/10.1177/0363546508326984
38. Toole BP (2004) Hyaluronan: From extracellular glue to pericellular cue. Nat Rev Cancer 4(7):528–539. https://doi.org/10.1038/nrc1391
39. Bollyky P, Bogdani M, Bollyky J, Hull R, Wight T (2012) The role of Hyaluronan and the extracellular matrix in islet inflammation and immune regulation. Curr Diab Rep 12(5):471–480. https://doi.org/10.1007/s11892-012-0297-0
40. Preston M, Sherman L (2011) Neural stem cell niches: critical roles for the hyaluronan-based matrix in neural stem cell proliferation and differentiation. Front Biosci 3:1165–1179
41. Balazs EA, Denlinger JL (1993) Viscosupplementation: a new concept in the treatment of osteoarthritis. J Rheumatol Suppl 39:3–9
42. Moreland LW (2003) Intra-articular hyaluronan (hyaluronic acid) and hylans for the treatment of osteoarthritis: mechanisms of action. Arthritis Res Ther 5(2):54–67
43. Lynen N, De Vroey T, Spiegel I, Van Ongeval F, Hendrickx NJ, Stassijns G (2017) Comparison of Peritendinous Hyaluronan injections versus extracorporeal shock wave therapy in the treatment of painful Achilles' Tendinopathy: a randomized clinical efficacy and safety study. Arch Phys Med Rehabil 98(1):64–71. https://doi.org/10.1016/j.apmr.2016.08.470
44. Araujo JP, Silva L, Andrade R, Pacos M, Moreira H, Migueis N, Pereira R, Sarmento A, Pereira H, Loureiro N, Espregueira-Mendes J (2016) Pain reduction and improvement of function following ultrasound-guided intra-articular injections of triamcinolone hexacetonide and hyaluronic acid in hip osteoarthritis. J Biol Regul Homeost Agents 30(4 Suppl 1): 51–62
45. Bannuru RR, Natov NS, Dasi UR, Schmid CH, McAlindon TE (2011) Therapeutic trajectory following intra-articular hyaluronic acid injection in knee osteoarthritis--meta-analysis. Osteoarthr Cartil/OARS Osteoarthr Res Soc 19(6):611–619. https://doi.org/10.1016/j.joca.2010.09.014
46. Wang CT, Lin J, Chang CJ, Lin YT, Hou SM (2004) Therapeutic effects of hyaluronic acid on osteoarthritis of the knee. A meta-analysis of randomized controlled trials. J Bone Joint Surg Am 86-a(3):538–545
47. Bellamy N, Campbell J, Robinson V, Gee T, Bourne R, Wells G (2006) Viscosupplementation for the treatment of osteoarthritis of the knee. Cochrane Database Syst Rev (2):Cd005321. https://doi.org/10.1002/14651858.CD005321.pub2
48. Karatosun V, Unver B, Ozden A, Ozay Z, Gunal I (2008) Intra-articular hyaluronic acid compared to exercise therapy in osteoarthritis of the ankle. A prospective randomized trial with long-term follow-up. Clin Exp Rheumatol 26(2):288–294
49. Migliore A, Bizzi E, Massafra U, Vacca F, Alimonti A, Iannessi F, Tormenta S (2009) Viscosupplementation: a suitable option for hip osteoarthritis in young adults. Eur Rev Med Pharmacol Sci 13(6):465–472
50. Monfort J, Rotes-Sala D, Segales N, Montanes FJ, Orellana C, Llorente-Onaindia J, Mojal S, Padro I, Benito P (2015) Comparative efficacy of intra-articular hyaluronic acid and corticoid injections in osteoarthritis of the first carpometacarpal joint: results of a 6-month single-masked randomized study. Joint Bone Spine 82(2):116–121. https://doi.org/10.1016/j.jbspin.2014.08.008
51. Porcellini G, Merolla G, Giordan N, Paladini P, Burini A, Cesari E, Castagna A (2015) Intra-articular glenohumeral injections of HYADD(R)4-G for the treatment of painful shoulder osteoarthritis: a prospective multicenter, open-label trial. Joints 3(3):116–121. https://doi.org/10.11138/jts/2015.3.3.116

52. Legre-Boyer V (2015) Viscosupplementation: techniques, indications, results. Orthop Trauma Surg Res OTSR 101(1s):S101–s108. https://doi.org/10.1016/j.otsr.2014.07.027

53. Braithwaite GJ, Daley MJ, Toledo-Velasquez D (2016) Rheological and molecular weight comparisons of approved hyaluronic acid products - preliminary standards for establishing class III medical device equivalence. J Biomater Sci Polym Ed 27(3):235–246. https://doi.org /10.1080/09205063.2015.1119035

54. Ayhan E, Kesmezacar H, Akgun I (2014) Intraarticular injections (corticosteroid, hyaluronic acid, platelet rich plasma) for the knee osteoarthritis. World J Orthop 5(3):351–361. https://doi. org/10.5312/wjo.v5.i3.351

55. Buda R, Vannini F, Cavallo M, Baldassarri M, Luciani D, Mazzotti A, Pungetti C, Olivieri A, Giannini S (2013) One-step arthroscopic technique for the treatment of osteochondral lesions of the knee with bone-marrow-derived cells: three years results. Musculoskelet Surg 97(2):145–151. https://doi.org/10.1007/s12306-013-0242-7

56. Doral MN, Bilge O, Batmaz G, Donmez G, Turhan E, Demirel M, Atay OA, Uzumcugil A, Atesok K, Kaya D (2012) Treatment of osteochondral lesions of the talus with microfracture technique and postoperative hyaluronan injection. Knee Surg Sports Traumatol Arthrosc Off J ESSKA 20(7):1398–1403. https://doi.org/10.1007/s00167-011-1856-7

57. Giannini S, Buda R, Battaglia M, Cavallo M, Ruffilli A, Ramponi L, Pagliazzi G, Vannini F (2013) One-step repair in talar osteochondral lesions: 4-year clinical results and t2-mapping capability in outcome prediction. Am J Sports Med 41(3):511–518. https://doi. org/10.1177/0363546512467622

58. Kon E, Mandelbaum B, Buda R, Filardo G, Delcogliano M, Timoncini A, Fornasari PM, Giannini S, Marcacci M (2011) Platelet-rich plasma intra-articular injection versus hyaluronic acid viscosupplementation as treatments for cartilage pathology: from early degeneration to osteoarthritis. Arthroscopy J Arthrosc Relat Surg Off Publ Arthrosc Assoc North Am Int Arthrosc Assoc 27(11):1490–1501. https://doi.org/10.1016/j.arthro.2011.05.011

59. Mason LW, Wilson-Jones N, Williams P (2014) The use of a cell-free chondroinductive implant in a child with massive cartilage loss of the talus after an open fracture dislocation of the ankle: a case report. J Pediatr Orthop 34(8):e58–e62. https://doi.org/10.1097/bpo.0000000000000198

60. Wong KL, Lee KB, Tai BC, Law P, Lee EH, Hui JH (2013) Injectable cultured bone marrow-derived mesenchymal stem cells in varus knees with cartilage defects undergoing high tibial osteotomy: a prospective, randomized controlled clinical trial with 2 years' follow-up. Arthroscopy J Arthrosc Relat Surg Off Publ Arthrosc Assoc North Am Int Arthrosc Assoc 29(12):2020–2028. https://doi.org/10.1016/j.arthro.2013.09.074

61. Colen S, Geervliet P, Haverkamp D, Van Den Bekerom MP (2014) Intra-articular infiltration therapy for patients with glenohumeral osteoarthritis: a systematic review of the literature. Int J Shoulder Surg 8(4):114–121. https://doi.org/10.4103/0973-6042.145252

62. Trellu S, Dadoun S, Berenbaum F, Fautrel B, Gossec L (2015) Intra-articular injections in thumb osteoarthritis: a systematic review and meta-analysis of randomized controlled trials. Joint Bone Spine 82(5):315–319. https://doi.org/10.1016/j.jbspin.2015.02.002

63. Abat F, Alfredson H, Cucchiarini M, Madry H, Marmotti A, Mouton C, Oliveira JM, Pereira H, Peretti GM, Romero-Rodriguez D, Spang C, Stephen J, van Bergen CJA, de Girolamo L (2017) Current trends in tendinopathy: consensus of the ESSKA basic science committee. Part I: biology, biomechanics, anatomy and an exercise-based approach. J Exp Orthop 4(1):18. https://doi.org/10.1186/s40634-017-0092-6

64. Frizziero A, Vittadini F, Barazzuol M, Gasparre G, Finotti P, Meneghini A, Maffulli N, Masiero S (2017) Extracorporeal shockwaves therapy versus hyaluronic acid injection for the treatment of painful non-calcific rotator cuff tendinopathies: preliminary results. J Sports Med Phys Fitness 57(9):1162–1168. https://doi.org/10.23736/S0022-4707.16.06408-2

65. Kumai T, Muneta T, Tsuchiya A, Shiraishi M, Ishizaki Y, Sugimoto K, Samoto N, Isomoto S, Tanaka Y, Takakura Y (2014) The short-term effect after a single injection of high-molecular-weight hyaluronic acid in patients with enthesopathies (lateral epicondylitis, patellar tendi-

nopathy, insertional Achilles tendinopathy, and plantar fasciitis): a preliminary study. J Orthop Sci 19(4):603–611. https://doi.org/10.1007/s00776-014-0579-2

66. Wu PT, Jou IM, Kuo LC, Su FC (2016) Intratendinous injection of hyaluronate induces acute inflammation: a possible detrimental effect. PLoS One 11(5):e0155424. https://doi.org/10.1371/journal.pone.0155424

67. Antunes JC, Oliveira JM, Reis RL, Soria JM, Gomez-Ribelles JL, Mano JF (2010) Novel poly(L-lactic acid)/hyaluronic acid macroporous hybrid scaffolds: characterization and assessment of cytotoxicity. J Biomed Mater Res A 94(3):856–869. https://doi.org/10.1002/jbm.a.32753

68. Forriol F, Longo UG, Duart J, Ripalda P, Vaquero J, Loppini M, Romeo G, Campi S, Khan WS, Muda AO, Denaro V (2015) VEGF, BMP-7, Matrigel(TM), hyaluronic acid, in vitro cultured chondrocytes and trephination for healing of the avascular portion of the meniscus. An experimental study in sheep. Curr Stem Cell Res Ther 10(1):69–76

69. Kon E, Filardo G, Robinson D, Eisman JA, Levy A, Zaslav K, Shani J, Altschuler N (2013) Osteochondral regeneration using a novel aragonite-hyaluronate bi-phasic scaffold in a goat model. Knee Surg Sports Traumatol Arthrosc. [Epub ahead of print]:1–13. https://doi.org/10.1007/s00167-013-2467-2

70. Anitua E, Sanchez M, De la Fuente M, Zalduendo MM, Orive G (2012) Plasma rich in growth factors (PRGF-Endoret) stimulates tendon and synovial fibroblasts migration and improves the biological properties of hyaluronic acid. Knee Surg Sports Traumatol Arthrosc Off J ESSKA 20(9):1657–1665. https://doi.org/10.1007/s00167-011-1697-4

71. Chen WH, Lo WC, Hsu WC, Wei HJ, Liu HY, Lee CH, Tina Chen SY, Shieh YH, Williams DF, Deng WP (2014) Synergistic anabolic actions of hyaluronic acid and platelet-rich plasma on cartilage regeneration in osteoarthritis therapy. Biomaterials 35(36):9599–9607. https://doi.org/10.1016/j.biomaterials.2014.07.058

72. Srinivasan PP, McCoy SY, Jha AK, Yang W, Jia X, Farach-Carson MC, Kirn-Safran CB (2012) Injectable perlecan domain 1-hyaluronan microgels potentiate the cartilage repair effect of BMP2 in a murine model of early osteoarthritis. Biomed Mater 7(2):024109. https://doi.org/10.1088/1748-6041/7/2/024109

73. Sanchez M, Azofra J, Anitua E, Andia I, Padilla S, Santisteban J, Mujika I (2003) Plasma rich in growth factors to treat an articular cartilage avulsion: a case report. Med Sci Sports Exerc 35(10):1648–1652. https://doi.org/10.1249/01.MSS.0000089344.44434.50

74. Tsaryk R, Gloria A, Russo T, Anspach L, De Santis R, Ghanaati S, Unger RE, Ambrosio L, Kirkpatrick CJ (2015) Collagen-low molecular weight hyaluronic acid semi-interpenetrating network loaded with gelatin microspheres for cell and growth factor delivery for nucleus pulposus regeneration. Acta Biomater 20:10–21. https://doi.org/10.1016/j.actbio.2015.03.041

Chapter 7
Semi-IPN- and IPN-Based Hydrogels

Nicole Zoratto and Pietro Matricardi

Abstract Semi-interpenetrating polymer networks (semi-IPNs) and interpenetrating polymeric networks (IPNs) have emerged as innovative materials for biomedical and pharmaceutical applications. The interest in these structures is due to the possibility of combining the favorable properties of each polymeric component of the IPNs or semi-IPNs leading to a new system with properties that often differ from those of the two single components. In this respect, polysaccharides represent an opportunity in this field, combining a general biocompatibility and a good availability. Moreover, the functional groups along the polymer chains allow chemical derivatization, widening the possibilities in semi-IPNs and IPNs building up. At the same time, materials based on proteins are often used in this field, due to their similarity to the materials present in the human body. All these overall properties allow tailoring new materials, thus designing desired properties and preparing new hydrogels useful in the biomedical field. In the present chapter, we chose to describe systems prepared starting from the most important and studied hydrogel-forming polysaccharides: alginate, hyaluronic acid, chitosan, dextran, gellan, and scleroglucan. Besides, systems based on proteins, such as gelatin, collagen, and elastin, are also described. With this chapter, we aim describing the routes already traveled in this field, depicting the state of the art and hoping to raise interest in designing new promising strategies useful in biomedical and pharmaceutical applications.

Keywords Semi-IPN · IPN · Hydrogels · Polysaccharides · Proteins

N. Zoratto · P. Matricardi (✉)
Departement of Drug Chemistry and Technologies, Sapienza University of Roma, Roma, Italy
e-mail: pietro.matricardi@uniroma1.it

© Springer International Publishing AG, part of Springer Nature 2018
J. M. Oliveira et al. (eds.), *Osteochondral Tissue Engineering*,
Advances in Experimental Medicine and Biology 1059,
https://doi.org/10.1007/978-3-319-76735-2_7

7.1   Semi-IPN and IPN Hydrogels

Semi-interpenetrating polymer networks (semi-IPNs) and interpenetrating polymeric networks (IPNs) have emerged as innovative materials for biomedical and pharmaceutical applications. The interest in these structures is due to the possibility of combining the favorable properties of each polymeric component of the IPNs or Semi-IPNs leading to a new system with properties that often differ from those of the two single components [1]. In this respect, tailoring of the resulting material thus reaching the desired properties represents an important route to widening the spectrum of materials useful in the biomedical field. Table 7.1 summarizes IPN and semi-IPN hydrogels recently investigated for tissue engineering applications and reported in the chapter.

Table 7.1 IPN and semi-IPN hydrogels recently investigated for tissue engineering applications and described in the chapter

Hydrogel composition		Type of network	Application	References
Polymer 1	Polymer 2			
Alginate (Alg)	Pluronic F127	IPN	Antiadhesive agent after surgery	[19]
	Fibrin	IPN	In vitro growth of ovarian follicles	[29]
	pHEMA	Semi-IPN	Biomedical applications	[30]
Methacrylated alginate (MAAlg)	Collagen	IPN	3D preosteoblast spreading and osteogenic differentiation	[38]
Hyaluronic Acid (HA)	Hyperbranched PEG-based copolymer	Semi-IPN	Potential scaffold	[42]
	dex-HEMA	Semi-IPN	Potential bioprintable scaffold	[43]
	Fibrin	IPN	Potential scaffold	[44]
Methacrylated hyaluronic Acid (MAHA)	Diacrylated PEG	IPN	Cartilage repair	[46]
HA glycidyl methacrylated (GMHA)	Collagen	IPN	Regenerative medicine	[47]
	Puramatrix	IPN	Neurite growth and extension	[48]
Chitosan	pNIPAAM	Semi-IPN	Potential scaffold	[50]
N-carboxyethyl Chitosan	2-pHEMA	Semi-IPN	Wound dressing	[52]
Dextran methacrylated and aldehyde bifunctionalized	Gelatin	IPN	Potential vascular scaffold	[56]

(continued)

Table 7.1 (continued)

Hydrogel composition		Type of network	Application	References
Polymer 1	Polymer 2			
Gelatin-graft-polyaniline	Oxidized dextran	IPN	Potential scaffold	[57]
Gellan gum methacrylated (GGMA)	Gelatin polyacrylamide	IPN	Regeneration of load-bearing tissue	[62]
Gellan gum (GG)	HA	Semi-IPN	Bone regeneration of osteochondral defects	[63]
		IPN sponge	Skin wound regeneration	[64]
Gelatin (Gel)	Diacrylated PEG	IPN	Cartilage tissue engineering	[75]
Gelatin methacrylamide (GelMA)	PEG	BioSIN	Potential scaffold	[76]
Gelatin	Silk fibroin	IPN	Potential scaffold	[77]
	PVA	Theta-gel	Cartilage regeneration	[78]
Collagen	HA	Semi-IPN	Potential scaffold	[80]
	Chitosan	IPN	3D–scaffold for carcinoma cells	[83]
	2-MPC	IPN	Corneal substitute	[84]
Collagen methacrylated	Chondroitin sulfate-HA	IPN	Cartilage tissue engineering	[81]

7.1.1 Definitions

The IPNs can be considered belonging to the class of polymer blends. An IPN is defined by IUPAC as "A polymer comprising two or more networks which are at least partially interlaced on a molecular scale but not covalently bonded to each other and cannot be separated unless chemical bonds are broken. A mixture of two or more pre-formed polymer networks is not an IPN" [2].

As Sperling reported, IPN systems are "rediscovered" several times over the years [3, 4]. In 1914, Aylsworth designed the first synthetic IPN, composed of a mixture of natural rubber, sulfur, and partly reacted phenol-formaldehyde resins [1]. Some years later, Staudinger and Hutchinson (1951) and Solt (1955) described the use of IPN systems, but only in 1960, the term "interpenetrating polymer networks" was coined by Millar. Millar cross-linked polystyrene and swelled this sample in solutions of styrene and divinylbenzene and then polymerized the monomers within the swollen network to form a polystyrene network within another polystyrene network. Thanks to his pioneering work, IPNs having both networks identical in chemical compositions have sometimes called Millar IPNs [5]. Only in 1969, Frisch [6] and Sperling [7] developed IPN composed of two different polymers independently.

Semi-IPNs differ from IPNs because the chains of the second polymer are dispersed only into the network formed by the first polymer without forming another network interpenetrated with the first one. The IUPAC definition referred to semi-IPN is "A polymer comprising one or more networks and one or more linear or branched polymer(s) characterized by the penetration on a molecular scale of at least one of the networks by at least some of the linear or branched macromolecules." In addition, a note was added in order to underline the difference between IPN and semi-IPN, reporting "semi-interpenetrating polymer networks are distinguished from interpenetrating polymer networks because the constituent linear or branched polymers can, in principle, be separated from the constituent polymer network(s) without breaking chemical bonds" [2].

7.1.2 Properties

In recent years, IPN and semi-IPN systems have attracted substantial interest, thanks to their favorable properties. Generally, an IPN or semi-IPN can combine the properties of their individual components with synergistic effects in many cases. Therefore, it is possible to tune the characteristics of the resulting materials by choosing properly the starting components of the IPN and semi-IPN systems, and in addition, thanks to the possibility of combining synthetic and natural polymers, the range of the reachable properties can be extended [1, 3]. Improved mechanical properties, thermal stability, and chemical resistance are some of the properties presented by IPN and semi-IPN systems. In addition, IPNs differ from other types of polymeric blends, as they swell without dissolving in solvents and creep and flow are suppressed [3, 8].

However, in most IPN systems, a phase separation occurs. Phase separation, which proceeds during the IPN formation, is due to the chemically different structure of the components forming the IPNs. This leads to the development of a heterogeneous structure. However, the process of separation proceeds very slowly due to the high viscosity of the system and to entanglements between chains. Clearly, the thermal behavior of IPN systems depends on the miscibility of the polymeric components and on the phase mixing [9]. In this respect, as far as the semi-IPN systems are concerned, they generally show a higher shift in transition temperature compared to the full IPNs. This effect was explained by the more complete phase separation in semi-IPNs as compared with full ones [9]. Two are the mechanisms of phase separation that may occur during the IPN formation: nucleation and growth and spinodal decomposition. In the nucleation and growth, kinetics spheres of the second phase are formed within the matrix of the first phase. These spheres grow by increasing their diameter. Spinodal decomposition tends to produce interconnected cylinders of the second phase within the matrix of the first phase. These cylinders grow by increasing their wave amplitude. Later, coarsening and coalescence may cause important changes. However, these changes may be hampered by cross-links, which keep the domains small. Spinodal decomposition is the most common mechanism of phase separation [10]. Generally, simultaneous IPNs are phase-separated by a spinodal mechanism, while sequential IPNs via a mechanism of nucleation and growth.

7.1.3 Characterization

IPNs are characterized mainly for their physico-mechanical, morphological, spectroscopic, and thermal properties [11]. The morphology of the IPN systems is deeply affected by the synthetic methodology adopted by the compatibility of the single components forming the IPN networks and by the relative rates of network formation. SEM and TEM are the most common techniques used for the investigation of the networks morphology. With these techniques, it is possible to determine the distribution of phase domains, the shape, the structure in terms of micro- and nanoscale, and the degree of mixing. In this way, a complete and clear elucidation of the IPN architecture may be obtained. Also, other techniques can be used to identify the morphology of IPNs. These include atomic force microscopy (AFM) analysis, confocal laser scanning microscopy (CLSM), X-ray scattering experiments, energy dispersive X-ray analysis (EDAX), and laser scanning confocal microscopy (LSCM) are other techniques for the morphological characterization of an IPN.

Differential scanning calorimetry (DSC), thermogravimetric analysis (TGA), and differential thermal analysis (DTA) are the widely used techniques for the thermal characterizations of IPNs and semi-IPNs. DSC analysis usually shows one or two Tg values for the IPN, allows understanding the interpenetration degree of the network components and, at the same time, the formation of any macro- or microphase separation. TGA analysis allows investigating the thermal stability of IPN networks.

Mechanical properties commonly used to characterize IPN are tensile strength, elongation at break, Young's modulus, and hardness. In order to determine these parameters, different techniques may be used. These include rheological analyses, tensile tests, extensiometry, tearing tests, and compression tests [12]. IPNs usually show mechanical properties intermediate of the single polymeric components, so the interaction of a less swellable, stiffer hydrogel with a more swellable, softer hydrogel can be used to tune the IPN mechanical properties and swellability. In many cases, IPNs show a higher toughness compared to the single constituents. Enhanced toughness occurs in double-network (DN) hydrogels, a special class of interpenetrated polymer, where the first network is more tightly cross-linked than the second network and the molar ratio of the second network to the first network is greater than ~ 5.

The most powerful method in the investigation of viscoelastic properties of polymeric systems is dynamic mechanical spectroscopy (DMS), which enables the estimation of the elastic moduli, mechanical losses, glass transition temperature, and relaxation characteristics of IPN and semi-IPNs.

Finally, spectroscopic techniques are widely used in the structure elucidation, such as, nuclear magnetic resonance (NMR), infrared (IR), and electron spin resonance (ESR) techniques. NMR spectroscopy is used to estimate the interpenetration of polymers in IPN at nanoscale level; moreover, information on the morphology, miscibility, microstructure, and mobility can be directly obtained by NMR studies. IR and FT-IR spectroscopic techniques have been widely used for identifying the components with their specific signals in IPNs.

7.2 Polysaccharides-Based IPN and Semi-IPN Hydrogels

Polysaccharides play a very important role in the IPN and semi-IPN hydrogels for the biomedical field due to the ability of some of them to form hydrogels by means of physical or chemical interactions with small biocompatible molecules and/or ions or among their chains, in special conditions. Moreover, polysaccharides are a favorable polymer system, thanks to their abundance in nature, their generally accepted biocompatibility, and the possibility to modify their chemical structure by means of easy and efficient reactions [1, 13]. In this chapter, we will describe some of the most important systems developed in recent years, in order to give to the reader a general overview of this field, classifying the paragraphs according to the main polysaccharide used in the system.

7.2.1 Alginate

Alginate (Alg) is a linear polysaccharide derived from brown algae or bacteria and composed of $1 \rightarrow 4$ linked β-D-mannuronic acid (M) and α-L-guluronic acid (G). The G and M residues are arranged in homopolymer blocks (MM, GG) interspersed by regions containing alternating blocks of M and G with different M/G ratios (MG-blocks) (Fig. 7.1). The wide variability of the composition of

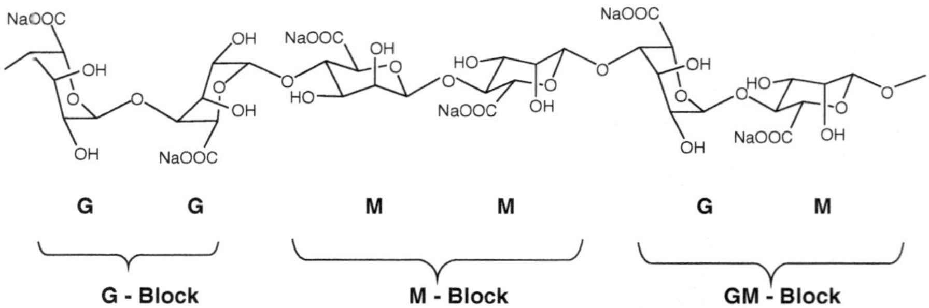

Fig. 7.1 Repeating units in alginate structure

the alternating zones and the M/G ratio affected the physicochemical properties of Alg. Applications of alginate are related to its gelling ability, thanks to the interactions with divalent cations. The selectivity in ion binding by alginate is highly dependent on the presence of G residues in the polysaccharide structure. These G/divalent cation interactions are responsible for the "egg box" model formation. Due to these characteristics, Alg has been widely used in food industry and in biotechnological and biomedical fields.

7.2.1.1 Smart Alginate IPNs and Semi-IPNs

Smart polymers or stimuli-responsive polymers are polymers that are able to be responsive to a number of stimuli, such as temperature, pH, electrical, or magnetic field. Several smart IPN and semi-IPN Alg-based systems are reported in literature. These polymeric networks are generally responsive to the temperature or pH but in most cases are responsive to both temperature and pH and thus defined as dual-stimuli-responsive gels. In the field of thermo-responsive hydrogels, hydrogels showing a lower critical solution temperature (LCST) polymers have been extensively studied for biomedical and pharmaceutical applications, and the attention was particularly focused on systems that show a LCST value near body temperature. It is well known that pNIPAAm hydrogels undergo a sharp and reversible phase transition near the body temperature (32–37 °C). Synthesis of multi-responsive IPN composite hydrogels, based on Alg and pNIPAAm, constitutes one of the strategies to increase the porosity of the resulting gels and thus to achieve networks with a faster release of the drug [14]. Both semi-IPN and full-IPN composed of Alg and pNIPAAm have been investigated [15–18]. Not only pNIPAAm but also other synthetic polymers have LCST values near body temperature, such as several Pluronics. Pluronics are block copolymers of poly(ethylene oxide)b-poly(propylene oxide)-b-poly(ethylene oxide) (PEO-PPO-PEO) with a LCST that depends on their composition (e.g., ratio and molecular weight of the PEO/PPO blocks and overall molecular weight). Some Pluronics exhibit a thermo-reversible gelation below the body temperature, and for this reason, they are often applied in the pharmaceutical field. Vong and co-workers have successfully developed an in situ semi-IPN device based on an Alg and Pluronic F127 for the ophthalmic release of the pilocarpine. Guardix SG™ is a clinically approved system successfully applied as an antiadhesive agent after the surgery [19]. It is composed of sodium Alg, poloxamer, and calcium chloride, and it is able to create a thermosensitive viscous gel in contact with the body temperature and to form a mechanical barrier that separates injured tissue. It is generally used in spine thyroid surgeries in order to reduce the incidence of postoperative adhesion [20].

Other natural polymers, such as gelatin, cellulose derivatives, and agarose, have been widely used in combination with Alg to form thermosensitive IPN and semi-IPN [21–23]. Choudhary and co-workers reported the preparation IPN hydrogels composed by alginate and hydrophobically modified ethyl hydroxyl ethyl cellulose (HMEHEC) [24]. They found that the rheology of these systems could be easily tuned because of the mechanical strength of the IPN is strongly dependent on the relative ratio of the polymers. These IPNs were able to entrap highly hydrophobic drugs, such as the prodrug NSAID (nonsteroidal anti-inflammatory drug) sulindac, increasing their solubility because of the presence of hydrophobic HMEHEC domains in the hydrogel matrix.

Ionic hydrogels are swollen polymer networks containing pendent groups, such as carboxylic or sulfate groups, which exhibit a sol/gel transition as a result of changing the environmental pH. In the case of the acidic groups, as the degree of the ionization is increased, the number of fixed charges is increased, resulting in increased electrostatic repulsion between the chains and consequently in a sol/gel transition. Ramesh Babu et al. prepared novel IPN microgels of sodium alginate and acrylic acid for the controlled release of ibuprofen [25]. As mentioned above, dual-stimuli-responsive hydrogels are deeply investigated systems. Among them, semi-IPNs based on Alg and pNIPAAm have been widely studied [26, 27]. Compared to pure pNIPAAm hydrogels, these combined networks showed a swelling profile that depends on the charge of Alg, which in turn depends on the pH of the medium [1]. Therefore, these Alg/pNIPAAM semi-IPN hydrogels are affected by the pH thanks to the presence of the carboxylic groups of the Alg and by the temperature due to the presence of the pNIPAAm. Shi and co-workers studied the effect of pH and temperature on the release of indomethacin, a nonsteroidal anti-inflammatory drug, from semi-IPN beads based on CaAlg and pNIPAAm [27].

7.2.1.2 Physically Cross-Linked Alginate IPNs

Lin et al. prepared IPN beads composed of a water-soluble derivative of chitosan (N,O-carboxymethyl chitosan, NOOC) and alginate [28]. Another interesting application of physically cross-linked Alg was reported by Shikanov and co-workers for the in vitro growth of ovarian follicles [29]. They developed an interpenetrating matrix based on Alg and fibrin aimed to tailor the mechanical properties of the hydrogel to fit the needs of the follicles during their development. The two biopolymers were simultaneously cross-linked by addition of Ca^{2+} ions and thrombin, respectively, leading to IPNs with mechanical properties suitable for tissue regeneration. After fibrin degradation, due to plasmin and various other proteases secreted by the cells, the Alg network assured the mechanical support to the hydrogel. Mechanical properties of the matrix emerged as significant regulators of follicle development, and the authors supposed that small two-layered follicles, cultured in a mechanically dynamic environment, could be able to mimic the in vivo environment and increase the rate of oocyte maturation.

7.2.1.3 Chemically Cross-Linked Alginate IPNs

Chemically cross-linked alginate IPNs were widely studied, thanks to the increased resistance to failure and mechanical strength that they possess compared to gels in which the networks are held together by physical/ionic interactions as in CaAlg gels

[1]. La Gatta et al. developed polyelectrolyte materials based on a semi-IPN matrix composed of Alg and a pHEMA [30]. In detail, after dissolution of the monomers in an aqueous solution of Alg, HEMA was copolymerized with the cationic monomer 2-methacryloxy ethyltrimethyl ammonium chloride (METAC). The mechanical properties as well as the swelling behavior of the resulting semi-IPN hydrogels were dependent on both charged polymers. In the studied systems, the Alg chains had a very positive effect on the biocompatibility of the systems as the gels showed better cell viability and cell adhesion properties than the control p(HEMA-co-METAC) hydrogels. Recently, Hina et al. formulated a novel Na-alginate/PVA hydrogels employing 2-Acylamido-2-methylpropane-sulfonic acid as monomer (AMPS) [31]. Figure 7.2 presents the reaction scheme for the preparation of Na-alginate/PVA-co-AMPS IPN hydrogel. Drug release characteristics from these IPN hydrogels as a

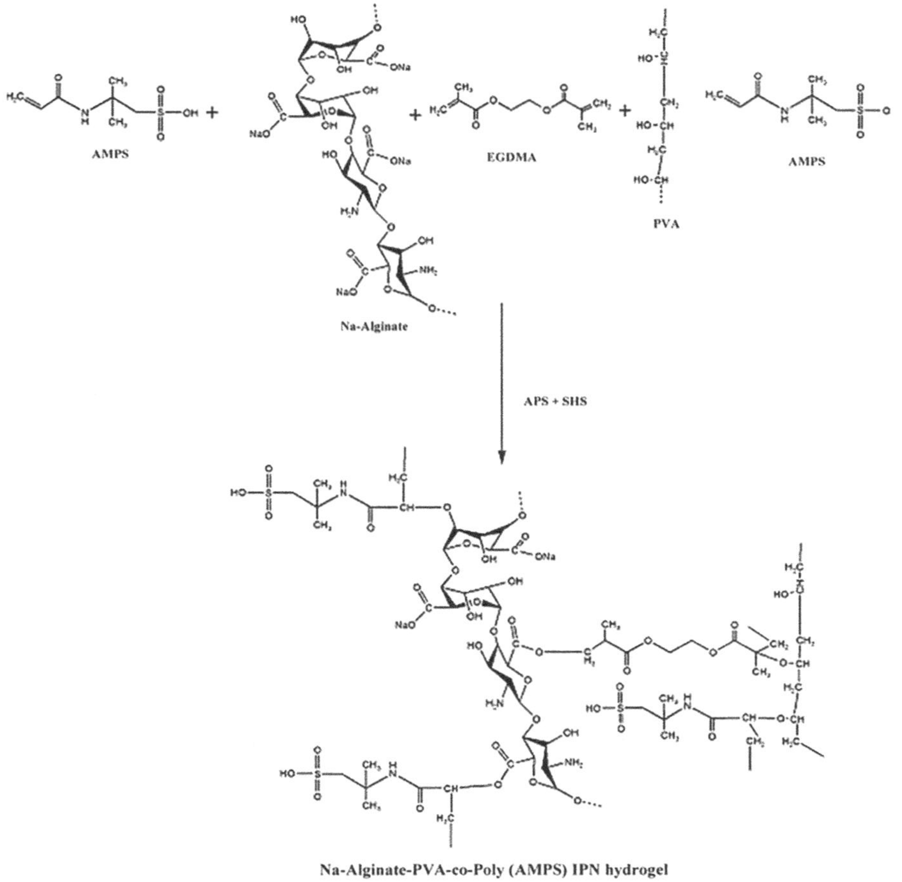

Fig. 7.2 Reaction scheme for the preparation of Na-Alginate/PVA-co-AMPS IPN hydrogel. For details, see Ref. [31]. (Figure reproduced with permission)

function of pH revealed pH-independent release at both pH 1.2 and 7.4, while varying polymer ratio and concentration of monomer to higher level resulted in prolonged drug delivery. Na-alginate-PVA-copoly(AMPS) hydrogel could be an interesting candidate alternate to conventional dosage forms for prolonged delivery of a variety of hydrophilic drugs.

7.2.1.4   Photopolymerized Alginate IPNs

Photopolymerization is an alternative approach for the formation of chemically cross-linked hydrogels and offers the possibility to obtain in situ hydrogel systems by means of UV or visible light irradiation [1]. In general, hydrophilic/water-soluble polymers with polymerizable groups, such as acrylate and methacrylate moieties, form a hydrogel when exposed to UV or visible light. Radicals, which initiate the polymerization, are generated when a so-called photoinitiator undergoes an homolytic bond cleavage upon exposure to UV/visible light. At present, several photoinitiators, having a good cytocompatibility, are available, allowing their in vivo use. Wang and co-workers used photopolymerization in order to improve the stability of calcium-Alg microcapsules by introducing additional polymers to provide covalent linkages via photopolymerization [32]. Photocross-linkable IPN beads based on Alg and polyethylene glycol were reported for the encapsulation of Langerhans islets [33]. Also, in situ cross-linkable IPN hydrogel composed of CaAlg and dextran methacrylate were widely investigated [34–37]. Recently, Sun and colleagues prepared a series of hydrogels based on IPNs based on methacrylated alginate (MAA) and collagen to support preosteoblast spreading and proliferation as well as osteogenic differentiation [38]. Compared to the pure MAA hydrogel, these hydrogels demonstrated higher mechanical moduli, lower swelling ratios, and denser network structures. Moreover, their properties could be fine-tuned by varying the ratio collagen/alginate. MC3T3-E1 cells in IPN hydrogels exhibited a rapid proliferation and spread gradually with prolonged culture time, and their osteogenic differentiation was greatly facilitated. These results should provide collagen-MAA IPN hydrogels as potential three-dimensional scaffolds for bone tissue engineering.

7.2.2   Hyaluronic Acid

Hyaluronic acid (or hyaluronan, HA) is a linear glycosaminoglycan composed of repeating disaccharide unit of D-glucuronic acid and N-acetyl-D-glucosamine, linked through alternating β-$(1 \rightarrow 4)$ and β-$(1 \rightarrow 3)$ glycosidic bonds (Fig. 7.3). HA is one of the major components of the extracellular matrix (ECM) and is present at high concentrations in all connective tissues such as cartilage, vitreous humor, and synovial fluids, where it performs structural and lubricant functions. The biological functions of HA have been widely investigated, and the enormous interest in HA is due to its

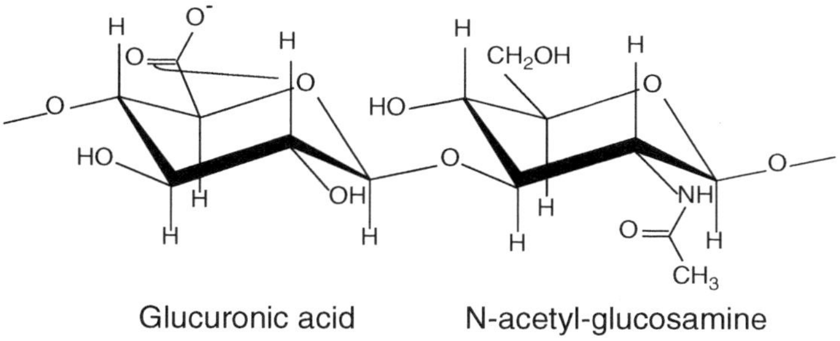

Fig. 7.3 Repeating unit of hyaluronic acid

biocompatibility and its capability to interact with specific cell receptors that can recognize and bind HA selectively. Three are the functional groups of HA that usually are chemically modified: the glucuronic acid group, the primary and secondary hydroxyl groups, and the amine group (after deacetylation of N-acetyl group). Actually, numerous HA derivatives are employed for tissue repair, wound healing, treatment of joint diseases, drug delivery, and as scaffolds for tissue engineering.

7.2.2.1 Temperature-Responsive Hyaluronic Acid IPNs and Semi-IPNs

A pH- and temperature-sensitive semi-IPN system was developed by Santos and co-workers by combining cross-linked thermosensitive pNIPAAm and HA [39]. The semi-interpenetrated polymer networks (semi-IPNs) hydrogel was prepared by mixing HA with pNIPAAm which was then cross-linked in the presence of N,N'-methylenebisacrylamide as cross-linking agent and tetramethylethylenediamine as catalyst. The LCST of the materials was measured, and it was found that it did not change with the introduction of HA. The swelling/deswelling behavior of the semi-IPN responded to pH and temperature changes with good reversibility: addition of HA led to an improvement in the deswelling process at 37 °C and an increase in both water uptake capability and swelling kinetics of the hydrogels at 25 °C. These favorable properties make the pNIPAAm-HA semi-IPN suitable for therapeutic and biomedical applications. Also, "Click" chemistries have been used to cross-link polymer chains within a hydrogel network capable to respond simultaneously or separately to different external stimuli [40]. The thermo-responsive pNIPAAm polymer was modified in order to obtain a new telechelic RAFT-generated pNIPAAm with a propargyl function at both ends, which in this way could be clicked together with the azido-grafted hyaluronic acid (HA) to create a polymer network. This hybrid system displays a multi-responsive behavior versus temperature, pH, and ionic strength and exhibited a distinct volume phase transition temperature (VPTT) between 32 and 34 °C as well as pH-dependent swelling behavior. Importantly, the pore size of the hydrogels could be controlled by varying the spacing of grafted azide functions on the HA chains or by varying the chain length of the pNIPAAm cross-links, which has great utility in designing scaffolds to mimic the extracellular matrix. Additionally, release of fluorescein isothiocyanate-labeled dextran from the gels was increased at temperatures above the VPTT, suggesting a potential dual use as a drug-releasing scaffold material. Degradability of the hydrogels by hyaluronidase was found to be controlled by degree of cross-linking.

Very recently, Joung et al. investigated a thermo-responsive semi-IPN made from hyaluronic acid and Pluronic F127 as a new intra-articular injectable hydrogel for the controlled release of piroxicam [41]. The use of the thermo-responsive surfactant properties of Pluronic F127 allowed to disperse the drug within the hydrogel but at the same time could erode the hydrogel in a very short time at physiological conditions. For this reason, HA was used, and the authors found that HA reduced the amount of Pluronic required for the gelation, thanks to the ability of high molecular weight HA to assist the intermicellar packing in the hydrogel structure (Fig. 7.4a, b) In addition, HA had also the advantage to enhance the mechanical strength of the semi-IPNs and to reduce critical gelation temperature value of the hydrogel compared to pure Pluronic F127 gels. The hydrogel exhibits both sustained drug release behavior and superior bioavailability in physiological conditions; thus, this IPN systems could be a promising hydrogel-based drug delivery platform for the treatment of arthritis.

Another thermo-responsive semi-IPN containing HA within the network was developed by Dong and co-workers [42]. They developed a physically and chemically in situ cross-linkable hydrogel system composed of a thermo-responsive hyperbranched PEG-based copolymer exhibiting a physical gelation around 37 °C. The branched copolymer was synthetized using one-pot and one-step in situ DE-ATRP reaction (deactivation enhanced atom transfer radical polymerization) of

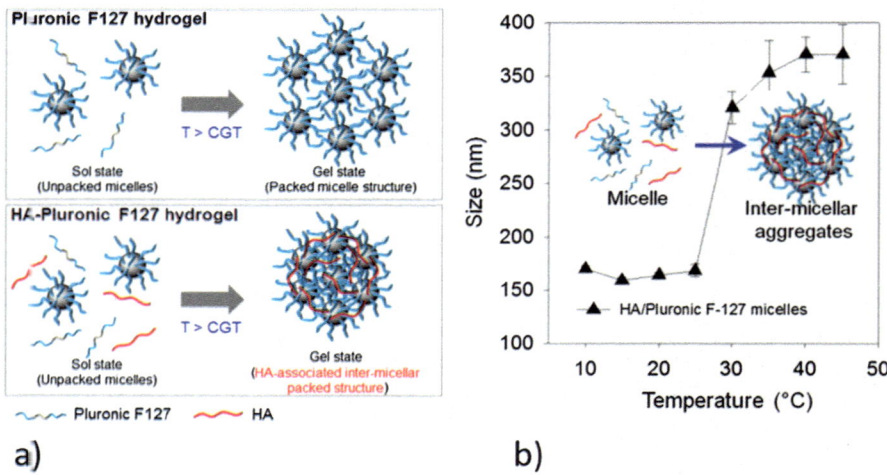

Fig. 7.4 (a) Schematic representation of bridged micellar packing formation of hydrogels. (b) Particle size analysis of HA-Pluronic F127 micelles as a function of temperature. For details, see Ref. [41]. (Figure reproduced with permission)

polyethylene glycol diacrylate, polyethylene glycol methyl ether methacrylate, and 2-methoxyethoxy ethyl methacrylate. Thanks to the several numbers of acrylated groups, the polymer was chemically cross-linked using pentaerythritol tetrakis (3-mercaptoproprionate) as cross-linker via thiol-ene Michael addition reaction. Finally, the resulting semi-IPN was obtained combined this cross-linked polymer with HA leading to an in situ cross-linkable hydrogel with high porosity, improved cell adhesion, and viability.

7.2.2.2 Chemically Cross-Linked Hyaluronic Acid Semi-IPN

An example of a chemically cross-linked semi-IPN based on hyaluronic acid was developed by Pescosolido and co-workers (Fig. 7.5), using a high molecular weight of HA and a dextran derivative, dex-HEMA (hydroxyethyl-methacrylate-derivatized dextran) that is a photocross-linkable polymer able to form stable hydrogels after UV irradiation [43]. Dex-HEMA was dissolved in HEPES buffer, and the photoinitiator (Irgacure 2959) was added. Then, HA was added, and the mixture was stirred before the UV irradiation. Kinetic studies of these semi-IPN hydrogels with different HA contents were performed evidencing that the cross-linking kinetics were almost instantaneous, as shown by the rapid increase of the storage modulus G′ after 10 s of UV exposure. Also, the printability of these systems was studied. The resulting 3D construct showed high porosity, and the construct architecture can be easily tuned by controlling the process parameters, such as fiber spacing and orientation, demonstrating the suitability of the HA/dex-HEMA systems for bioprinting applications in tissue engineering.

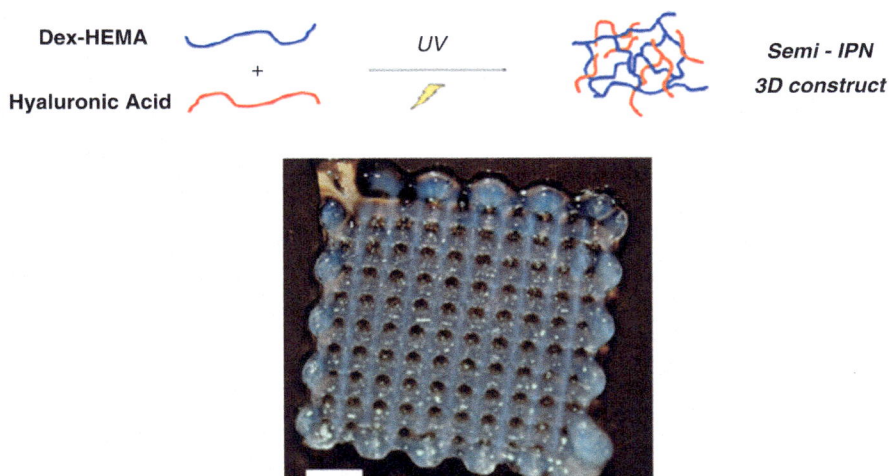

Fig. 7.5 Schematic representation of the 3D hydrogel formation and the picture of the resulting 3D printed hydrogel. For details, see Ref. [43]. (Figure reproduced with permission)

Recently, another interesting IPN system based on fibrin and disulfide cross-linked HA was described by Zhang and co-workers [44]. Fibrin is formed during the physiological coagulation cascade after thrombin-mediated cleavage of fibrinogen in the presence of Ca^{2+}. In their study, HA was dually functionalized with both 2-dithiopyridyl (-SSPy) and hydrazide (-hy) groups, providing the HA-hy-SSPy derivative. The complementary tiolated HA was also prepared (HA-SH). The IPN system was then obtained through the mixing of two mixtures: A and B. The mixture A was composed of a fibrinogen solution containing HA-hy-SSPy, while the mixture B was composed of thrombin and HA-SH. They suggested that the formation of a disulfide network between thiolated HA (HA-SH) and HA modified with reactive 2-dithiopyridyl (SSPy) group should be chemoselective toward enzymatically catalyzed steps of fibrin formation. Synthesis and structures of the HA-fibrin network is show in Fig. 7.6. Because of the polypeptide chains in fibrinogen are linked together via disulfide bonds, the authors demonstrated that the fibrin can be formed even in the presence of thiol and 2-dithiopyridyl-modified HA derivatives. In fact, the reactive groups of the HA derivatives do not interfere with the thrombin-mediated activation of fibrinogen, thanks to the very fast reaction of HA-SH with HA-hy-SSPy for the formation of a disulfide HA network via thiol-disulfide exchange reaction. The mechanical characterization of this novel IPN showed an increased stiffness in comparison to that of the pure fibrin gel, and this is probably due to the entanglements between fibrin and HA networks. In addition, the IPN network presented a lower degradation rate compared to the pure fibrin gel, suggesting that this new material could be a useful scaffold for tissue engineering.

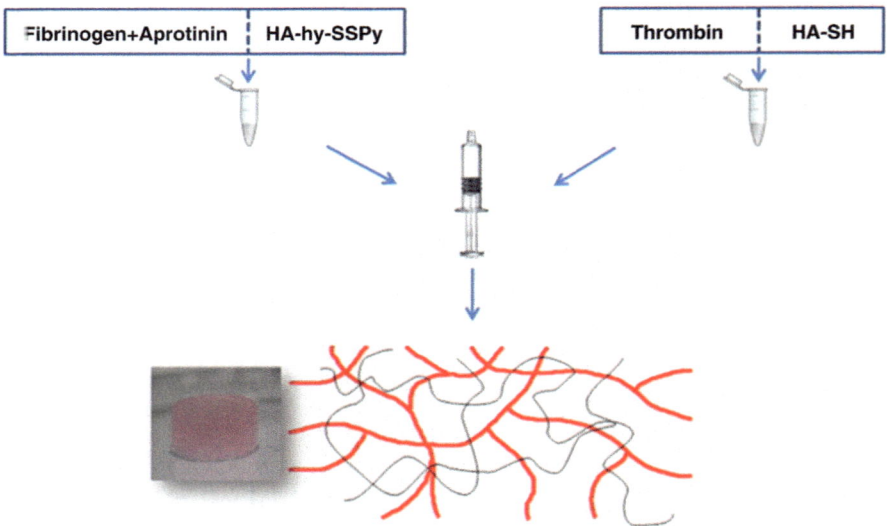

Fibrin-HA double interpenetrating network

Fig. 7.6 Synthesis and structures of the HA-fibrin network developed by Zhang et al. [44]. (Figure reproduced with permission)

7.2.2.3   Photopolymerized Hyaluronic Acid IPNs and Semi-IPNs

The most common chemical modifications on HA polymers for the IPN formation
are the methacrylation and the acrylation, because these derivatives can be prepared
using very mild conditions, leading to generally nontoxic compounds. In 1994, the
Italian company Fidia Farmaceutici SpA prepared HA derivatives based on methacrylated moieties, claiming that those HA or semi-synthetic HA derivatives, used in
the IPNs or semi-IPNs formation, are nontoxic and non-carcinogenic synthetic
materials. In their patent, they reported that these hyaluronan or hyaluronan-
derivative IPNs can be used in the "biomedical and sanitary fields, including derma-
tology, urology, orthopedics, otologic microsurgery, otoneurology, functional,
post-traumatic and rhinosinusal endoscopic microsurgery, plastic surgery, and in the
cardiovascular system" [45]. Many examples are reported in the literature of IPN
systems based on methacrylate or acrylate HA polymers. D'arrigo et al. [37]
reported the development of a semi-IPN network based on methacrylate derivative
of hyaluronic acid (HAMA) and calcium alginate (AlgCa). Particularly, they first
added $CaCl_2$ to a solution of HAMA and AlgNa in order to obtain a semi-IPN structure, thanks to the formation of a AlgCa hydrogel in the presence of HAMA chains;
then, they added a photoinitiator and irradiated the semi-IPN. In this way, a photo-
chemical cross-linking of the methacrylate moieties of HA was obtained. The
resulting network was characterized, exhibiting completely new mechanical properties compared to the two polymeric starting materials. In addition, drug release studies showed that the IPN could act as a depot system for modified release of drugs or
proteins, showing a retention of the activity of proteins embedded within the new
hydrogels. Park and co-workers synthetized an IPN system based on methacrylated
HA (HA-Ac) and diacrylated PEG (PEG-DA) for cartilage repair [46].
Experimentally, RGD residues were introduced into PEG-DA through a Michael
addition between the cysteine residue of the peptides and the acrylated groups of the
polymer. In this way, the RGD residues provided a matrix suitable for fibroblast
adhesion and proliferation on the gels. IPN hydrogels were then obtained through
the photopolymerization of solutions of HA-Ac, previously synthetized, and
peptide-modified-PEG-DA in the presence of eosin Y as visible light sensitizer and
triethanolamine as initiator. The properties of the systems were characterized varying the derivatization degree of HA-Ac and of the PEG-DA concentration; an
increase in these two parameters led to an increase in the G' modulus of the gels and
a decrease in the swelling degree. The hydrogels obtained using this technique were
demonstrated to be still degradable by hyaluronidase, and the presence of RDG
peptides was found to be fundamental for cell adhesion and proliferation.

Collagen, along with HA, is a major component of ECM, and this protein forms
gels without chemical modification. Suri and co-workers developed novel
photocross-linkable semi-IPN and IPN gels made of glycidyl methacrylated HA
(GMHA) and collagen [47]. Particularly, they prepared a semi-IPN in which only
the collagen was in a network form, while HA chains were entangled in the collagen
network without being photocross-linked. Then, they compared this semi-IPN net-

work to the IPN network formed after the exposure of the previous semi-IPN to UV light in order to have both networks entangled within each other. SEM images and rheological data revealed that IPNs are denser than semi-IPNs, which results in their molecular reinforcement. In addition, the degradation of the collagen-HA IPNs was slower than the semi-IPNs because of the presence of the cross-linked HA network. Cytocompatibility of IPNs was confirmed by Schwann cell and dermal fibroblasts adhesion and proliferation studies, confirming that these hydrogels can be employed as potential candidates for regenerative medicine applications.

Recently, another interesting interpenetrating polymer network based on glycidyl methacrylate hyaluronic acid (GMHA) and "Puramatrix™ (PM) was developed obtaining a material with tunable properties that could influence the neurite growth and proliferation [48]. PM is a self-assembling peptide scaffold structurally composed of 99% water and synthetic peptide (1% w/v) that is widely used to recreate the 3D microenvironments useful for cell growth in the absence of animal-derived materials and pathogens. In this study, the core of the IPN system is obtained by the photopolymerization of a solution of GMHA and PM in the presence of the photoinitiator. This IPN hydrogel is then surrounded by a photocross-linkable polyethylene glycol (PEG), and through a UV light drawing defined geometries onto the photocross-linkable substrates, it is possible to irradiate PEG and the IPN throughout the gel. In this way, a dual hydrogel system with a 3D microenvironment was created. Rheological and morphological analyses were performed on the IPNs showing that simply by controlling the degree of methacrylation of HA is possible to tune the mechanical properties. This model provides a simple, in vitro environment to generate different mechanical properties and study neurite growth and extension.

7.2.3   Chitosan

Chitosan is a linear cationic polysaccharide obtained by partial deacetylation of the insoluble naturally available chitin, derived from exoskeletons of crustaceans, fungi, and insects. However, chitin is not readily used as such due to its high level of acetylated groups and its rigid structure as well as its poor solubility in aqueous solutions. Therefore, deacetylation of chitin leads to an increase in the number of amino groups and thus in an enhancement of the water solubility of the polymer. Chitosan is composed of D-glucosamine and N-acetyl-D-glucosamine units linked by β-(1 → 4) glycosidic bonds. Chitosan is an interesting polysaccharide, thanks to the presence of the amino groups, which could be modified in order to modify the physicochemical properties of the polymer, thus improving important characteristic such as solubility and bioadhesivity [49]. Among the various features, Chitosan has generated considerable interest because of its permeation enhancer and mucoadhesive properties.

7.2.3.1 Temperature and pH-Responsive Chitosan IPNs and Semi-IPNs

Chitosan can be combined with specific responsive polymers in IPNs or semi-IPNs with the aim to be responsive to different properties such as temperature, pH, and enzymatic activity. Several studies focused on the formation of thermo-responsive IPNs based on chitosan and N-isopropylacrylamide. Fernández-Gutiérrez and co-workers synthesized a semi-IPN based on poly(N-isopropylacrylamide) (pNIPAAm) and chitosan and studied the effect of pH and temperature on their rheological and swelling properties [50]. The semi-IPNs were prepared by free radical polymerization of NIPAAm in the presence of chitosan using ammonium persulfate as radical initiator. TGA analyses performed on the gels revealed that the presence of chitosan in the network increases the thermal stability of the semi-IPN compared to the networks composed of pure pNIPAAm, probably thanks to the ability of chitosan to stabilize the free radicals generated by the thermal decomposition of polymeric chains, through transfer reactions. Morphological analyses suggested the presence of large pores, with size ranging from 40 to 50 μm up to more than 100 μm, useful for the diffusion and the proliferation of cells in tissue engineering applications. Tze-Wen Chung and co-workers prepared and investigated the drug delivery characteristics of new thermosensitive IPN gel composed of poloxamer (P) and chitosan (CS) cross-linked with various concentrations of glutaraldehyde (GA) for the release of 5-fluorouracil [51]. Two different poloxamer were used for the IPNs preparation, F127 and F68. The LCST value of the poloxamer-chitosan gels was slightly affected by the concentrations of chitosan. In contrast, the presence of GA affected deeply not only the thermosensitivity of the resulting IPN but also the viscosity and swelling ratios. Particularly, high amount of GA led to the formation of a denser CS network because of higher reaction rates and thus to a higher viscosity of the system. In a same way, high swelling ratios of GA-containing gels may be associated with the CS network that interpenetrates the gels and binds the aggregated P micelles/gels in compartments during hydration, thereby preventing the rapid dissolution of the gels. Also, the release of 5-FU was investigated, revealing that the presence of a CS network in the P-CS/GA gels strongly affected the release of 5-FU from the gels.

7.2.3.2 Chemically Modified Chitosan-Based IPNs and Semi-IPNs

Recently, the hydrogels based on chitosan and synthetic polymers were investigated extensively due to their special properties and thus the potential exploitation in the field of biomedicine and pharmaceutics. However, during the preparation of these hydrogels, an aqueous acid medium was usually used to dissolve chitosan or chitosan derivatives, which inevitably led to the presence of small amount of residual acid. These residues, even very small, may be harmful when it is applied onto wounded human skin or tissues. To overcome this problem, several chemical modifications of chitosan macromolecules were described in literature attempting to make the polymer soluble in aqueous media. On this basis, Zhou and colleagues synthetized N-carboxyethyl chitosan (CECS) through Michael addition reaction making chitosan water-soluble in this way [52]. Then, a semi-IPN was prepared by UV polymerization of an aqueous solution of CECS, 2-hydroxyethyl methacrylate (HEMA) in the presence of a photoinitiator. Some properties of the hydrogels including swelling properties, thermal stability, mechanical properties, and cytotoxicity were investigated. These analyses showed that the hydrogels were sensitive to the pH of the medium and had good mechanical properties in their wet state. In addition, the cytotoxicity studies indicated a good biocompatibility of the gels, suggesting a potential use of the CECS/poly (HEMA) hydrogels as drug delivery matrix or wound dressing materials. O-carboxymethyl derivative of chitosan was also investigated for IPNs formation. Particularly, Yin and co-workers [53] developed a superporous hydrogels containing poly(acrylic acid-co-acrylamide)/carboxymethyl chitosan interpenetrating polymer networks (SPH-IPNs) with the aim of enhancing the mechanical strength, in vitro muco-adhesive force, and drug loading capacity of SPHs. Even Ching Chen et al. [54] developed a pH-sensitive hydrogel useful for protein drug delivery, composed of a water-soluble chitosan derivative (N,O-carboxymethyl chitosan, NOCC), and alginate blended with a naturally occurring cross-linking agent (genipin) to form a semi-IPN. Guo and co-workers have reported an interesting approach to obtain thermo- and pH-responsive semi-IPN polyampholyte hydrogels based on carboxymethyl chitosan and poly(2-(dimethylamino)ethyl methacrylate) (PDMAEM) [55]. The swelling behavior and the mechanical properties of the systems suggested that CM-CS/PDMAEMA semi-IPN polyampholyte hydrogels could be used as a pH-/temperature-responsive drug delivery system.

7.2.4 Dextran

Dextran is a nontoxic, bacterial hydrophilic polysaccharide consisting of consecutive α-$(1 \rightarrow 6)$ linked D-glucopyranose units with a low percentage of α-$(1 \rightarrow 2)$, α-$(1 \rightarrow 3)$ and α-$(1 \rightarrow 4)$ side chains. The biocompatibility of dextran is well known; thus, dextran has been extensively explored in tissue engineering and biomedical applications. It is currently used in medicine as an antithrombotic agent to reduce blood viscosity and as a plasma expander. In addition, the hydroxyl groups of dextran can be chemically modified to graft various functional groups, and, as a result, the polymer properties can be tuned with specific characteristics and can be modified in several ways to obtain chemically cross-linkable polymers.

Pescosolido and co-workers have developed a semi-IPN networks based on dextran hyaluronic acid, as reported above [43]. The same authors reported also an in situ IPN and semi-IPN systems based on chemically modified dextran, methacrylate dextran (Dex-MA), and Ca/Alginate. Particularly, the semi-IPNs were formed by dex-HEMA chains interpenetrated with the Alg-Ca network. This semi-IPN behaves like a weak gel and can be easily injected. The semi-IPN can be then transformed in an IPN network through UV polymerization of dex-HEMA. Thanks to the hydrolytic nature of the ester bonds, the IPN was fully biodegradable and thus suitable for biomedical and pharmaceutical applications. Cytocompatibility studies showed that the cells remain viable within the IPN and were able to differentiate depending on the composition of the IPN. Liu and co-workers have synthesized a methacrylate- and aldehyde-bifunctionalized dextran (Dex-MA-AD) and prepared an interpenetrating network of Dex-MA-AD and gelatin [56]. The IPN was formed by UV cross-linking between the methacrylate pendant groups of Dex-MA-AD and Shiff base reaction between Dex-MA-AD and gelatin. The synthetized IPN was characterized, and also compressive, swelling, and mechanical properties were measured showing a rather high elastic modulus. The Dex-MA-AD imparted to the hydrogels elastic properties, while the gelatin component provided cell adhesive and enzymatically degradable properties to the IPN and also significantly increased the compressive modulus and strength through the Schiff base contribution to the cross-link density. The mechanical properties of this new hydrogel coupled with 2D and 3D biocompatibility with vascular cells make this a promising material for 3D scaffolds for vascular tissue engineering. Recently, electrical conductive hydrogels are emerging as promising material for tissue engineering by combining conductivity properties with the possibility of the formation of a three-dimensional network. These hydrogels can promote cell adhesion, proliferation, and differentiation. On this basis, Li and colleagues focused on the development of a novel IPN system exhibiting conductive properties based on gelatin-graft-polyaniline and carboxymethyl-chitosan, which were cross-linked with oxidized dextran [57]. Particularly, three kinds of natural polymers were modified to prepare this biodegradable and biocompatible hydrogel: gelatin, which was modified with various contents of polyaniline on the side chains to improve conductivity; chitosan modified with carboxymethyl groups; and dextran, whose hydroxyl groups were oxidized into aldehyde groups. The synthesis of this novel IPN conductive hydrogels is shown in Fig. 7.7. The amino groups of modified gelatin and carboxymethyl chitosan were cross-linked with the aldehyde groups of oxidized dextran by simple mixing and heating at

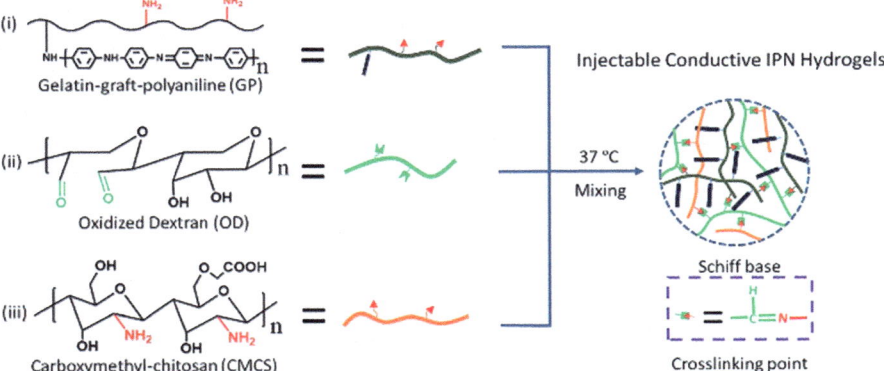

Fig. 7.7 Synthesis scheme of the novel IPN conductive hydrogels composed of gelatin-graft-polyaniline and carboxymethyl-chitosan. For details, see Ref. [57]. (Figure reproduced with permission)

37 °C, thus forming an IPN at body temperature. The storage modulus of the gelatin-grafted polyaniline/carboxymethyl-chitosan/oxidized dextran hydrogels was greatly improved compared to other injectable conductive hydrogel. The conductivity, swelling ratio, and pore size of the hydrogels depend on the polyaniline content. The hydrogels showed also good cytocompatibility and biocompatibility, thus showing a great potential for tissue engineering.

Several studies on IPNs and semi-IPNs focused also on the use of dextran for the development of drug delivery systems in form of microspheres or beads [37, 58, 59, 60].

7.2.5 Gellan Gum

Gellan gum is a bacterial exopolysaccharide of the "sphingans" family, based on a tetrasaccharide repeating unit composed of two molecules of D-glucose, one of L-rhamnose, and one of D-glucuronic acid. In its native form, gellan is partially esterified with acyl substituents. The acetyl groups in native gellan gum can be removed by alkaline treatment to produce deacetylated gellan gum. Gellan is structurally a double helix, formed by two intertwined left-handed, threefold helical chains. This helical geometry is promoted by the $(1 \rightarrow 3)$ linkage in the gellan repeating unit. It is well known that gellan gum can form gels in the presence of monovalent and divalent cations because of the electrostatic interaction of these ions and the carboxylate groups of the polymer. Thanks to its characteristics, gellan gum is used for a great variety of applications in the field of food, cosmetics, pharmaceutics, and biomedicine.

Amici and colleagues [61] studied a new type of IPN based on gellan gum and agarose by simple mixing. The two components appeared to form their own individual ordered conformations at the appropriate temperatures. However, there is a strong evidence from the microscopy and turbidity data that the two networks interpenetrate on a molecular length scale, one network essentially passing through the pores of the other. Shin and co-workers [62] developed a double-network (DN) hydrogel composed of gellan gum methacrylate (GGMA) and gelatin methacrylamide (GelMA). The DN hydrogels were fabricated by a two-step procedure (Fig. 7.8): first, the GGMA solutions were photocross-linked to form the first

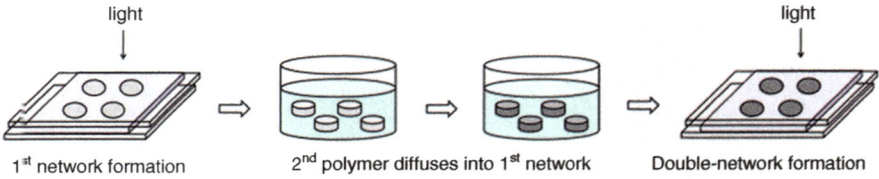

Fig. 7.8 A two-step procedure for the formation of DN gels developed by Shin et al. [62]. (Figure reproduced with permission)

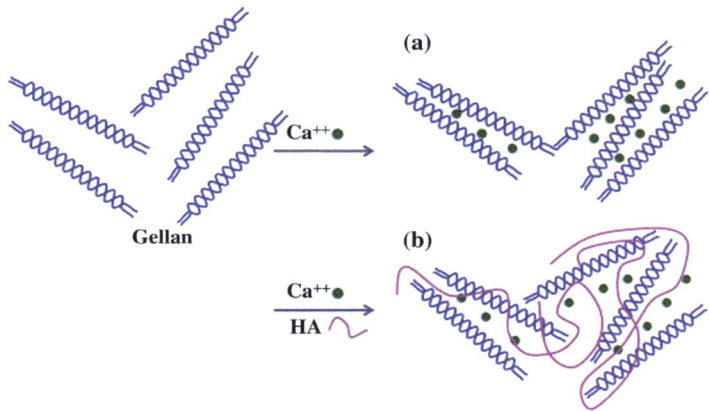

Fig. 7.9 Schematic representation of the HA-Ca-G hydrogels formation. For details, see Ref. [63]. (Figure reproduced with permission)

network and then the GGMA hydrogels were then immersed in GelMA solution so that the GelMA molecules diffused into the GGMA hydrogels and the resulting gels were exposed to the UV light, thus forming the second network. DN hydrogels exhibited higher strength with respect the single components, approaching closer to the strength of cartilage. The DN showed a good cell encapsulation; thus, DN hydrogels made by those photocross-linkable macromolecules could be useful for the regeneration of load-bearing tissues.

Always within the field of tissue engineering, Bellini et al. developed an in situ gelling semi-IPN composed of hyaluronic acid (HA), gellan gum (GG), and calcium chloride (Ca) for bone regeneration of osteochondral defects [63]. In their study, the authors investigated the disturbing effect of HA on the GG-Ca hydrogels. As shown in Fig. 7.9, when HA was mixed with GG, the association of the GG chains during the Ca-mediated gelling process were disturbed, and, thus, weaker junction zones were formed. For the samples preparation, a solution of HA/Ca was poured in a becker, and the solution of G was poured onto the first solution and the resulting mixture was left at room temperature until complete gelation. The resulting HA-Ca-GG hydrogels were characterized through rheological, dynamo-mechanical, and swelling-degradation analyses. The results show that when higher HA concentrations were used weaker hydrogels with a more rapid degradation were obtained, while stronger and more stable hydrogels were formed in presence of high amount of calcium. Also in vitro adhesion tests were carried out by using pig bones and human primary osteoblasts, confirming that this semi-IPN promoted cell survival and osteoblastic progression. The composition of the semi-IPN was crucial for the in situ gelation of the hydrogel: in fact, the use of low HA concentrations leads to a low viscosity solution, which can leak out of the defect, when covered with the viscous G solution; on the other side, high calcium concentrations lead to a too rapid gelation, hindering the polymer interpenetration, thus forming a rapid hydrogel with low adhesion properties.

Cerqueira et al. developed a tissue-engineered construct based on a prevascularized gellan gum and hyaluronic acid (GG-HA) spongy-like hydrogel with the aim to promote full-thickness skin wound regeneration by improving neotissue [64]. These sponges were formed by transferred a solution of GG-HA in a plate and adding divalent cations in order to obtain an ionic cross-linking. This hydrogel was able to meet quality regeneration parameters such as fast wound closure and reepithelialization, a distinct dermal matrix remodeling, and improved neovascularization. Hoosain and co-workers prepared a semi-interpenetrating polymer network xerogel matrix system for the controlled delivery of sulpiride by using epichlorohydrin as cross-linker [65]. The ability of the xerogel to sustain drug release was determined by in vitro and in vivo drug release experiments. Recently, a large number of IPN microspheres have been developed using a combination of gellan gum with other polymer for drug delivery purpose [66–68].

7.2.6 Scleroglucan

Scleroglucan is a bacterial polysaccharide produced by microorganisms, especially by fungi of genus sclerotium. It consists of a main chain of $(1 \rightarrow 3)$-linked β-D-glucopyranosyl units where every third unit has a $(1 \rightarrow 6)$-linked β-D-glucopyranosyl unit. In an aqueous solution, this polysaccharide adopts a stable triple-stranded helical conformation held together by hydrogen bonds. Due to the rod-like character of native scleroglucan in an aqueous solution and resistance to hydrolysis and temperature, it has several commercial applications. Aalaje and co-workers [69] developed a novel semi-IPN hydrogel by cross-linking of partially hydrolyzed polyacrylamide and scleroglucan aqueous solutions using chromium triacetate. Oscillatory shear rheological data showed that the limiting storage modulus of the semi-IPN gels increased with the increase of scleroglucan concentration. In addition, increasing the scleroglucan content, the loss factor decreased slightly, thus indicating that the viscous properties of this gelling system decrease more than its elastic properties. The swelling tests showed that the equilibrium swelling ratios of the semi-IPN networks decreased with increase in the scleroglucan content due to the decrease in the ionic hydrophilic groups of the semi-IPN network. Corrente et al. [70] synthetized a novel injectable and in situ cross-linkable hydrogels composed of methacrylate dextran (DEX-MA) and scleroglucan, in its native form or carboxymethylated. Both unmodified and modified scleroglucan tuned the mechanical properties of DEX-MA hydrogels that became harder but also more elastic. All the hydrogels investigated were able to swell in contact with biological fluids showing a consistence similar to natural tissues. Drug delivery properties were also tested on the hydrogels DEX500-MA, scleroglucan, and DEX500-MA/Scl-CM. The systems released very fast small molecules but were able to modulate the release of

VitB$_{12}$ and to behave as depot delivery systems. Matricardi and co-workers [71] prepared a semi-IPN based on scleroglucan, Alg, and borax and studied its mechanical properties and release behavior of a model drug. The semi-IPN was thermo-irreversible in the range of the temperature investigated, and the presence of Alg led to an increase in the storage modulus (gel strength) of the resulting network. Circular dichroic data showed that no significant conformational changes of Alg occurred, suggesting a possible explanation was the synergistic effect between the Sclg/borax hydrogel network elasticity and the rigidity of entangled Alg chains stretched by the electrostatic repulsions among the carboxylic groups on the backbone.

7.3 Protein-Based IPN and Semi-IPN Gels

7.3.1 Gelatin

Gelatin is a product of the denaturation and structural degradation of collagen. However, gelatin exhibits a behavior that is more similar to those of rigid-chain synthetic polymers rather than the native collagen. Gelatin shows a wide molecular weight distribution and in aqueous solutions, and at high temperature, it has the conformation of a statistical coil. It is possible to tune the characteristics and the physicochemical properties of gelatin by modifying its molecular structure and exploiting both acidic and basic functional groups along the macromolecules. In addition, gelatin forms in solutions at low temperatures a triple-stranded helical structure. The rate of formation of this structure depends on many factors such as the presence of covalent cross bonds, gelatin molecular weight, the presence of amino acids, and the gelatin concentration in the solution. A further important aspect of this biopolymer is its specific interaction with water that leads to drastic changes of its physico-mechanical properties depending on the moisture content [72].

Gelatin have been widely used in the formation of IPN and semi-IPN [73, 57], especially in cartilage tissue engineering applications, thanks to its relatively low antigenicity and bioactivity signals, such as the Arg-Gly-Asp (RGD) sequence. In this context, Zhang et al. developed a macroporous IPN hydrogels composed of gelatin and PEG diacrylate in order to fulfill the required mechanical properties of the scaffolds and improve cell function for cartilage tissue engineering [74]. Particularly, they dissolved PEGDA in the presence of saturated NaCl solution, N,N,N',N'-tetramethylethylenediamine (TEMED), ammonium persulfate (APS), and NaCl particles. Gelatin was also added to this precursor solution in order to have a range of gelatin concentrations, from 0 to 10% (w/v). Each mixture was cross-linked for 10 minutes, and then deionized water was added in order to remove NaCl particles and excess precursor. Finally, glutaraldehyde (0.5% w/V) was added to cross-link the gelatin at 4 °C for 24 h, and the resulting scaffolds were washed with deionized water to remove excess of glutaraldehyde. SEM analyses revealed that the IPN scaffolds had highly interconnected structures with pores of ca. 80 μm in diameter. Confocal microscope analyses performed by fluorescence-labeling showed that the gelatin was distributed regularly only in 5% and 10% gelatin-IPN, leading to a complete pore structures. The Young's modulus of both lyophilized and swollen IPN hydrogels increased with increasing the amount of gelatin. These IPNs were also found to facilitate cell-cell interaction and improve cell attachment, cell proliferation, cartilage-specific gene expression, and ECM accumulation. In fact, chondrocytes in such IPN hydrogels were elongated and had fibroblast morphologies, thanks to the presence of gelatin. Gelatin can induce cell adhesion and has been proven to enhance cell function as native collagen. Cell adhesion receptors on chondrocytes can recognize binding motifs in gelatin and bind to form extensive cytoskeletons. Daniele et al., recently, developed a new bio/synthetic interpenetrating network (BioSINx), based on modified gelatin methacrylamide (GelMA), and PEG [75]. The covalently cross-linked PEG network was formed by thiol-ene coupling, while the bioactive GelMA was integrated using a concurrent thiol-ene coupling reaction (Fig. 7.10). This new BioSINx can be considered a new generation of hydrogel-based tissue scaffold showing tunable mechanical properties and long-

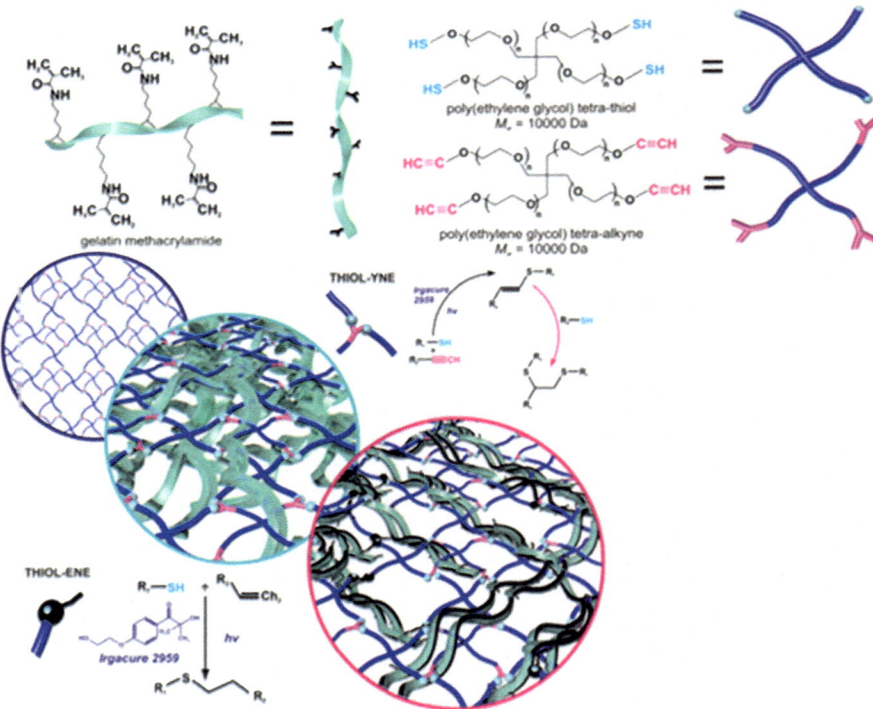

Fig. 7.10 Fabrication of bio/synthetic interpenetrating network (BioSIN) of (a) gelatin methacrylamide and poly(ethylene glycol) via concurrent photoinitiation of thiol-ene and thiolyne coupling. (Reproduced with permission [75])

term cell anchorage points and supporting cell attachment and proliferation in the 3D environment.

GelMA was also used by Xiao and colleagues [76] for the synthesis of a new photocross-linkable IPNs hydrogel with silk fibroin (SF). Subsequently, exposure to aqueous methanol was used to test the induction of SF crystallization in polymerized GelMA-SF hydrogels to form crystalline sheets, which acted as a reinforcement component. Further variation of the concentration of SF demonstrated the tunability of the resulting series of IPN hydrogels from biophysical, structural, and cell compatibility point of view. The IPN showed a lower swelling ratio, higher compressive modulus, and lower degradation rate compared to the GelMA and semi-IPN hydrogels, where only GelMA was cross-linked. These differences were due to a higher degree of overall cross-linking due to the presence of crystallized SF in the IPN hydrogels. Finally, they investigated the fabrication of three-dimensional (3D) microscaffolds and assessed their cytocompatibility and found that NIH-3 T3 fibroblasts readily attached to spread and proliferated on the surface of IPN hydrogels, as demonstrated by F-actin staining and analysis of mitochondrial activity (MTT). In addition, photolithography combined with lyophilization techniques was used to fabricate three-dimensional micropatterned and porous microscaffolds from GelMA-SF IPN hydrogels, favoring their versatility in various microscale tissue engineering applications. Not only methacrylation of gelatin was used for the development of new IPN and semi-IPN networks with the aim to have a cross-linking among methacrylate groups of gelatin, thus obtaining the formation of a network of gelatin. In this contest, Liu et al. [56] developed a very interesting IPN hydrogel based on dextran and gelatin for vascular tissue engineering. In this study, dextran was biofunctionalized with methacrylate and aldehyde groups (Dex-MA-AD), and the resulting IPN hydrogel was formed by ultraviolet cross-linking of methacrylate moieties on Dex-MA-AD and a Schiff base reaction between Dex-MA-AD and gelatin, as previously reported in 2.4. Within the framework of tissue engineering, also IPN theta-gels™ based on gelatin were developed. As Miao and colleagues reported, theta-gels are hydrogels that form during the solidification and phase separation of two dislike polymers, in which a low molecular weight polymer behaves as a porogen and is removed through dialysis [77]. In their work, PEG was used as porogen, and the IPN was obtained from the physical cross-linking of PVA and gelatin. The thermal gelation of both PVA and gelatin in the presence of PEG created a macro-porous IPN. The short-chain, hydrophilic PEG molecules behaved as porogens, aggregating into large domains. After cooling, the PVA-gelatin hydrogel was dialyzed for 5 days, allowing the soluble, nucleated PEG molecules to escape the hydrogels, creating a large interconnected porous structure (Fig. 7.11). The resulting IPN showed an increased storage and elastic moduli compared to PVA-gelatin scaffold controls and supported chondrogenic differentiation of MSCs and cartilage matrix deposition in the presence of chondrogenic media.

Fig. 7.11 Schematic
illustrating the formation
of a PVA and gelatin
theta-gel, through the
physical cross-linking of
PVA (solid black lines) and
gelatin (dashed black
lines), respectively, in the
presence and subsequent
removal of PEG. For
details, see Ref. [77].
(Figure reproduced with
permission)

PVA, PEG and gelatin
solution, autoclaved
for 1 hour.

Solution was cooled to RT
and formed a physically
crosslinked theta-gel
(before dialysis).

PEG was removed, which
formed a macro-porous
hydrogel (after dialysis).

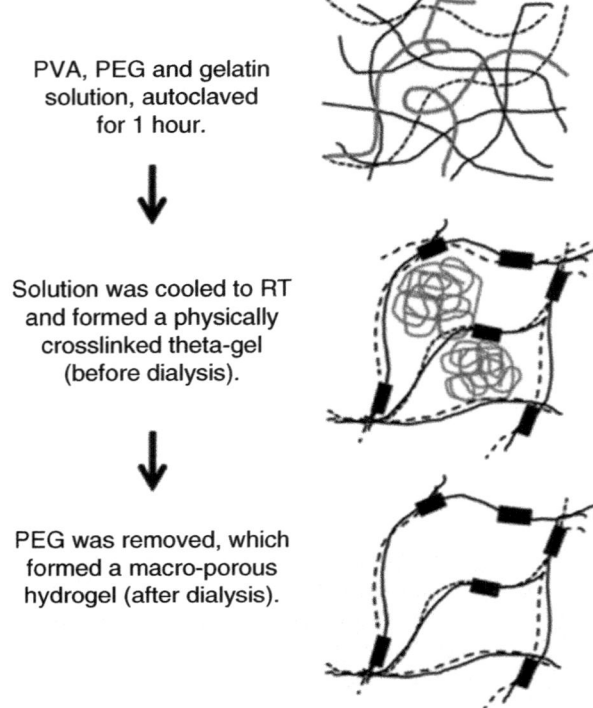

7.3.2   Collagen

Collagen is the main fibrous protein of the body, and it is possible to find it in bones, cartilages, and skins. Nowadays, more than 27 different types of collagen have been identified; however, type I collagen is the most widely occurring collagen in connective tissue. Collagen is composed of three α-chains intertwined in the so-called collagen triple-helix. This structure is stabilized by intra- and inter-chain hydrogen bonding, thanks to presence of a repeating sequence of the Gly-X-Y-, where X is mostly proline and Y is mostly hydroxyproline. Native collagen is insoluble in water and almost in every organic solvent; therefore, it must be pretreated before to be converted into a form suitable for extraction. A chemical pretreatment of collagen breaks non-covalent bonds so as to disorganize the protein structure, thus producing adequate swelling and collagen solubilization. Subsequently, heat treatment cleaves the hydrogen and covalent bonds to destabilize the triple helix, resulting in helix-to-coil transition and conversion into soluble gelatin. The degree of collagen conversion into gelatin is related to the severity of both the pretreatment and the warmwater extraction process, as a function of pH, temperature, and extraction time [78].

Collagen is the most abundant protein of the ECM and favors cell adhesion by interacting with cell surface integrins. Being a ubiquitous ECM component, collagen has low antigenicity and excellent biocompatibility and biodegradability. Thanks to these features, collagen has been widely investigated in tissue engineering for several biomedical applications. In most of the studies, collagen was used in association with hyaluronic acid to obtain hydrogel with enhanced mechanical properties [47, 79]. It is worth noticing that both polymers are components of the ECM, presenting a range of useful properties such as biocompatibility, participation in cell signaling events, and native biodegradability. On these bases, Guo et al. prepared a new IPN based on collagen (Col)/chondroitin sulfate (CS)/hyaluronan (HA) [80]. First, CS and HA were chemically modified by reaction with methacrylated acid (MA), and then the IPN hydrogels were obtained via concomitant self-assembly of collagen and free radical polymerization of CSMA and HAMA under physiological conditions. Mechanical properties of the resulting IPN hydrogels were related to the two networks formed, collagen and CSMA/HAMA, respectively, and it was observed that the IPN hydrogels prepared with a higher methacrylation of CSMA showed higher compressive modulus, slower degradation rate, and denser network structures compared to the semi-IPN hydrogel. The results of in vitro cell culture indicated that the IPN hydrogels had good cytocompatibility, and therefore, they could be excellent candidates for three-dimensional cell culture studies. Moreover, the authors suggested that these systems might have potential applications in cartilage tissue engineering.

Among natural polymers, also alginate was used in association with collagen for the formation of a fully IPN by Branco da Cunha et al. Sodium alginate polymeric backbone presents no intrinsic cell-binding domains but can be used to regulate gel mechanical properties; collagen I presents specific peptide sequences recognized by cell surface receptors and provides a substrate for cell adhesion that better recreates many in vivo contexts [81]. The alginate network was cross-linked by divalent cations of calcium that preferentially intercalate between the guluronic acid residues. Extensive characterization of the microarchitecture of the alginate/collagen-I IPNs revealed that the degree of calcium cross-linking did not change gel architecture or porosity, as expected as the polymer concentration in the system was kept constant. However, the extent of cross-linking with calcium influenced the storage modulus of the resulting IPNs with values ranging from 50 to 1200 Pa. Furthermore, these IPNs showed viscoelastic behavior. Tuning the storage modulus of the alginate/collagen-I IPN also induced different wound healing-related genetic profiles in dermal fibroblasts. Another interesting IPN scaffold based on collagen and chitosan was developed by Shanmugasundaramet et al. [82], using glutaraldehyde as cross-linking agent. In this study, IPN composed of different ratio of collagen and chitosan was investigated and characterized by Fourier transform infrared spectroscopy (FT-IR), differential scanning calorimetry (DSC), and thermogravimetric analysis (TGA), and also the swelling behavior was investigated. Thermodynamic investigations showed that the IPN of collagen-chitosan composite is more thermally stable than the individual networks. The IPN composed of 60:40 collagen/chitosan showed a maximum swelling suggesting a high surface area/volume ratio, and thus the cells can attach and grow in a three-dimensional scaffold. Finally, in vitro analyses showed that this scaffold can be useful in culture of HEp-2 cells and as an in vitro model to test various anticancer drugs.

In the field of tissue engineering, an interesting collagen-based IPN was developed as corneal substituted by Liu and colleagues [83]. In their study, they prepared an IPN network composed of collagen and 2-methacryloyloxyethyl phosphorylcholine (MPC). MPC networks were initially developed as a biocompatible, antifouling material with a unique, high water-holding capacity. Despite these favorable properties, MPC showed antiadhesive properties that inhibited cell adhesion. For this reason, the authors used collagen for the formation of the IPN network. Both medical grade porcine collagen and the human recombinant collagen (RHCIII) were examined as IPN scaffolds for artificial corneas. Collagen network was realized by cross-linking with 1-ethyl-3-(3-dimethyl aminopropyl) carbodiimide (EDC) and N-hydroxysuccinimide (NHS), while the MPC network was cross-linked by poly(ethylene glycol) diacrylate (PEGDA) initiated by ammonium persulfate (APS) or a photoinitiator, Irgacure 2959. The mechanical characterization of the IPN showed that hydrogels fabricated from RHCIII are significantly stronger than those of extracted porcine collagen. On the other hand, porcine collagen hydrogels are more elastic than RHCIII hydrogels. In vitro degradation tests showed that porcine collagen-based hydrogels were more rapidly degraded by the high concentrations of collagenase used in the degradation assay, while RHCIII collagen hydrogels were more robust but still fully degraded after approximately 20 h. These hydrogels retained the properties of collagen in promoting corneal cells and nerve regeneration, showing optical properties comparable to those of the human cornea. In addition, the glucose and albumin permeability were comparable to those of human corneas. Finally, they found that collagen could be substituted with recombinant human collagen, resulting in a fully synthetic implant that is free from the potential risks of disease transmission (e.g., prions) present in animal source materials.

7.3.3   Elastin

Elastin is an extracellular matrix protein that is known for providing elasticity to tissues/organs. As a result, elastin is most abundant in organs where elasticity is of major importance, like in blood vessels, in elastic ligaments, in the lung, and in the skin [84]. Elastin has an uncommon amino acid composition, with about 75% hydrophobic residues (Gly, Val, Ala) and is highly insoluble due to interchain cross-links. However, insoluble elastin may also be hydrolyzed to obtain soluble elastin preparations. Incorporation of elastin in biomaterials is especially significant when its elasticity or biological effects can be exploited. However, several problems may occur when using elastin as a biomaterial such as calcification [84]. Not several studies focused on the use of elastin and collagen for biomedical application and as drug delivery systems.

7.4 Conclusions

The opportunity to obtain IPNs and semi-IPNs with a wide range of mechanical properties and the possibility to tune these properties by combining different kinds of natural and synthetic polymers lead to an increasing interest in these materials. Hence, literature reported a wide range of applications of these systems in various fields. In addition, the patent literature reveals that there are many products based on IPN and semi-IPN systems, including adhesive, optically smooth surfaces, damping materials, and ion exchange resins [8]. However, the most important applications of IPN and semi-IPN hydrogels are their use as drug delivery systems and as biomaterials in tissue engineering.

The most important requirements in designing efficient drug or protein delivery systems are the active control of site/kinetics of drug release and the improvement of the stability of the drug before the delivery [85]. IPNs can provide these requirements, exhibiting, by an appropriate choice of the components, a lower toxicity in comparison to other systems. They can be design to be endowed with various physical and biological properties such as enhanced solubility of hydrophobic drugs, excellent swelling capacity, stable drug formulations, biodegradability, biocompatibility, weak antigenicity, and specific tissue targeting, which make them suitable for drug delivery applications. Several factors like nature and molecular weight of the drug, density, pore size of the matrix, degree of cross-linking, and type of solvent govern the release of a drug from an IPN hydrogel.

Over the past years, hydrogels have evolved as promising candidates for engineered tissue scaffolds due to their biocompatibility and similarities to native extracellular matrix.

However, controlling the porosity of a hydrogel is a challenging issue. The ability to control porosity and the microarchitectural features in hydrogels are vital in creating engineered tissues with structure and functions similar to native tissues. The importance of the porosity in cell proliferation, survival, and migration are widely reported, and an efficient method to obtain reinforced porous hydrogels is the accomplishment of an IPN. IPN hydrogels based on fully synthetic polymers present the advantage to maintain over the time the elasticity and an enhanced strength coupled with high failure stress and stiffness; however, they are not the best candidates for cell culture due to their low degradability and toxicity. Therefore, biologically derived IPNs, in particular those from collagen and hyaluronic acid, have been widely explored for tissue engineering applications. Natural biomaterials based IPNs can be design to show dynamic cell-responsive mechanical properties which can provide a favorable environment in supporting tissue growth. In the middle, IPNs derived from synthetic and a natural polymer could result in materials, which combine the mechanical properties of the synthetic component with the biological properties of the natural one, representing, in some cases, the best choice to fulfill a specific need.

References

1. Matricardi P, Di Meo C, Coviello T, Hennink WE, Alhaique F (2013) Interpenetrating polymer networks polysaccharide hydrogels for drug delivery and tissue engineering. Adv Drug Deliv Rev 65:1172–1187

2. Jenkins AD, Kratochvìl P, Stepto RFT, Suter UW (1996) Glossary of basic terms in polymer science (IUPAC recommendations 1996). Pure Appl Chem 68:2287–2311

3. Banerjee S, Ray S, Maiti S, Sen KK, Bhattacharyya UK, Kaity S, Ghosh A (2010) Interpenetrating polymer network (IPN): a novel biomaterial. Int J Appl Pharm 2:28–34

4. Sperling LH (1977) Interpenetrating polymer networks and related materials. J Polym Sci Macromol Rev 12:141–180

5. Klempner D, Frisch KC (eds) (1980) Polymer alloys II: blends, blocks, grafts and interpenetrating networks. Plenum, New York

6. Frisch HL, Klempner D, Frisch KC (1969) A topologically interpenetrating elastomeric network. J Polym Sci B Polym Phys 7:775–779

7. Sperling LH, Friedman DW (1969) Synthesis and mechanical behavior of interpenetrating polymer networks: poly(ethyl acrylate) and polystyrene. J Polym Sci A-2 Polym Phys 7:425–427

8. Mathew AP (2013) Interpenetrating polymer networks: processing, properties and applications. In: Ochsner A, FM SL, Altenbach H (eds) Advances in elastomers I. Springer, Berlin, pp 283–301

9. Lipatov YS, Alekseeva T (2007) Phase-separated interpenetrating polymer networks. Adv Polym Sci 208:1–227

10. Sperling LH (1994) Interpenetrating polymer networks: an overview. In: Utracki (ed) Interpenetrating polymer networks. ACS, Washington, DC, pp 3–38

11. James J, Thomas GV, Akhina H, Thomas S (2016) Micro- and nano-structured interpenetrating polymer networks: state of the art, new challenges,m and opportunities. In: Thomas SD, Grande D, Cvelbar U, Raju KVSN, Narayan R, Thomas SP (eds) Micro- and Nano-structured interpenetrating polymer networks: from design to applications. John Wiley & Sons, Inc., pp 1–27

12. Lapasin R (2015) Rheological characterization of hydrogels. In: Matricardi P, Alhaique F, Coviello T (eds) Polysaccharide hydrogels: characterization and biomedical applications. Pan Stanford Publishing Pte. Ltd, Singapore, pp 83–137

13. Coviello T, Matricardi P, Marianecci C, Alhaique F (2007) Polysaccharide hydrogels for modified release formulations. J Control Release 119:5–24

14. Dragan ES (2014) Design and applications of interpenetrating polymer network hydrogels. A review. Chem Eng J 243:572–590

15. Hernàndez R, Mijangos C (2009) In situ synthesis of magnetic iron oxide nanoparticles in thermally responsive alginate-poly(N-isopropylacrylamide) semi-interpenetrating polymer networks. Macromol Rapid Commun 30:176–181

16. Dumitriu RP, Oprea AM, Vasile C (2009) A drug delivery system based on stimuli-responsive alginate/N-isopropylacryl amide hydrogel. Cellul Chem Technol 43:251–262

17. Guilherme MR, De Moura MR, Radovanovic E, Geuskens G, Rubira AF, Muniz EC (2005) Novel thermo-responsive membranes composed of interpenetrated polymer networks of alginate-Ca^{2+} and poly(N-isopropylacrylamide). Polymer 46:2668–2674

18. Reddy M, Ramesh Babu V, Krishna Rao KSV, Subha MCS, Chowdoji Rao K, Sairam M, Aminabhavi TM (2008) Temperature sensitive semi-IPN microspheres from sodium algi-

nate and N-isopropylacrylamide for controlled release of 5-fluorouracil. Appl Polym Sci 107:2820–2829

19. Choi B, Loh XJ, Tan A, Loh CK, Ye E, Kyung Joo M, Jeong B (2015) Introduction to in situ forming hydrogels for biomedical applications. In: Loh XJ (ed) In-Situ Gelling Polymers. Spinger, Singapore, pp 5–35

20. Marta Szekalska M, PuciBowska A, Szymanska E, Ciosek P, Winnicka K (2016) Alginate: current use and future perspectives in pharmaceutical and biomedical applications. Int J Polym Sci:7697031. 17 pages

21. Liu Z, Li J, Nie S, Liu H, Ding P, Pan W (2006) Study of an alginate/HPMC-based in-situ gelling ophthalmic delivery system for gatifloxacin. Int J Pharm 315:12–17

22. Nochos A, Douroumis D, Bouropoulos N (2008) In vitro release of bovine serum albumin from alginate/HPMC hydrogel beads. Carbohydr Polym 74:451–457

23. Karewicz A, Zasada K, Szczubialka K, Zapotoczny S, Lach R, Nowakowska M (2010) "Smart" alginate–hydroxypropylcellulose microbeads for controlled release of heparin. Int J Pharm 385:163–169

24. Choudhary S, White J, Stoppel WL, Roberts S, Bhatia S (2011) Gelation behavior of polysaccharide-based interpenetrating polymer network (IPN) hydrogels. Rheol Acta 50:39–52

25. Ramesh Babu V, Krishna Rao KSV, Sairam M, Naidu VK, Hosamani KM, Aminabhavi TM (2006) pH sensitive interpenetrating network microgels of sodium alginate-acrylic acid for the controlled release of ibuprofen. J Appl Polym Sci 99:2671–2678

26. Ju HK, Kim SY, Lee YM (2001) pH/temperature-responsive behaviors of semi-IPN and comb-type graft hydrogels composed of alginate and poly(N-isopropylacrylamide). Polymer 42:6851–6857

27. Shi J, Alves NM, Mano JF (2006) Drug release of pH/temperature-responsive calcium 1109 alginate/poly(N-isopropylacrylamide) semi-IPN beads. Macromol Biosci 6:358–363

28. Lin YH, Liang HF, Chung CK, Chen MC, Sung HW (2005) Physically crosslinked alginate/N,O-carboxymethyl chitosan hydrogels with calcium for oral delivery of protein drugs. Biomaterials 26:2105–2113

29. Shikanov A, Xu M, Woodruff TK, Shea LD (2009) Interpenetrating fibrin–alginate matrices for in vitro ovarian follicle development. Biomaterials 30:5476–5485

30. La Gatta A, Schiraldi C, Esposito A, D'Agostino A, De Rosa A (2009) Novel poly(HEMA-co-METAC)/alginate semi-interpenetrating hydrogels for biomedical applications: synthesis and characterization. J Biomed Mater Res 90A:292–302

31. Anwar H, Ahmad M, Minhas MU, Rehman S (2017) Alginate-polyvinyl alcohol based interpenetrating polymer network for prolonged drug therapy, optimization and in-vitro characterization. Carbohydr Polym 166:183–194

32. Wang MS, Childs RF, Chang PL (2005) A novel method to enhance the stability of alginate-poly-L-lysine-alginate microcapsules. J Biomater Sci Polymer Edn 16:91–113

33. Desai NP, Sojomihardjo A, Yao Z, Ron N, Soon-Shiong P (2000) Interpenetrating polymer networks of alginate and polyethylene glycol for encapsulation of islets of Langerhans. J Microencapsul 17:677–690

34. Pescosolido L, Miatto S, Di Meo C, Cencetti C, Coviello T, Alhaique F, Matricardi P (2010) Injectable and in situ gelling hydrogels for modified protein release. Eur Biophys J Biophys 39:903–909

35. Matricardi P, Pontoriero M, Coviello T, Casadei MA, Alhaique F (2008) In situ cross-linkable novel alginate–dextran methacrylate IPN hydrogels for biomedical applications: mechanical characterization and drug delivery properties. Biomacromolecules 9:2014–2020

36. Pescosolido L, Piro T, Vermonden T, Coviello T, Alhaique F, Hennink WE, Matricardi P (2011) Biodegradable IPNs based on oxidized alginate and dextran-HEMA for controlled release of proteins. Carbohydr Polym 86:208–213

37. D'Arrigo G, Di Meo C, Pescosolido L, Coviello T, Alhaique F, Matricardi P (2012) Calcium alginate/dextran methacrylate IPN beads as protecting carriers for protecting carriers for protein delivery. J Mater Sci Mater Med 23:1715–1722

38. Sun J, Xiao W, Tang Y, Li K, Fan H (2012) Biomimetic interpenetrated polymer network hydrogels based on methacrylated alginate and collagen for 3D pre-osteoblast spreading and osteogenic differentiation. Soft Matter 8:2398–2404

39. Santos JR, Alves NM, Mano JF (2010) New thermo-responsive hydrogels based on poly(N-isopropylacrylamide)/hyaluronic acid semi-interpenetrated polymer networks: swelling properties and drug release studies. J Bioact Compat Polym 25:169–184

40. Pasale SK, Cerroni B, Ghugare SV, Paradossi G (2014) Multiresponsive Hyaluronan-p(NiPAAm) "click"-linked hydrogels. Macromol Biosci 14:1025–1038

41. Jung YS, Park W, Park H, Leeb DK, Naa K (2017) Thermo-sensitive injectable hydrogel based on the physical mixing of hyaluronic acid and Pluronic F-127 for sustained NSAID delivery. Carbohydr Polym 156:403–408

42. Dong Y, Hassan W, Zheng Y, Saeed AO, Cao H, Tai H, Pandit A, Wang W (2002) Thermoresponsive hyperbranched copolymer with multi acrylate functionality for in situ cross-linkable hyaluronic acid composite semi-IPN hydrogel. J Mater Sci Mater Med 23:25–35

43. Pescosolido L, Schuurman W, Malda J, Matricardi P, Alhaique F, Coviello T, van Weeren PR, Dhert WJA, Hennink WE, Vermonden T (2011) Hyaluronic acid and dextran-based semi-IPN hydrogels as biomaterials for bioprinting. Biomacromolecules 12:1831–1838

44. Zhang Y, Heher P, Hilborn J, Redl H, Ossipov DA (2016) Hyaluronic acid-fibrin interpenetrating double network hydrogel prepared in situ by orthogonal disulfide cross-linking reaction for biomedical applications. Acta Biomater 38:23–32

45. Giusti P, Callegaro L (1994) Hyaluronic acid and derivatives thereof in interpenetrating polymer networks (IPN). Wo9401468 (A1)

46. Park YD, Tirelli N, Hubbell JA (2003) Photopolymerized hyaluronic acid-based hydrogels and interpenetrating networks. Biomaterials 24:893–900

47. Suri S, Schmidt CE (2009) Photopatterned collagen–hyaluronic acid interpenetrating polymer network hydrogels. Acta Biomater 5:2385–2397

48. Khoshakhlagh P, Moore MJ (2015) Photoreactive interpenetrating network of hyaluronic acid and Puramatrix as a selectively tunable scaffold for neurite growth. Acta Biomater 16:23–34

49. Ahmadi F, Oveisi Z, Samani SM, Amoozgar Z (2015) Chitosan based hydrogels: characteristics and pharmaceutical applications. Res Pharm Sci 10:1–16

50. Fernández-Gutiérrez M, Fusco S, Mayol L, San Román J, Borzacchiello A, Ambrosio LJ (2016) Stimuli-responsive chitosan/poly (N-isopropylacrylamide) semi-interpenetrating polymer networks: effect of pH and temperature on their rheological and swelling properties. Mater Sci Mater Med 27:109

51. Chung TW, Lin SY, Liu DZ, Tyan YC, Yang JS (2009) Sustained release of 5-FU from Poloxamer gels interpenetrated by crosslinking chitosan network. Int J Pharm 382:39–44

52. Zhou Y, Yang D, Ma G, Tan H, Jin Y, Nie J (2008) A pH-sensitive water-soluble N-carboxyethyl chitosan/poly(hydroxyethyl methacrylate) hydrogel as a potential drug sustained release matrix prepared by photopolymerization technique. Polym Adv Technol 19:1133–1141

53. Yin L, Fei L, Cui F, Tang C, Yin C (2007) Superporous hydrogels containing poly(acrylic acid-co-acrylamide)/O-carboxymethyl chitosan interpenetrating polymer networks. Biomaterials 28:1258–1266

54. Chen SC, Wu YC, Mi FL, Lin YH, Yu LC, Sung H-WJ (2004) Novel pH-sensitive hydrogel composed of N,O-carboxymethyl chitosan and alginate cross-linked by genipin for protein drug delivery. Control Release 96:285–300

55. Guo B, Yuan J, Yao L, Gao Q (2007) Preparation and release profiles of pH/temperature-responsive carboxymethyl chitosan/P(2-(dimethylamino) ethyl methacrylate) semi-IPN amphoteric hydrogel. Colloid Polym Sci 285:665–671

56. Liu Y, Chan-Park MB (2009) Hydrogel based on interpenetrating polymer networks of dextran and gelatin for vascular tissue engineering. Biomaterials 30:196–207

57. Li L, Ge J, Ma PX, Guo B (2015) Injectable conducting interpenetrating polymer network hydrogels from gelatin-graft-polyaniline and oxidized dextran with enhanced mechanical properties. RSC Adv 5:92490–92498

58. Rokhade AP, Patil SA, Aminabhavi TM (2007) Synthesis and characterization of semi-interpenetrating polymer network microspheres of acrylamide grafted dextran and chitosan for controlled release of acyclovir. Carbohydr Polym 67:605–613
59. Sullad AG, Manjeshwar LS, Aminabhavi TM (2011) Novel semi-interpenetrating microspheres of dextran-grafted-acrylamide and poly(vinyl alcohol) for controlled release of Abacavir sulfate. Ind Eng Chem Res 50:11778–11784
60. Ahmed A, Al-Kahtani AA, Sherigara BS (2009) Controlled release of theophylline through semi-interpenetrating network microspheres of chitosan-(dextran-g-acrylamide). J Mater Sci Mater Med 20:1437–1445
61. Amici E, Clark AH, Normand V, Johnson NB (2000) Interpenetrating network formation in Gellan-agarose gel composites. Biomacromolecules 1:721–729
62. Shin H, Olsen BD, Khademhosseini A (2012) The mechanical properties and cytotoxicity of cell-laden double-network hydrogels based on photocrosslinkable gelatin and gellan gum biomacromolecules. Biomaterials 33:3143–3152
63. Bellini D, Cencetti C, Meraner J, Stoppoloni D, Scotto D'Abusco A, Matricardi P (2015) An *in situ* gelling system for bone regeneration of osteochondral defects. Eur Polym J 72:642–650
64. Cerqueira MT, Da Silva LP, Santos TC, Pirraco RP, Correlo VM, Reis RL, Marques AP (2014) Gellan gum-hyaluronic acid spongy-like hydrogels and cells from adipose tissue synergize promoting Neoskin vascularization. ACS Appl Mater Interfaces 6:19668––19679
65. Famida G, Hoosain FG, Choonara YE, Kumar P, Tomar LK, Tyagi C, du Toit LC, Pillay V (2016) In vivo evaluation of a PEO-Gellan gum semi-interpenetrating polymer network for the oral delivery of Sulpiride. AAPS PharmSciTech 18:654–670
66. Mundargi RC, Shelke NB, Babu VR, Patel P, Rangaswamy V, Aminabhavi TM (2010) *Novel thermo-responsive semi-interpenetrating network microspheres of gellan gum-poly(N-isopropylacrylamide) for controlled release of atenolol.* J Appl Polym Sci 116:1832–1841
67. Agnihotri SA, Tejraj M, Aminabhavi TM (2005) Development of novel interpenetrating network Gellan Gum-Poly(vinyl alcohol) hydrogel microspheres for the controlled release of carvedilol. Drug Dev Ind Pharm 31:491–503
68. Kulkarni RV, Mangond BS, Mutalik S, Sa B (2011) Interpenetrating polymer network microcapsules of gellan gum and egg albumin entrapped with diltiazem–resin complex for controlled release application. Carbohydr Polym 83:1001–1007
69. Aalaiea J, Rahmatpourb A, Vasheghani-Farahani E (2009) Rheological and swelling behavior of semi-interpenetrating networks of polyacrylamide and scleroglucan. Polym Adv Technol 20:1102–1106
70. Corrente F, Abu Amara HM, Pacelli S, Paolicelli P, Casadei MA (2013) Novel injectable and in situ cross-linkable hydrogels of dextran methacrylate and scleroglucan derivatives: preparation and characterization. Carbohydr Polym 92:1033–1039
71. Matricardi P, Onorati I, Masci G, Coviello T, Alhaique F (2007) Semi-IPN hydrogel based on scleroglucan and alginate: drug delivery behaviour and mechanical characterisation. J Drug Del Sci Tech 17:193–197
72. Kozlov PV (1983) The structure and properties of solid gelatin and the principles of their modification. Polymer 24:651–666
73. Hago EE, Li X (2013) Interpenetrating polymer network hydrogels based on gelatin and PVA by biocompatible approaches: synthesis and characterization. Adv Mater Sci Eng ID 328763. 1–8
74. Zhang J, Wang J, Zhang H, Lin J, Ge Z, Zou X (2016) Macroporous interpenetrating network of polyethylene glycol (PEG) and gelatin for cartilage regeneration. Biomed Mater 11. https://doi.org/10.1088/1748-6041/11/3/035014
75. Daniele MA, Adams AA, Naciri J, North SH, Ligler FS (2014) Interpenetrating networks based on gelatin methacrylamide and PEG formed using concurrent thiol click chemistries for hydrogel tissue engineering scaffolds. Biomaterials 35:1845–1856
76. Xiao W, He J, Nichol JW, Wang L, Hutson CB, Wang B, Du Y, Fan H, Khademhosseini A (2011) Synthesis and characterization of photocrosslinkable gelatin and silk fibroin interpenetrating polymer network hydrogels. Acta Biomater 7:2384–2393

77. Miao T, Miller EJ, McKenzie C, Oldinski RA (2015) Physically crosslinked polyvinyl alcohol and gelatin interpenetrating polymer network theta-gels for cartilage regeneration. J Mater Chem B 3:9242–9249

78. Gómez-Guillén MC, Giménez B, López-Caballero ME, Montero MP (2011) Functional and bioactive properties of collagen and gelatin from alternative sources: a review. Food Hydrocoll 25:1813–1827

79. Brigham MD, Bick A, Lo E, Bendali A, Burdick JA, Khademhosseini A (2008) Mechanically robust and bioadhesive collagen and Photocrosslinkable hyaluronic acid semi-interpenetrating networks. Tissue Eng Part A 15:1645–1653

80. Guo Y, Yuan T, Xiao Z, Tang P, Xiao Y, Fan Y, Zhang X (2012) Hydrogels of collagen/chondroitin sulfate/hyaluronan interpenetrating polymer network for cartilage tissue engineering. J Mater Sci Mater Med 23:2267–2279

81. Branco da Cunha C, Klumpers DD, Li WA, Koshy ST, Weaver JC, Chaudhuri O, Granja PL, Mooney D (2014) Influence of the stiffness of three-dimensional alginate/collagen-I interpenetrating networks on fibroblast biology. J Biomater Dent 35:8927–8936

82. Shanmugasundaram N, Ravichandran P, Reddy NP, Ramamurty N, Pal S, Rao KP (2001) Collagen-chitosan polymeric scaffolds for the in vitro culture of human epidermoid carcinoma cells. Biomaterials 22:1943–1951

83. Liu W, Deng C, McLaughlin CR, Fagerholm P, Lagali NS, Heyne B, Scaiano JC, Watsky MA, Kato Y, Munger R, Shinozaki N, Li F, Griffith M (2009) Collagen–phosphorylcholine interpenetrating network hydrogels as corneal substitutes. Biomaterials 30:1551–1559

84. Daamen WF, Veerkamp JH, van Hest JCM, van Kuppevelt TH (2007) Elastin as a biomaterial for tissue engineering. Biomaterials 28:4378–4398

85. Raveendran R, Sharma CP (2016) Applications of interpenetrating polymer networks. In: Thomas S, Grande D, Cvelbar U, Raju KVSN, Narayan R, Thomas SP (eds) Micro- and Nano-structured interpenetrating polymer networks: from design to applications. John Wiley & Sons, Inc., pp 383–397

Chapter 8
Promising Biomolecules

Isabel Oliveira, Ana L. Carvalho, Hajer Radhouani, Cristiana Gonçalves, J. Miguel Oliveira, and Rui L. Reis

Abstract The osteochondral defect (OD) comprises the articular cartilage and its subchondral bone. The treatment of these lesions remains as one of the most problematic clinical issues, since these defects include different tissues, requiring distinct healing approaches. Among the growing applications of regenerative medicine, clinical articular cartilage repair has been used for two decades, and it is an effective example of translational medicine; one of the most used cell-based repair strategies includes implantation of autologous cells in degradable scaffolds such as alginate, agarose, collagen, chitosan, chondroitin sulfate, cellulose, silk fibroin, hyaluronic acid, and gelatin, among others. Concerning the repair of osteochondral defects, tissue engineering and regenerative medicine started to design single- or bi-phased scaffold constructs, often containing hydroxyapatite-collagen composites, usually used as a bone substitute. Biomolecules such as natural and synthetic have been explored to recreate the cartilage-bone interface through multilayered biomimetic scaffolds. In this chapter, a succinct description about the most relevant natural and synthetic biomolecules used on cartilage and bone repair, describing the procedures to obtain these biomolecules, their chemical structure, common modifications to improve its characteristics, and also their application in the biomedical fields, is given.

I. Oliveira · A. L. Carvalho · H. Radhouani (✉) · C. Gonçalves
3B's Research Group – Biomolecules, Biodegradables and Biomimetics, University of Minho, Headquarters of the European Institute of Excellence on Tissue Engineering and Regenerative Medicine, Barco, Guimarães, Portugal

ICVS/3B's – PT Government Associate Laboratory, Braga/Guimarães, Portugal
e-mail: hajer.radhouani@dep.uminho.pt

J. M. Oliveira · R. L. Reis
3B's Research Group – Biomolecules, Biodegradables and Biomimetics, University of Minho, Headquarters of the European Institute of Excellence on Tissue Engineering and Regenerative Medicine, Barco, Guimarães, Portugal

ICVS/3B's – PT Government Associate Laboratory, Braga/Guimarães, Portugal

The Discoveries Centre for Regenerative and Precision Medicine, Headquarters at University of Minho, Barco/Guimarães, Portugal

© Springer International Publishing AG, part of Springer Nature 2018
J. M. Oliveira et al. (eds.), *Osteochondral Tissue Engineering*,
Advances in Experimental Medicine and Biology 1059,
https://doi.org/10.1007/978-3-319-76735-2_8

Keywords Synthetic biomolecules · Natural biomolecules · Bone repair ·
Cartilage repair · Osteochondral defects

8.1   Introduction

Tissue engineering field has been responsible for the promising application development to regenerate and repair osteochondral defects. The biomolecules studied
and applied for cartilage and bone repair are selected considering the knowledge of
the anatomical complexity of both structures. The increase of knowledge in the
biomolecular area, with progress in new technology such as cellular, molecular
biology and biochemistry, offer a good opportunity to create biomolecules with
stimulating and precise properties [1, 2].

The main properties sought in a biomolecule to be applied for cartilage and bone
regeneration are biocompatibility, bioactivity, biomimetic skill, and biodegradability, being bio-responsive, highly porous, suitable for cell attachment (as well as
proliferation and differentiation), osteoconductive, non-cytotoxic, flexible and elastic, and nonantigenic [3, 4]. Currently, several natural and synthetic polymers, with
most of these characteristics, have been studied as therapy for cartilage repair. The
natural polymers, more used to osteochondral defects, are collagen, fibrin, alginate,
silk, and chitosan. These biopolymers have been investigated as bioactive scaffolds
for bone engineering such as alginate, agarose, fibrin, hyaluronic acid (HA), collagen, gelatin, chitosan, chondroitin sulfate, and cellulose [5, 6]. Regarding synthetic
polymers, they are often considered to be used for proteins' and growth factors'
delivery with or without cells locally to enhance tissue repair and regeneration such
as poly(ethylene glycol) (PEG), poly(lactic-co-glycolic acid) (PLGA), poly-L-
lactic acid (PLLA), polycaprolactone (PCL), poly(N-isopropylacrylamide)
(PolyNiPAAm), and polycarbonate [7, 8]. At this point, it is important to clear up
that both natural and synthetic materials can be considered biomolecules, since that
they can be equally used to medical and surgical purposes.

In the following section, promising trends in the development of biomolecules
for cartilage and bone repair are described.

8.2   Biomolecules

Bone and cartilage are two different tissues with specific structural and mechanical
properties belonging to osteochondral interface [9].

The biodegradable scaffolds, with or without cells and/or growth factors, have
been widely used for cartilage and bone repair. Cartilage tissue shows single conjunction of nonlinear tensile and compressive properties due to arranged collagen
fibrils, proteoglycans, and proteins. The structural and mechanical properties of

natural tissues can be mimicked by engineered bio-nanocomposites through the use of polymers and nanoparticles [10].

It is important to highlight that classical scaffolds cannot be together for both chondrogenesis and osteogenesis. Scaffolds with two layers have more potential, being organized by different space structures and mechanical forces. One layer is responsible by the repair of the cartilage, and the other layer supports the regeneration of the subchondral bone [11].

8.2.1 Natural Materials

Nature has been using natural polymers long before the creation of plastics and other synthetic materials. For that reason, the natural biomolecules' development is not truly a scientific area. Nonetheless this not turns it less important, in fact the use of this products is currently resurging in the biomedical field. Natural polymers remain attractive compared with synthetic polymers since they can avoid waste disposal problems that are associated with traditional synthetic polymers. Moreover, these polymers present more advantages related to the possibility of several chemical modifications, potential to be degradable and biocompatible due to their natural origin [12].

Over the past few years, this area have gained special attention, and a large number of biomolecules have been reported as a great option to apply on osteochondral defects repair and regeneration [13–20].

8.2.1.1 Silk Fibroin

Silk fibroin (SF) is a natural polymeric biomolecule composed of two proteins: hydrophobic fibroin and hydrophilic sericin. This biopolymer shows several interesting characteristics for tissue engineering, such as mammalian cell compatibility, remarkable oxygen and water vapor permeability, biodegradability, and suitable mechanical properties, with a unique combination of elasticity and strength [21]. Recently, the attention for silk biomolecules has increasing, since this biomolecule has shown to be a great option as a therapeutic material like biomimetic scaffolds for tissue engineering [22, 23].

Silk fibroin presents crystalline regions, with an amino acid sequence (Table 8.1) due to their heavy chains (350 kDa), and non-repetitive amorphous regions. Moreover, SF also presents three crystal structures denoted as silk I, silk II, and silk III [24, 25].

Silk proteins, present in glands of silk, can be obtained from silkworms, spiders, scorpions, mites, and bees and spun into fibers during their metamorphosis [26]. The cannibalistic nature of spiders represents a difficulty on the commercial production of spider silk. For that reason, silkworm's silk, an established fiber, has been the option for textile but also for medical research area. In fact, the yield of fiber

Table 8.1 Structures of the main biomolecules used in osteochondral tissue engineering

Natural biomolecules	Synthetic biomolecules
Collagen	*Polycaprolactone*
Silk fibroin	*Poly(lactic-co-glycolic acid).*
Chitosan	*Poly(ethylene glycol)*
Cellulose	*Poly(N-isopropylacrylamide).*
Agarose	
Hyaluronic acid	
High acyl gellan gum	
Chondroitin sulfate.	

from a single silk cocoon is 600–1500 m, compared to only about 137 m from the ampullate gland of a spider and about 12 m from the spider web [27]. Silk-based biomaterials can be easily prepared from *Bombyx mori* silkworm silk. The raw silk fibers (cocoons) are degummed with Na_2CO_3 solution at 100 °C, rinsed with distilled water in order to extract the highly immunogenic sericin protein (other protein secreted by silk worm) and then air-dried. SF thus extracted is dissolved in a ternary solvent containing $CaCl_2$, CH_3CH_2OH, and H_2O, at 70 °C for 1 h, with continuous stirring. The final regenerated SF solution is obtained following dialysis against distilled water and filtration at room temperature [28].

As refereed above, SF fibers have been extensively used in biomedical field, namely, as surgical sutures, not presenting side effects for humans [24]. In fact, due to its excellent properties, SF is considered one of the best biomolecule for tissue engineering, more specifically for skeletal and osteochondral applications [26, 29]. Another add value of SF is the fact that besides the ways it is processed (films, sponges, or hydrogels), it is able to sustain cell adhesion and proliferation and even shown capable of ECF production in vivo and in vitro with a low inflammatory potential [30].

Several research studies have shown the potential of SF for osteochondral defect regeneration. In fact, silk fibroin scaffolds showed to be osteoinductive with good integration of stem cell-silk biomolecule, which is elementary in the bone or cartilage area [31].

8.2.1.2 Collagen

Collagen is the primary structural material of vertebrates and is the most abundant protein in mammalian, accounting for about 20–30% of the whole-body protein content [32]. In the human body, collagen is mostly found in fibrous tissues such as tendons, the skin, ligaments, and also tissues (as cartilage, bone, and intervertebral disc), and mostly collagen in the body is of type I [33]. In fact, it is in the skin that collagen is more abundant, and about 70% of the material, besides water that are present in dermis of skin and tendon, is collagen [34]. Collagen is synthesized by fibroblasts, which usually originate from pluripotential adventitial cells or reticulum cells [35]. There are different collagen types, differentiated by their complexity and diversity in their structure, their splice variants, the presence of additional, non-helical domains, their assembly, and their function. Types I and V collagen fibrils are the main contribution for the structural backbone of the bone, while type II and XI collagens predominantly contribute to the fibrillar matrix of articular cartilage [36].

Collagen is gaining popularity not only for the easier way of fabrication since it can be processed, through chemical and biochemical modifications, into a variety of forms including cross-linked films, steps, sheets, beads, meshes, fibers, and sponges. Furthermore, it presents a unique structural, physical, chemical, and immunological properties, being biodegradable, biocompatible, and non-cytotoxic, supporting cellular growth [37, 38].

Collagen can be extracted from bovine skin, pig skin, and chicken waste, but these kinds of sources could have religious and ethnic constraints. Taking this into account many reports have been showed to extract collagen from marine sources as alternative source and have used to screen their potential industrial applications [39, 40].

Collagen molecules are structural macromolecule of the extracellular matrix (ECM) comprising three polypeptide chains. These chains, aligned in a parallel manner and coiled in triple helix, wrap around each other by interstrand hydrogen bonds in a stabilized form [38]. Collagen production is quite simple, and a diversity of matrix systems, such as meshes, hydrogels, scaffolds, injectable solutions, and dispersions, among others, can be achieved [41]. Several studies showed that collagen-based scaffolds provide a suitable scaffold for cartilage and bone regeneration, as it supports the adhesion, migration, and proliferation of cells in vitro [18, 42].

8.2.1.3   Chitosan

Chitosan polysaccharide is the only pseudo-natural cationic polymer and an important derivative of chitin. Chitin is a homopolymer comprised of 2-acetamido-2-deoxy-b-D-glucopyranose units [43] (Table 8.1). It has application on the biomedical field and many others, due to its unique cationic character (protonation of the amino group) [44]. Extensive research has been conducted in order to improve chitosan characteristics for specific applications. This polysaccharide mostly finds its application in wound dressings and scaffolds (on physical cross-linking) and as antimicrobial agent [45, 46]. Moreover, although it has several interesting properties (such as biodegradability and non-toxicity) [47], this biopolymer has poor mechanical properties and high swelling ability, getting easily deformed, which is generally improved by blending with other polymers [48].

8.2.1.4   Alginate

Alginate is a linear and anionic natural polysaccharide derived from brown seaweed such as *Laminaria hyperborea*, *Macrocystis pyrifera*, and *Ascophyllum nodosum*. This biopolymer is composed of alternating blocks of α-1, 4-l-guluronic acid (G) and β-1,4-d-mannuronic acid (M) units [49] (Table 8.1). The monomer sequence can diverge depending on algal species and different tissues of the same species. The ratio between monomers and the block structure has a significant effect on the physicochemical and rheological properties of alginates. These biopolymers with more industrial significance usually show high G content [50, 51]. Although it can be produced by bacterial sources, it is commercially available from algae in the form of salt, e.g., sodium alginate. This biomolecule presents excellent properties, namely, their biodegradability, low toxicity, and chemical versatility, but also its unique property to form stable gel in aqueous media and mild condition by addition of multivalent cations makes this biopolymer very useful for drug delivery and cell immobilization [52].

The main industrial applications of alginates are linked to their ability to cross-link with ions, retain water, gelling, viscosity booster, and stabilizing properties [53].

Alginate's extraction procedure from brown algae has suffered an optimization to obtain more industrially and economically sustainable products, with controlled properties as to envisage different therapeutic applications [50, 54].

Alginate is widely used on biotechnological industry, due to its ability to efficiently bind divalent cations, leading to hydrogel formation, being also a perfect candidate for chemical functionalization. By forming alginate derivatives through chemical functionalization, the various properties such as solubility, hydrophobicity, and physicochemical and biological characteristics may be improved. Several studies reported the alginate chemical modification of hydroxyl groups through different techniques such as oxidation [55], reductive amination of oxidized alginate [56], sulfation [57], copolymerization, or cyclodextrin-linked alginate [58] and also chemical modification of carboxyl groups using other techniques as esterification [59], Ugi reaction [60], or amidation [61, 62]. Alginate has been extensively used for many biomedical applications. It could form scaffolds through the use of ionic cross-linking, allowing for encapsulation of cells.

8.2.1.5 Cellulose

Cellulose is the most widely spread natural raw material with total production of 10^{11}–10^{12} tons/year. It is a cheap, biodegradable, and renewable polymer, being fibrous and the main constituent of the cell wall of green plants. Cellulose has a versatile production and exists in a wide range of forms and shapes, e.g., as membrane sponges, microspheres, and non-woven, woven, or knitted textiles [33]. Commercial sources of cellulose include mainly wood or cotton. However, cellulose can also be extracted from different parts of plants and other sources [33]. Usually, purification and isolation processes almost engender the degradation of cellulose and also permit the cellulose to undergo oxidation by reaction with both acids and bases [63].

Cellulose is one of the many beta-glucan compounds; it is a polycarbohydrate composed of a series of cellobiose units, formed by two anhydroglucose subunits (Table 8.1). This polysaccharide presents unique properties and cannot be synthesized or hydrolyzed due to its intricate hydrogen bond network. This complex is also responsible for the good mechanical properties of cellulose, important for its function in nature [63]. Purification and isolation of cellulose comprises several steps including a pulping process, partial hydrolysis, dissolution, re-precipitation, and extraction with organic solvents. Cellulose polymer allows the proliferation of chondrocytes and also has shown an interesting biocompatibility [64]. Moreover, it was found that a product based on hydroxypropyl methylcellulose hydrogel may be used for articular cartilage repair [16]. This versatile compound could also be modified for bone defects, due to its good mechanical properties.

8.2.1.6 Agarose

Agarose is a neutral linear polysaccharide extracted from red algae (*Rhodophyceae*) and is the major component of agar, being the other component of agaropectin. It is composed of alternating β-D-galactopyranose and anhydro-α-L-galactopyranose [15] (Table 8.1). Agarose gel characteristics are determined by two parameters: temperature and concentration, being obtained through changing temperatures. Cross-linked alginate matrix can be formed via ionic bonding in the presence of Ca^{2+} [42]. Agarose hydrogels are at present well characterized and have been studied in different biomedical fields. The hydrogels may be polymerized in situ, reducing invasiveness of the surgery and allowing the hydrogel to acquire the required shape [65].

These kind of hydrogels have been extensively studied for chondrocyte culture and cartilage tissue engineering, by its mechanical and cell-seeding density properties [66–68]. Moreover, agarose also demonstrated to be a great option for cell therapy with chondrocytes or mesenchymal stem cells (chondrocyte- or MSC-laden hydrogels) in osteochondral defects. This approach ensure physiologically relevant mechanical properties and allows the formation of a repaired tissue containing collagens and proteoglycans [69].

8.2.1.7 Hyaluronic Acid

Hyaluronic acid (HA) is a polysaccharide that can be found in all tissues and body fluids of vertebrates. It can also be found in some bacteria such as *Streptococcus* genus. This polysaccharide, a linear polymer composed of N-acetylglucosamine and glucuronic acid (Table 8.1), is especially abundant in loose connective tissue. HA is a polyanion with a pKa around 2.9. It is a pseudoplastic material, and in aqueous solutions, it has shear-thinning properties, behaving like gelatin [70]. HA has a remarkable hydrodynamic characteristics, especially in terms of its viscosity and its ability to retain water, thus having an important role in tissue homeostasis and biomechanical integrity [71]. HA is produced through bacterial fermentation of *streptococcus* species or extracted from rooster combs, umbilical cords, synovial fluids, the skin, or vitreous humor for commercial purposes. The usual sources for its industrial production are bovine vitreous humor, bovine synovial liquid, and rooster crest, with an increasing interest to bacterial cultivations [72].

Although the HA extraction protocols have been developed and optimized over the years, these protocols are still limited to low yields. The HA products are mostly from animal origins, which have several risks of proteins and viruses' contaminations. But the risks can be reduced if there is a special care in choosing tissues from healthy animals [73].

Alternatively, HA biomolecule could be functionalized in order to obtain a more rigid and stable, hydrophobic, and less susceptible enzyme decomposition. These functionalizations include sulfation, carbodiimide-mediated modification, esterification, hydrazide modification, and cross-linking with polyfunctional epoxides, divinyl sulfone, and glutaraldehyde, among others [70].

HA presents several advantages, namely, for adhesion of cells, extracellular matrix deposition, transport of gases and nutrients, and metabolic product release. It offers a good interface of material-cell function mainly due to the HA 3D structure with a significant porosity, surface, and space area. This polymer has been reported as an important role in joint lubrication, nutrition, and preserving cartilage properties since it can help in the control of water balance [70, 74, 75]. All these evidences turn HA a good option for articular cartilage, besides the fact that it is also able to maintain normal growth of cartilage cells and promote the integration of transplanted chondrocytes and damaged cartilage [76, 77].

8.2.1.8 Gellan Gum

Gellan gum, a bacterial polysaccharide, is a linear and anionic exopolysaccharide, with the repeating unit consisting of α-1-rhamnose, β-d-glucose, and β-d-glucuronate, in molar ratios of 1:2:1 (Table 8.1). Gellan is produced by the bacteria *Sphingomonas elodea*, and the process efficiency is dependent on many factors such as media composition, pH, temperature, agitation rate, and available oxygen. Native form of gellan contains two types of acyl substituents, namely, l-glyceryl and acetyl. Usually, alkaline hydrolysis is used to remove both of the residues and gives deacetylated gellan, also called low acetyl or low acyl [78].

Native low acyl gellan can forms hydrogels in the presence of mono-, di-, and trivalent cations [79]. With native gellan, it is possible to obtain a soft, easily deformable gels, while with deacetylated one, it results to rigid and brittle gels. The production of gellan gum is temperature dependent, at least 70 °C is needed, with subsequent cooling (to room temperature) to change the conformation of polymer chains [79–81]. Commonly, gellan gum is used as agar substitute since it can be clarified by filtration of hot deacetylated gellan gum [78]. In fact, due to the advantages of GG, it has received both US FDA and EU approval, being widely used in many food, cosmetics, pharmacy, and medicine applications [79, 82]. It has also being applied in tissue engineering, mostly as a material for cartilage reconstruction [79, 82]. Moreover, some researchers have modified gellan gum with methacrilated groups to improve its physical and mechanical properties, without affecting its biocompatibility, being relevant in a widespread of tissue engineering applications [83]. Moreover, gellan gum gel also showed to be a promising material for cartilage tissue engineering [84].

8.2.1.9 Chondroitin Sulfate

Chondroitin sulfate, member of the glycosaminoglycan family, is a complex sulfated polysaccharide containing repetitive units of glucuronic acid and galactosamine extensively distributed in human and other mammals and invertebrates, as well as some bacteria [85, 86] (Table 8.1). It is negatively charged and responsible for the water retention of the cartilage, which is essential for pressure resistance

[87]. Moreover, chondroitin sulfate has a molecular mass of 20–80 kDa [88] depending with source, being always heterogeneous with respect to size. The sources for extraction can be bovine, chicken and porcine, and also marine species such as bony fishes, whale, shark, squid, and salmon [86, 89, 90]. This polysaccharide is largely used as a biomacromolecule in the treatment of osteoarthritis via oral administration alone or in combination with other active ingredients [90]. The isolation process of from cartilage has been defined for many years and generally includes four steps: (1) chemical hydrolysis of cartilage, (2) breakdown of proteoglycan core, (3) elimination of proteins and recovery, and (4) purification [86]. Recently, chondroitin sulfate was shown to enhance resistance to apoptosis in vascular cells [91].

8.2.2 Synthetic Materials

Synthetic polymers are very attractive candidates as their material properties are typically more flexible than those of natural materials, being also possible to control the mechanical and chemical properties of synthetic polymers. Since it is possible to have a superior control on the production, synthetic materials can be non-toxic, available, and inexpensive to produce and can be compatible with cells. For these reasons, synthetic polymers are being used for tissue engineering and regenerative medicine; however, they do not have the inherent qualities that can promote desirable cell responses. The most common synthetic polymers are poly-lactic acid (PLA, which is present in both L and D forms), poly-glycolic acid (PGA), and their copolymer poly(lactic-co-glycolic acid) (PLGA) [86, 90]. Some examples of synthetic biomolecules will be discussed in the following topics.

8.2.2.1 Polycaprolactone

Polycaprolactone (Table 8.1) is a hydrophobic and semicrystalline polymer. Its crystallinity tends to decrease with increasing molecular weight. Extensive research or its application in the biomedical field has been performed due to its amazing properties, for instance, good solubility, low melting point, and exceptional blend compatibility. This polymer can be prepared by either ring-opening polymerization of ε-caprolactone using a variety of anionic, cationic, and coordination catalysts or also via free-radical ring-opening polymerization of 2-methylene-1-3-dioxepane. Polycaprolactone has many advantages over other polymers such as tailorable degradation kinetics and mechanical properties (ease molding and manufacturing enabling suitable pore sizes conducive to tissue in-growth and controlled delivery of drugs). Furthermore, functional groups can be added to turn the polymer more hydrophilic, adhesive, or biocompatible that enabled favorable cell responses [90, 91]. Recent research demonstrated that the PCL-HA scaffolds loaded with bone marrow cells improved chondrogenesis, and implantation of these scaffolds for osteochondral repair enhanced integration with host bone [92].

8.2.2.2 Poly(Lactic-Co-Glycolic Acid)

Poly(lactic-co-glycolic acid) (Table 8.1) is a copolymer of poly-lactic acid and poly-glycolic acid. This compound contains an asymmetric α-carbon which is typically described as the D or L form in classical stereochemical terms and sometimes as R and S form, respectively. The enantiomeric forms of the polymer poly-lactic acid are poly-D-lactic acid and poly-L-lactic acid. This polymer is the best defined material available for drug delivery regarding to shape and performance. In fact, poly(lactic-co-glycolic acid) can have any shape and size and can encapsulate molecules of any size. It is soluble in wide range of common solvents including chlorinated solvents, tetrahydrofuran, acetone, ethyl acetate, and water (by hydrolysis of its ester linkages). Due to the hydrolysis of this polymer, parameters of a solid formulation can change, such as the glass transition temperature, molecular weight, and moisture content. The change in poly(lactic-co-glycolic acid) properties during polymer biodegradation influences the release and degradation rates of incorporated drug molecules. Mechanical strength of the poly(lactic-co-glycolic acid) is related to physical properties such as molecular weight and polydispersity index that can affect the ability of drug delivery device and may control the device degradation rate and hydrolysis. Mechanical strength, swelling behavior, capacity to suffer hydrolysis, and biodegradation rate of the polymer are influenced by the degree of crystallinity of poly(lactic-co-glycolic acid) [86, 93]. Several researches showed the use of this biomolecules to treat osteochondral defects [94, 95].

8.2.2.3 Poly(Ethylene Glycol)

Poly(ethylene glycol) (Table 8.1) is a linear or branched neutral polyether that has molecular weights less than 1000. This biomolecule is viscous and soluble in water and also in most organic solvents. The melting point of the solid is proportional to molecular weight, approaching 67 °C. Poly(ethylene glycol) is normally prepared by an anionic initiation process with few chain transfer and terminal steps. This polymer has much interest in biomedical community because it is nontoxic, soluble, and high mobile and does not harm active proteins or cells. This biomolecule is weakly immunogenic and FDA approved for internal consumption. Furthermore, it interacts with cell membranes and partitioning controlled by making derivatives. If covalently linked, poly(ethylene glycol) will solubilize other molecules, render proteins nonimmunogenic and toleragenic, change electroosmotic flow, change render surfaces protein-rejecting, move molecules across cell membranes, and change pharmacokinetics [96, 97]. Hui and co-workers showed that the rehydrated freeze-dried oligo[poly(ethylene glycol)fumarate] hydrogel can enhance formation of hyaline-fibrocartilaginous mixed repair tissue of osteochondral defects in a small model [98].

8.2.2.4 Poly(N-Isopropylacrylamide)

Poly(N-isopropylacrylamide) (Table 8.1) is a biocompatible and stimuli-responsive polymer with potential pharmaceutical applications, including controlled drug delivery, artificial muscles, cell adhesion mediators, and precipitation of proteins. This polymer can be synthesized from N-isopropylacrylamide, which is commercially available and can be polymerized through different methods. This polymer, water soluble at room temperature, is able to suffer transition above 32 °C (the low critical solution temperature, LCST). This temperature of such thermally sensitive poly(N-isopropylacrylamide) can be adjusted to a desired temperature range by copolymerization with a more hydrophilic comonomer, which raises the LCST, or a more hydrophobic comonomer, which lowers the LCST. When temperature is increased above its LCST (hydrophobic), the hydrophobic isopropyl groups are exposed to the water interface and so insoluble, suffering a collapse into gel form. Therefore researches reported the modifications of poly(N-isopropylacrylamide) which showed a great interest to tailor the LCST of poly(N-isopropylacrylamide) systems for drug delivery [99, 100]. Moreover, it has been shown that hydrogel obtained by covalently grafting poly(N-isopropylacrylamide) to hyaluronan is biocompatible and does not interfere with the intrinsic healing response of osteochondral defects in a rabbit model [101].

8.3 Final Remarks

The biomolecules discussed in this chapter covered some of the most current materials that can be used in osteochondral defects treatment. The biomolecule production procedures, their chemical structures, and also their modifications as well as their applications on the bone and cartilage were described for each one of them, as well as its characteristic favorable and the unfavorable properties. To overcome these unfavorable characteristics, several methodologies have been employed and are under research nowadays. In fact, physical and chemical modifications could be performed to overcome these problems.

The field of cartilage tissue engineering has developed novel biological solutions; there is still a paucity of clinical options for treatment. Although the field has concentrated on finding therapies for local lesions, it has now developed sufficiently to begin considering the challenge of finding novel solutions for the extensive joint damage osteochondral defects. In fact, the development of novel biomolecules to apply on osteochondral defects is an extremely active and challenging area of research, due to the complexity of treating two different tissues. The best biomolecules should possess several properties and characteristics, which actively participate both on cartilage and bone repair and regeneration.

References

1. Mano J, Reis R (2007) Osteochondral defects: present situation and tissue engineering approaches. J Tissue Eng Regen Med 1(4):261–273
2. Mellor LF et al (2015) Extracellular calcium modulates chondrogenic and osteogenic differentiation of human adipose-derived stem cells: a novel approach for osteochondral tissue engineering using a single stem cell source. Tissue Eng A 21(17–18):2323–2333
3. James HP et al (2014) Smart polymers for the controlled delivery of drugs–a concise overview. Acta Pharm Sin B 4(2):120–127
4. Liechty WB et al (2010) Polymers for drug delivery systems. Ann Rev Chem Biomol Eng 1:149–173
5. Singh J (2016) Natural polymers based drug delivery systems. World J Pharm Pharm Sci 5(4):805–816
6. Malafaya PB, Silva GA, Reis RL (2007) Natural–origin polymers as carriers and scaffolds for biomolecules and cell delivery in tissue engineering applications. Adv Drug Deliv Rev 59(4):207–233
7. Sokolsky-Papkov M et al (2007) Polymer carriers for drug delivery in tissue engineering. Adv Drug Deliv Rev 59(4):187–206
8. Rodriguez F et al (2014) Principles of polymer systems. CRC Press, Boca Raton
9. Martin I et al (2007) Osteochondral tissue engineering. J Biomech 40(4):750–765
10. Slotkin JR et al (2017) Biodegradable scaffolds promote tissue remodeling and functional improvement in non-human primates with acute spinal cord injury. Biomaterials 123:63–76
11. Neto BA, Carvalho PH, Correa JR (2015) Benzothiadiazole derivatives as fluorescence imaging probes: beyond classical scaffolds. Acc Chem Res 48(6):1560–1569
12. Francis R, Joy N, Sivadas A (2016) Relevance of natural degradable polymers in the biomedical field. In: Biomedical applications of polymeric materials and composites. Wiley-VCH Verlag GmbH & Co. KGaA, 303–360
13. Temenoff JS, Mikos AG (2000) Review: tissue engineering for regeneration of articular cartilage. Biomaterials 21(5):431–440
14. Bonzani IC, George JH, Stevens MM (2006) Novel materials for bone and cartilage regeneration. Curr Opin Chem Biol 10(6):568–575
15. Doulabi AH, Mequanint K, Mohammadi H (2014) Blends and nanocomposite biomaterials for articular cartilage tissue engineering. Materials 7:5327–5355
16. Vinatier C et al (2009) Cartilage tissue engineering: towards a biomaterial-assisted mesenchymal stem cell therapy. Curr Stem Cell Res Ther 4(4):318–329
17. Johnstone B et al (2013) Tissue engineering for articular cartilage repair--the state of the art. Eur Cell Mater 25:248–267
18. Cao Z, Dou C, Dong S (2014) Scaffolding biomaterials for cartilage regeneration. J Nanomater 2014:1
19. Lee EJ, Kasper FK, Mikos AG (2014) Biomaterials for tissue engineering. Ann Biomed Eng 42(2):323–337
20. Chajra H et al (2008) Collagen-based biomaterials and cartilage engineering. Application to osteochondral defects. Biomed Mater Eng 18(1 Suppl):S33–S45
21. Koh L-D et al (2015) Structures, mechanical properties and applications of silk fibroin materials. Prog Polym Sci 46:86–110
22. Kambe Y et al (2016) Silk fibroin sponges with cell growth-promoting activity induced by genetically fused basic fibroblast growth factor. J Biomed Mater Res A 104(1):82–93
23. Liu H et al (2015) Composite scaffolds of nano-hydroxyapatite and silk fibroin enhance mesenchymal stem cell-based bone regeneration via the interleukin 1 alpha autocrine/paracrine signaling loop. Biomaterials 49:103–112
24. Hashimoto T et al (2015) Changes in the properties and protein structure of silk fibroin molecules in autoclaved fabrics. Polym Degrad Stab 112:20–26

25. Lai GJ et al (2014) Composite chitosan/silk fibroin nanofibers for modulation of osteogenic differentiation and proliferation of human mesenchymal stem cells. Carbohydr Polym 111:288–297

26. Kundu B et al (2013) Silk fibroin biomaterials for tissue regenerations. Adv Drug Deliv Rev 65(4):457–470

27. Lewis R (1996) Unraveling the weave of spider silkOne of nature's wondrous chemical structures is being dissected so that it can be used in human inventions. Bioscience 46(9):636–638

28. Rockwood DN et al (2011) Materials fabrication from Bombyx mori silk fibroin. Nat Protoc 6(10):1612–1631

29. Jin SH et al (2014) The effects of tetracycline-loaded silk fibroin membrane on proliferation and osteogenic potential of mesenchymal stem cells. J Surg Res 192(2):e1–e9

30. Foss C et al (2013) Silk fibroin/hyaluronic acid 3D matrices for cartilage tissue engineering. Biomacromolecules 14(1):38–47

31. Saha S et al (2013) Osteochondral tissue engineering in vivo A comparative study using layered silk fibroin scaffolds from mulberry and nonmulberry silkworms. PLoS One 8(11):e80004

32. Miranda-Nieves D, Chaikof EL (2017) Collagen and elastin biomaterials for the fabrication of engineered living tissues. ACS Biomater Sci Eng 3(5):694–711

33. Ong KL, Lovald S, Black J (2015) Orthopaedic biomaterials in research and practice, 2nd edn. CRC Press, Boca Raton, p 476

34. Gelse K, Pöschl E, Aigner T (2003) Collagens—structure, function, and biosynthesis. Adv Drug Deliv Rev 55(12):1531–1546

35. Gaharwar AK et al (2013) Nanocomposite polymer: biomaterials for tissue repair of bone and cartilage: a material science perspective. In: Gaharwar AK et al (eds) Nanomaterials in tissue engineering: fabrication and applications. Woodhead Publishing, Cambridge, p 468

36. Lee JC, Volpicelli EJ (2017) Bioinspired collagen scaffolds in cranial bone regeneration: from bedside to bench. Adv Healthc Mater 6(17)

37. Chaudhari AA et al (2016) Future prospects for scaffolding methods and biomaterials in skin tissue engineering: a review. Int J Mol Sci 17(12)

38. Chattopadhyay S, Raines RT (2014) Review collagen-based biomaterials for wound healing. Biopolymers 101(8):821–833

39. Abedin MZ et al (2013) Isolation and characterization of pepsin-solubilized collagen from the integument of sea cucumber (Stichopus vastus). J Sci Food Agric 93(5):1083–1088

40. Potaros T et al (2009) Characteristics of collagen from Nile Tilapia (oreochromis niloticus) skin isolated by two different methods. Kasetsart J 43:584–593

41. Gorgieva S, Kokol V (2011) Collagen- vs. gelatine-based biomaterials and their biocompatibility: review and perspectives. In: Pignatello R (ed) Biomaterials applications for nanomedicine. INTECH Open Access Publisher, Rijeka.

42. Zhang L, Hu J, Athanasiou KA (2009) The role of tissue engineering in articular cartilage repair and regeneration. Crit Rev Biomed Eng 37(1–2):1–57

43. Sionkowska A et al (2014) The influence of UV-irradiation on thermal and mechanical properties of chitosan and silk fibroin mixtures. J Photochem Photobiol B 140:301–305

44. Bhardwaj N et al (2011) Potential of 3-D tissue constructs engineered from bovine chondrocytes/silk fibroin-chitosan for in vitro cartilage tissue engineering. Biomaterials 32(25):5773–5781

45. Je JY, Kim SK (2012) Chitosan as potential marine nutraceutical. Adv Food Nutr Res 65:121–135

46. Zhang K et al (2013) Repair of an articular cartilage defect using adipose-derived stem cells loaded on a polyelectrolyte complex scaffold based on poly(l-glutamic acid) and chitosan. Acta Biomater 9(7):7276–7288

47. Bano I et al (2017) Chitosan: a potential biopolymer for wound management. Int J Biol Macromol 102:380–383

48. Bhardwaj N, Kundu SC (2011) Silk fibroin protein and chitosan polyelectrolyte complex porous scaffolds for tissue engineering applications. Carbohydr Polym 85(2):325–333

49. Dragan ES (2014) Design and applications of interpenetrating polymer network hydrogels. A review. Chem Eng J 243:572–590
50. Draget KI, Smidsrød O, Skjåk-Bræk G (2005) Alginates from algae. In: Biopolymers online. Wiley Online Libray
51. Fertah M et al (2015) Extraction and characterization of sodium alginate from Moroccan *Laminaria digitata* brown seaweed. Arab J Chem 8(1):1–142
52. Cardoso MJ, Costa RR, Mano JF (2016) Marine origin polysaccharides in drug delivery systems. Mar Drugs **14**(2)
53. Nalamothu N, Potluri A, Muppalla MB (2014) Review on marine alginates and its applications. Indo Am J Pharm Res 4(10):4006–4015
54. Silva TH et al (2012) Materials of marine origin: a review on polymers and ceramics of biomedical interest. Int Mater Rev 2012
55. Boontheekul T, Kong HJ, Mooney DJ (2005) Controlling alginate gel degradation utilizing partial oxidation and bimodal molecular weight distribution. Biomaterials 26(15):2455–2465
56. Li C et al (2009) Preparation and drug release of hydrophobically modified alginate. Chemistry 1:93–96
57. Alban S, Schauerte A, Franz G (2002) Anticoagulant sulfated polysaccharides: part I. Synthesis and structure-activity relationships of new pullulan sulfates. Carbohydr Polym 47(3):267–276
58. Pluemsab W, Sakairi N, Furuike T (2005) Synthesis and inclusion property of alpha-cyclodextrin-linked alginate. Polymer 46(23):9778–9783
59. Pelletier S et al (2001) Amphiphilic derivatives of sodium alginate and hyaluronate for cartilage repair: rheological properties. J Biomed Mater Res 54(1):102–108
60. Bu HT et al (2006) Interaction of unmodified and hydrophobically modified alginate with sodium dodecyl sulfate in dilute aqueous solution - Calorimetric, rheological, and turbidity studies. Colloids Surf A Physicochem Eng Asp 278(1–3):166–174
61. Yang J-S, Xie Y-J, He W (2010) Research progress on chemical modification of alginate: a review. Carbohydr Polym 84(1):33–39
62. Galant C et al (2006) Altering associations in aqueous solutions of a hydrophobically modified alginate in the presence of beta-cyclodextrin monomers. J Phys Chem B 110(1):190–195
63. Suhas et al (2016) Cellulose: a review as natural, modified and activated carbon adsorbent. Bioresour Technol 216:1066–1076
64. Muller FA et al (2006) Cellulose-based scaffold materials for cartilage tissue engineering. Biomaterials 27(21):3955–3963
65. Varoni E et al (2012) Agarose gel as biomaterial or scaffold for implantation surgery: characterization, histological and histomorphometric study on soft tissue response. Connect Tissue Res 53(6):548–554
66. Yodmuang S et al (2015) Silk microfiber-reinforced silk hydrogel composites for functional cartilage tissue repair. Acta Biomater 11:27–36
67. Khanarian NT et al (2012) A functional agarose-hydroxyapatite scaffold for osteochondral interface regeneration. Biomaterials 33(21):5247–5258
68. Zignego DL et al (2014) The mechanical microenvironment of high concentration agarose for applying deformation to primary chondrocytes. J Biomech 47(9):2143–2148
69. Rackwitz L et al (2014) Functional cartilage repair capacity of de-differentiated, chondrocyte-and mesenchymal stem cell-laden hydrogels in vitro. Osteoarthritis Cartilage 22(8):1148–1157
70. Collins MN, Birkinshaw C (2013) Hyaluronic acid based scaffolds for tissue engineering--a review. Carbohydr Polym 92(2):1262–1279
71. Zhao F et al (2014) The application of polysaccharide biocomposites to repair cartilage defects. Int J Polym Sci 2014:9
72. Murado MA et al (2012) Optimization of extraction and purification process of hyaluronic acid from fish eyeball. Food Bioprod Process 90(C3):491–498
73. Liu L et al (2011) Microbial production of hyaluronic acid: current state, challenges, and perspectives. Microb Cell Fact 10:99

74. Kang JY et al (2009) Novel porous matrix of hyaluronic acid for the three-dimensional culture of chondrocytes. Int J Pharm 369(1–2):114–120

75. Kim IL, Mauck RL, Burdick JA (2011) Hydrogel design for cartilage tissue engineering: a case study with hyaluronic acid. Biomaterials 32(34):8771–8782

76. Unterman SA et al (2012) Hyaluronic acid-binding scaffold for articular cartilage repair. Tissue Eng Part A 18(23–24):2497–2506

77. Park YB et al (2017) Single-stage cell-based cartilage repair in a rabbit model: cell tracking and in vivo chondrogenesis of human umbilical cord blood-derived mesenchymal stem cells and hyaluronic acid hydrogel composite. Osteoarthritis Cartilage 25(4):570–580

78. Prajapati VD et al (2013) An insight into the emerging exopolysaccharide gellan gum as a novel polymer. Carbohydr Polym 93(2):670–678

79. Osmalek T, Froelich A, Tasarek S (2014) Application of gellan gum in pharmacy and medicine. Int J Pharm 466(1–2):328–340

80. da Silva RMP et al (2008) Poly(N-Isopropylacrylamide) surface-grafted chitosan membranes as a new substrate for cell sheet engineering and manipulation. Biotechnol Bioeng 101(6):1321–1331

81. da Silva LP et al (2014) Engineering cell-adhesive gellan gum spongy-like hydrogels for regenerative medicine purposes. Acta Biomater 10(11):4787–4797

82. Kang D, Zhang F, Zhang H (2015) Fabrication of stable aqueous dispersions of graphene using gellan gum as a reducing and stabilizing agent and its nanohybrids. Mater Chem Phys 149-150:129–139

83. Coutinho DF et al (2010) Modified Gellan Gum hydrogels with tunable physical and mechanical properties. Biomaterials 31(29):7494–7502

84. Tang Y et al (2012) An improved complex gel of modified gellan gum and carboxymethyl chitosan for chondrocytes encapsulation. Carbohydr Polym 88(1):46–53

85. Thalla PK et al (2014) Chondroitin sulfate coatings display low platelet but high endothelial cell adhesive properties favorable for vascular implants. Biomacromolecules 15(7):2512–2520

86. Shi YG et al (2014) Chondroitin sulfate: extraction, purification, microbial and chemical synthesis. J Chem Technol Biotechnol 89(10):1445–1465

87. Jerosch J (2011) Effects of glucosamine and chondroitin sulfate on cartilage metabolism in OA: outlook on other nutrient partners especially Omega-3 fatty acids. Int J Rheumatol 2011:17

88. Lai JY et al (2012) Nanoscale modification of porous gelatin scaffolds with chondroitin sulfate for corneal stromal tissue engineering. Int J Nanomedicine 7:1101–1114

89. Fardellone P et al (2013) Comparative efficacy and safety study of two chondroitin sulfate preparations from different origin (avian and bovine) in symptomatic osteoarthritis of the knee. Open Rheumatol J 7:1–12

90. Vazquez JA et al (2013) Chondroitin sulfate, hyaluronic acid and chitin/chitosan production using marine waste sources: characteristics, applications and eco-friendly processes: a review. Mar Drugs 11(3):747–774

91. Charbonneau C et al (2011) Stimulation of cell growth and resistance to apoptosis in vascular smooth muscle cells on a chondroitin sulfate/epidermal growth factor coating. Biomaterials 32(6):1591–1600

92. Wei B et al (2015) Three-dimensional polycaprolactone-hydroxyapatite scaffolds combined with bone marrow cells for cartilage tissue engineering. J Biomater Appl 30(2):160–170

93. Place ES et al (2009) Synthetic polymer scaffolds for tissue engineering. Chem Soc Rev 38(4):1139–1151

94. Solchaga LA et al (2005) Repair of osteochondral defects with hyaluronan- and polyester-based scaffolds. Osteoarthritis Cartilage 13(4):297–309

95. Kang SW et al (2006) The use of poly(lactic-co-glycolic acid) microspheres as injectable cell carriers for cartilage regeneration in rabbit knees. J Biomater Sci Polym Ed 17(8):925–939

96. Schmaljohann D (2006) Thermo-and pH-responsive polymers in drug delivery. Adv Drug Deliv Rev 58(15):1655–1670

97. Woodruff MA, Hutmacher DW (2010) The return of a forgotten polymer—Polycaprolactone in the 21st century. Prog Polym Sci 35(10):1217–1256
98. Hui JH et al (2013) Oligo[poly(ethylene glycol)fumarate] hydrogel enhances osteochondral repair in porcine femoral condyle defects. Clin Orthop Relat Res 471(4):1174–1185
99. Emami J et al (2015) Formulation and optimization of celecoxib-loaded PLGA nanoparticles by the Taguchi design and their in vitro cytotoxicity for lung cancer therapy. Pharm Dev Technol 20(7):791–800
100. Harris JM (2013) Poly (ethylene glycol) chemistry: biotechnical and biomedical applications. Springer Science & Business Media
101. D'Este M et al (2016) Evaluation of an injectable thermoresponsive hyaluronan hydrogel in a rabbit osteochondral defect model. J Biomed Mater Res A 104(6):1469–1478

Part III
Technological Advances in Osteochondral Tissue Engineering

Chapter 9
Nanoparticles-Based Systems for Osteochondral Tissue Engineering

Isabel Oliveira[*], Sílvia Vieira[*], J. Miguel Oliveira, and Rui L. Reis

Abstract Osteochondral lesions represent one of the major causes of disabilities in the world. These defects are due to degenerative or inflammatory arthritis, but both affect the articular cartilage and the underlying subchondral bone. Defects from trauma or degenerative pathology frequently cause severe pain, joint deformity, and loss of joint motion. Osteochondral defects are a significant challenge in orthopedic surgery, due to the cartilage complexity and unique structure, as well as its exposure to high pressure and motion. Although there are treatments routinely performed in the clinical practice, they present several limitations. Tissue engineering can be a suitable alternative for osteochondral defects since bone and cartilage engineering had experienced a notable advance over the years. Allied with nanotechnology, osteochondral tissue engineering (OCTE) can be leveled up, being possible to create advanced structures similar to the OC tissue. In this chapter, the current strategies using nanoparticles-based systems are overviewed. The results of the studies herein considered confirm that advanced nanomaterials will undoubtedly play a crucial role in the design of strategies for treatment of osteochondral defects in the near future.

[*]These authors contributed equally to this chapter.

I. Oliveira · S. Vieira
3B's Research Group – Biomolecules, Biodegradables and Biomimetics,
University of Minho, Headquarters of the European Institute of Excellence
on Tissue Engineering and Regenerative Medicine, Barco, Guimarães, Portugal

ICVS/3B's – PT Government Associate Laboratory, Braga/Guimarães, Portugal
e-mail: silvia.vieira@dep.uminho.pt

J. M. Oliveira (✉) · R. L. Reis
3B's Research Group – Biomolecules, Biodegradables and Biomimetics,
University of Minho, Headquarters of the European Institute of Excellence on Tissue
Engineering and Regenerative Medicine, Barco, Guimarães, Portugal

ICVS/3B's – PT Government Associate Laboratory, Braga/Guimarães, Portugal

The Discoveries Center for Regenerative and Precision Medicine,
Headquarters at University of Minho, Barco/Guimarães, Portugal
e-mail: miguel.oliveira@dep.uminho.pt

© Springer International Publishing AG, part of Springer Nature 2018   209
J. M. Oliveira et al. (eds.), *Osteochondral Tissue Engineering*,
Advances in Experimental Medicine and Biology 1059,
https://doi.org/10.1007/978-3-319-76735-2_9

Keywords Nanoparticles · Articular cartilage · Subchondral bone · Osteochondral defects

9.1 Introduction

Osteochondral tissue defects affect both the articular cartilage and the underlying subchondral bone (Fig. 9.1). Usually, these defects are due to degenerative or inflammatory arthritis and represent one of the major causes of disabilities in the world [1, 2].

Osteochondral lesions lead to the formation of fibrocartilage with different biomechanical properties from the native hyaline cartilage and do not protect the subchondral bone from further degeneration. Damage from trauma or degenerative pathology frequently causes severe pain, joint deformity, and loss of joint motion [3]. Osteochondral tissue repair requires an advanced knowledge of how bone and cartilage interact, that is, comprehension of the osteochondral interface and its combined yet separate mechanical strengths, structure, and biology [4, 5].

Osteochondral defects are a big challenge in orthopedic surgery, due to the cartilage complexity and unique structure, as well as its exposure to high pressure and motion. Current clinical treatments comprise palliative methods (arthroscopic debridement), intrinsic repair enhancement (microfracture), tissue transplantation (osteochondral autograft), and cell-based tissue repair (autologous chondrocyte implantation) [6].

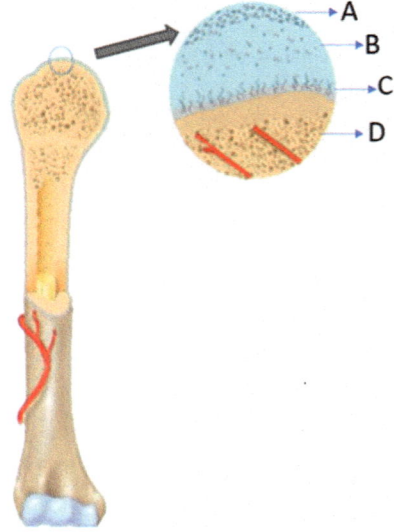

Fig. 9.1 Osteochondral tissue structure. (A) Superficial zone, (B) middle zone, (C) deep zone and calcified cartilage, and (D) subchondral bone. (Adapted with permission [10])

Despite the progress in clinical treatment, there are several limitations, such as the amount of material available, the donor site morbidity, and the difficulty to match the topology of the grafts with the injured site [7]. Furthermore, the absence of a scaffold able to guide cell differentiation and support the secretion of a structurally coherent ECM, simultaneously with the inherent predisposition of chondrocytes to dedifferentiate into fibroblast-like phenotypes upon ex vivo expansion, represents a big limitation of current clinical treatments [8]. Clinical treatments generally need long postoperative treatments, with limited mechanical loading, until tissue remodeling is achieved at the defect site, and, although patients initially demonstrate improvement, studies show that the functionality of the tissue is not improved in the long term [9].

Tissue engineering takes advantage of biomaterials, scaffolds, and cells to regenerate injured tissue, and it is a suitable alternative for osteochondral defects. The study of both bone engineering and cartilage engineering has experienced a great advance over the years, and it is improving the understanding of tissue engineering materials and biology [10].

Strategies to osteochondral repair using tissue engineering typically involve nanoparticles-based system. These systems play a key role in the repair of osteochondral defects and must have a number of criteria such as possess similar mechanical properties as the native tissue; support the growth and proliferation of cells; be biocompatible, non-immunogenic in vivo; and remain integrated in the defect while subjected to repetitive physiological loads, till the tissue repair is complete [11].

Herein, the current osteochondral tissue engineering (OCTE) strategies using nanoparticles-based systems are overviewed.

9.2 Nanoparticles-based Systems for Osteochondral Repair

Nanotechnology opens up a new era for tissue engineering (TE), with nanoparticles (NPs) being applied in several approaches, from scaffold construction to drug delivery or cell tracking. This versatility is a result of the several NP available and countless functionalization reactions that can be used to tailor NP for a desired application. These include antibodies, labeling probes, hydrophobic or hydrophilic molecules, DNA, and/or oligonucleotides [12].

Considering its nature, NP can be divided as organic and inorganic. Organic NP can include liposomes and polymeric NP. These are mostly used as delivery systems or reservoirs. On the other hand, inorganic NPs comprise silica and metallic NP, bioceramic and bioactive glass NP, carbon nanotubes, and quantum dots (QD) [13]. Although all of them are of inorganic nature, they have distinct properties and, consequently, different applications that will be further discussed in this chapter.

9.2.1 NP for Scaffolding Applications

Anatomically, the bone can be considered as a hierarchical complex nanocomposite, with an organic extracellular matrix, strengthened by inorganic calcium phosphate NPs, namely, hydroxyapatite (HA) crystals [14]. Hence, bioceramic NPs, like HA or tricalcium phosphate (TCP), are broadly applied in biomedical field alone or combined with natural [15, 16] or synthetic polymers [17, 18]. This combination results into a nanocomposite [19], usually with a similar structure to the one observed in bone. These biomaterials usually hold superior mechanical properties [20] which are very attractive for bone tissue engineering strategies. Indeed, bone-inspired hybrid scaffolds, comprising an organic and an inorganic portion with similar size and functionality as natural nanosized inorganic ceramic particles, have been considered as a potential approach to mimic the bone part of OC region.

Following this rationale, Nowicki et al. [18] combined fused deposition modeling technique (FDM) and nanocrystalline HA (nHA) to design a scaffold with tunable porosity and improved bone marrow human mesenchymal stem cell (hMSC) adhesion, growth, and osteochondral differentiation. The nanocomposite scaffold was constituted by a mix of poly(ethylene glycol) (PEG) and PEG diacrylate, as organic phase, and an nHA equivalent of 60% wt of PEG-DA as inorganic part. Bare scaffolds, i.e., without inorganic phase, were also studied to mimic the cartilaginous part of the OC region. Using a supplemented media with chondrogenic factors as culture media, it was possible to observe that the presence of nHA enhanced the osteogenic differentiation of hMSC but inhibited chondrogenic differentiation. On the other hand, cells cultured in scaffolds void of nHA showed enhanced chondrogenic differentiation rather than osteogenic.

In another approach, Amadori et al. [16] combined gelatin, a natural polymer, with nHA, using a gradient of these NPs to better mimic the OC anatomy. The scaffold is obtained using a bottom-up strategy where different layers of gelatin, combined with a decreasing gradient of nHA (50% wt., 30% wt., and 0% wt.), are glued together to obtain a multilayer scaffold. Such scaffolds showed interconnected porosity, which was not affected by the presence of nHA. Conversely, the mechanical performance of the resulting scaffolds was greatly improved by the presence of an inorganic phase. As expected, the presence of nHA pushed hMSC, seeded on these scaffolds, into an osteogenic lineage, while on the layer without nanoparticles, cells differentiated into a chondrogenic lineage.

Another interesting method was considered by Mellor et al. [21]. After concluding that an elevated extracellular concentration in calcium promoted the osteogenic differentiation of human adipose stem cells (hASC) but inhibited chondrogenic differentiation, the authors engineered a scaffold with site-specific calcium concentrations. For that, the authors used polylactic acid nanofibers, obtained by electrospinning, containing either 0% or 20% of TCP nanoparticles corresponding to low and elevated calcium concentration, respectively. Scaffolds were then seeded with hASC and cultured in presence of chondrogenic differentiation medium. Under such conditions, hASC differentiated locally into cartilage in the layers with no TCP and generated calcified tissue in layers containing 20% TCP, as anticipated.

9.2.2 NP for Imaging

Another application of NP is as imaging tracking agents. Either incorporated within a scaffold or internalized by cells, NPs appear as a good alternative for noninvasive tracking of TE constructs [13]. Different NPs can be used as contrast agents, being the most common: the magnetic nanoparticles [22, 23], gold NP [24], mesoporous silica NP [25], and quantum dots (QDs) [26]. The detection method is dependent on the type of NP used and can include magnetic resonance (MRI), computed tomography (CT), and photo-acoustic imaging.

Magnetic nanoparticles (MNP) are composed by magnetic elements such as iron, nickel, cobalt, and their oxides [27, 28]. Typically, MNP are detected using MRI techniques due to their magnetism and strong contrast enhancement effects [27]. Since MNP can be functionalized at surface level, these particles can be modified to recognize specific targets or to include fluorescent probes for multimodal imaging. Thence, these NPs are a very versatile tool for imaging processes, cell tracking and isolation, biosensors, guided drug and gene delivery, 3D cell organization, and hyperthermia [28]. Nevertheless, care must be taken when a biological application of NP is envisaged, as particles can be toxic for cells or may alter their phenotype [29].

To access whether MNP can be used together with rabbit chondrocytes, Su et al. [30] labeled these cells with commercial iron oxide magnetic nanoparticles and further analyzed cell proliferation, viability, and differentiation capacity. After finding an optimal concentration (250 μg/ml), the authors concluded that, at this concentration, MNP did not affect cell morphology, viability, or phenotype. Additionally, the authors took advantage of the magnetic properties of NP to guide cells inside of a biphasic scaffold made by type II collagen-chitosan/PLGA.

9.2.3 NP as Delivery Vehicles

Nanoparticles-based systems have the potential to provide more effective tissue regeneration when compared to the existing therapies. Systems containing nanoparticles loaded with bioactive agents can be used for their local delivery, enabling site-specific pharmacological effects, for instance, the induction of cell proliferation and differentiation and, therefore, neo-tissue formation, such as cartilage repair and bone regeneration [31].

As aforementioned, magnetic NP can be used to take advantage of their inherent magnetic properties, but they can also be functionalized to deliver specific molecules to cells. Taking advantage of this, Zhang et al. [32] developed a new Fe_3O_4 magnetic nanoparticle coated with nanoscale graphene oxide to label stem cells and delivery growth factors. These nanoparticles were successfully fabricated, with an average diameter of 10 nm, and exhibited a core-shell structure and kept high-saturation magnetization values. The Fe_3O_4 magnetic nanoparticle coated with nanoscale graphene oxide did not affect the viability and proliferation of dental-pulp stem cells (DPSCs), and nanoparticle-labeled cells could be organized via magnetic force to form multilayered cell sheets with different patterns. When compared to traditional Fe_3O_4 nanoparticles, the graphene oxide coating gives abundance of carboxyl groups to bind and deliver growth factors. With these Fe_3O_4 magnetic nanoparticle coated with nanoscale graphene oxide, bone-morphogenetic-protein-2 (BMP2) is favorably included into DPSC sheets to promote more bone formation. Moreover, an integrated osteochondral complex is also constructed using a combination of DPSCs/TGF-β3 and DPSCs/BMP2. This study showed that Fe_3O_4 magnetic nanoparticle coated with nanoscale graphene oxide supports a novel magnetically controlled vehicle for stem cells and growth factors to construct protein-immobilized cell sheets, and they have promising potential for future use in regenerative medicine.

Other recent study developed a bilayer scaffold, in order to promote the regeneration of osteochondral tissue within a single integrated construct. For the subchondral bone layer, organic type I collagen and the inorganic hydroxyapatite were used to mimic the bone matrix. A composite scaffold was made through mineralization of hydroxyapatite nanocrystals, with oriented growth on collagen fibrils. For that, a multi-shell NP system with different layers was developed, comprised by a calcium phosphate core and a DNA/calcium phosphate shell conjugated with polyethyleneimine to operate as nonviral vectors for delivery of plasmid DNA encoding BMP2 and TGF-β3. For this, it was used microbial transglutaminase as a cross-linking agent to cross-link the bilayer scaffold. The results showed that the produced scaffold has the capacity to promote transfection of human mesenchymal stem cells (hMSCs) and the functional osteochondral tissue formation. Moreover, the sustained release of plasmids, BMP2 and TGF-b3, from gene-activated matrix could induce prolonged transgene expression and stimulate hMSC differentiation into osteogenic and chondrogenic lineages by spatial and temporal control manner. This system may increase the functionalization composite graft to accelerate healing process for osteochondral tissue regeneration [33]. This work is a great example of the versatility of NP, since they were used not only to mimic the natural anatomy of the subchondral bone layer but also to simultaneously deliver nonviral vectors, improving the outcome of the system.

Polymeric NPs have been also greatly used in recent years, owing to their bulk physical properties, tunable architecture, and biodegradability. Synthesis of polymeric NP is commonly an easy and flexible process, allowing the incorporation of a wide range of molecules [12]. It is possible to produce NP with different shapes including nanofibers [34] and spherical NPs [35], either nanocapsules or nanospheres. Typically, this type of particles is used for transport and/or delivery of molecules, since they are characterized by a high drug-loading capacity and can be easily functionalized to perform an active targeting. Considering the above, Wang et al. [36] studied chitosan nanoparticles and electrospun fiber scaffolds as a sustained release system of Nel-like molecule-1(Nell-1) growth factor, protecting its bioactivity. Then, the effect and process of Nell-1 on inducing human bone MSC (hBMSC) differentiate toward chondrocytes were analyzed. The results showed that scaffolds mimicked the oriented structure of native articular cartilage by regulating cell adhesion and distribution. In release and bioactivity protection study, the incorporation of Nell-1 into chitosan nanoparticles significantly prolonged the release time and increased the bioactivity of the released growth factor, as compared to the scaffolds with free Nell-1. Furthermore, in vitro chondrogenesis study proved that Nell-1 increases hBMSC chondrogenic differentiation and extracellular matrix production. This study showed the potential ability of Nell-1 integrated dual release scaffold for cartilage tissue engineering.

9.3 Conclusions

In this chapter, several alternatives that can be potentially used for the treatment osteochondral defects, rather than traditional treatments, are presented. Many techniques using nanoparticles have been developed for the treatment osteochondral tissues. The results of recent studies allow confirming that the use of nanoparticle systems is very promising, due to their physicochemical and biological properties for OCTE.

Undoubtedly, nanotechnology is a breakthrough in the TE field, and OCTE is not an exception. Being a very complex tissue, the application of NP can be justified by different aims, which is only possible because of the large variability concerning NP synthesis, nature, and properties.

As part of the OC tissue is composed by bone, this field of TE particularly benefits of NP usage, as they can mimic the nanosized inorganic phase present in the bone. The reinforced scaffolds present not only better mechanical properties but also a commitment for the osteogenic lineage. Despite the encouraging results of most the studies herein discussed, it is still a great challenge to pass these nanosystems to clinical applications. Yet it is possible to state that advanced nanomaterials, in the future, will certainly play a crucial role on the design of strategies for treatment of osteochondral defects.

References

1. Castro NJ, Hacking SA, Zhang LG (2012) Recent progress in interfacial tissue engineering approaches for osteochondral defects. Ann Biomed Eng 40(8):1628–1640
2. Khan MS, Vishakante GD, Siddaramaiah H (2013) Gold nanoparticles: a paradigm shift in biomedical applications. Adv Colloid Interface Sci 199–200:44–58. https://doi.org/10.1016/j.cis.2013.06.003
3. Panseri S, Russo A, Cunha C et al (2012) Osteochondral tissue engineering approaches for articular cartilage and subchondral bone regeneration. Knee Surg SportTraumatol Arthrosc 20(6):1182–1191
4. Amini AR, Adams DJ, Laurencin CT, Nukavarapu SP (2012) Optimally porous and biomechanically compatible scaffolds for large-area bone regeneration. Tissue Eng Part A 18:1376. https://doi.org/10.1089/ten.tea.2011.0076

5. Nukavarapu SP, Amini AR (2011) Optimal scaffold design and effective progenitor cell identification for the regeneration of vascularized bone. In: Proceedings of the Annual International Conference of the IEEE Engineering in Medicine and Biology Society, EMBS, Boston, MA, USA.
6. Nejadnik H, Daldrup-Link HE (2012) Engineering stem cells for treatment of osteochondral defects. Skeletal Radiol 41:1
7. Martin I, Miot S, Barbero A et al (2007) Osteochondral tissue engineering. J Biomech 40(4):750–765
8. Makris EA, Gomoll AH, Malizos KN et al (2014) Repair and tissue engineering techniques for articular cartilage. Nat Rev Rheumatol 11:21. https://doi.org/10.1038/nrrheum.2014.157
9. Camarero-Espinosa S, Cooper-White J (2017) Tailoring biomaterial scaffolds for osteochondral repair. Int J Pharm 523:476. https://doi.org/10.1016/j.ijpharm.2016.10.035
10. Nukavarapu SP, Dorcemus DL (2013) Osteochondral tissue engineering: current strategies and challenges. Biotechnol Adv 31:706. https://doi.org/10.1016/j.biotechadv.2012.11.004
11. Degoricija L, Bansal PN, Söntjens SHM et al (2008) Hydrogels for osteochondral repair based on photocrosslinkable carbamate dendrimers. Biomacromolecules 9:2863. https://doi.org/10.1021/bm800658x
12. Nicolas J, Mura S, Brambilla D et al (2013) Design, functionalization strategies and biomedical applications of targeted biodegradable/biocompatible polymer-based nanocarriers for drug delivery. Chem Soc Rev 42:1147–1235. https://doi.org/10.1039/C2CS35265F
13. Vieira S, Vial S, Reis RL, Oliveira JM (2017) Nanoparticles for bone tissue engineering. Biotechnol Prog 33:590–611. https://doi.org/10.1002/btpr.2469
14. Alves Cardoso D, Jansen JA, Leeuwenburgh SCG (2012) Synthesis and application of nanostructured calcium phosphate ceramics for bone regeneration. J Biomed Mater Res Part B Appl Biomater 100B:2316–2326. https://doi.org/10.1002/jbm.b.32794
15. Yan L-P, Silva-Correia J, Correia C et al (2012) Bioactive macro/micro porous silk fibroin/nano-sized calcium phosphate scaffolds with potential for bone-tissue-engineering applications. Nanomedicine 8:359–378. https://doi.org/10.2217/nnm.12.118
16. Amadori S, Torricelli P, Panzavolta S et al (2015) Multi-layered scaffolds for osteochondral tissue engineering: in vitro response of co-cultured human mesenchymal stem cells. Macromol Biosci 15:1535–1545. https://doi.org/10.1002/mabi.201500165
17. Gaharwar AK, Dammu SA, Canter JM et al (2011) Highly extensible, tough, and elastomeric nanocomposite hydrogels from poly(ethylene glycol) and hydroxyapatite nanoparticles. Biomacromolecules 12:1641–1650. https://doi.org/10.1021/bm200027z
18. Nowicki MA, Castro NJ, Plesniak MW, Zhang LG (2016) 3D printing of novel osteochondral scaffolds with graded microstructure. Nanotechnology 27:414001. https://doi.org/10.1088/0957-4484/27/41/414001
19. Pina S, Oliveira JM, Reis RL (2015) Natural-based nanocomposites for bone tissue engineering and regenerative medicine: a review. Adv Mater 27:1143–1169. https://doi.org/10.1002/adma.201403354
20. Verma S, Domb AJ, Kumar N (2011) Nanomaterials for regenerative medicine. Nanomedicine (Lond) 6:157–181. https://doi.org/10.2217/nnm.10.146
21. Mellor LF, Mohiti-Asli M, Williams J et al (2015) Extracellular calcium modulates Chondrogenic and osteogenic differentiation of human adipose-derived stem cells: A novel approach for osteochondral tissue engineering using a single stem cell source. Tissue Eng Part A 21:2323. https://doi.org/10.1089/ten.tea.2014.0572
22. Fan J, Tan Y, Jie L et al (2013) Biological activity and magnetic resonance imaging of superparamagnetic iron oxide nanoparticles-labeled adipose-derived stem cells. Stem Cell Res Ther 4:44. https://doi.org/10.1186/scrt191
23. Lalande C, Miraux S, Derkaoui SM et al (2011) Magnetic resonance imaging tracking of human adipose derived stromal cells within three-dimensional scaffolds for bone tissue engineering. Eur Cell Mater 21:341–354

24. Meir R, Motiei M, Popovtzer R (2014) Gold nanoparticles for in vivo cell tracking. Nanomedicine 9:2059–2069. https://doi.org/10.2217/nnm.14.129
25. Shen Y, Shao Y, He H et al (2013) Gadolinium3+−doped mesoporous silica nanoparticles as a potential magnetic resonance tracer for monitoring the migration of stem cells in vivo. Int J Nanomedicine 8:119–127. https://doi.org/10.2147/IJN.S38213
26. Wegner KD, Hildebrandt N (2015) Quantum dots: bright and versatile in vitro and in vivo fluorescence imaging biosensors. Chem Soc Rev 44:4792. https://doi.org/10.1039/C4CS00532E
27. Shin T-H, Choi Y, Kim S, Cheon J (2015) Recent advances in magnetic nanoparticle-based multi-modal imaging. Chem Soc Rev 44:4501. https://doi.org/10.1039/C4CS00345D
28. Colombo M, Carregal-Romero S, Casula MF et al (2012) Biological applications of magnetic nanoparticles. Chem Soc Rev 41:4306
29. Pandey RK, Prajapati VK (2018) Molecular and immunological toxic effects of nanoparticles. Int J Biol Macromol 107(Pt A):1278–1293
30. Su JY, Chen SH, Chen YP, Chen WC (2017) Evaluation of magnetic nanoparticle-labeled chondrocytes cultivated on a type II collagen–chitosan/poly(lactic-co-glycolic) acid biphasic scaffold. Int J Mol Sci 18. https://doi.org/10.3390/ijms18010087
31. Monteiro N, Martins A, Reis RL, Neves NM (2015) Nanoparticle-based bioactive agent release systems for bone and cartilage tissue engineering. Regen Ther 1:109. https://doi.org/10.1016/j.reth.2015.05.004
32. Zhang W, Yang G, Wang X et al (2017) Magnetically controlled growth-factor-immobilized multilayer cell sheets for complex tissue regeneration. Adv Mater 29:1703795. https://doi.org/10.1002/adma.201703795
33. Lee Y-H, Wu H-C, Yeh C-W et al (2017) Enzyme-crosslinked gene-activated matrix for the induction of mesenchymal stem cells in osteochondral tissue regeneration. Acta Biomater 63:210–226. https://doi.org/10.1016/J.ACTBIO.2017.09.008
34. Yoo HS, Kim TG, Park TG (2009) Surface-functionalized electrospun nanofibers for tissue engineering and drug delivery. Adv Drug Deliv Rev 61:1033–1042. https://doi.org/10.1016/j.addr.2009.07.007
35. Rao JP, Geckeler KE (2011) Polymer nanoparticles: Preparation techniques and size-control parameters. Prog Polym Sci 36:887–913. https://doi.org/10.1016/j.progpolymsci.2011.01.001
36. Wang C, Hou W, Guo X et al (2017) Two-phase electrospinning to incorporate growth factors loaded chitosan nanoparticles into electrospun fibrous scaffolds for bioactivity retention and cartilage regeneration. Mater Sci Eng C 79:507. https://doi.org/10.1016/j.msec.2017.05.075

Chapter 10
Stem Cells for Osteochondral Regeneration

Raphaël F. Canadas, Rogério P. Pirraco, J. Miguel Oliveira, Rui L. Reis, and Alexandra P. Marques

Abstract Stem cell research plays a central role in the future of medicine, which is mainly dependent on the advances on regenerative medicine (RM), specifically in the disciplines of tissue engineering (TE) and cellular therapeutics. All RM strategies depend upon the harnessing, stimulation, or guidance of endogenous developmental or repair processes in which cells have an important role. Among the most clinically challenging disorders, cartilage degeneration, which also affects subchondral bone becoming an osteochondral (OC) defect, is one of the most demanding. Although primary cells have been clinically applied, stem cells are currently seen as the promising tool of RM-related research because of its availability, in vitro proliferation ability, pluri- or multipotency, and immunosuppressive features. Being the OC unit, a transition from the bone to cartilage, mesenchymal stem cells (MSCs) are the main focus for OC regeneration. Promising alternatives, which can also be obtained from the patient or at banks and have great differentiation potential toward a wide range of specific cell types, have been reported. Still, ethical concerns and tumorigenic risk are currently under discussion and assessment. In this book chapter, we revise the existing stem cell-based approaches for engineering bone and

R. F. Canadas · R. P. Pirraco
3B's Research Group – Biomaterials, Biodegradables, and Biomimetics,
University of Minho, Headquarters of the European Institute of Excellence on Tissue Engineering and Regenerative Medicine, Barco, Guimarães, Portugal

ICVS/3B's - PT Government Associate Laboratory, Braga/Guimarães, Portugal

J. M. Oliveira · R. L. Reis · A. P. Marques (✉)
3B's Research Group – Biomaterials, Biodegradables, and Biomimetics,
University of Minho, Headquarters of the European Institute of Excellence on Tissue Engineering and Regenerative Medicine, Barco, Guimarães, Portugal

ICVS/3B's - PT Government Associate Laboratory, Braga/Guimarães, Portugal

The Discoveries Centre for Regenerative and Precision Medicine, Headquarters at the University of Minho, Barco, Guimarães, Portugal
e-mail: apmarques@dep.uminho.pt

© Springer International Publishing AG, part of Springer Nature 2018
J. M. Oliveira et al. (eds.), *Osteochondral Tissue Engineering*,
Advances in Experimental Medicine and Biology 1059,
https://doi.org/10.1007/978-3-319-76735-2_10

cartilage, focusing on cell therapy and TE. Furthermore, 3D OC composites based or cell co-cultures are described. Finally, future directions and challenges still to be faced are critically discussed.

Keywords Skeletogenesis · Stem cells · Bone · Cartilage · Osteochondral constructs

10.1   Introduction

Osteochondral (OC) lesion is an injury or defect of the articular cartilage extended to the subchondral bone [1]. A disorder affecting the OC interface usually results in osteoarthritis (OA), which is not just associated to disability but also to other conditions, such as neuropathic pain, depression, and sleep disorders [2, 3]. Some assessments of disease burden suggest that OA is even an important cause of premature death [4, 5].

Alterations in the tissues and cells surrounding articular joint of osteoarthritic patients such as the synovium are nowadays considered a predictive factor of the disease progression [6]. The activity and phenotype of the cell population's resident within the synovium are crucial to keep a healthy joint. However, might also accelerate the OA symptoms when unstable, indicating that synovial inflammation further than being a feature in early disease, might be the initiator of degenerative cascades that lead to tissue destruction [7]. On the other hand, the synovium is a natural origin of repair responses involving the endogenous progenitor cells present in the tissue [8]. Still, full-thickness articular cartilage damage does not resolve spontaneously.

Clinically applied operative techniques for OC repair are usually based on bone marrow (BM) stimulation for promotion of tissue restoration by recruitment of stromal cells. These treatments provide acceptable clinical results over midterm follow-up periods but often fail in the long term, resulting in fibrous tissue covering the lesion [9, 10], which then leads to scar tissue and biomechanical insufficiency [11, 12]. The most used techniques like abrasion arthroplasty, subchondral drilling, microfracture, autologous matrix-induced chondrogenesis [13], and mosaicplasty also frequently face drawback. Formation of fibrocartilage rather than hyaline cartilage [14, 15], limited donor site availability or morbidity, and poor integration [16, 17] are among the most common marks of unsuccessful regeneration.

Cartilage repair has evolved at a rapid pace from marrow stimulation techniques to articular chondrocyte transplantation (ACI). Taking this into consideration, the field started using techniques that apply specialized cells together with invading stems cells resulting from marrow stimulation. ACI and its later evolution, matrix-induced autologous chondrocyte implantation (MACI) [18, 19], offered great promise when compared to other approaches with 80% of patients showing improved results at 10 years [20]. However, despite the improvements, hyaline-like cartilage repair keep experiencing complications not only as the failure of graft integration in the host tissue but also periosteal hypertrophy and delamination [21, 22]. In addition, it has also been reported that chondrocytes may lose their phenotype during expansion in vitro [23, 24]. Recent reports are now projecting the future focused on stem cell-based strategies for cartilage repair [25].

There is, therefore, a growing interest in exploring stem cells features to regenerate OC lesions either through the direct injection of the cells into the bloodstream or tissues or the combination of the cells with supporting scaffolds. Stem cell-based strategies are therefore seen as the possibility to improve the limitations experienced with primary chondrocytes in ACI therapies. Unlike primary differentiated cells, stem cells have the ability to divide and specialize into specific cell types, depending on the differentiation potential and source. Pluripotent stem cells (PSCs) are often harvested from embryonic niches and can develop into any type of cell derived from the three germ layers whereas multipotent stem cells are generally isolated from adult tissues and have limited differentiation ability [25]. Mesenchymal stem cells (MSCs), not hampered by availability and donor site morbidity, are a form of multipotent cells that may offer an alternative to cartilage repair techniques [25]. In vitro generated chondral or OC grafts using stem cells are currently in a queue for FDA approval for clinical studies [26]. However, the use of stem cell therapies into the clinic is a form of translational research that still is associated with regulatory issues that are currently under discussion under different perspectives worldwide [27, 28].

Current investigation using MSCs from bone marrow (BM), adipose or embryonic tissues for cell therapies, and TE approaches on bone and cartilage regeneration are overviewed in this book chapter. Specific issues that are slowing down the fast progress observed over the last 30 years are also analyzed. Since stem cells niche has been proved to be an important factor when considering regeneration of bone and cartilage disorders, the works applying stem cells from varied origins are described. Furthermore, 3D co-culture systems are revised as the basis to discuss the future of stem cell-based approaches for osteochondral regeneration.

10.2 Progenitor Cells, Stem Cell Sources, and Cell Recruitment on Endogenous Skeletogenesis

Skeletogenesis is the process encompassing the formation of several components of the skeleton such as bones, cartilage, and joints. These components are the result of the action of specialized cells such as chondrocytes and osteoblasts that give rise, respectively, to the cartilage and bone. During embryonic development, each of these cells is derived from skeletogenic mesenchymal progenitors of mesodermal or ectodermal origin, depending on the skeletal site [29]. These progenitors home at prospective skeletal formation sites and condense in structures that then give rise to the different skeletal components [30]. In the case of the bone, two processes are possible for ossification of the skeleton: endochondral and intramembranous ossification [30]. The former consists on the production by chondrocytes of a cartilaginous template that is later mineralized and is the way through which most bone tissue is formed. In intramembranous ossification, through which flat bones are formed, the bone is directly produced by osteoblasts without an intermediate cartilage anlage.

The coordination of all these events is made through direct cell-cell communication and by the action of several key signaling molecules that regulate the cell recruitment and the associated patterning of the skeleton. Among the most significant signaling molecules are fibroblast growth factors (FGF) [31], bone morphogenetic proteins (BMPs) [32], sonic hedgehog (SHH) [33], and notch [34] and Wnt [35] ligands. The coordinated action of these molecules will determine the location of skeletogenic mesenchymal cells and orchestrate the timing of condensation. At this stage, skeletogenic mesenchymal progenitor cells express both Sox9 [36] and Runx2 [37] transcription factors which are determinant for the differentiation into the chondrogenic and osteogenic lineages, respectively. The commitment to either one is achieved by the action of signaling pathways such as Wnt and Indian hedgehog (IHH) that mostly down- or upregulate Sox9 and Runx2 [30].

Critically, mesenchymal progenitors are present in adult vertebrates as what are known as MSCs. It is not clear if MSCs directly derive from the same skeletogenic mesenchymal progenitors responsible for tissue development during embryogenesis. MSCs are multipotent stem cells that can differentiate in specific lineages of tissues such as the bone, cartilage, fat, muscle, etc. and present no immunogenicity after transplantation [38]. These cells are present in almost all tissues of the human body [39], and therefore it is hypothesized that they have the ultimate involvement in tissue repair [40]. In the skeletal system, MSCs have a direct role in injury repair and tissue regeneration, where some of the embryonic events of skeletal formation are recapitulated. MSCs are recruited to the injury site and undergo differentiation by the action of signaling molecules such as BMPs, FGF, TGF-β, and SDF-1 [41]. Such signaling molecules can be released from the matrix or secreted by other cells present at the injury site. The low oxygen tension typically found at the injury site due to blood vessel disruption is also a major factor in MSC recruitment and differentiation [42]. Under such type of stimuli, MSCs not only home to the injury site but also engage in proliferation and differentiation, ultimately repairing or regenerating the injured tissue.

MSCs have been isolated and cultured in vitro and were found to possess clinically relevant proliferation ability [43]. This is important since typically, fully differentiated cells such as osteoblasts present decreased proliferation rates in vitro, which undermines their potential to be used in, e.g., a tissue engineering strategy. Furthermore, MSCs have been found to positively modulate the response of immune system cells, which besides being important on itself for injury resolution also suggests these cells might be tolerated in allogeneic approaches [44]. It is then clear that MSCs are a very attractive source of cells to induce postnatal skeletogenesis in TE strategies. Being present on multiple tissues means that several potential sources of MSCs for therapeutic strategies have been projected [45]. A suitable source of

MSCs should allow easy accessibility, minimally invasive harvesting, and yield usable numbers of cells. BM is the most explored source of MSC in a myriad to TE strategies [46]. However, the isolation of these cells encompasses a significant degree of morbidity to the patient, and therefore alternative sources, namely, perinatal tissues such as the umbilical cord [47], teeth [48], and adipose tissue [49], have been sought. The latter, in particular, has been the focus of increasing attention since fat can be obtained in relatively noninvasive procedures, and adipose-derived stem cells (ASCs) can be isolated from discarded fat residues from plastic surgeries. While these cells appear to have more tendency for adipogenic regeneration, several works demonstrate their feasibility for bone [50] and cartilage [51] TE. Recently, it has been hypothesized that the perivascular niche might also be a suitable source of cells for bone and cartilage TE [52]. In fact, some researchers even claim that MSCs and pericytes might be one and the same entity [53] although this is still quite controversial at the moment. Nevertheless, it is unquestionable that MSCs have been the prime choice for developing TE strategies for bone and cartilage tissues.

Other examples of stem cells that can be considered for postnatal skeletogenic strategies are embryonic stem cells (ESCs) [54]. These cells are pluripotent, i.e., they have a broader differentiation potential than MSCs and similar, if not better, proliferation ability. However, they bring attached several ethical constraints since they are derived from human embryos. Furthermore, they have a significant tumorigenic risk associated. Due to these concerns, great promise has been placed in induced pluripotent stem cells (iPSCs) as proposed by Yamanaka and others [55]. These cells can be obtained by introducing specific reprogramming factors into terminally differentiated somatic cells. This avoids the need of embryo destruction and yields pluripotent stem cells from a specific patient, allowing in principle the development of personalized therapies [56].

While all these cells have been used to obtain bone and cartilage cells for tissue engineering, the question as to which is the best cell source for postnatal skeletogenesis is still open.

10.3   Stem Cells for Engineering Bone

As stated above, BM-MSCs have been the top choice as a MSC source for bone TE. This is mainly due to their high potential for osteogenic differentiation, which is probably related with their tissue of origin [57]. These cells have been combined with 3D scaffolds to deliver a myriad of bone tissue engineering strategies that ultimately resulted in the formation of bone tissue after transplantation [46, 58–60]. However, these cells present several limitations. The isolation procedure is painful for the patient as it normally consists of an invasive procedure to aspirate the marrow in the iliac crest. Other methods exist where BM is accessed from patients undergoing hip or knee replacement [61]. These are however exceptions to the rule. Furthermore, the yield of stem cells obtained from BM is low, estimated at 0.001% of colony forming unit cells per nucleated cells [38]. This means that these cells need to be heavily expanded to reach therapeutic numbers, which ultimately results in reduced differentiation capacity and therapeutic efficiency [62]. Finally, the potency of each population of BM-MSC greatly depends on the donor's age and general condition, which further complicates their applicability [63]. The quantity of MSCs declines with age; approximately 1 MSC per 10,000 marrow cells are found in a newborn, whereas 1 MSC per 400,000 marrow cells is found in a 50-year-old adult [64, 65].

Adipose-derived stem cells (ASCs) are an increasingly used cell source for bone TE. Through a relatively simple and painless procedure, adipose tissue can be harvested from a patient in generous amounts, and ASCs can be isolated from the stromal vascular fraction (SVF) by employing straightforward enzymatic protocols [66]. Importantly, the frequency of stem cells in adipose tissue is 100–1000 times bigger than for BM, which coupled with the relatively large volumes of adipose tissue that can be harvested, and makes this cell source even more appealing [67]. Furthermore, the potential of these cells to undergo osteogenic differentiation is robust although a higher tendency for adipogenic differentiation has been found [68]. The increasing number of bone TE strategies using these cells attempts to capitalize on these advantages [52, 69–71]. It is interesting to note that the SVF from which ASCs are isolated also contains angiogenic cell populations such as endothelial cells, endothelial progenitor cells, and pericytes [72]. These constitute a perfect cocktail of cells to create functional vasculature in bone TE constructs, therefore addressing one of the main current obstacles in TE [73, 74]. The possibility of using the whole fraction both for osteogenesis and angiogenesis is extremely exciting and will allow from one simple procedure to obtain patient-specific cell population for a complete bone TE strategy.

Other adult MSCs sources have been considered for bone TE. Dental pulp MSCs (DP-MSCs) are isolated from the pulp of definitive or deciduous teeth [75]. While accessibility and ease of isolation are strong points in favor of DP-MSCs, the fact is that only a small number of cells can be isolated from each tooth. The ensuing need of comprehensive cell expansion until therapeutic numbers are achieved is negative due to potential impact in cell potency. Nevertheless, these cells have been extensively tested and characterized [76]. Several works describe their similarity with BM-MSCs in term of cell markers and multilineage differentiation potential while having superior proliferative ability [75, 76]. Accordingly, these cells are being increasingly proposed for bone TE applications [77–80]. Cells from perinatal tissues such as the umbilical cord blood, umbilical cord, and amniotic membrane and fluid have been found to have all the hallmarks of MSCs while retaining some traits of embryonic cells such as robust proliferation ability and differentiation pluripotency [81–83]. Furthermore, such cells present low tumorigenic risk and appear to retain the immune privileged nature of MSCs [84]. The volume of bone TE research using perinatal cells trails behind that of other stem cells [85, 86] and the coming years will reveal if this is a cell source that will be embraced by tissue engineers, particularly in the case of bone tissue.

ESCs, as stated before, are pluripotent stem cells isolated from embryonic tissues that can differentiate in cells from every germ layer. Osteogenic cells have been obtained from ESCs using well-developed protocols [54]. Therefore, many bone TE strategies have been proposed using cells derived from ESCs [54, 87–90]. However, several constraints are associated with the use of these cells. Since they are isolated from the inner cell mass of blastocysts, they encompass the destruction of the latter which raises a series of ethical issues. Furthermore, these cells have a high tumorigenic potential in vivo which is obviously a very serious safety concern for clinical applications. This tumorigenic risk can be partially mitigated using a combination of culture and cell sorting techniques [91]. However, the ethical concerns are unavoidable. In order to overcome this issue, iPSCs were developed and famously proposed by Yamanaka and colleagues [55]. These cells are pluripotent stem cells vastly similar to ESCs, but they are derived from adult cells. The technique to achieve this derivation is based on the delivery of pluripotency-associated factors such as cMyc, Oct4, Sox2, and Klf4 to fully dedifferentiate cells in order to effectively reprogram those cells into pluripotent cells. This strategy avoids the ethical issues of ESCs while adding a very significant benefit: the ability to produce patient-specific pluripotent stem cells [56]. This is very important since it allows the combination of the advantages of adult stem cell sources such as BM-MSCs and ASCs, with the pluripotency of ESCs. While some concerns exist related with potential tumorigenicity of the original reprogramming factors used in iPSC protocols, other protocols have been put forward using less worrisome factors [92]. Therefore, numerous bone TE strategies using iPSCs have been proposed by first inducing differentiation to the mesenchymal and finally to the osteogenic lineages [93–96]. However, the safety of these cells is still a major issue, and more clinical research is needed to confirm these cells potential for bone TE.

10.4   Stem Cells for Engineering Cartilage

To date, only few studies directly compared MSCs and chondrocytes for cartilage repair. Interestingly, no significant differences were found on the histological scores [97, 98]. Moreover, the efficacy of MSCs in vitro pre-differentiated into chondrocytes or undifferentiated was equally superior in comparison to the untreated condition, but without significant differences among themselves [98]. Clinically, the intra-articular injection of MSCs favorably influenced the progression of lesions larger than 109 mm^2 associated with subchondral cysts, in patients older than 50 years, thus hampering the development of an associated degenerative disease [1]. Therefore, many uncertainties are still to be clarified regarding the mode of action and the efficacy of MSCs to engineer cartilage tissue.

MSCs were proposed as a valid option for the treatment of cartilage defects because of their ability to differentiate into chondrocytes, among other cell lineages [99–101]. These cells are also immune privileged [102–104] and have the ability to modulate inflammatory cytokines including interferon gamma, tumor necrosis factor-α, and interleukin-1α and interleukin-1β [105]. While these features might be of major therapeutic interest when inflammatory conditions, such as OA, are associated to cartilage degeneration, it also reinforces the possibility of following allogeneic approaches as demonstrated for BM-MSCs [106, 107] and ASCs [108] in different animal models. In fact, the need for cartilage repair is greatly associated to age; thus autologous approaches are significantly compromised by its retrograde effect on the cell's intrinsic properties and regenerative capability [109].

While adipose tissue could be considered an alternative source, ASCs have less chondrogenic potential than BM-MSCs [110]. The infrapatellar fat pad also known as Hoffa's body, an extra-synovial tissue placed in the knee under the patella [111] that is commonly harvested and discarded in arthroscopic surgeries [112], was shown to possess a higher percentage of stromal cells than subcutaneous fat [113, 114]. In addition, the regenerative capacity of the cells obtained from Hoffa's body in an OA model was demonstrated [115], thus reinforcing the positioning of this source for cartilage engineering.

As an alternative to the abovementioned sources of MSCs, peripheral blood [116], periosteum [117], and synovium [118] have been explored to engineer cartilage in particular. However, there are no reports showing clear evidences of superiority of one over the other.

Interestingly enough, the debate on the possibility of following an allogeneic approach comes up again when considering Wharton's jelly-derived MSCs (WJ-MSCs). The umbilical cord, which is discarded at birth, provides an unlimited source of cells which are believed to be more primitive and proliferative and to have broader multipotency than adult MSCs [119, 120]. WJ-MSCs express markers of both MSCs and ESCs, but in opposition to ESCs, these cells do not induce teratoma formation [121]. In addition, the composition of Wharton's jelly extracellular matrix, rich in aggrecan and type II collagen, is very similar to that of cartilage [122]. hWJ-MSCs also express growth factors, chemokines, and cytokines at levels similar to those of cartilage cells [123]. Altogether, these indications might be sufficiently supportive to consider Wharton's jelly as an ethically noncontroversial source of MSCs to engineer and regenerate cartilage.

At a different stage of development but with promising expectations, a multipotent subpopulation of muscle-derived stem cells (MDSCs) isolated from mouse skeletal muscle was genetically engineered to express BMP-4 as a way to enhance chondrogenesis [124]. MDSCs transduced with retroviral vectors CLBMP-4 and CLLacZ can potentially be used to locally secrete BMP-4 for cartilage repair.

As mentioned before, one of the great promises of iPSCs is the possibility to generate autologous cells from adult somatic fully differentiated cells that can be expanded and then differentiated into the cells of interest. iPSCs derived from human chondrocytes biopsied from osteoarthritic knees [125] and from fetal human neural stem cells [126] were successfully differentiated into the chondrogenic lineage. Despite this, the lack of expression of pluripotency-associated markers is not fully achieved until sometime in culture [127] which raises some concerns regarding the control of cell phenotype and cell fate. In alternative, some studies have been supporting the generation of intermediate MSC-like cells and the purification of chondrogenic progenitors, such as neural crest cells (NCCs) [128], to eliminate residual PSCs. NCCs, which have an ectoderm origin, are known to give rise to many craniofacial tissues including the bone and cartilage [129], but NCC-derived cells have been also detected in the BM of limb tubular bones [130]. MSC-like cells derived from intermediate NCCs upon chondrogenic differentiation exhibited X collagen upregulation, features of hypertrophic chondrogenesis [131]. Still, further investigation on this cells potential for bone and cartilage regeneration is needed.

While the use of pluripotent stem cells is still hampered, alternative approaches that take advantage of stem cells intrinsic signaling moieties such as extracellular vesicles/exosomes are starting to be explored. As the efficacy of many MSC-based therapies has been attributed to paracrine secretion, exosomes have been posed as a "cell-free" strategy [132]. A weekly intra-articular injection in rat OC defects led to the formation of cartilage similar to hyaline cartilage and underlying subchondral bone resembling that of age-matched native control.

Despite an extensive preclinical research and promising clinical results, there are yet some drawbacks related to the harvesting and culture of stem cells to be addressed. Cell seeding number, serum conditions, or even the plastic surface of the expansion systems can affect cell phenotype. A forced selection happens during cell expansion, which can affect, for example, MSC properties by their technical preparation [149–151]. Cell culture was not a concern for long time, being poorly controlled over most of the published studies. As a specific example, chondrogenic differentiation requires a 3D environment, but MSCs are commonly expanded on 2D plastic surfaces. Furthermore, as two-stage procedures involving cell culture are expensive and cumbersome, there is an increasing push toward a single stage stem cell treatment, which focuses again on autologous strategies. In this situation, there is some supportive preclinical data [133–136], but a direct comparison between fresh MSC concentrates and MSCs expanded in vitro is not available [25]. The schematic represented in Fig. 10.1 summarizes the use of stem cell strategies for regenerative approaches.

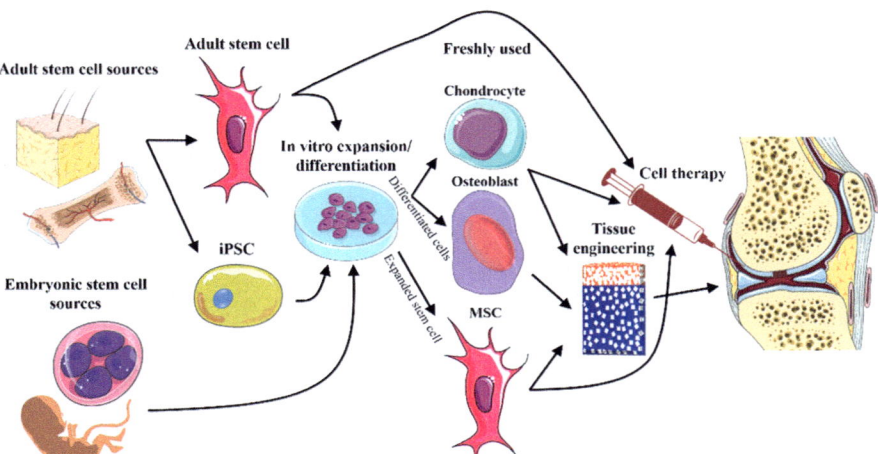

Fig. 10.1 Stem cell-based strategies for bone and cartilage engineering. Cell isolation has been relying on several sources, from adult to embryonic tissues. After isolation, cells can be freshly used or in vitro expanded. During in vitro phase, stem cells can be directly used taking advantage of their stemless or differentiated. When applied, these cells can be directly injected at the defect site using carriers or not

10.5 Engineering Osteochondral Composites Using Stem Cells

A persistent challenge in the field of RM has been the ability to engineer complex tissues comprised of multiple cell types and organizational features. This is the case of the OC unit in which a particular gradient interface integrates the cartilage and the bone parts in which chondrocytes and osteoblasts can be recognized as characteristic cells, respectively. Furthermore, OC tissue also represents a gradient of ECM constituents through it. The control over spatial patterning of tissue development represents in this way the main challenge for the advance toward engineering functional OC grafts. This control has been approached following different strategies such as heterogeneous 3D structures providing physical cues for stem cell differentiation [137], which are represented in Fig. 10.2. In addition to several materials engineering strategies, the spatial controlled delivery of differentiation factors as GFs [138] and gene vectors [139], and the use of parallel culture medium flow for osteo- and chondrogenic conditioning by custom-made dual-chamber bioreactors [140] or microfluidic devices [141], has been applied.

In any of the previous cases, to build a heterogeneous construct, cells either from different lineages or differently differentiated (ending up in two different phenotypes) are applied at the seeding moment. In coordination with the previous strategies, different cell types from primary osteoblasts co-cultured with primary chondrocytes [142], primary chondrocytes co-cultured with stem cells [143], or co-differentiation of stem cells toward osteo- and chondrogenesis [144] have been tested envisioning the enhancement of OC phenotype by the synergistic action of the different factors produced by each cell type. Envisioning these strategies, both pre-differentiation [145] and in situ differentiation [146] of the stem cells were tested. The full set of previous materials and cell combinations is summarized in Fig. 10.3.

The challenge of controlling osteogenic and chondrogenic differentiations in a single system has been addressed with the independent prematuration of the bone

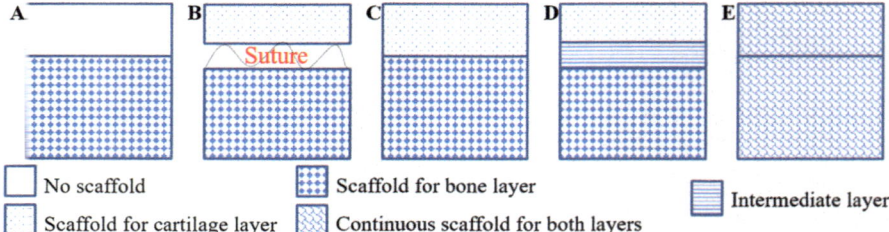

Fig. 10.2 Strategies based on materials engineering strategies to be coordinated with cellular strategies for OC TE. (**A**) Scaffold-free cartilage layer combined with bony scaffold, (**B**) two separated constructs sutured together, usually involving individual osteo- and chondrogenic prematuration, (**C**) two different integrated scaffolds merged together, (**D**) tri-layer structure composed of two different scaffolds for the bone and cartilage and an intermediate layer in the interface, (E) continuous scaffold for both bone and cartilage layers, which can be different in material properties or cell type

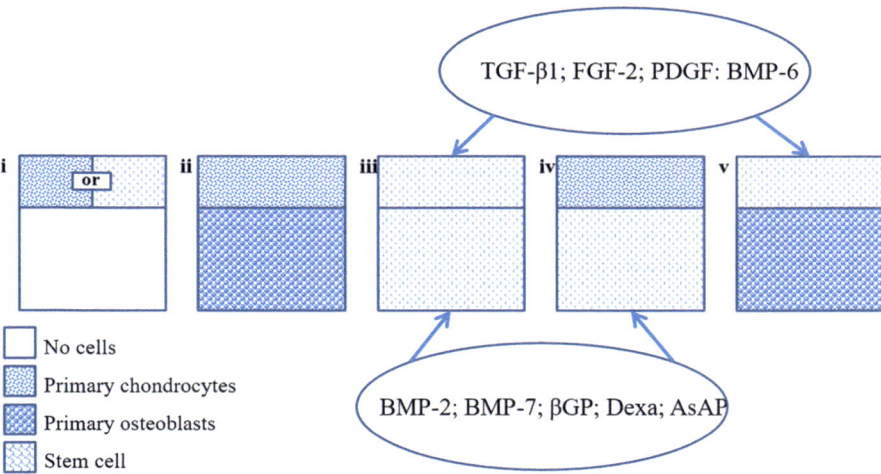

Fig. 10.3 Strategies-based cellular approached coordinated with scaffolds for OC TE. (i) Cell-free approach for bone region combined with primary chondrocytes or stem cells for cartilage layer; (ii) primary chondrocytes co-cultured with primary osteoblasts; (iii) single stem cell population for co-differentiation; (iv) primary chondrocytes co-cultured with stem cells for the cartilage and bone, respectively; (v) stem cells co-cultured with primary osteoblasts for the cartilage and bone, respectively

and cartilage parts. A hydrogel phase made of chitosan and a cancellous bone part were seeded with hASCs for chondrogenesis and osteogenesis, respectively [147]. After 2 weeks of differentiation period, constructs were sutured together and cultured for further 2 weeks under static or dynamic conditions. The dynamic culture improved the resulting interface connection and enhanced homogeneous nutrient transfer. However, this technique keeps lacking integrity and a robust interface integration between cartilage and bone tissues.

In a functional OC unit, the interface needs to recapitulate the transition between the bone and cartilage without compromising a strong integration between the layers. A tri-layer structure has been proposed as a way to improve layer's integrity and to have a better control over cell phenotype by avoiding cell migration from one region to the other due to the interface layer. Like the sutured approach, primary chondrocytes embedded in a RGD-modified peptide amphiphilic nanofibrous hydrogel, and MSCs seeded in a silk scaffold, were respectively and independently cultured in chondrogenic and osteogenic media. After 2 weeks the structures were combined with a soft silk scaffold as an interface layer and cultured under an OC cocktail medium without the addition of GFs [148]. The presence of chondrocytes in the co-cultures significantly increased the osteogenic differentiation potential of the MSCs. On the other hand, the effect of hMSCs on chondrogenic phenotype was less. A similar tri-layer approach was compared with the respective independent part. A layer of pure gelatin was assembled with layers composed of varied amounts of gelatin and hydroxyapatite. Gelatin was also used as a glue to stick all the layers together. The multilayered strategy showed an increased expression of bone markers but not of the chondrogenic ones, in relation to the respective single constructs. Authors reported that each layer enhanced cell phenotype but difficulties associated to the simultaneous use of very different media must be also highlighted [149].

This issue has been addressed with the use of microfluidics and bioreactors with special interfaced designs. Based on a microfluidic approach, Goldman and Barabino [141] described a process for the formation of integrated cartilaginous and bony tissues by culturing a single cell source, bovine BM-MSC, within a single structure. Two independent microfluidic networks supplying inductive media independently for osteo- and chondrogenesis were designed to achieve co-differentiation as shown by the differential gene expression of chondrogenic and osteogenic markers through the 3D constructs. As an alternative, Kuiper et al. [140] used a custom-made bioreactor to perform a co-culture of primary osteoblasts, seeded in a tricalcium phosphate and poly-L-lactic acid composite, and chondrocytes encapsulated in Extracel® hydrogel. The cell-laden hydrogel was cast on top of the composite in a dual-chamber perfusion bioreactor system and compared to a static condition. The authors demonstrated that the dual-chamber bioreactor positively influenced the co-culture of chondrocytes and osteoblasts in the scaffolds in terms of gene expression and matrix deposition markers for the bone and cartilage.

In addition to the different biochemical requirements to achieve concomitant osteogenic and chondrogenic differentiation within an OC construct, the spatial control of cells distribution in between layers is also critical. This has been addressed by playing with the architectures of the supporting structures. An integrated scaffold made of one layer with monodispersed porosity of 38 um pore size and coated with hydroxyapatite particles, and of a second layer with 200 um pores coated with hyaluronan, was proposed to recreate, respectively, the bone and cartilage parts of an OC construct. After seeded with MSCs and chondrocytes differentiated from MSCs, respectively, constructs were cultured over 4 weeks in vitro in the absence of soluble GFs. Concurrent deposition of ECM typical to each cell type was promoted in standard-based medium [150]. A similar approach used pre-differentiated rat BM-MSCs into the chondrogenic and osteogenic lineages, respectively, to populate a sponge composed of a hyaluronan derivative and of calcium phosphate ceramic. Both layers were joined together with fibrin sealant to form an OC graft. After 6 weeks of in vivo implantation, the heterogeneous structures were integrated in the host OC tissue. However, fibrocartilage but not hyaline cartilage was identified in the sponge. Although collagen type II was predominant in the neo-cartilage, collagen type I was found in both bone and cartilage regions [151]. In a different approach, the control over 3D OC differentiation has been targeted by a defined local displaying of GFs or genetic vectors. Brunger et al. [152] hypothesized that by genetically modifying MSCs in a spatially restricted fashion would differentially determine cell fate and ECM deposition. To prove this, the authors tested whether two interfaced tissues and an OC composite could be derived from a single stem cell type [153]. The localized differentiation factors, such as BMP-2 or RUNX2 and TGFb-3, drove concomitant mineral and GAG production in a spatially defined fashion, showing the potential of this strategy to be further explored.

Although there are several approaches to recreate the OC tissue in vitro promoting the control over cell phenotypes under co-culture systems, the exact phenotype changes of subchondral bone during inflammation and OC repair are still poorly understood. In addition, inconsistent outcomes have been reported. Studies focus on GFs and stem cell source effectiveness in vitro, and implantation of autologous chondrocytes and MSCs in vivo, creating mismatching conclusions [16]. The current state of the art of 3D engineered grafts is characterized by tailoring stem cells phenotype, such as advanced scaffold strategies with multiple layers, which were eventually combined with GFs or bioreactor designs to provide chemical and physical stimuli. However, there is a gap on vascularized constructs, which will demand one extra level of complexity. A vascularized bone and non-vascularized cartilage are required. Specific pathways and incorporation of GFs have to be explored for connecting the graft to the host vasculature, and co-cultures of stem cells with endothelial cells have to be optimized under a whole and integrated bone and cartilage-like 3D construct.

10.6   Conclusion and Future Directions

Over the last years, great progress has been made to validate cell therapy and TE strategies for OC regeneration. Approaches that mimic physiological stimuli, support cell proliferation, and warrant adequate differentiation, thus avoiding hypertrophic cartilage maturation and osteogenic induction in cartilage region, are a demand. Knowing and taking into consideration the principles of skeletogenesis is believed to be fundamental. Although MSCs are known to be key players, alternatives to BM are required mainly because this is an age-dependent and limited cell source. For example, infrapatellar fat pad is an interesting source of stem cells to be applied on knee OC disorders, since Hoffa's body has to be harvested for surgical arthroscopy in most of the cases. ASCs are much more available than BM-MSCs, but the coming years will reveal if iPSC and perinatal stem cells like WJ-MSCs can be used to replace adult multipotent stem cells, overcoming the current ethical and tumorigenic issues and becoming the cell sources under the focus of the tissue engineers for OC repair. Having this in mind, the ideal stem cell source might be dependent on the target tissue and the right strategy dependent, for example, on donor's age. However, further investigations relating to optimal cell strategy, such as autologous versus allogeneic, freshly isolated versus expanded, stem cells versus differentiated cells, as well as the use of pluripotent stem cells in a consistent way and long-term efficacy, are the main challenges still remaining.

Beyond the stem cell source, another issue that is a matter of controversy is the dose of cells that should be used. Huge ranges of cell densities have been applied in the different phases of investigation. In in vitro studies aiming for cell therapy, numbers vary from few thousands to several millions of cells per milliliter. Preclinical animal studies have been applying from a thousand to a billion of cells per milliliter, depending on the animal model, while in clinical trials, the reported densities range from 1.2 to 24 million cells/mL, which is, in fact, a more concise range but still too broad. Although some studies state that higher cell number leads to a better repair [25], the most appropriate cell dose remains unclear as the huge reported range of cell densities evidences. In fact, cell saturation is also linked to limited cell survival [154]. Thus, there is certainly a maximum cell number to aid repair [155]. This uncertainty has inevitably direct implications in the costs of each process, but even bigger consequences on the costs associated to false-positive results at the level of clinical trials [156].

Moreover, in vitro models have been changing from 2D to 3D, increasing the number of varying parameters affecting stem cell differentiation and asking for further optimization on cell densities. Studies focusing on the analysis of the cell phenotype after in vitro expansion, the cell density for cell therapy or for engineering a graft, the long-term viability of a graft after implantation, and the maintenance of a stable cartilaginous phenotype are required. The frequently varying outcomes of cartilage repair from cell implantation approaches are likely the consequence of different cell sources and differentiation strategies, GF liberation patterns, scaffold properties, model-specific defect environments, and, when applied, the animal model itself.

Regarding the strategies applied for OC unit development, the most promising strategies seem to rely on gradient structures supporting one stem cell type population under two different culture media for its maturation. This has been requiring the design and use of specific bioreactors. Some efforts have been made to create grafts using these, while in vitro models could be developed using microfluidic approaches. In any case, this concept was just born 5 years ago, and several improvements are required. Independently of the generation of OC TE constructs, either by the individual or concurrent development of the bone and cartilage counterparts, the vascularization of the engineered bone is under developed. Co-cultures can be introduced for bone vascularization processes, and SVF fraction is one of the most promising cell cocktail sources that should be further explored. In addition to the development of a vascularized subchondral bone, the mimicking of OA inflammatory state is still to be achieved.

In conclusion, the current methods for engineering physiologically relevant bone, cartilage, and OC composites from stem cells ask for a better understanding of the complex in vivo healing environment, its maturation, and homeostasis. For the future, patient-specific approaches have to be the focus, and big data can foster optimal variant selections for this goal. So, the availability of compound libraries and high-throughput screening technologies will play a key role in the selection process of the most appropriate approach in view of a stable clinical outcome.

Acknowledgments The authors would like to thank H2020-MSCA-RISE program, as this work is part of developments carried out in BAMOS project, funded from the European Union's Horizon 2020 research and innovation program under grant agreement N° 734156. Thanks are also due to the Portuguese Foundation for Science and Technology (FCT) for the distinction attributed to J. M. Oliveira (IF/00423/2012 and IF/01285/2015) and to Rogério Pirraco (IF/00347/2015) under the Investigator FCT program. The authors also thank FCT for the Ph.D. scholarship provided to R. F. Canadas (SFRH/BD/92565/2013).

References

1. Kim YS, Park EH, Kim YC, Koh YG (2013) Clinical outcomes of mesenchymal stem cell injection with arthroscopic treatment in older patients with osteochondral lesions of the talus. Am J Sports Med 41(5):1090–1099
2. Barry F, Murphy M (2013) Mesenchymal stem cells in joint disease and repair. Nat Rev Rheumatol 9(10):584–594
3. Gore M, Tai K-S, Sadosky A, Leslie D, Stacey BR (2011) Clinical comorbidities, treatment patterns, and direct medical costs of patients with osteoarthritis in usual care: a retrospective claims database analysis. J Med Econ 14(4):497–507
4. Michaud CM, McKenna MT, Begg S, Tomijima N, Majmudar M, Bulzacchelli MT et al (2006) The burden of disease and injury in the United States 1996. Popul Health Metr 4:11
5. McKenna MT, Michaud CM, Murray CJL, Marks JS (2005) Assessing the burden of disease in the United States using disability-adjusted life years. Am J Prev Med 28(5):415–423
6. Ayral X, Pickering EH, Woodworth TG, Mackillop N, Dougados M (2005) Synovitis: a potential predictive factor of structural progression of medial tibiofemoral knee osteoarthritis – results of a 1 year longitudinal arthroscopic study in 422 patients. Osteoarthr Cartil 13(5):361–367
7. Cascão R, Vidal B, Lopes IP, Paisana E, Rino J, Moita LF et al (2015) Decrease of CD68 synovial macrophages in celastrol treated arthritic rats. Ng LFP, editor. PLoS One 10(12):e0142448
8. Roelofs AJ, Zupan J, Riemen AHK, Kania K, Ansboro S, White N et al (2017) Joint morphogenetic cells in the adult mammalian synovium. Nat Commun 8:15040
9. Hangody L, Kish G, Módis L, Szerb I, Gáspár L, Diószegi Z et al (2001) Mosaicplasty for the treatment of osteochondritis dissecans of the talus: two to seven year results in 36 patients. Foot Ankle Int 22(7):552–558
10. Kono M, Takao M, Naito K, Uchio Y, Ochi M (2006) Retrograde drilling for osteochondral lesions of the talar dome. Am J Sports Med 34(9):1450–1456
11. Baltzer AWA, Arnold JP (2005) Bone-cartilage transplantation from the ipsilateral knee for chondral lesions of the talus. Arthrosc J Arthrosc Relat Surg 21(2):159–166
12. Imhoff AB, Ottl GM, Burkart A, Traub S (1999) Autologous osteochondral transplantation on various joints. Orthopade 28(1):33–44
13. Jacobi M, Villa V, Magnussen RA, Neyret P (2011) MACI - a new era? Sport Med Arthrosc Rehabil Ther Technol SMARTT 3:10
14. Grässel S, Lorenz J (2014) Tissue-engineering strategies to repair chondral and osteochondral tissue in osteoarthritis: use of mesenchymal stem cells. Curr Rheumatol Rep 16(10):452
15. Gupta PK, Das AK, Chullikana A, Majumdar AS (2012) Mesenchymal stem cells for cartilage repair in osteoarthritis. Stem Cell Res Ther 3(4):25
16. Reyes R, Pec MK, Sanchez E, del Rosario C, Delgado A, Evora C (2013) Comparative, osteochondral defect repair: stem cells versus chondrocytes versus bone morphogenetic protein-2, solely or in combination. Eur Cell Mater 25:351–365. discussion 365
17. Khan WS, Johnson DS, Hardingham TE (2010) The potential of stem cells in the treatment of knee cartilage defects. Knee 17(6):369–374
18. Peterson L, Menche D, Grande D, Pitman M (1984) Chondrocyte transplantation: an experimental model in the rabbit. Trans Orthop Res Soc 9(218)
19. Brittberg M, Lindahl A, Nilsson A, Ohlsson C, Isaksson O, Peterson L (1994) Treatment of deep cartilage defects in the knee with autologous chondrocyte transplantation. N Engl J Med 331(14):889–895
20. Bentley G, Bhamra JS, Gikas PD, Skinner JA, Carrington R, Briggs TW (2013) Repair of osteochondral defects in joints – How to achieve success. Injury 44(Supplement 1):S3–10
21. Wood JJ, Malek MA, Frassica FJ, Polder JA, Mohan AK, Bloom ET et al (2006) Autologous cultured chondrocytes: adverse events reported to the United States Food and Drug Administration. J Bone Joint Surg Am 88(3):503–507

22. Peterson L, Minas T, Brittberg M, Nilsson A, Sjogren-Jansson E, Lindahl A (2000) Two- to 9-year outcome after autologous chondrocyte transplantation of the knee. Clin Orthop Relat Res 374:212–234

23. Benya PD, Shaffer JD (1982) Dedifferentiated chondrocytes reexpress the differentiated collagen phenotype when cultured in agarose gels. Cell 30(1):215–224

24. Takata N, Furumatsu T, Ozaki T, Abe N, Naruse K (2011) Comparison between loose fragment chondrocytes and condyle fibrochondrocytes in cellular proliferation and redifferentiation. J Orthop Sci 16(5):589–597

25. Goldberg A, Mitchell K, Soans J, Kim L, Zaidi R (2017) The use of mesenchymal stem cells for cartilage repair and regeneration: a systematic review. J Orthop Surg Res 12:39

26. Makris EA, Gomoll AH, Malizos KN, Hu JC, Athanasiou KA (2015) Repair and tissue engineering techniques for articular cartilage. Nat Rev Rheumatol 11(1):21–34

27. Keramaris NC, Kanakaris NK, Tzioupis C, Kontakis G, Giannoudis PV (2008) Translational research: from benchside to bedside. Injury 39(6):643–650

28. Woolf SH (2008) The meaning of translational research and why it matters. JAMA 299(2):211–213

29. Lefebvre V, Bhattaram P (2010) Vertebrate skeletogenesis. Curr Top Dev Biol 90:291–317

30. Yang Y (2013) Skeletal morphogenesis and embryonic development. In: Primer on the metabolic bone diseases and disorders of mineral metabolism. John Wiley & Sons, Inc., Ames, pp 1–14

31. Su N (2008) FGF signaling: its role in bone development and human skeleton diseases. Front Biosci 13(13):2842

32. Rosen V (2009) BMP2 signaling in bone development and repair. Cytokine Growth Factor Rev 20(5–6):475–480

33. Kim HJ, Rice DP, Kettunen PJ, Thesleff I (1998) FGF-, BMP- and Shh-mediated signalling pathways in the regulation of cranial suture morphogenesis and calvarial bone development. Development 125(7):1241–1251

34. Zanotti S, Canalis E (2012) Notch regulation of bone development and remodeling and related skeletal disorders. Calcif Tissue Int 90(2):69–75

35. Day TF, Yang Y (2008) Wnt and hedgehog signaling pathways in bone development. J Bone Joint Surg Am 90(Suppl 1):19–24

36. Akiyama H (2002) The transcription factor Sox9 has essential roles in successive steps of the chondrocyte differentiation pathway and is required for expression of Sox5 and Sox6. Genes Dev 16(21):2813–2828

37. Komori T (2010) Regulation of bone development and extracellular matrix protein genes by RUNX2. Cell Tissue Res 339(1):189–195

38. Pittenger MF, Mackay AM, Beck SC, Jaiswal RK, Douglas R, Mosca JD et al (1999) Multilineage potential of adult human mesenchymal stem cells. Science 284(5411):143–147

39. da Silva Meirelles L (2006) Mesenchymal stem cells reside in virtually all post-natal organs and tissues. J Cell Sci 119(11):2204–2213

40. Bielby R, Jones E, McGonagle D (2007) The role of mesenchymal stem cells in maintenance and repair of bone. Injury 38(1):S26–S32

41. Devescovi V, Leonardi E, Ciapetti G, Cenni E (2008) Growth factors in bone repair. Chir Organi Mov 92(3):161–168

42. Maes C, Carmeliet G, Schipani E (2012) Hypoxia-driven pathways in bone development, regeneration and disease. Nat Rev Rheumatol 8(6):358–366

43. Caplan AI (1991) Mesenchymal stem cells. J Orthop Res 9(5):641–650

44. Le Blanc K (2003) Immunomodulatory effects of fetal and adult mesenchymal stem cells. Cytotherapy 5(6):485–489

45. Hass R, Kasper C, Böhm S, Jacobs R (2011) Different populations and sources of human mesenchymal stem cells (MSC): a comparison of adult and neonatal tissue-derived MSC. Cell Commun Signal [Internet] 9(1):12. Available from: http://www.pubmedcentral. nih.gov/articlerender.fcgi?artid=3117820&tool=pmcentrez&rendertype=abstract

46. Polymeri A, Giannobile W, Kaigler D (2016) Bone marrow stromal stem cells in tissue engineering and regenerative medicine. Horm Metab Res 48(11):700–713
47. Romanov YA (2003) Searching for alternative sources of postnatal human mesenchymal stem cells: candidate MSC-like cells from umbilical cord. Stem Cells 21(1):105–110
48. Pierdomenico L, Bonsi L, Calvitti M, Rondelli D, Arpinati M, Chirumbolo G et al (2005) Multipotent mesenchymal stem cells with immunosuppressive activity can be easily isolated from dental pulp. Transplantation 80(6):836–842
49. Zuk P, Zhu M, Ashjian P, De Ugarte DA, Huang JI, Mizuno H et al (2002) Human adipose tissue is a source of multipotent stem cells. Mol Biol Cell [Internet] 13:4279–4295. Available from: http://www.molbiolcell.org/cgi/content/abstract/13/12/4279
50. Mesimäki K, Lindroos B, Törnwall J, Mauno J, Lindqvist C, Kontio R et al (2009) Novel maxillary reconstruction with ectopic bone formation by GMP adipose stem cells. Int J Oral Maxillofac Surg 38(3):201–209
51. Dragoo JL, Carlson G, McCormick F, Khan-Farooqi H, Zhu M, Zuk PA et al (2007) Healing full-thickness cartilage defects using adipose-derived stem cells. Tissue Eng 13(7):1615–1621
52. Crisan M, Yap S, Casteilla L, Chen C-W, Corselli M, Park TS et al (2008) A perivascular origin for mesenchymal stem cells in multiple human organs. Cell Stem Cell 3(3):301–313
53. Caplan AI, Correa D (2011) The MSC: an injury drugstore. Cell Stem Cell 9(1):11–15
54. Marolt D, Campos IM, Bhumiratana S, Koren A, Petridis P, Zhang G et al (2012) Engineering bone tissue from human embryonic stem cells. Proc Natl Acad Sci 109(22):8705–8709
55. Takahashi K, Yamanaka S (2006) Induction of pluripotent stem cells from mouse embryonic and adult fibroblast cultures by defined factors. Cell 126(4):663–676
56. Yamanaka S (2010) Patient-specific pluripotent stem cells become even more accessible. Cell Stem Cell 7(1):1–2
57. Bianco P, Riminucci M, Gronthos S, Robey PG (2001) Bone marrow stromal stem cells: nature, biology, and potential applications. Stem Cells 19(3):180–192
58. Papadimitropoulos A, Piccinini E, Brachat S, Braccini A, Wendt D, Barbero A et al (2014) Expansion of human mesenchymal stromal cells from fresh bone marrow in a 3D scaffold-based system under direct perfusion. Ivanovic Z, editor. PLoS One 9(7):e102359
59. Di Maggio N, Piccinini E, Jaworski M, Trumpp A, Wendt DJ, Martin I (2011) Toward modeling the bone marrow niche using scaffold-based 3D culture systems. Biomaterials 32(2):321–329
60. Pirraco RP, Obokata H, Iwata T, Marques AP, Tsuneda S, Yamato M et al (2011) Development of osteogenic cell sheets for bone tissue engineering applications. Tissue Eng Part A 17(11–12):1507–1515
61. Juneja SC, Viswanathan S, Ganguly M, Veillette C (2016) A simplified method for the aspiration of bone marrow from patients undergoing hip and knee joint replacement for isolating mesenchymal stem cells and in vitro chondrogenesis. Bone Marrow Res 2016:1–18
62. Agata H, Asahina I, Watanabe N, Ishii Y, Kubo N, Ohshima S et al (2010) Characteristic change and loss of in vivo osteogenic abilities of human bone marrow stromal cells during passage. Tissue Eng Part A 16(2):663–673
63. Galipeau J (2013) The mesenchymal stromal cells dilemma—does a negative phase III trial of random donor mesenchymal stromal cells in steroid-resistant graft-versus-host disease represent a death knell or a bump in the road? Cytotherapy 15(1):2–8
64. Caplan AI (2007) Adult mesenchymal stem cells for tissue engineering versus regenerative medicine. J Cell Physiol [Internet] 213(2):341–347. Available from: http://www.ncbi.nlm.nih.gov/pubmed/17620285
65. Mahmoud EE, Tanaka Y, Kamei N, Harada Y, Ohdan H, Adachi N et al (2017) Monitoring immune response after allogeneic transplantation of mesenchymal stem cells for osteochondral repair. J Tissue Eng Regen Med 12(1):e275-e286
66. Zuk PA, Zhu M, Mizuno H, Huang J, Futrell JW, Katz AJ et al (2001) Multilineage cells from human adipose tissue: implications for cell-based therapies. Tissue Eng 7(2):211–228

67. Aust L, Devlin B, Foster SJ, Halvorsen YDC, Hicok K, du Laney T et al (2004) Yield of human adipose-derived adult stem cells from liposuction aspirates. Cytotherapy 6(1): 7–14

68. Zhu M, Kohan E, Bradley J, Hedrick M, Benhaim P, Zuk P (2009) The effect of age on osteogenic, adipogenic and proliferative potential of female adipose-derived stem cells. J Tissue Eng Regen Med 3(4):290–301

69. Zanetti AS, Sabliov C, Gimble JM, Hayes DJ (2013) Human adipose-derived stem cells and three-dimensional scaffold constructs: a review of the biomaterials and models currently used for bone regeneration. J Biomed Mater Res Part B Appl Biomater 101B(1):187–199

70. Ko E, Yang K, Shin J, Cho S-W (2013) Polydopamine-assisted osteoinductive peptide immobilization of polymer scaffolds for enhanced bone regeneration by human adipose-derived stem cells. Biomacromolecules 14(9):3202–3213

71. Kakudo N, Shimotsuma A, Miyake S, Kushida S, Kusumoto K (2008) Bone tissue engineering using human adipose-derived stem cells and honeycomb collagen scaffold. J Biomed Mater Res Part A 84A(1):191–197

72. Koh YJ, Koh BI, Kim H, Joo HJ, Jin HK, Jeon J et al (2011) Stromal vascular fraction from adipose tissue forms profound vascular network through the dynamic reassembly of blood endothelial cells. Arterioscler Thromb Vasc Biol 31(5):1141–1150

73. Costa M, Cerqueira MT, Santos TC, Sampaio-Marques B, Ludovico P, Marques AP et al (2017) Cell sheet engineering using the stromal vascular fraction of adipose tissue as a vascularization strategy. Acta Biomater 55:131–143

74. Costa M, Pirraco RP, Cerqueira MT, Reis RL, Marques AP (2016) Growth factor-free pre-vascularization of cell sheets for tissue engineering. Methods Mol Biol:219–226

75. Gronthos S, Mankani M, Brahim J, Robey PG, Shi S (2000) Postnatal human dental pulp stem cells (DPSCs) in vitro and in vivo. Proc Natl Acad Sci U S A 97(25):13625–13630

76. Kawashima N (2012) Characterisation of dental pulp stem cells: a new horizon for tissue regeneration? Arch Oral Biol 57(11):1439–1458

77. d'Aquino R, De Rosa A, Lanza V, Tirino V, Laino L, Graziano A et al (2009) Human mandible bone defect repair by the grafting of dental pulp stem/progenitor cells and collagen sponge biocomplexes. Eur Cell Mater 18:75–83

78. Ito K, Yamada Y, Nakamura S, Ueda M (2011) Osteogenic potential of effective bone engineering using dental pulp stem cells, bone marrow stem cells, and periosteal cells for osseointegration of dental implants. Int J Oral Maxillofac Implants 26(5):947–954

79. Yamada Y, Ito K, Nakamura S, Ueda M, Nagasaka T (2011) Promising cell-based therapy for bone regeneration using stem cells from deciduous teeth, dental pulp, and bone marrow. Cell Transplant 20(7):1003–1013

80. Tatullo M, Marrelli M, Shakesheff KM, White LJ (2015) Dental pulp stem cells: function, isolation and applications in regenerative medicine. J Tissue Eng Regen Med 9(11):1205–1216

81. Degistirici Ö, Jäger M, Knipper A (2008) Applicability of cord blood-derived unrestricted somatic stem cells in tissue engineering concepts. Cell Prolif 41(3):421–440

82. Atala A, Murphy SV (eds) (2014) Perinatal stem cells, vol 9781493911. Springer New York, New York, pp 1–373

83. Perin L, Sedrakyan S, Da Sacco S, De Filippo R (2008) Characterization of human amniotic fluid stem cells and their pluripotential capability. Methods Cell Biol:85–99

84. Deuse T, Stubbendorff M, Tang-Quan K, Phillips N, Kay MA, Eiermann T et al (2011) Immunogenicity and immunomodulatory properties of umbilical cord lining mesenchymal stem cells. Cell Transplant 20(5):655–667

85. Baba K, Yamazaki Y, Takeda A, Uchinuma E (2014) Bone regeneration using Wharton's jelly mesenchymal stem cells. In: Atala A, Murphy SV (eds) Perinatal stem cells. Springer New York, New York, pp 299–311

86. Kim J, Ryu S, Ju YM, Yoo JJ, Atala A (2014) Amniotic fluid-derived stem cells for bone tissue engineering. In: Atala A, Murphy SV (eds) Perinatal stem cells. Springer New York, New York, pp 107–114

87. Handschel J, Naujoks C, Depprich R, Lammers L, Kübler N, Meyer U et al (2011) Embryonic stem cells in scaffold-free three-dimensional cell culture: osteogenic differentiation and bone generation. Head Face Med 7(1):12
88. Jukes JM, Both SK, Leusink A, Sterk LMT, C a v B, de Boer J (2008) Endochondral bone tissue engineering using embryonic stem cells. Proc Natl Acad Sci 105(19):6840–6845
89. Marcos-Campos I, Marolt D, Petridis P, Bhumiratana S, Schmidt D, Vunjak-Novakovic G (2012) Bone scaffold architecture modulates the development of mineralized bone matrix by human embryonic stem cells. Biomaterials 33(33):8329–8342
90. Kang H, Wen C, Hwang Y, Shih Y-RV, Kar M, Seo SW et al (2014) Biomineralized matrix-assisted osteogenic differentiation of human embryonic stem cells. J Mater Chem B Mater Biol Med 2(34):5676–5688
91. Blum B, Benvenisty N et al (2008) Adv Cancer Res 100:133–158
92. Hou P, Li Y, Zhang X, Liu C, Guan J, Li H et al (2013) Pluripotent stem cells induced from mouse somatic cells by small-molecule compounds. Science (80-) 341(6146):651–654
93. Wang P, Liu X, Zhao L, Weir MD, Sun J, Chen W et al (2015) Bone tissue engineering via human induced pluripotent, umbilical cord and bone marrow mesenchymal stem cells in rat cranium. Acta Biomater 18:236–248
94. Jeon OH, Panicker LM, Lu Q, Chae JJ, Feldman RA, Elisseeff JH (2016) Human iPSC-derived osteoblasts and osteoclasts together promote bone regeneration in 3D biomaterials. Sci Rep 6(1):26761
95. TheinHan W, Liu J, Tang M, Chen W, Cheng L, Xu HHK (2013) Induced pluripotent stem cell-derived mesenchymal stem cell seeding on biofunctionalized calcium phosphate cements. Bone Res 1(4):371–384
96. de Peppo GM, Marcos-Campos I, Kahler DJ, Alsalman D, Shang L, Vunjak-Novakovic G et al (2013) Engineering bone tissue substitutes from human induced pluripotent stem cells. Proc Natl Acad Sci 110(21):8680–8685
97. Yan H, Yu C (2007) Repair of full-thickness cartilage defects with cells of different origin in a rabbit model. Arthrosc J Arthrosc Relat Surg 23(2):178–187
98. Dashtdar H, Rothan HA, Tay T, Ahmad RE, Ali R, Tay LX et al (2011) A preliminary study comparing the use of allogenic chondrogenic pre-differentiated and undifferentiated mesenchymal stem cells for the repair of full thickness articular cartilage defects in rabbits. J Orthop Res 29(9):1336–1342
99. Beris AE, Lykissas MG, Papageorgiou CD, Georgoulis AD (2005) Advances in articular cartilage repair. Injury 36(4, Supplement):S14–S23
100. Galois L, Freyria A-M, Herbage D, Mainard D (2005) Ingénierie tissulaire du cartilage: état des lieux et perspectives. Pathol Biol 53(10):590–598
101. Oreffo ROC, Cooper C, Mason C, Clements M (2005) Mesenchymal stem cells. Stem Cell Rev 1(2):169–178
102. Potian JA, Aviv H, Ponzio NM, Harrison JS, Rameshwar P (2003) Veto-like activity of mesenchymal stem cells: functional discrimination between cellular responses to alloantigens and recall antigens. J Immunol 171(7):3426 LP–3423434
103. Glennie S, Soeiro I, Dyson PJ, Lam EW-F, Dazzi F (2005) Bone marrow mesenchymal stem cells induce division arrest anergy of activated T cells. Blood 105(7):2821 LP–2822827
104. Corcione A, Benvenuto F, Ferretti E, Giunti D, Cappiello V, Cazzanti F et al (2005) Human mesenchymal stem cells modulate B-cell functions. Blood 107(1):367 LP–367372
105. Aggarwal S, Pittenger MF (2005) Human mesenchymal stem cells modulate allogeneic immune cell responses. Blood 105(4):1815 LP–1811822
106. Kang S-H, Chung Y-G, Oh I-H, Kim Y-S, Min K-O, Chung J-Y (2014) Bone regeneration potential of allogeneic or autogeneic mesenchymal stem cells loaded onto cancellous bone granules in a rabbit radial defect model. Cell Tissue Res 355(1):81–88
107. Arinzeh TL, Peter SJ, Archambault MP, van den Bos C, Gordon S, Kraus K et al (2003) Allogeneic mesenchymal stem cells regenerate bone in a critical-sized canine segmental defect. J Bone Joint Surg Am 85–A(10):1927–1935

108. Mei L, Shen B, Ling P, Liu S, Xue J, Liu F et al (2017) Culture-expanded allogenic adipose tissue-derived stem cells attenuate cartilage degeneration in an experimental rat osteoarthritis model. Lammi MJ, editor. PLoS One 12(4):e0176107

109. Zhou S, Greenberger JS, Epperly MW, Goff JP, Adler C, LeBoff MS et al (2008) Age-related intrinsic changes in human bone marrow-derived mesenchymal stem cells and their differentiation to osteoblasts. Aging Cell 7(3):335–343

110. Danisovic L, Varga I, Polak S, Ulicna M, Hlavackova L, Bohmer D et al (2009) Comparison of in vitro chondrogenic potential of human mesenchymal stem cells derived from bone marrow and adipose tissue. Gen Physiol Biophys 28(1):56–62

111. Saddik D, McNally EG, Richardson M (2004) MRI of Hoffa's fat pad. Skelet Radiol 33(8):433–444

112. Staeubli HU, Bollmann C, Kreutz R, Becker W, Rauschning W (1999) Quantification of intact quadriceps tendon, quadriceps tendon insertion, and suprapatellar fat pad: MR arthrography, anatomy, and cryosections in the sagittal plane. Am J Roentgenol 173(3):691–698

113. Pires de Carvalho P, Hamel KM, Duarte R, King AGS, Haque M, Dietrich MA et al (2014) Comparison of infrapatellar and subcutaneous adipose tissue stromal vascular fraction and stromal/stem cells in osteoarthritic subjects. J Tissue Eng Regen Med 8(10):757–762

114. Koh Y-G, Choi Y-J (2012) Infrapatellar fat pad-derived mesenchymal stem cell therapy for knee osteoarthritis. Knee 19(6):902–907

115. Jurgens WJFM, van Dijk A, Doulabi BZ, Niessen FB, Ritt MJPF, van Milligen FJ et al (2009) Freshly isolated stromal cells from the infrapatellar fat pad are suitable for a one-step surgical procedure to regenerate cartilage tissue. Cytotherapy 11(8):1052–1064

116. Saw K-Y, Anz A, Siew-Yoke Jee C, Merican S, Ching-Soong Ng R, Roohi SA et al (2013) Articular cartilage regeneration with autologous peripheral blood stem cells versus hyaluronic acid: a randomized controlled trial. Arthrosc J Arthrosc Relat Surg [Internet] 29(4):684–694. Available from: http://www.sciencedirect.com/science/article/pii/S0749806312018993

117. Fukumoto T, Sperling JW, Sanyal A, Fitzsimmons JS, Reinholz GG, Conover CA et al (2003) Combined effects of insulin-like growth factor-1 and transforming growth factor-β1 on periosteal mesenchymal cells during chondrogenesis in vitro. Osteoarthr Cartil 11(1):55–64

118. De Bari C, Dell'Accio F, Tylzanowski P, Luyten FP (2001) Multipotent mesenchymal stem cells from adult human synovial membrane. Arthritis Rheum 44(8):1928–1942

119. Baksh D, Yao R, Tuan RS (2007) Comparison of proliferative and multilineage differentiation potential of human mesenchymal stem cells derived from umbilical cord and bone marrow. Stem Cells [Internet] 25(6):1384–92. Available from: https://doi.org/10.1634/stemcells.2006-0709

120. Chen M-Y, Lie P-C, Li Z-L, Wei X (2009) Endothelial differentiation of Wharton's jelly-derived mesenchymal stem cells in comparison with bone marrow–derived mesenchymal stem cells. Exp Hematol [Internet] 37(5):629–640. Available from: http://www.sciencedirect.com/science/article/pii/S0301472X09000514

121. Fong C-Y, Chak L-L, Biswas A, Tan J-H, Gauthaman K, Chan W-K et al (2011) Human Wharton's jelly stem cells have unique transcriptome profiles compared to human embryonic stem cells and other mesenchymal stem cells. Stem Cell Rev Reports [Internet] 7(1):1–16. Available from: https://doi.org/10.1007/s12015-010-9166-x

122. Ahmed TAE, Hincke MT (2014) Mesenchymal stem cell-based tissue engineering strategies for repair of articular cartilage. Histol Histopathol 29(6):669–689

123. Berg LC, Koch TG, Heerkens T, Bessonov K, Thomsen PD, Betts DH (2009) Chondrogenic potential of mesenchymal stromal cells derived from equine bone marrow and umbilical cord blood. Vet Comp Orthop Traumatol 22(5):363–370

124. Adachi N, Sato K, Usas A, Fu FH, Ochi M, Han C-W et al (2002) Muscle derived, cell based ex vivo gene therapy for treatment of full thickness articular cartilage defects. J Rheumatol 29(9):1920 LP–1921930

125. Wei Y, Zeng W, Wan R, Wang J, Zhou Q, Qiu S et al (2012) Chondrogenic differentiation of induced pluripotent stem cells from osteoarthritic chondrocytes in alginate matrix. Eur Cell Mater 23:1–12

126. Medvedev SP, Grigor'eva EV, Shevchenko AI, Malakhova AA, Dementyeva EV, Shilov AA et al (2010) Human induced pluripotent stem cells derived from fetal neural stem cells successfully undergo directed differentiation into cartilage. Stem Cells Dev 20(6):1099–1112

127. Nguyen D, Hägg DA, Forsman A, Ekholm J, Nimkingratana P, Brantsing C et al (2017) Cartilage tissue engineering by the 3D bioprinting of iPS cells in a Nanocellulose/alginate bioink. Sci Rep 7:658

128. Menendez L, Kulik MJ, Page AT, Park SS, Lauderdale JD, Cunningham ML et al (2013) Directed differentiation of human pluripotent cells to neural crest stem cells. Nat Protoc 8(1):203–212

129. Ishii M, Arias AC, Liu L, Chen Y-B, Bronner ME, Maxson RE (2012) A stable cranial neural crest cell line from mouse. Stem Cells Dev 21(17):3069–3080

130. Nagoshi N, Shibata S, Kubota Y, Nakamura M, Nagai Y, Satoh E et al (2017) Ontogeny and multipotency of neural crest-derived stem cells in mouse bone marrow, dorsal root ganglia, and whisker pad. Cell Stem Cell 2(4):392–403

131. Chijimatsu R, Ikeya M, Yasui Y, Ikeda Y, Ebina K, Moriguchi Y et al (2017) Characterization of mesenchymal stem cell-like cells derived from human iPSCs via neural crest development and their application for osteochondral repair. Stem Cells Int 2017:1960965

132. Zhang S, Chu WC, Lai RC, Lim SK, Hui JHP, Toh WS (2017) Exosomes derived from human embryonic mesenchymal stem cells promote osteochondral regeneration. Osteoarthr Cartil 24(12):2135–2140

133. Solchaga LA, Gao J, Dennis JE, Awadallah A, Lundberg M, Caplan AI et al (2002) Treatment of osteochondral defects with autologous bone marrow in a hyaluronan-based delivery vehicle. Tissue Eng 8(2):333–347

134. Betsch M, Schneppendahl J, Thuns S, Herten M, Sager M, Jungbluth P et al (2013) Bone marrow aspiration concentrate and platelet rich plasma for osteochondral repair in a porcine osteochondral defect model. PLoS One 8(8):e71602

135. Oshima Y, Watanabe N, Matsuda K, Takai S, Kawata M, Kubo T (2017) Fate of transplanted bone-marrow-derived mesenchymal cells during osteochondral repair using transgenic rats to simulate autologous transplantation. Osteoarthr Cartil 12(10):811–817

136. Saw K-Y, Hussin P, Loke S-C, Azam M, Chen H-C, Tay Y-G et al (2017) Articular cartilage regeneration with autologous marrow aspirate and hyaluronic acid: an experimental study in a goat model. Arthroscopy 25(12):1391–1400

137. Oliveira JM, Rodrigues MT, Silva SS, Malafaya PB, Gomes ME, Viegas CA et al (2006) Novel hydroxyapatite/chitosan bilayered scaffold for osteochondral tissue-engineering applications: Scaffold design and its performance when seeded with goat bone marrow stromal cells. Biomaterials [Internet] 27(36):6123–6137. Available from: http://www.ncbi. nlm.nih.gov/pubmed/16945410

138. Wang X, Wenk E, Zhang X, Meinel L, Vunjak-Novakovic G, Kaplan DL (2009) Growth factor gradients via microsphere delivery in biopolymer scaffolds for osteochondral tissue engineering. J Control Release 134(2):81–90

139. Lee Y-H, Wu H-C, Yeh C-W, Kuan C-H, Liao H-T, Hsu H-C et al (2017) Enzyme-crosslinked gene-activated matrix for the induction of mesenchymal stem cells in osteochondral tissue regeneration. Acta Biomater 63(Supplement C):210–226

140. Kuiper NJ, Wang QG, Cartmell SH (2014) A perfusion co-culture bioreactor for osteochondral tissue engineered plugs. J Biomater Tissue Eng 4(2):162–171

141. Goldman SM, Barabino GA (2016) Spatial engineering of osteochondral tissue constructs through microfluidically directed differentiation of mesenchymal stem cells. Biores Open Access 5(1):109–117

142. Jiang J, Nicoll SB, Lu HH (2005) Co-culture of osteoblasts and chondrocytes modulates cellular differentiation in vitro. Biochem Biophys Res Commun 338(2):762–770

143. Sheehy EJ, Vinardell T, Buckley CT, Kelly DJ (2013) Engineering osteochondral constructs through spatial regulation of endochondral ossification. Acta Biomater 9(3):5484–5492

144. Lee W, Park J (2016) 3D patterned stem cell differentiation using thermo-responsive methyl-cellulose hydrogel molds. Sci Rep 6:29408

145. Lam J, Lu S, Meretoja V V, Tabata Y, Mikos AG, Kasper FK (2014) Generation of osteochondral tissue constructs with chondrogenically and osteogenically predifferentiated mesenchymal stem cells encapsulated in bilayered hydrogels. Acta Biomater 10(3):1112–1123

146. Mellor LF, Mohiti-Asli M, Williams J, Kannan A, Dent MR, Guilak F et al (2015) Extracellular calcium modulates Chondrogenic and osteogenic differentiation of human adipose-derived stem cells: a novel approach for osteochondral tissue engineering using a single stem cell source. Tissue Eng Part A 21(17–18):2323–2333

147. Song K, Li W, Wang H, Zhang Y, Li L, Wang Y, Wang H, Wang L, Liu T (2016) Development and fabrication of a two-layer tissue engineered osteochondral composite using hybrid hydrogel-cancellous bone scaffolds in a spinner flask. Biomed Mater 11(6):65002

148. Cakmak S, Cakmak AS, Kaplan DL, Gumusderelioglu M (2016) A silk fibroin and peptide amphiphile-based co-culture model for osteochondral tissue engineering. Macromol Biosci 16(8):1212–1226

149. Amadori S, Torricelli P, Panzavolta S, Parrilli A, Fini M, Bigi A (2015) Multi-layered scaffolds for osteochondral tissue engineering: in vitro response of co-cultured human mesenchymal stem cells. Macromol Biosci 15(11):1535–1545

150. Galperin A, Oldinski RA, Florczyk SJ, Bryers JD, Zhang M, Ratner BD (2013) Integrated bi-layered scaffold for osteochondral tissue engineering. Adv Healthc Mater 2(6):872–883

151. Gao J, Dennis JE, Solchaga LA, Awadallah AS, Goldberg VM, Caplan AI (2001) Tissue-engineered fabrication of an osteochondral composite graft using rat bone marrow-derived mesenchymal stem cells. Tissue Eng 7(4):363–371

152. Brunger JM, Huynh NPT, Guenther CM, Perez-Pinera P, Moutos FT, Sanchez-Adams J et al (2014) Scaffold-mediated lentiviral transduction for functional tissue engineering of cartilage. Proc Natl Acad Sci U S A 111(9):E798–E806

153. Brunger JM, Huynh NPT, Moutos FT, Guilak F, Gersbach CA (2017) 407. Biomaterial-mediated lentiviral gene delivery for osteochondral tissue engineering. Mol Ther 22:S155

154. Zhao Q, Wang S, Tian J, Wang L, Dong S, Xia T et al (2013) Combination of bone marrow concentrate and PGA scaffolds enhance bone marrow stimulation in rabbit articular cartilage repair. J Mater Sci Mater Med 24(3):793–801

155. Agung M, Ochi M, Yanada S, Adachi N, Izuta Y, Yamasaki T et al (2006) Mobilization of bone marrow-derived mesenchymal stem cells into the injured tissues after intraarticular injection and their contribution to tissue regeneration. Knee Surg Sport Traumatol Arthrosc 14(12):1307–1314

155. Food and Drug Administration (FDA) (2004). Innovation or stagnation: challenge and opportunity on the critical path to New Medical Products [Internet]. Available from: http://www.fda.gov/ScienceResearch/SpecialTopics/CriticalPathInitiative/CriticalPathOpportunitiesReports/ucm077262.htm#intro

Chapter 11
PRP Therapy

Ibrahim Fatih Cengiz, J. Miguel Oliveira, and Rui L. Reis

Abstract Osteochondral lesions remain as a clinical challenge despite the advances in orthopedic regenerative strategies. Biologics, in particular, platelet-rich plasma, has been applied for the reparative and regenerative effect in many tissues, and osteochondral tissue is not an exception. Platelet-rich plasma is an autologous concentrate prepared from the collected blood; thus, this safe application is free of immune response or risk of transmission of disease. It has a high potential to promote regeneration, thanks to its content, and can be applied alone or can reinforce a tissue engineering strategy. The relevant works making use of platelet-rich plasma in osteochondral lesions are overviewed herein. The practical success of platelet-rich plasma is uncertain since there are many factors involved including but not limited to its preparation and administration method. Nevertheless, today, the issues and challenges of platelet-rich plasma have been well acknowledged by researchers and clinicians. Thus, it is believed that a consensus will be built it, and then with high-quality randomized controlled trials and standardized protocols, the efficacy of platelet-rich plasma therapy can be better evaluated.

I. F. Cengiz (✉)
3B's Research Group–Biomaterials, Biodegradables and Biomimetics, University of Minho, Headquarters of the European Institute of Excellence on Tissue Engineering and Regenerative Medicine, Barco, Guimarães, Portugal

ICVS/3B's–PT Government Associate Laboratory, Braga/Guimarães, Portugal
e-mail: fatih.cengiz@dep.uminho.pt

J. M. Oliveira · R. L. Reis
3B's Research Group–Biomaterials, Biodegradables and Biomimetics, University of Minho, Headquarters of the European Institute of Excellence on Tissue Engineering and Regenerative Medicine, Barco, Guimarães, Portugal

ICVS/3B's–PT Government Associate Laboratory, Braga/Guimarães, Portugal

The Discoveries Centre for Regenerative and Precision Medicine, Headquarters at University of Minho, Barco/Guimarães, Portugal

© Springer International Publishing AG, part of Springer Nature 2018   241
J. M. Oliveira et al. (eds.), *Osteochondral Tissue Engineering*,
Advances in Experimental Medicine and Biology 1059,
https://doi.org/10.1007/978-3-319-76735-2_11

Highlights

- The need of treating the osteochondral lesions has not been yet met in the clinics.
- Thanks to being an autologous source of growth factors, interleukins, and other cytokines and relative ease of clinical application, i.e., during a single-step surgical procedure, the use of platelet-rich plasma is of great interest.
- The high theoretical potential of the role of platelet-rich plasma in the regeneration process of osteochondral lesions is known, and the efficiency needs to be confirmed by high-quality randomized controlled trials for a robust position in the treatments of osteochondral lesions in the clinics.

Keywords Platelet-rich plasma · Osteochondral · Regeneration · Tissue engineering

11.1   Platelet-Rich Plasma (PRP): What, Why, and How?

Platelet-rich plasma (PRP) is one of the important biologics and widely employed in the field of orthopedics. PRP (also named as platelet gel, platelet-enriched plasma, and platelet-rich concentrate [1]) is an autologous blood-derived concentrate of platelets that are a source of growth factors, interleukins, and other cytokines [2, 3]. Platelets are small (with a diameter of 2–3 μm) cytoplasmic fragments of megakaryocytes, which can be found in the peripheral blood. Many bioactive proteins are found in the α-granules of the platelets that function in tissue healing and hemostasis upon secretion [4]. Qureshi et al. [5] identified a total of 1507 unique proteins in platelets upon analysis of 10 independent human samples. The platelet concentration in PRP is typically around 10^6 platelets per μL while the baseline that is around 2×10^5 platelets per μL [6–8]. Thus, PRP is a pool of biologically active autologous components which can be contributors to tissue healing and regeneration [6, 9]. PRP has been of keen interest in orthopedics thanks to their potential regenerative effect [10–13].

PRP is prepared from anticoagulated blood [14], typically by the centrifuge of the collected peripheral blood of the patient, and activated by clotting [15] (Fig. 11.1). Activation can be done by several means including calcium chloride, thrombin, and soluble collagen type I [4, 15]. Being autologous makes it possible to be free of immune response and risk of transmission of disease or cannot lead to a mutation since the growth factors of PRP do not go into the cell or into cell nucleus, but just bind to the external surface of the cell membrane [8]. Moreover, they involve a relative ease of clinical administration, i.e., during a single-step surgical procedure in the operating room, and regulatory convenience owing to being considered as a "minimally manipulated tissue." Nevertheless, as concluded in almost all PRP review paper, recently, the inconsistencies in the clinical studies were highlighted in the systematic review of Chahla et al. [16].

Growth factors are polypeptides that affect the function of cells (Table 11.1) such as proliferation, matrix synthesis, adhesion, or differentiation. Other bioactive components of the platelets such as cytokines, chemokines, and metabolites

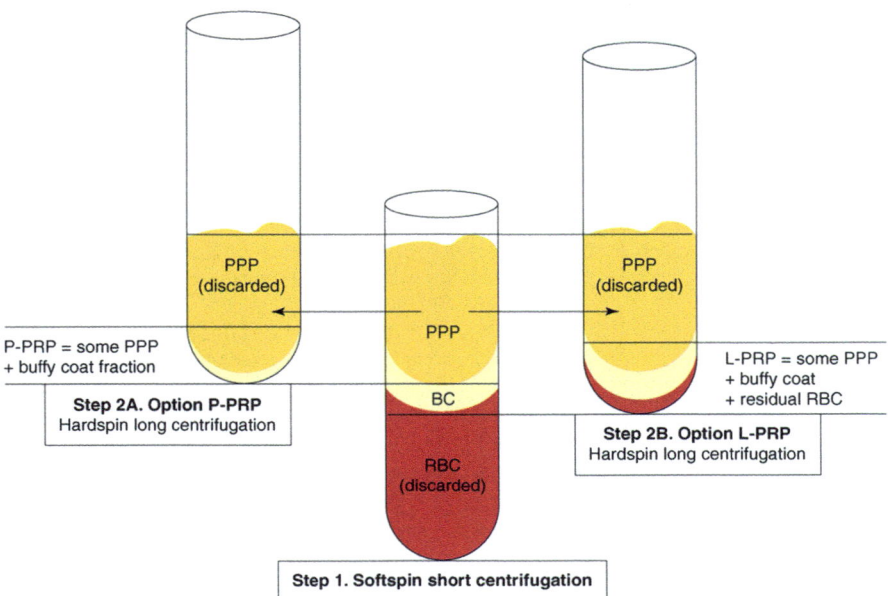

Fig. 11.1 A conventional manual way of PRP preparation through a two-step centrifugation based on density separation. Step 1: The blood is collected from the patient and, in the presence of an anticoagulant, centrifuged for a short time with a low force (soft spin). As a result of the soft spin, three layers are obtained from bottom to top: (i) red blood cells (RBCs) layer, (ii) the "buffy coat" (BC) layer, and (iii) the platelet-poor plasma (PPP) layer. The BC layer has typically whitish color and contains the major proportion of the platelets and leukocytes. Step 2A: To obtain pure PRP (P-PRP), PPP and the superficial BC are transferred to another tube for the second centrifuge at high-force (hard spin) centrifugation. After the step 2A, most of the PPP layer is discarded, and most of the leukocytes are not collected. The final P-PRP concentrate consists of an undetermined fraction of BC with a large number of platelets suspended in some fibrin-rich plasma. Step 2B: For the production of leukocyte-rich PRP (L-PRP), PPP, the entire BC layer, and some residual RBCs are transferred to another tube. After a hard spin centrifugation, the PPP is discarded. The final L-PRP contains the entire BC that has most of the platelets, leukocytes, and the residual RBCs suspended in some fibrin-rich plasma. Thus, the final version depends on the how the BC is collected. Since this is a manual method, using a robust commercial system would critically reproducibility from [20]

contribute to the function of growth factors [17]. The growth factors within PRP include platelet-derived growth factor (PDGF), transforming growth factor (TGF-β), platelet-derived epidermal growth factor (PDEGF), vascular endothelial growth factor (VEGF), insulin-like growth factor 1 (IGF-1), fibroblastic growth factor (FGF), and epidermal growth factor (EGF) [4, 18]. While almost all growth factors are secreted in the first hour post-clotting, most of them are secreted within the first 10 minutes post-clotting; and more growth factors can be synthesized by the platelets for around 7 days [7, 8]. Thus, it is critical to optimize the mean and timing of the activation for a better therapeutic outcome. Besides, it is critically important to consider that α-granule secreted proteins can have opposite roles such as in

Table 11.1 Possible effects of growth factors [4]

Growth factor	Effect
Platelet-derived growth factor	Angiogenesis, macrophage activation Fibroblasts: Proliferation, chemotaxis, collagen synthesis Enhances the proliferation of bone cells
Transforming growth factor-β	Proliferation of fibroblasts Collagen type I and fibronectin synthesis Induction of bone matrix deposition, inhibition of bone resorption
Platelet-derived epidermal Growth factor	Stimulation of epidermal regeneration Promotion of wound healing by the proliferation of keratinocytes and dermal fibroblasts stimulation Enhances the production and effects of other growth factors
Vascular endothelial growth factor	Vascularization by stimulating vascular endothelial cells
Insulin-like growth factor 1	Chemotactic for fibroblasts and stimulates protein synthesis Enhances bone formation
Platelet factor 4	Stimulation of the initial influx of neutrophils into wounds Chemoattractant for fibroblasts
Epidermal growth factor	Cellular proliferation and differentiation

coagulation, angiogenesis, and proteolysis (Fig. 11.2). Thus, the balance between catabolism and anabolism is of great importance [10].

Many ready-to-use products are commercially available to prepare PRP based on different techniques and providing different outcomes [1, 13, 19–21], including ACP-DS (Arthrex, Naples, Florida), FIBRINET (Cascade; Musculoskeletal Foundation, New Jersey), GPS III (Biomet, Warsaw, Indiana), Magellan (Medtronic, Minneapolis, Minnesota), SmartPReP2 (Harvest Technologies, Plymouth, Massachusetts), GenesisCS/Exactech (Gainesville, Florida), Symphony II (DePuy, Warsaw, Indiana), EmCyte (Fort Myers, Florida), and Haemonetics Cell Saver 5 (Haemonetics, Leeds, UK) [13]. Their platelet recovery varies between 31% and 78%, and the obtained PRP's concentration may go up to 6x of the baseline concentration [4]. Moreover, given that PRP is an autologous concentrate, its features are donor-dependent such as the growth factor concentrations [22].

11.2 PRP Studies and Applications

Despite the advances in orthopedic regenerative strategies, osteochondral lesions remain as a clinical challenge [23, 24]. Thanks to aforementioned features of PRP, there is a firm rationale to expect a benefit for the treatment of osteochondral lesions. However, the literature on the effect of PRP is contradictory, while some study designs do not allow to deduct specific effect of PRP due to lack of a control group. In an in vitro study [25], chondrogenic and osteogenic differentiation of human adipose-derived stem cells were found to be superior in the presence of PRP and insulin. Marmotti et al. [26] studied a one-step osteochondral repair in a rabbit

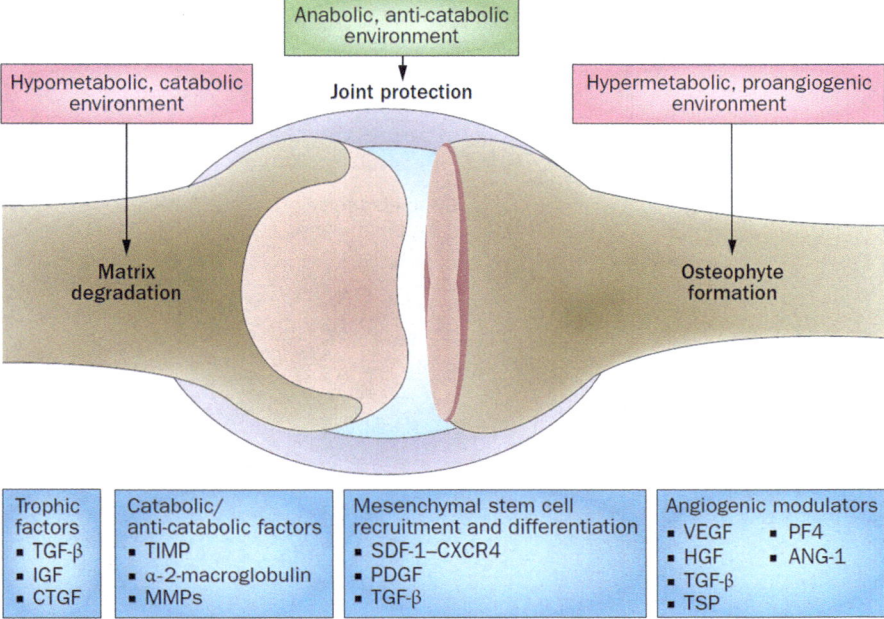

Fig. 11.2 PRP can modulate the synthesis and degradation processes. The balance between catabolism and anabolism is of great importance for joint function and homeostasis. Trophic factors can positively influence cell metabolism and activate anabolic pathways. Processes of angiogenesis, mesenchymal stem cell migration, self-renewal, and differentiation depend on the PRP content. However, trophic and pro-angiogenic factors can promote osteophyte formation. The effects of cytokines depend on the timing, dose, and context. Abbreviations in alphabetical order: ANG-1, angiopoietin-1; CTGF, connective tissue growth factor; CXCR4, CXC chemokine receptor 4; HGF, hepatocyte growth factor; IGF, insulin-like growth factor; MMPs, metalloproteinases; PDGF, platelet-derived growth factor; PF4, platelet factor 4; SDF-1, stromal cell-derived factor; TGF-β, transforming growth factor β; TIMP, tissue inhibitor of metalloproteinases; TSP1, thrombospondin 1; VEGF, vascular endothelial growth factor. (Adapted with permission [10])

model. As depicted in Fig. 11.3a, rabbit knee cartilage fragments with or without human fibrin glue (FG) combined with hyaluronic acid (HA)-based membranes and PRP. At 6 months, the use of cartilage fragments provided superior results, while human fibrin glue provided inferior results [26]. Since all treated groups received PRP, additional effect of PRP could not be detected. Lee et al. [27] treated osteochondral lesions of rabbits with an injectable PRP gel (Fig. 11.3b, c) with and without encapsulated mesenchymal stem cells. At 24 weeks, the treatment with PRP + cells provided a superior outcome: the lesion was resurfaced with cartilage, subchondral bone was restored, and higher glycosaminoglycan content, safranin-O staining, and collagen type II immunostaining were obtained [27].

In another rabbit study, the effect of PRP on osteochondral lesion healing was studied [28]. PRP intra-articular injection was performed in one knee of each rabbit,

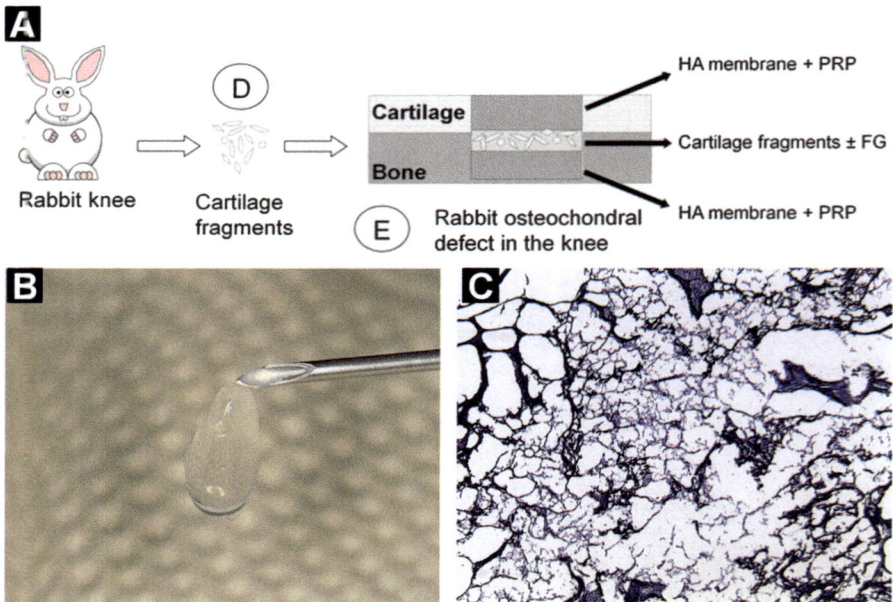

Fig 11.3 A strategy for leporine osteochondral tissue regeneration using rabbit knee cartilage fragments with or without fibrin glue (FG), hyaluronic acid (HA)-based membranes, and PRP (**A**). (Adapted with permission [26]). Gross appearance of PRP gel (**B**) and its cryosection's hematoxylin and eosin staining (**C**) (original magnification × 100). (Adapted with permission [27])

while saline as a placebo was injected into the other knee. No significant difference was found in International Cartilage Repair Society (ICRS) macroscopic scores between the groups. However, subjective macroscopic evaluation indicated PRP treatment provided a greater tissue infill with fewer fissures and a more similar to native tissue appearance in PRP-treated knees. The PRP group has higher ICRS histological scores than the placebo group with more glycosaminoglycan and type II collagen in the repair tissue [28]. PRP can be combined with conventional treatments as an adjunct therapy. For instance, the use of PRP with mosaicplasty provided a better healing response and integration of the adjacent surfaces and superior histological scores 3 weeks as compared to mosaicplasty without PRP treatment in the rabbit osteochondral lesions [29]. A similar rationale was also tried in human [30–32]. Guney et al. [31] reported that PRP could enhance the functional score of arthroscopic microfracture treatment for osteochondral lesions of the talus at a mean of 16.3 months of follow-up. Guney et al. [32] compared mosaicplasty and arthroscopic microfracture with and without PRP and reported that PRP provided no benefit at a median follow-up time of 44 months, and mosaicplasty was superior to microfracture in pain relief, while all groups had a similar American Orthopedic Foot and Ankle Society score [32]. In the randomized clinical trial by Gormeli et al. [30], the influence of intra-articular injections of hyaluronic acid and

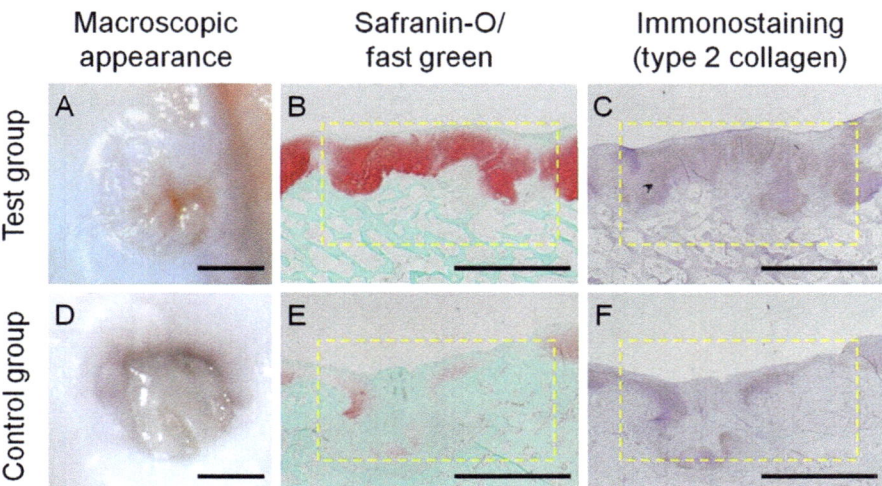

Fig. 11.4 Macroscopic and histological images of the equine osteochondral lesions at 4 months post-surgery. The surface of the lesions of the test group (gelatin/β-tricalcium phosphate scaffolds with PRP, mesenchymal stem cells, and bone morphogenetic protein-2 with a synovial flap cover) (**A**) was smoother than those in the test group (gelatin/β-tricalcium phosphate scaffolds with PRP, mesenchymal stem cells, and bone morphogenetic protein-2 without a synovial flap cover) (**D**). Histology confirmed no remaining implant material and no inflammatory reactions in or near the lesions. The upper part of the lesion in the test group that showed more positive for safranin-O staining was more positive in the test group (**B**) than that in the control group (**E**). The test group (**C**) had more collagen type II immunostained than the control group (**F**). The scale bars indicate 5 mm. (From [34])

PRP was compared as an adjunct therapy to arthroscopic debridement and microfracture treatment for talar osteochondral lesions at a mean follow-up time of 15.3-month post-surgery. It was reported that both of hyaluronic acid and PRP injections augmented the outcomes, while the American Orthopedic Foot and Ankle Society score was highest in the PRP group; thus, PRP was recommended as an adjunct therapy [30].

Seo et al. [33] studied bilayer gelatin/β-tricalcium phosphate scaffolds that were seeded with one layer of mesenchymal stem cells and bone morphogenetic protein-2 (for the subchondral bone layer) and another layer with chondrocytes and PRP (for the cartilage layer) for the treatment of osteochondral lesions in horses, and bilayer scaffolds provided superior results than control. In a later study of Seo et al. [34], they investigated whether the use of a synovial flap cover would improve the results of the regeneration of osteochondral lesions in horses (Fig. 11.4). It was shown that the lesions treated with gelatin/β-tricalcium phosphate scaffolds with PRP, mesenchymal stem cells, and bone morphogenetic protein-2 with a synovial flap cover were superior to the control group that lacks the synovial flap cover [34]. However, since in that study, both the test and control group received PRP, the PRP-specific effects were not detectable due to the study design.

Platelet-rich Plasma   Control

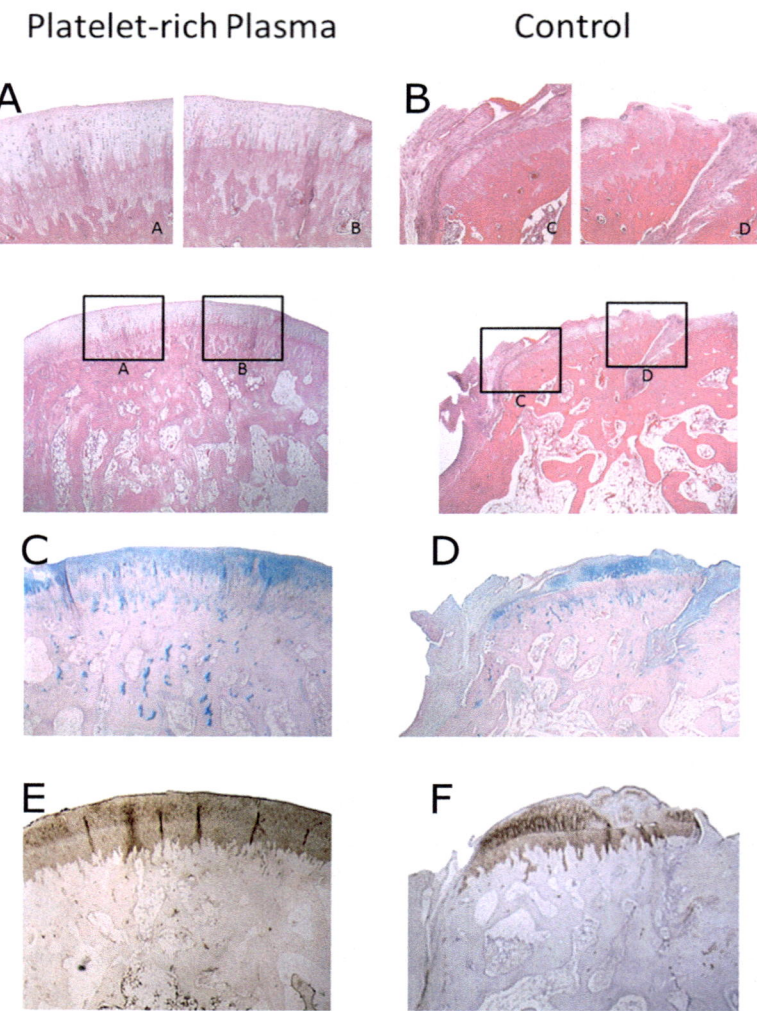

Fig. 11.5 Hematoxylin and eosin staining (**A**, **B**), alcian blue staining (**C**, **D**), and immunohisto-chemistry images for collagen type II (**E**, **F**) of the osteochondral grafts with PRP (left column) and without PRP (right column) at 12 weeks post-surgery (the ones with superior results are presented). Magnification is × 200 in the top row of (**A**) and (**B**) and × 20 in the rest of the images. (Adapted from [35] that is accessible of http://journals.lww.com/jbjsjournal/subjects/Knee/Abstract/2013/12180/The_Effect_of_Platelet_Rich_Plasma_on_Autologous.4.aspx)

Application of PRP was also performed with autologous osteochondral transplantation in a rabbit model [35]. Groups with and without PRP were similarly based on the macroscopic evaluation, while the PRP group has higher International Cartilage Repair Society score and better graft integration [35] (Fig. 11.5). Scaffolds (Trufit BGS, Smith and Nephew, USA) with bone marrow aspiration concentrate or

PRP had superior outcome than the scaffolds alone when used for an osteochondral repair in a mini-pig model at 26 weeks post-surgery, while the use of both biologics together did not further improve the outcome [36]. Addition of PRP to a polylactic-glycolic acid scaffold improved the osteochondral healing in a rabbit model and has promising results indicating a resurfaced lesion and restored subchondral bone [37]. Interestingly, Kon et al. [38] reported that the use of PRP decreased the osteochondral regeneration of collagen-hydroxyapatite scaffolds used for osteochondral lesions in sheep at 6 months post-surgery. Scaffolds without PRP led to significantly better bone regeneration and cartilage surface reconstruction compared to than the ones with PRP [38]. In the similar direction, van Bergen et al. [39] reported that inclusion of PRP to allogeneic demineralized bone matrix did not enhance the healing of osteochondral lesions of talus in goats at 24 weeks post-surgery assessed by micro-computed tomography, histology, histomorphometry, fluorescence microscopy, and macroscopic evaluation [39]. Xie et al. [40] treated the osteochondral lesions in rabbits with PRP scaffolds seeded with mesenchymal stem cells from either bone marrow or adipose tissue. PRP scaffolds were suitable for osteochondral regeneration. Bone marrow-derived stem cells in PRP scaffolds had superior subchondral bone healing, cartilage-specific protein and gene expressions, and histological and immunohistochemical outcomes and better gross appearance compared to PRP scaffold seeded with adipose-derived cells [40].

In the study of Krych et al. [41], a total of 46 patients were treated with one of three different strategies, polylactide-co-glycolide osteochondral scaffold alone (11 patients) and with either PRP (23 patients) or with bone marrow aspirate concentrate (12 patients), and were evaluated with magnetic resonance imaging at 12 months post-surgery. Augmentation with PRP or bone marrow aspirate concentrate improved the tissue fill better than the scaffold alone. While the scaffold with PRP group had T2 relaxation time value similar to the scaffold-alone group, the scaffold with bone marrow aspirate concentrate group has higher value that is close to that of the top layer of the native cartilage [41]. Gu et al. [42] implanted autograft with PRP to the Hepple stage V osteochondral lesions of the talus of 14 patients, and a significant functional improvement and pain relief were achieved at a mean follow-up time of 18 months post-surgery. However, since there was no control group and all patients received an autograft with PRP treatment, it is impossible to deduct any PRP-specific effects. In the randomized controlled trial by Mei-Dan et al. [43], osteochondral lesions of the talus in human were treated with PRP or hyaluronic acid intra-articular injections by which a decrease in pain scores and an increase in function for minimum of 6 months were achieved by both approaches. However, PRP therapy functioned significantly better than hyaluronic acid therapy under the design of this study.

The outcome of PRP therapies are context-dependent [10] and thus contra-dictory [12, 44]. A meta-analysis of randomized trials and prospective cohort studies by Sheth et al. [45] indicated the absence of standardization of proto-cols, and there is no certain evidence of clinical benefit of PRP in orthopedics; similarly there is no evidence-based medicine data that favors the use of PRP in arthroscopic surgery [46].

11.3 Final Remarks

The autologous nature of PRP is a critical factor that the outcome of the PRP therapy becomes patient-dependent, since (i) the applied PRPs are not the same PRP and (ii) the recipients are not same; thus there is always a variation, and the statistical error can be minimized by highly increasing the number of patients that are being taken into account. This is an indication of a need for statistically powered studies. "It is also substantial here to discuss that in some amount of the original papers on PRP, the transition from the "results" to the "conclusion" is not robust or evidenced-based but rather sentimental, having a tendency to favor the use of PRP due to the rationale of PRP usage; therefore, in the conclusion part of the papers, it is taken somewhat as a tradition to say "PRP *could* be beneficial for..." (which is sometimes not very correct based on the presented results).

PRP science is beyond a simple procedure for the blood collection, the use of ready-to-use products, and the application; but also the basic science is crucial. For instance, the half-life of released bioactive proteins and the life span of platelets is to be considered when defining the time points of the studies (i.e., shall we still expect a clinical benefit in the patients months after the therapy?). Moreover, the secretion of bioactive proteins is critical. The proteins in the α-granules of platelets are in fact not complete unless they are soluble, and at the time of the secretion, the proteins get completed by the inclusion of specific side chains. This means that proteins cannot be secreted if platelets are damaged (e.g., when the PRP was prepared), so the PRP therapy may not be beneficial at the end.

The evaluation of the PRP therapy alone or adjunct requires well-organized test and control groups, which lacks some studies due to some reasons. Nevertheless, today, the issues and challenges of platelet-rich plasma have been well acknowledged by researchers and clinicians and expressed in many review papers by the experts. Thus, we believe that a consensus will be built around it, and then with high-quality randomized controlled trials, the efficacy of platelet-rich plasma therapy can be better evaluated. Addressing the outstanding issues (e.g., PRP preparation method, platelet and growth factor concentration evaluation, and delivery/administration strategy) would minimize the uncertainty that is detectable by the systematic reviews and meta-analyses, and we can be able to deliver a robust PRP therapy to patients, perhaps in a patient-specific manner by analyzing the blood of the patient and optimizing the therapy timing, dose, interval, administration, and the role in the regenerative tissue engineering strategy, i.e., injections or as gel in an open surgery or introduced within a biomaterial such as scaffolds or grafts.

Acknowledgments This work is a result of the project FROnTHERA (NORTE-01-0145-FEDER-000023), supported by Norte Portugal Regional Operational Programme (NORTE 2020), under the Portugal 2020 Partnership Agreement, through the European Regional Development Fund (ERDF). IFC thanks the Portuguese Foundation for Science and Technology (FCT) for the PhD scholarship (SFRH/BD/99555/2014). JMO also thanks the FCT for the funds provided under the program Investigador FCT 2012 and 2015 (IF/00423/2012 and IF/01285/2015).

References

1. Dhillon MS, Patel S, John R (2017) PRP in OA knee–update, current confusions and future options. SICOT J 3:27
2. Andia I, Abate M (2017) Platelet-rich plasma: combinational treatment modalities for musculoskeletal conditions. Front Med:1–14. https://rd.springer.com/article/10.1007/s11684-017-0551-6. DOI:10.1007/s11684-017-0551-6
3. Nguyen RT, Borg-Stein J, McInnis K (2011) Applications of platelet-rich plasma in musculoskeletal and sports medicine: an evidence-based approach. PM R 3(3):226–250
4. Engebretsen L, Steffen K, Alsousou J, Anitua E, Bachl N, Devilee R, Everts P, Hamilton B, Huard J, Jenoure P (2010) IOC consensus paper on the use of platelet-rich plasma in sports medicine. Br J Sports Med 44(15):1072–1081
5. Qureshi AH, Chaoji V, Maiguel D, Faridi MH, Barth CJ, Salem SM, Singhal M, Stoub D, Krastins B, Ogihara M (2009) Proteomic and phospho-proteomic profile of human platelets in basal, resting state: insights into integrin signaling. PLoS One 4(10):e7627
6. Everts PA, Knape JT, Weibrich G, Schonberger J, Hoffmann J, Overdevest EP, Box HA, van Zundert A (2006) Platelet-rich plasma and platelet gel: a review. Journal of ExtraCorporeal. Technology 38(2):174
7. Marx RE (2001) Platelet-rich plasma (PRP): what is PRP and what is not PRP? Implant Dent 10(4):225–228
8. Marx RE (2004) Platelet-rich plasma: evidence to support its use. J Oral Maxillofac Surg 62:489–496
9. Pietrzak WS, Eppley BL (2005) Platelet rich plasma: biology and new technology. J Craniofac Surg 16(6):1043–1054
10. Andia I, Maffulli N (2013) Platelet-rich plasma for managing pain and inflammation in osteoarthritis. Nat Rev Rheumatol 9(12):721–730
11. Cengiz IF, Oliveira JM, Ochi M, Nakamae A, Adachi N, Reis RL (2017) "Biologic" treatment for meniscal repair. In: Injuries and health problems in football. Springer. Berlin, Heidelberg. 679–686
12. Piuzzi NS, Chughtai M, Khlopas A, Harwin SF, Miniaci A, Mont MA, Muschler GF (2017) Platelet-rich plasma for the treatment of knee osteoarthritis: a review. J Knee Surg 30(07):627–633
13. Utku B, Dönmez G, Büyükdoğan K, Karanfil Y, Tolevska RD, Korkusuz F, Doral MN (2015) Platelet-rich plasma: from laboratory to the clinic. Sports injuries: prevention, diagnosis, treatment and. Rehabilitation:3223–3250
14. Araki J, Jona M, Eto H, Aoi N, Kato H, Suga H, Doi K, Yatomi Y, Yoshimura K (2011) Optimized preparation method of platelet-concentrated plasma and noncoagulating platelet-derived factor concentrates: maximization of platelet concentration and removal of fibrinogen. Tissue Eng Part C Methods 18(3):176–185
15. Cavallo C, Roffi A, Grigolo B, Mariani E, Pratelli L, Merli G, Kon E, Marcacci M, Filardo G (2016) Platelet-rich plasma: the choice of activation method affects the release of bioactive molecules. Biomed Res Int 2016:1
16. Chahla J, Cinque ME, Piuzzi NS, Mannava S, Geeslin AG, Murray IR, Dornan GJ, Muschler GF, LaPrade RF (2017) A call for standardization in platelet-rich plasma preparation protocols and composition reporting: a systematic review of the clinical orthopaedic literature. JBJS 99(20):1769–1779
17. Navani A, Li G, Chrystal J (2017) Platelet rich plasma in musculoskeletal pathology: a necessary rescue or a lost cause? Pain Physician 20(3):E345
18. Foster TE, Puskas BL, Mandelbaum BR, Gerhardt MB, Rodeo SA (2009) Platelet-rich plasma from basic science to clinical applications. Am J Sports Med 37(11):2259–2272
19. Castillo TN, Pouliot MA, Kim HJ, Dragoo JL (2011) Comparison of growth factor and platelet concentration from commercial platelet-rich plasma separation systems. Am J Sports Med 39(2):266–271

20 Ehrenfest DMD, Rasmusson L, Albrektsson T (2009) Classification of platelet concentrates: from pure platelet-rich plasma (P-PRP) to leucocyte-and platelet-rich fibrin (L-PRF). Trends Biotechnol 27(3):158–167

21 Riboh JC, Saltzman BM, Yanke AB, Fortier L, Cole BJ (2016) Effect of leukocyte concentration on the efficacy of platelet-rich plasma in the treatment of knee osteoarthritis. Am J Sports Med 44(3):792–800

22 Fréchette J-P, Martineau I, Gagnon G (2005) Platelet-rich plasmas: growth factor content and roles in wound healing. J Dent Res 84(5):434–439

23 Cengiz IF, Oliveira JM, Reis RL (2014) Tissue engineering and regenerative medicine strategies for the treatment of osteochondral lesions. In: 3D multiscale physiological human. Springer, London pp 25–47

24 Yan L-P, Oliveira JM, Oliveira AL, Reis RL (2015) Current concepts and challenges in osteochondral tissue engineering and regenerative medicine. ACS Biomater Sci Eng 1(4):183–200

25 Scioli MG, Bielli A, Gentile P, Cervelli V, Orlandi A (2017) Combined treatment with platelet-rich plasma and insulin favours chondrogenic and osteogenic differentiation of human adipose-derived stem cells in three-dimensional collagen scaffolds. J Tissue Eng Regen Med 11(8):2398–2410

26 Marmotti A, Bruzzone M, Bonasia D, Castoldi F, Rossi R, Piras L, Maiello A, Realmuto C, Peretti G (2012) One-step osteochondral repair with cartilage fragments in a composite scaffold. Knee Surg Sports Traumatol Arthrosc 20(12):2590–2601

27 Lee J-C, Min HJ, Park HJ, Lee S, Seong SC, Lee MC (2013) Synovial membrane–derived mesenchymal stem cells supported by platelet-rich plasma can repair osteochondral defects in a rabbit model. Arthroscopy: The Journal of Arthroscopic & Related Surgery 29(6):1034–1046

28 Smyth NA, Haleem AM, Ross KA, Hannon CP, Murawski CD, Do HT, Kennedy JG (2016) Platelet-rich plasma may improve osteochondral donor site healing in a rabbit model. Cartilage 7(1):104–111

29 Altan E, Aydin K, Erkocak O, Senaran H, Ugras S (2014) The effect of platelet-rich plasma on osteochondral defects treated with mosaicplasty. Int Orthop 38(6):1321–1328

30 Gormeli G, Karakaplan M, Gormeli CA, Sarikaya B, Elmali N, Ersoy Y (2015) Clinical effects of platelet-rich plasma and hyaluronic acid as an additional therapy for talar osteochondral lesions treated with microfracture surgery: a prospective randomized clinical trial. Foot Ankle Int 36(8):891–900

31 Guney A, Akar M, Karaman I, Oner M, Guney B (2015) Clinical outcomes of platelet rich plasma (PRP) as an adjunct to microfracture surgery in osteochondral lesions of the talus. Knee Surg Sports Traumatol Arthrosc 23(8):2384–2389

32 Guney A, Yurdakul E, Karaman I, Bilal O, Kafadar IH, Oner M (2016) Medium-term outcomes of mosaicplasty versus arthroscopic microfracture with or without platelet-rich plasma in the treatment of osteochondral lesions of the talus. Knee Surg Sports Traumatol Arthrosc 24(4):1293–1298

33 J-p S, Tanabe T, Tsuzuki N, Haneda S, Yamada K, Furuoka H, Tabata Y, Sasaki N (2013) Effects of bilayer gelatin/β-tricalcium phosphate sponges loaded with mesenchymal stem cells, chondrocytes, bone morphogenetic protein-2, and platelet rich plasma on osteochondral defects of the talus in horses. Res Vet Sci 95(3):1210–1216

34 J-p S, Kambayashi Y, Itho M, Haneda S, Yamada K, Furuoka H, Tabata Y, Sasaki N (2015) Effects of a synovial flap and gelatin/β-tricalcium phosphate sponges loaded with mesenchymal stem cells, bone morphogenetic protein-2, and platelet rich plasma on equine osteochondral defects. Res Vet Sci 101:140–143

35 Smyth NA, Haleem AM, Murawski CD, Do HT, Deland JT, Kennedy JG (2013) The effect of platelet-rich plasma on autologous osteochondral transplantation: an in vivo rabbit model. JBJS 95(24):2185–2193

36. Betsch M, Schneppendahl J, Thuns S, Herten M, Sager M, Jungbluth P, Hakimi M, Wild M (2013) Bone marrow aspiration concentrate and platelet rich plasma for osteochondral repair in a porcine osteochondral defect model. PLoS One 8(8):e71602

37. Sun Y, Feng Y, Zhang C, Chen S, Cheng X (2010) The regenerative effect of platelet-rich plasma on healing in large osteochondral defects. Int Orthop 34(4):589–597

38. Kon E, Filardo G, Delcogliano M, Fini M, Salamanna F, Giavaresi G, Martin I, Marcacci M (2010) Platelet autologous growth factors decrease the osteochondral regeneration capability of a collagen-hydroxyapatite scaffold in a sheep model. BMC Musculoskelet Disord 11(1):220

39. van Bergen CJ, Kerkhoffs GM, Özdemir M, Korstjens CM, Everts V, van Ruijven LJ, van Dijk CN, Blankevoort L (2013) Demineralized bone matrix and platelet-rich plasma do not improve healing of osteochondral defects of the talus: an experimental goat study. Osteoarthr Cartil 21(11):1746–1754

40. Xie X, Wang Y, Zhao C, Guo S, Liu S, Jia W, Tuan RS, Zhang C (2012) Comparative evaluation of MSCs from bone marrow and adipose tissue seeded in PRP-derived scaffold for cartilage regeneration. Biomaterials 33(29):7008–7018

41. Krych AJ, Nawabi DH, Farshad-Amacker NA, Jones KJ, Maak TG, Potter HG, Williams RJ III (2016) Bone marrow concentrate improves early cartilage phase maturation of a scaffold plug in the knee: a comparative magnetic resonance imaging analysis to platelet-rich plasma and control. Am J Sports Med 44(1):91–98

42. Gu W, Li T, Shi Z, Mei G, Xue J, Zou J, Wang X, Zhang H, Xu H (2017) Management of Hepple stage V osteochondral lesion of the talus with a platelet-rich plasma scaffold. Biomed Res Int 2017:1

43. Mei-Dan O, Carmont MR, Laver L, Mann G, Maffulli N, Nyska M (2012) Platelet-rich plasma or hyaluronate in the management of osteochondral lesions of the talus. Am J Sports Med 40(3):534–541

44. Metcalf KB, Mandelbaum BR, McIlwraith CW (2013) Application of platelet-rich plasma to disorders of the knee joint. Cartilage 4(4):295–312

45. Sheth U, Simunovic N, Klein G, Fu F, Einhorn TA, Schemitsch E, Ayeni OR, Bhandari M (2012) Efficacy of autologous platelet-rich plasma use for orthopaedic indications: a meta-analysis. J Bone Joint Surg 94(4):298–307

46. Nourissat G, Mainard D, Kelberine F, French Arthroscopic Society (SFA) (2013) Current concept for the use of PRP in arthroscopic surgery. Orthop Traumatol Surg Res 99(8):S407–S410

Chapter 12
Enhancing Biological and Biomechanical Fixation of Osteochondral Scaffold: A Grand Challenge

Maryam Tamaddon and Chaozong Liu

Abstract Osteoarthritis (OA) is a degenerative joint disease, typified by degradation of cartilage and changes in the subchondral bone, resulting in pain, stiffness and reduced mobility. Current surgical treatments often fail to regenerate hyaline cartilage and result in the formation of fibrocartilage. Tissue engineering approaches have emerged for the repair of cartilage defects and damages to the subchondral bones in the early stage of OA and have shown potential in restoring the joint's function. In this approach, the use of three-dimensional scaffolds (with or without cells) provides support for tissue growth. Commercially available osteochondral (OC) scaffolds been studied in OA patients for repair and regeneration of OC defects. However, some controversial results are often reported from both clinical trials and animal studies. The objective of this chapter is to report the scaffolds clinical requirements and performance of the currently available OC scaffolds that have been investigated both in animal studies and in clinical trials. The findings have demonstrated the importance of biological and biomechanical fixation of the OC scaffolds in achieving good cartilage fill and improved hyaline cartilage formation. It is concluded that improving cartilage fill, enhancing its integration with host tissues and achieving a strong and stable subchondral bone support for overlying cartilage are still grand challenges for the early treatment of OA.

Keywords Osteoarthritis · Cartilage · Osteochondral scaffold · Biomaterials · Tissue engineering

M. Tamaddon · C. Liu (✉)
Institute of Orthopaedics & Musculoskeletal Science, Division of Surgery & Interventional Science, University College London, Royal National Orthopaedic Hospital, Stanmore, UK
e-mail: m.tamaddon@ucl.ac.uk; Chaozong.liu@ucl.ac.uk

© Springer International Publishing AG, part of Springer Nature 2018
J. M. Oliveira et al. (eds.), *Osteochondral Tissue Engineering*,
Advances in Experimental Medicine and Biology 1059,
https://doi.org/10.1007/978-3-319-76735-2_12

12.1 Introduction

In joints, the articular cartilage, calcified cartilage and subchondral bone form a biocomposite system, referred to as the osteochondral (OC) unit, which has the unique capability of transferring loads during joint motion [1]. Repetitive overloading to this unit could result in cartilage damage and changes in the subchondral bone, leading to mechanical instability of the joints and loss of joint function [2, 3]. If left untreated, the OC defects will lead to the development of osteoarthritis (OA) [3], where the composition and structure of this unit undergo significant alterations [1]. During the process of OA, thinning and degradation of articular cartilage, joint-space narrowing, osteophytes formation and subchondral bone remodelling [4–6] take place. Cartilage destruction results from an unbalanced relationship between matrix synthesis by chondrocytes and matrix degradation [6]. Other pathological processes including microfractures, microedema or microbleeding within the subchondral bone could lead to subchondral bone defects such as subchondral cyst formation [5]. If the OC defect has progressed to the stage where the patient's quality of life has significantly reduced and non-surgical treatments are no longer effective, then a joint replacement has to be performed. This major surgical procedure often does not restore the full function of joints and have high long-term complication rates.

Between 2003 and 2013, there were 1.296 m joint replacements performed in England and Wales including 620,400 hip procedures and 676,082 knee procedures predominantly for OA (over 93%) [7]. OA is a major contributor to functional impairment and reduced independence in older adults [8] and represents an enormous socioeconomic challenge [9]. Regeneration of the tissues affected by OA in early, mid or late stages of the disease can enhance the quality of life and delay or avoid the need for total joint replacement, thereby reducing the costs.

12.1.1 Progression of OA and Available Treatments

In early stages of OA, there is an increase in water content of cartilage, resulting in swelling of the matrix and an increase in metabolic activity of chondrocytes (Fig. 12.1). These changes are accompanied by the appearance of surface fibrillations characterised by microscopic cracks in the superficial zone of the articular cartilage. In the subchondral bone, increased remodelling of the cortical bone plate usually leads to increased porosity [1]. In this stage, pain and stiffness dominate the other symptoms, and the goal of the treatments is therefore to reduce pain and physical disability and some attempt to control structural deterioration in the affected joints [4, 10], using physical therapy [11], analgesics and NSAIDs [9]. Intra-articular injection of long-acting glucocorticoids is an effective treatment of inflammatory flares of OA. Hyaluronic acid has varying effectiveness when used for intra-articular injections for the treatment of OA of the knee [10]. With the progression of OA, loss

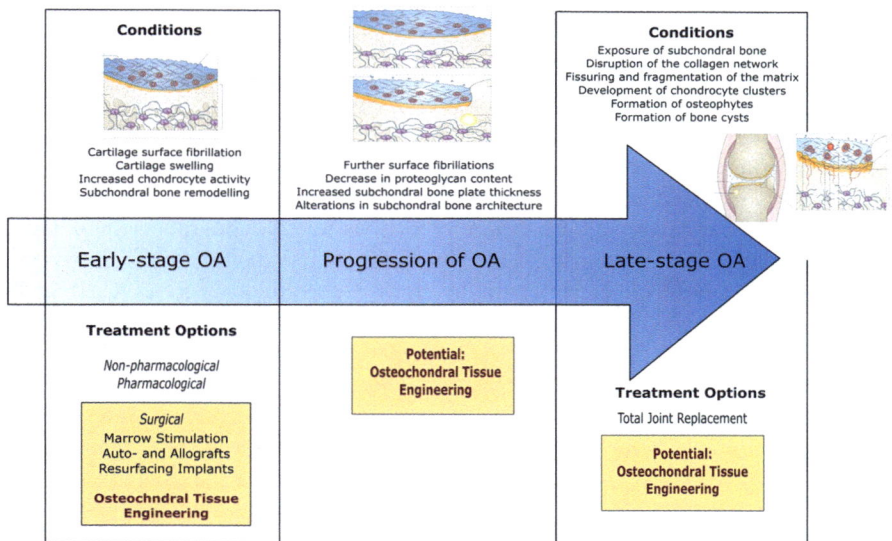

Fig. 12.1 Progression of OA: conditions and treatments in each stage. Pictures of osteochondral units. (Adapted from Ref. [1])

of cartilage matrix proteoglycans and erosion of the collagen network lead to the development of deep fissures and partial delamination of the cartilage, while in the subchondral bone, cortical plate thickness gradually increases [1] (Fig. 12.1). At this stage of the disease, where the cartilage defect is still small (area <2–3 cm^2), microfracture (MF) marrow stimulation is considered a medically necessary treatment. MF is a minimally invasive procedure which seeks to repair cartilage damage through releasing mesenchymal stem cells (MSCs) from the underlying bone which then differentiate to become chondrocytes and create new cartilage. It involves removing the damaged cartilage and then drilling into or otherwise puncturing the surface of the underlying bone in order to allow blood and bone marrow to come through to the bone/cartilage interface, where the MSCs contribute to the formation and repair of the cartilage and bone. However, the regenerated cartilage is mainly fibrocartilage and is not expected to have the same durability as the articular hyaline cartilage. This type of cartilage is mostly type I collagen, fibrocytes and a disorganised matrix that lacks the biomechanical and viscoelastic characteristics of normal hyaline cartilage [12] and can fail with high shear forces in the joint, leading to an ongoing articular surface irregularity and subsequent secondary arthritic change [13]. This was demonstrated by the high 5-year post-microfracture reoperation rates, which are between 30% and 50% [14].

Osteochondral (OC) autografts or allografts [14], scaffolds and focal knee resurfacing implants are among the approaches that have been explored for treatment of small- to mid-sized lesions [15]. OC autografts have been proposed to provide an immediate reliable tissue transfer of a viable OC unit in a single-stage procedure.

This procedure exploits the regenerative potential of bone and bone-to-bone healing, since the cartilage has a limited healing capacity [14]. For example, Hangody et al. [15] analysed the results of mosaicplasty (where multiple autografts are used) in 82 athletes with signs of OA. They reported significant improvements after the procedure, although slight radiographic degeneration in one-third of the patients at mid- to long-term follow-up was observed [14, 16].

Fresh OC allografts provide the surgeon with more freedom regarding the size of the defect that can be treated. Common indications for OC allograft include large, focal chondral defect, osteochondritis dissecans and unicompartmental arthritis [17]. However, apart from general complications of open joint surgery, OC allograft transplantation is also associated with a risk of disease transmission from the allograft and subchondral collapse due to inadequate integration. The latter is responsible for a majority of graft-related failures [14].

Tissue engineering (TE) approaches have been developed as a potential solution for repair and regeneration of OC defects. In this approach, scaffolds are designed and fabricated to provide a physical environment to support cellular activities and prompt tissue regeneration. OC scaffolds can be implanted by arthroscopy or mini-arthrotomy and fixed by press fit. Some cases may require additional fixation through sutures, pins or fibrin glue. Currently, lesion size range from 2 to 8 cm^2 can be treated using osteochondral scaffolds which are available in predetermined sizes or patches that can be shaped and sized at the time of implantation [14]. Commercially available scaffolds such as Chondromimetic (Tigenix NV), MaioRegen (Finceramica) and TruFit® BGS Plugs (Smith & Nephew) have been used, with or without cells, in clinical trials for treatment of small cartilage and osteochondral defects (OCDs) (<1.5 cm^2). However, limited success was reported, and none of these scaffolds have achieved satisfactory durable clinical results.

In late-stage OA, chondrocytes clustering and apoptosis are evident in the cartilage. In the deeper zones of the cartilage, chondrocytes undergo hypertrophic differentiation, and the calcified cartilage expands and advances into the overlying hyaline articular cartilage, with duplication of the tidemark (Fig. 12.1). This process is initiated by the penetration of vascular elements into the osteochondral junction. In addition to the development of osteophytes at the joint margins and cysts within the subchondral bone, the subchondral cortical plate becomes flattened and deformed, a process referred to as "bone attrition". In this stage of the disease, underlying zones of calcified cartilage and subchondral bone are exposed [1]. In these advanced cases, joint prosthesis surgery is often required [9].

To date, OC tissue engineering approaches have mainly focussed on regeneration of small OC defects mostly in early stages of OA. However, with the right scaffold, treatment of large, late-stage OC defects could become possible. The idea of a "smart" scaffold which provides an appropriate biomechanical environment to support healthy cell growth and promote OC regeneration has been reported as the Holy Grail in the last decades in the treatment of both early and late stages of OA. However, this has been achieved only in early stages of OA, and with limited success. In this paper, we discuss the requirements of an OC scaffold and insights from the studies of OC scaffolds performance, both in vivo and in clinical settings, in the light of similar events observed during the development of OA. The effect of biomechanical and biological fixations of the scaffold on the healthy regeneration of OC tissue has become increasingly apparent. The results discussed in this study would provide us with the essential knowledge for the successful development of future clinical OC scaffolds.

12.2   Osteochondral Tissue Engineering

Tissue engineering (TE) is a discipline that applies the knowledge of materials science, cell biology and bioengineering to construct tissue templates and restore the function of an injured tissue (Fig. 12.2). It may involve a cell-free approach by using a scaffold only, or it may involve taking the cells from the patient, seeding the cells onto a scaffold and culture this whole in a bioreactor system and then transplanting it back into the patient once the tissue has matured. In either processes, the three-dimensional porous scaffold plays an important role in supporting the (seeded/resident) cells growth and guiding new tissue formation [20]. Due to the unique structure and property of OC tissue unit, the concept of the simultaneous regeneration of articular cartilage and underlying bone (OC defect) to develop a well-defined tissue-to-tissue interface [21] has drawn considerable attention, especially as a technique for promoting superior cartilage integration and a treatment for OC defects as often observed in osteoarthritic joints [18].

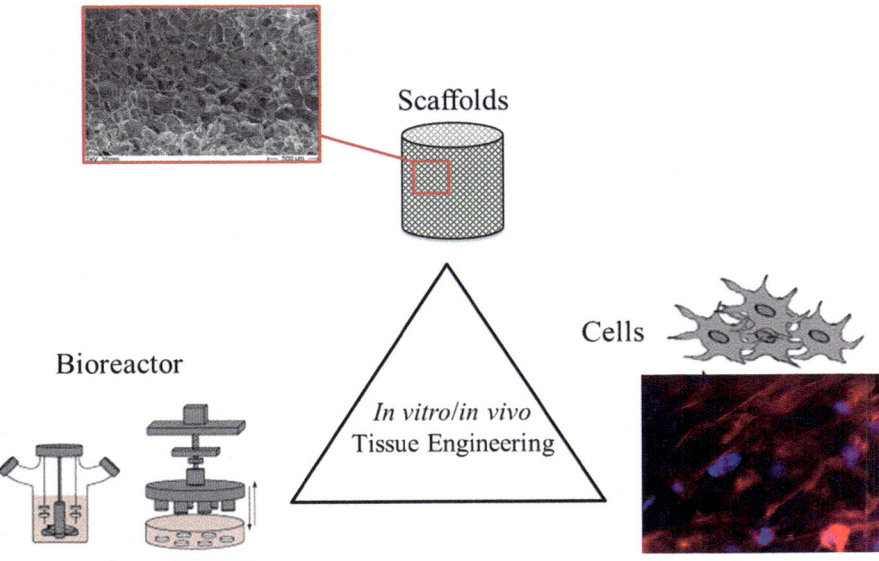

Fig. 12.2 Three key ingredients of tissue engineering are scaffolds, cells and bioreactors. (Adapted from Refs. [18, 19])

In general, cartilage-to-cartilage interfaces do not integrate well because of the dense avascular nature of cartilage and scarcity of cells; however, it is often observed that host bone integrates well with the grafted bone tissue and other implant materials. To improve the integration of the engineered cartilage tissue with the host tissue, an OC implant can be used where the bony region serves as an anchor for the implant. Thus a successful OC scaffold needs to address both regions concurrently. The cartilage and bone regions of the OC composite scaffold require different physical and mechanical properties to mimic the gradient mechanical property, structure and functionality of the OC unit. A major challenge is thus generating the natural gradation in porosity, composition and biomechanical properties associated with both tissues (i.e. bone and cartilage) as well as the integration of the two types of tissues [22].

12.2.1 Scaffold Design Considerations: Mimicking the Nature

The OC tissue is composed of cartilage and subchondral bone, each with their own specific hierarchical structure and biological property [23]. Therefore, to design a biomimetic scaffold, an understanding of the OC unit, including its composition, structure and function, is essential.

12.2.1.1 Cartilage–Bone Junction

Articular cartilage – the top layer of an OC unit – is vital for facilitating a smooth motion within joints and absorbing impact. It consists of chondrocytes embedded in an extracellular matrix (ECM) mainly comprising collagen (60% dry weight [24], 90–95% type II [25]), proteoglycans and non-collagenous proteins. The structure and composition of the cartilage are organised on two levels, which are determined according to the distance from the cartilage surface and in relation to its distance from the cells. Typically, articular cartilage is divided into four zones based on the distance from the surface: superficial, middle, deep and calcified zones [25]. The latter is directly below the deep zone containing hypertrophic chondrocytes embedded in a densely mineralised matrix which constitutes the OC interface [26]. Calcified cartilage is separated from the deep zone by a discrete band of mineralised cartilage called "tidemarks". This line represents the mineralisation front of the calcified cartilage and provides a gradual transition between the two dissimilar regions of cartilage (non-calcified and calcified). Immediately below the calcified zone lies subchondral bone plate – a bony lamella (cortical endplate, 1–3 mm thick [27]), which is separated by a "cement line" from the calcified cartilage. Together with the supporting trabeculae and subarticular spongiosa, they form the subchondral bone unit [5]. While the tidemark is crossed by collagen fibrils between the articular cartilage and calcified cartilage resulting in a strong link between these two zones, the cement line marks the separation of the cartilage and underlying bone. It is presumed to be a region of weakness since no collagen fibres are continuous between the calcified cartilage and subchondral bone plate [5, 28]. The steep stiffness gradient between cartilage and subchondral bone unit may be one of causes of cartilage delamination from the bone due to shear stresses [5]. Different magnitudes of strain, internal pressure and fluid flow are developed in each of the osteochondral layers during loading [29], the convex joint surfaces can be exposed to large lateral forces which may lead to a variety of shear-induced lesions in the osteochondral region [28]. Therefore, one of the important considerations in designing bi-/multilayered scaffolds [30] for OC TE is to avoid abrupt and large changes in mechanical properties of different layers.

12.2.1.2 Role of Subchondral Bone in Maintenance of Cartilage

Subchondral bone is essential in function and maintenance of articular cartilage. From biomechanical point of view, the joints can withstand about 2.5–5 times of the body weight caused by the dynamic loading generated during walking. Subchondral bone enhances the load bearing capacity: normal subchondral bone attenuates about 30% of the loads through joints; only 1–3% is attenuated by cartilage [5]. Cartilage and bone act in concert by performing a biomechanical function of the joints, the former as a bearing and the latter as a structural girder and shock absorber [31]. As such, simultaneous repair and regeneration of cartilage and subchondral bone is a key concept in osteochondral tissue engineering.

It is hypothesised that the thickness of cartilage is dependent on the joint congruency [5] and local stresses [32]. A joint with a high congruency has thin cartilage, whereas a joint with a low congruency is covered by a thick layer of cartilage so that it can more easily deform thereby increasing the load-bearing area and decreasing the stress per unit area [5, 33]. It is intuitive that heavily stressed regions have thicker cartilage. As for the subchondral bone, regional differences in mineralisation can also be recognised, and greater density is usually found in the more heavily loaded regions of the joint surface [5, 32]. Therefore, it can be suggested at places within the joint where the stress is assumed to be greatest, the subchondral bone mineralisation is higher and the cartilage is thicker [5]. Therefore, design of the scaffold may need to be adjusted in terms of the thickness and mineralisation/stiffness of layers according to the location of the defect within the joint.

The subchondral bone also plays an important role from nutritive point of view. The subchondral bone plate has a high number of vessels and hollow spaces invading the cement line into the calcified cartilage, and they are mainly concentrated in heavily stressed zones, providing a rich blood supply to subchondral bone and nutrients to the cartilage [5]. Whereas the superficial zone of cartilage is mainly dependent on diffusion via synovial fluid as its nutritive source, the subchondral circulation may make a significant contribution to the nutrition in deep and calcified cartilage [34, 35]. In fact it has been shown that more than 50% of the glucose, oxygen and water requirements of cartilage are provided by perfusion from these subchondral vessels [36, 37]. The abrogation of contact between the subchondral bone and cartilage leads to degeneration of cartilage in the long run [38]. This emphasises the importance of subchondral bone regeneration and vascularisation in OC tissue engineering.

Highlighting the significance of vasculature in bone formation is the fact that the metabolically active cells are no more than 100 μm away from a capillary for supply of oxygen and nutrients [39, 40]. Often, this poses a problem for tissue engineering (especially cell-free scaffolds – because in the in vitro construct this can be alleviated by using bioreactors), since the resident cells may not be able to migrate deep into the scaffold due to diffusion constraints of oxygen and nutrients, only cells close to the surface are able to survive. However, the mineralisation at the periphery of the scaffold actually block further diffusion and mass transfer to the interior of the scaffold, leading to growth of only thin cross-sections of tissue (<500 μm) [41]. This needs to be taken into account when designing a scaffold for OC tissue engineering, for example, by devising internal channels [42] in the OC scaffold.

12.2.2 Scaffold Structure and Properties

When the calcified zone of cartilage is surgically removed, such as in the case when the basis of a chondral or OC defect is prepared prior to microfracture or scaffold implantation, the bone marrow which contains mesenchymal stem cells (MSCs) may enter the debrided defect via the blood vessel or channels [5]. Controlling the cellular behaviour of bone marrow mesenchymal stem cells (BMMSCs) is vital for achieving correct type of regenerated tissue (e.g. hyaline cartilage as opposed to fibrocartilage). The scaffold provides physical environment to support BMMSC growth and plays a vital role in controlling BMMSC fate [43] through scaffold–cell interactions to regulate cell phenotype, cytoskeleton spreading, proliferation, gene expression and ECM secretion through metabolic activity, cell–matrix and cell–cell contact [44].

The microenvironmental factors affecting stem cell behaviour [45] include scaffold surface characteristics (e.g. wettability and charge – cell attachment), material (cell attachment and differentiation), microstructure (porosity, pore size and shape – cell adhesion, migration and differentiation) and stiffness (cell differentiation).

12.2.2.1 Scaffold Surface and Material Characteristics Affect Cell Attachment and Differentiation

When a scaffold is implanted into an OCD, the biological fluid (e.g. from bone marrow, synovial fluid, etc.) from the patient will be in contact with the surface of the scaffold. The surface wettability affects protein absorption and consequently cell attachment to it. Therefore, understanding the molecular mechanism of cell adhesion on biomaterials is important to manipulate the scaffold–cell interaction.

When an anchorage-dependent cell (such as MSCs) comes in contact with a surface, it must adhere to the surface to remain viable, proliferate or differentiate. Attachment to the surface can be non-receptor-mediated via weak chemical bonding, such as electrostatic, hydrogen or ionic bonding; however, this type of adhesion does not guarantee the transmission of signals from the microenvironment to the cells, which is necessary to prompt the secretion of ECM molecules from cells, and without it the cells may go into apoptosis. By contrast, receptor-mediated adhesion through ECM molecules such as fibronectin or collagen allows signal transmission. These ECM molecules adsorb onto the surface of biomaterial from the surrounding environment and then bind cell integrins through their specific amino acid sequences. These specific amino acids are called ligands, and the minimum adhesion motif on ECM molecules should contain at least three amino acids, which are often symbolised by Arg-Gly-Asp (RGD). On the other hand, integrins are glycoproteins consisting of one α and one β chain. There are about 16 recognised α subunits and eight β subunits with various combinations, which results in receptors with preferential binding affinity to certain ECM molecules [46, 47].

Once ligand binds to the surface, integrins are formed into dot-like or streak-like "focal adhesions". In these focal adhesions, integrins communicate with structural and signalling molecules, such as talin, α-actinin, filamin, paxillin or vinculin, which link integrin receptors to cytoplasmatic actin cytoskeleton (Fig. 12.3). External signals are then transmitted from the microenvironment to the nuclei of cells [49], thereby influencing intracellular transport processes and the secretion of various molecules and determining the cellular activities such as cell proliferation and differentiation or apoptosis [47]. The extent and strength of cell adhesion to a surface affects its migration and its decision to switch between proliferation and differentiation behaviour [47, 49, 50]. For example, cells with large adhesion are usually dormant in migration and proliferation and more active in the expression of differentiation markers [50]. The optimum adhesion of cells generally occurs on a moderately hydrophilic and positively charged scaffold because the adhesion molecules are adsorbed in a favourable geometric conformation, making ligands available to bind with cell receptors [50]. In short, cell attachment requires specific binding proteins to be present on the material. The surface morphology and chemistry of the scaffold regulate protein adsorption and influence cell attachment and alignment [51].

Biomaterials used in tissue engineering of OCDs are usually categorised into four major groups: natural polymers, synthetic polymers, metallic materials and inorganic materials such as ceramics and bioactive glasses. Multicomponent systems can be designed to generate composites of enhanced performance [3]. Naturally derived polymers such as collagen, alginate, gelatin and chitosan have the advantage of native biological function, enhancing cellular attachment, proliferation and function [52, 53]. As explained earlier, cells primarily interact with scaffolds via ligands on the material surface. Scaffolds synthesised from natural extracellular materials (e.g. collagen) naturally possess these ligands in the form of RGD-binding sequences, whereas scaffolds made from synthetic materials may require deliberate incorporation of these ligands through, for example, protein adsorption [54]. The main disad-

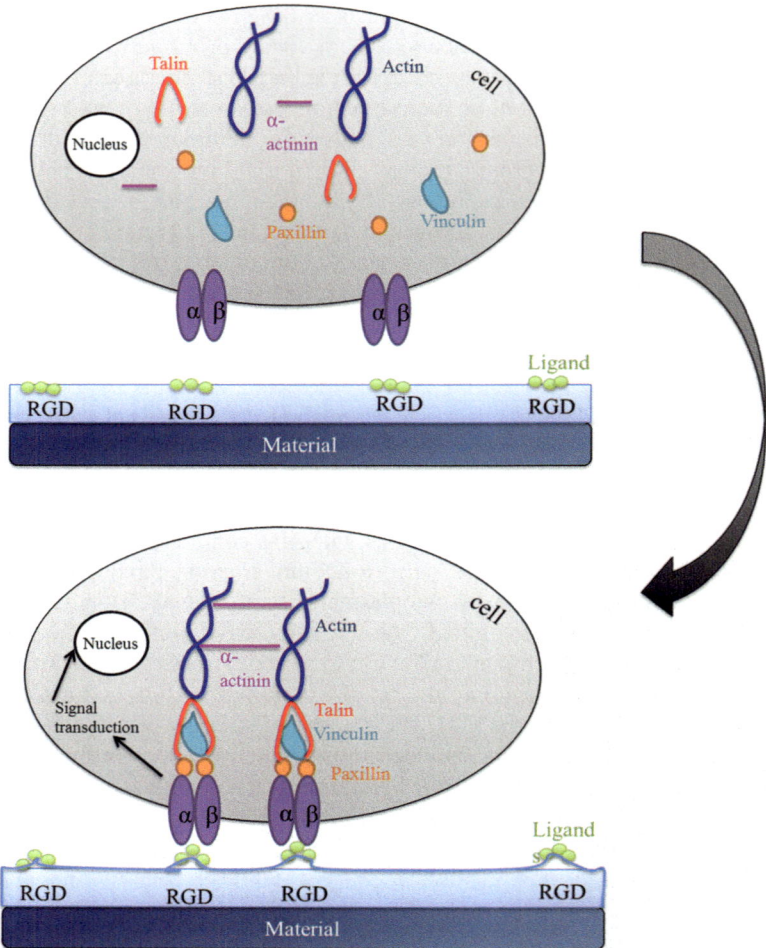

Fig. 12.3 Cell–scaffold interaction. (Modified from Refs. [46, 48])

vantages of these naturally derived biomaterials are batch-to-batch variability and low mechanical strength. With synthetic polymers (e.g. PCL, PLA, PLGA) on the other hand, it is possible to precisely control the mechanical properties and tailor the structure and apply surface modifications. However, they exhibit poor cell adhesion due to their intrinsic hydrophobicity and lack of natural ligand binding sites [52].

Bioceramics, such as calcium phosphates, are known for their excellent osteo-conductivity [3, 55]. The most common types of calcium phosphates for bone TE scaffolds are hydroxyapatite ($Ca_{10}(PO_4)_6(OH)_2$), tricalcium phosphate (TCP), biphasic calcium phosphates and multiphasic bioglasses [56]. The physical properties of the calcium phosphate ceramics, such as degradation rate, modulus and processability, can be controlled by altering their composition [57].

12.2.2.2   Scaffold Microstructure Affects Cell Adhesion, Migration and Differentiation

Once attach to a substrate, the cells need to migrate into the scaffold three-dimensional space, proliferate and differentiate into the appropriate tissue type there. Cell migration requires scaffold to be porous [58, 59] and to have an interconnected pore structure to allow for healthy cellular invasion and growth, nutrition delivery [60] to the cells inside the scaffold as well as removal of metabolic waste from the cells. Vascularisation – as explained in the design considerations – is not therefore possible without porosity to allow oxygen and nutrition diffusion and vasculature formation [61]. In fact, absence of any bone formation on a solid scaffold that lacked porosity demonstrates the importance of this factor in tissue formation [62].

Cell migration is also affected by another microstructural factor: pore size. Cells use a bridging mechanism when migrating through a porous scaffold; that is, they use neighbouring cells as support to bridge across pores larger than their diameters. If the pore dimension greatly exceeds the size of a cell, then the cell can only spread along the walls of the pore, instead of bridging a pore, a phenomenon that influences cells' migration ability and speed in general [60]. If pores are too small, cell migration is limited; the cells trying to bridge the small pores block the way for migration into the centre of the scaffold, resulting in the formation of a cellular capsule around the edges of the scaffold. This in turn can limit diffusion of nutrients and removal of waste resulting in necrotic regions within the construct. However, cells travelling through larger pores may migrate slower, but their directional movement allow them to travel further into the scaffold increasing cell migration and scaffold infiltration [60].

Differentiation of MSCs can be influenced by pore size as well as the pore shape. This can be attributed to the particular cellular events prerequisite in the chondrogenic and osteogenic differentiation processes of MSCs. Chondrogenesis of MSCs is associated with morphologic changes from fibroblast to spherical morphology, in which the fibroblastic morphology is formed through cell–matrix interactions during migration and proliferation, and develops into spherical morphology in the process of condensation. The chondrogenic differentiation of MSCs occurs in highly regulated stages. Aggregation of mesenchymal cells into pre-cartilage condensations is crucial for chondrogenesis. Condensation occurs through cell–cell contacts, which is controlled by the association of cell adhesion molecules of the adjacent cells, formation of gap junctions and changes in the cytoskeletal architecture, subsequently activating intracellular signalling pathways to initiate the transition from chondroprogenitor cells to a fully committed chondrocyte [44, 63, 64]. The ability of MSCs to aggregate in the larger pores in a porous scaffold, coupled with proliferation of cells within the scaffold, might facilitate chondrogenic condensation process of MSCs [44], showing the effect of pore size in controlling MSCs differentiation.

Cell shape is a potent regulator of cell growth and physiology [65], and many events related to embryonic development (e.g. single and collective cell migration, dorsal closure, etc. [66]) and stem cell differentiation are influenced by cell shape [43]. An example can be seen in Fig. 12.4, where the effects of cell spreading and focal adhesions on viability, migration, proliferation and differentiation of cells are shown.

In bone and cartilage development, flattened and spherical cell morphologies are the most relevant, where the spherical morphology of chondrocytes is closely related to their chondrogenic potential [67]. A direct comparison of cell and nuclear shape of BMMSCs shows that a more rounded nuclear shape is associated with the greatest expression of molecular markers associated with chondrogenesis [68] and a change in the cell shape profoundly alters the organisation of the actin cytoskeleton and the assembly of focal adhesions [69]. As shown in Fig. 12.4, the extent and strength of cell adhesion to a surface affect its migration and its decision to switch between proliferation and differentiation behaviour [47, 49, 50]. For example, cells with large adhesion are usually dormant concerning migration and proliferation and more active in the expression of differentiation markers [50]. It is therefore possible to influence MSCs fate artificially through control of their shape by synthetic extracellular matrices [43], such as scaffold pore shape or surface chemistry, and hence influencing differentiation of MSCs to chondrocytes and osteoblasts.

There are a number of techniques available to produce porous scaffolds, depending on the scaffold material. Pore-inducing techniques for synthetic polymers include solvent casting in conjunction with particulate leaching, phase separation, gas foaming, melt-moulding and fibre bonding [40, 70], all of which involve high

Cell behaviour/ cell spreading	Viability	Migration	Proliferation	Differentiation
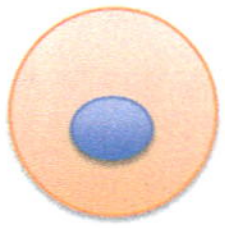	↓	↓	↓	↓
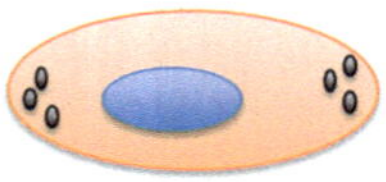	↑	↑	↑	↓
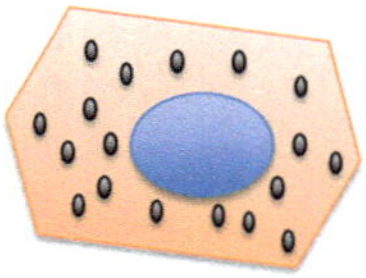	↑	↓	↓	↑

Fig. 12.4 Effect of cell spreading/adhesion on subsequent cell behaviour. Blue refers to cell nuclei, and black dots represent focal adhesions. (Adapted from Ref. [47])

temperatures, the use of chemicals or pH levels unsuitable for protein-based natural polymers. Consequently, the number of methods to generate pores in natural polymer is quite limited. Two of the most commonly used methods are freeze-drying [71] and critical point drying.

The use of 3D printing has gained considerable attention in recent years. This technique is especially fitting to generate OC scaffolds, since this tissue has a complex graded structure where biological, physiological and mechanical properties vary significantly over the full thickness of osteochondral unit [72]. "Solid free-form" technologies including 3D printing provide us with tools to closely control the design and shape (including the distinct curvatures of joints) in the final products; hence producing tailorable scaffolds has become a reality. Different techniques of 3D printing are extensively discussed in Do et al. [73], O'Brien et al. [74], and Sachlos and Czernuszka [40]. These include direct 3D printing, indirect 3D printing [75], bioplatter printing (using a "bioink" or cell-laden gels) [76, 77], fused deposition modelling (FDM) [78], selective laser sintering (SLS) [79] and stereolithography (SLA).

12.2.2.3 Scaffold Stiffness Affects Cell Migration and Differentiation

Scaffold stiffness is another factor affecting stem cell fate [80, 81]. Naive mesenchymal stem cells are shown to specify lineage and commit to phenotypes with extreme sensitivity to tissue-level elasticity [81]. At the cellular scale, normal tissue cells probe elasticity as they anchor and pull on their surroundings [80] (Fig. 12.5). The cells' response to mechanical properties of their matrix involves a feedback loop, where cells exert contractile forces on the matrix (e.g. it has been shown that each fibroblast on a scaffold exerts 1nN force to the matrix [82], which can contract it depending on the stiffness of the matrix). The contractile forces generate strain in the substrate. This strain can be detected by other cells, and they respond to this by adjusting their cytoskeleton and overall state [60, 80, 83]. Engler et al. [81] showed that human MSCs favoured differentiation into neuron-like cells on soft substrates, into myogenic lineage on substrates with moderate elasticity and into osteogenic lineage on rigid substrates [81] (Fig. 12.5). Similarly, the matrix stiffness can affect cell migration. The contact between fibroblasts, epithelial cells, smooth muscle and the matrix reduces on softer substrates, and cells migrate from softer regions to stiffer regions when subjected to gradient stiffness [45]. Tailoring the stiffness of the scaffold layers to induce chondrogenesis and osteogenesis is an important consideration in achieving the correct type of regenerated tissue.

Regardless of microstructure, the intrinsic resistance of a solid to a stress is measured by the solid's elastic modulus E, which is obtained by the linear slope of stress–strain curve when subjected to a force [80]. A scaffold's mechanical properties, on the other hand, are derived from both its composition (solid elastic modulus), microstructure (%porosity) and topology (connectivity and shape of the pores) [60, 84]. This is why the bulk modulus of a porous scaffold may be many times lower than that of the local elastic modulus facing an individual cell (e.g [85, 86]).

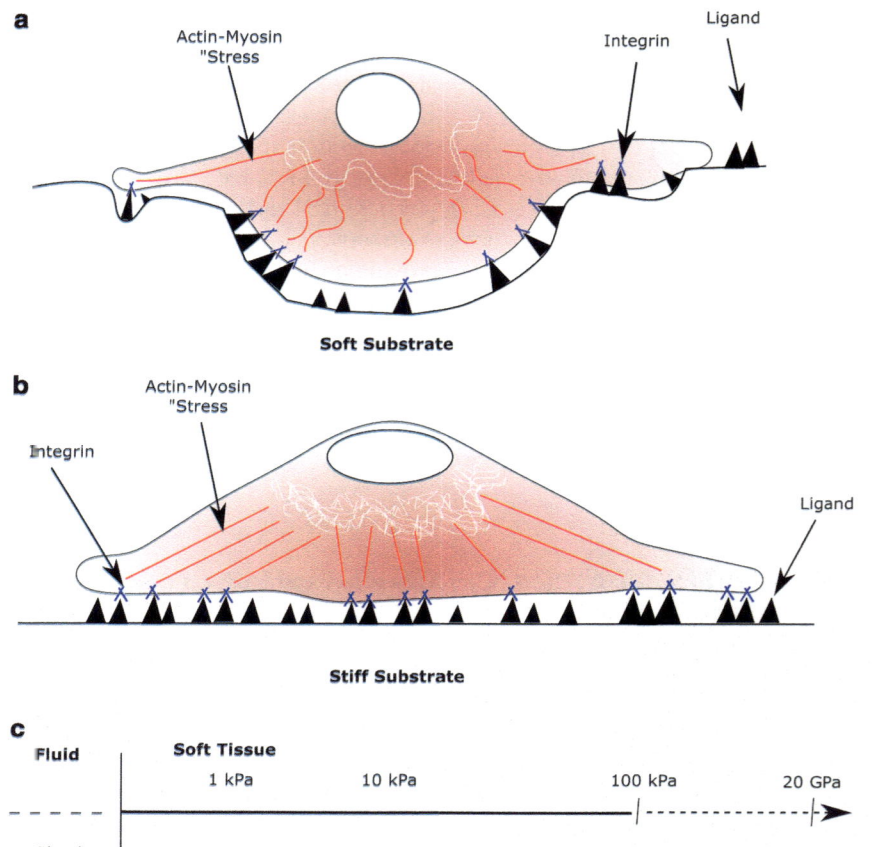

Fig. 12.5 Cell–scaffold interaction, (**a**) in soft and (**b**) in stiff substrates and (**c**) the resultant cytoskeletal formations; redrawn from [60]

Scaffolds can be developed as cell-free matrices, or as tissue-engineered construct before implantation. In the former, the fabricated scaffold is implanted either in a single-step (with no cell) or a two-step (with autologous cells) procedure. In the latter, the scaffold is seeded with autologous cells from the patient in the laboratory, the tissue is grown in vitro and the neo-tissue is implanted in the patient (two-step procedure). In this method, utilising a bioreactor could be beneficial. During physiological loading, a range of mechanical stimuli are developed in cartilage such as compressive and shear strain, stress, hydrostatic pressure and fluid flow. Bioreactors help to emulate these conditions in vitro by providing environmental, biochemical and mechanical cues to the cells [20]. The beneficial effects include differentiation of progenitor cells towards chondrogenic and osteogenic lineages [29].

There are several types of bioreactors suitable for osteochondral tissue engineering, including "rotating" and "perfusion" bioreactors. In the rotating bioreactors, scaffolds are suspended between two cylinders in the cell culture medium, and gas flow occurs via a silicone membrane. Although mass transport to the surface of the construct is enhanced by this method, it still only takes place by diffusion within the scaffold. The perfusion bioreactors include perfusion cartridges and perfusion chambers, with the latter being capable of applying 3D mechanical loading to the construct. This method is designed in a way to provide interstitial fluid flow through the scaffold [20] and is an effective way for initial cell seeding as well as subsequent mass transport [87]. Perfusion bioreactors have been used for generation of osteoinductive [88] grafts and cartilage constructs [89] and simultaneous construction of both tissues in a double-chamber (hydrostatic pressure but no mechanical stimulation) perfusion bioreactor [90].

12.3   Performance of OC Scaffolds in Animal and Clinical Studies

A great number of scaffolds have been fabricated and explored for osteochondral tissue engineering. These scaffolds have been developed specifically to reproduce bone and cartilage either with or without the addition of cells [91]. Of those, only a small number has been advanced into clinical trials [92].

12.3.1   In Vivo Performance of Osteochondral Scaffolds: Animal Model Studies

OCDs involve articular cartilage and associated subchondral bone. Multilayered osteochondral scaffolds have been developed to mimic the native architecture of the osteochondral tissue unit. There are many research groups around the world that have developed different osteochondral scaffolds from a range of biomaterials and their combinations. Tables 12.1 and 12.2 summarise the typical osteochondral scaffolds that have been evaluated in in vivo animal studies (Table 12.1) and in clinical studies (Table 12.2).

Gotterbarm et al. have reported a composite scaffold of collagen I/III for the upper cartilage layer and tricalcium phosphate (TCP) as bone section. They incorporated growth factors such as bone morphogenetic protein (BMP) and transforming growth factor beta (TGF-β) in their scaffold and evaluated the performance of the scaffolds in a minipig model. Fifty-two weeks post-operation results showed that the scaffolds were fully degraded. However, only approximately 32% of the defect area was restored with the lamellar trabecular bone. They concluded that although the use of growth factors assisted the rate of resorption, it

Table 12.1 Properties and performance of scaffolds used for in vivo animal studies

Scaffolds in animal studies

Cartilage material	Bone material	Mechanical pro	Other properties – notes	Cell/single vs double step	No. of samples and duration	Findings	References
Collagens I and II (porcine) + hyaluronic acid	Collagen I (bovine) + hydroxyapatite / collagens I and II + hydroxyapatite	0.95 kPa (bottom), 0.35 kPa (intermediate layer), 0.3 kPa (top)	>97% porosity	Single	8 rabbits (femoral condyle), 12 weeks	ICRS I score of regenerated cartilage: grade II (nearly normal); bone volume/total volume of 0.4 ± 0.05 compared to 0.35 ± 0.03 in control, evidence of tidemark	[93, 94]
Collagens I/III scaffold (insoluble bovine tendon type I collagen (4% w/v) with type III)	b-tricalcium phosphate (TCP) MedArtis AG	TCP – compressive stiffness 1 MPa	TCP = 65–80% (pore diameter of 600 μm) – size 7 mm × 5.45 mm	Growth factors: BMP 2, 3, 4, 6, 7, TGF-β1, –2, –3, FGF-1, osteocalcin and osteonectin	27 and 36 minipigs	The TCP layer significantly increased new bone formation 29.8 ± 9.68% at 6 weeks and 40.09 ± 4.76% at 12 weeks; at 52 weeks scaffolds almost fully degraded and 31.28 ± 5% defect was restored with trabecular lamellar bone; use of GF increased resorption of TCP but did not increase new bone formation	[95, 96]
Gelatin–chondroitin sulfate–sodium hyaluronate (GCH)	Gelatin–ceramic bovine bone (GCBB)	5.86 ± 0.77 cartilage and 13.44 ± 0.89 MPa bone	In rabbit > 3 mm, defect is large	Two-step: group A, with cells; group B, no cells; and group C, empty	36 rabbits, patella defects 15 × 10 × 15 mm, press fit	Group A showed hyaline cartilage formation at 6 and 12 weeks, while fibrous tissue was observed in groups B and C, by RT-PCR and type II collagen staining	[97]
Chitosan–gelatin + TGF-β1	HAp–chitosan–gelatin +BMP-2		Layers glues with fibrin glue	BMMSCs – 4 groups: gene-activated, DNA-free scaffold, gene-activated monolayers	36 rabbits, 4 × 5 mm, at 4, 8 and 12 weeks	In gene activated: at week 4, new trabecular bone formation; at 8 weeks, stain for Alcian blue and col. II; at 12 weeks, integration with surrounding tissue/hyaline cartilage/indistinguishable	[98]
Calcibon granules (200–220um) + chondrocytes	Calcibon: injectable in situ setting bone cement, consists of a powder (α-TCP, CaHPO4, CaCO3, precipitated HAp) and a liquid (Na2HPO4)	50 MPa compressive strength	Edges were cut before implantation to provide fresh surface	Autologous chondrocytes, culture for 3 weeks before implantation	8 minipigs, femoral condyle, 26 and 52 weeks 2 × 4.5 mm	Well integration of cartilage and scaffold No degradation/remodelling of the carrier Absence of subchondral cyst which was seen in TruFit	[99]

Silk scaffold ±TGF-ß	Silk scaffold ± BMP-2 → two layers	–	Layers attached by fibrin glue	No cells TGF-ß and BMP-2	20 rats, 1.8 mm × 1 mm, 8 weeks	Formation of neo-OC tissue Good biointegration Mulberry silk favours osseous differentiation (col I), non-mulberry chondrogenic differentiation (proteoglycan + col. I and col. II)	[100]
PolyGraft (PLGA + calcium sulfate): softer by increasing porosity	PolyGraft (PLGA + calcium sulfate)	95 MPa vs 150 MPa vs 172 MPa for subchondral bone		–	24 sheep, 7.45 × 10 mm, femoral condyle, 3 and 6 months	Scaffold stiffness affected subchondral bone formation, but at 6 months mechanical properties of regenerated cartilage were the same in stiff and softer Defects in some animals (rabbit) have more healing potential Stiffness change with porosity but porosity itself affects the behaviour, so not the effect of stiffness alone	[101]
CG = 1% collagen I bovine tendon +0.044% chondroitin-6-sulfate + cross-link EDC/NHS → for bone not cartilage	CCP = 0.5% col. I bovine tendon immersed in biphasic calcium phosphate	1 ± 0.27 kPa and 10.3 ± 6 kPa	Separate layers implanted, once cell-free and once cell-seeded for 28 days (TE)	Rat BMMSCs	90 rats, 7 mm calvarial defect, 4 and 8 weeks	CCP promoted more bone healing compared to CG due to mineral phase TE scaffolds prompted M1 pro-inflammatory macrophage-mediated response from the host tissue, while cell-free M2 pro-remodelling macrophage → over-engineering can impair in vivo healing A major barrier to clinical success is capsule formation and avascular necrosis in the centre of scaffold due to barrier to macrophage activity and thus remodelling	[102]

Table 12.2 Properties and performance of scaffolds used in clinical trials

Trade name	Martials	Mechanical properties	Fixation	Clinical	No. of samples	Duration	Findings	References
Cell-free clinical scaffolds								
MaioRegen	Cross-linked self-assembled equine atelocollagen I (30%, 60%) + magnesium–hydroxyapatite (70%, 40%)		Press fit/ no fixation	Clinical	79 patients, 82 defects (41 medial, 26 femoral, 15 trochlea) grades III and IV, age 31 ± 11.3, included previous surgeries and concurrent procedure, size = 3.2 ± 2 cm²	24 months	IKDC: 82.2% improved symptoms at 2 years; 47.4 ± 17.1 to 72.1 ± 18.9 to 76.2 ± 19.6) Tegner: improved from 2.9 ± 2 to 3.8 ± 1.6 to 4.4 ± 1.9 MRI: MOCART score increased significantly Those with OCD better than degenerative lesions	[103]
			Press fit/ no fixation/ fibrin glue	Clinical	27 patients (age 34.9 ± 10.2), 32 defects (7 medial femoral condyles, 5 lateral femoral condyles, 11 patellae, 7 trochleae, and 2 tibial plateaus) defect size 1.5–6 cm²	2 and 5 years	IKDC: 40.0 ± 15.0 to 76.5 ± 14.5 (2 years) and 77.1 ± 18.0 (5 years) Tegner: 1.6 ± 1.1 to 4.0 ± 1.8 (2 years) and 4.1 ± 1.9 (5 years) MRI: significant improvement in MOCART score and subchondral bone status from 2 to 5 years. At 5 years, complete filling of the cartilage in 78.3%; complete integration of the graft in 69.6; the repair tissue surface was intact in 60.9%; and the structure of the repair tissue was homogeneous in 60.9% of the cases No control, combined surgeries	[104, 105]
	Bilayered and tri-layered		Press fit	Pilot in vivo	2 horses: chondral defects in lateral condyle, OC defects in medial condyle	2 and 6 months	Second-look arthroscopy at 2 months: no inflammatory One OC filled with fibrocartilage Tide mark in one OC region Cartilage region filled with fibrocartilage → aim is hyaline not fibro	[106]

–	Clinical	3 complex cases: 46-year-old former athlete, 50-year-old woman and an Olympic level athlete	12 months, 24 months, and 4 years	Due to complexity different approaches were combined Good results in terms of MRI, level of pain and level of activity	[107–109]
Press fit	Clinical	2007: 13 patients (15 defects) 4 medial femoral condyle, 2 lateral femoral condyles, 5 patellae and 4 trochleae. The mean defect size 2.8 cm² (range: 1.5–5.9 cm²), grades IV and V according to ICRS	6 months	Attachment: 13 of 15 sites complete graft attachment, 2 partial detachment Filling and integration: at 6 months, 10 complete filling of lesion, 8 full integration MOCART: 73 ± 19.7 (6 months) Biopsy: 2 patients required additional surgery. Fibrocartilage with no proteoglycans, mostly type I collagen	[110]
Press fit and fibrin glue	Human cadaveric knee	Scaffolds in human cadaveric knee subjected to continuous passive motion (CPM) → significantly improved fixation with fibrin glue compared to only press fitting			[111]
	Clinical	23 patients with OCD, average size of 3.5 ± 1.43 cm3, 9 mm depth defect created, grade III or IV	12 and 24 months	2 failed cases; ICRS score: 50.9 3 ± 20.6 to 76.44 ± 18.03 ($p < 0.0005$) to 82.23 ± 17.36; EQ-VAS: significant increase at 2 years; Tegner, significant increase at 2 years from 2.34 ± 0.64 to 5.60 ± 1.72 but didn't reach pre-op: 80% complete filling of lesion, scaffold at bone still detectable after 2 years	[112]

(continued)

Table 12.2 (continued)

Trade name	Martials	Mechanical properties	Fixation	Clinical	No. of samples	Duration	Findings	References
				Clinical	49 patients (2009–2011), defect size of 4.35 ± 1.26 cm2, 6 mm depth, age, 37 ± 14 years	1 and 2 years	IKDC: significant increase ($p < 0.001$) pre-op to 1 and 2 years, 45.45 ± 19.29 to 70.86 ± 18.08 to 75.42 ± 19.31 Significant improvement ($p < 0.005$) in VAS from 6.69 ± 1.88 to 1.96 ± 2.47 Significant increase ($p < 0.001$) in Tegner score 5 biopsies: 4 from failed grafts Age, type of defect (OCD better outcome) and level of previous activity affected the outcome	[113]
TruFit	Bilayered, semiporous polylactide–co-glycolide copolymer (75:25), polyglycolic acid (PGA) and calcium sulfate (licensed for chondral and OC defects in Europe but only for bone void filling in the US)	Modulus of elasticity = 50–80 MPa based on diameter (7–11 mm) [114]	–	Clinical	26 patients, 2 with OC defect, 15 medial femoral condyle, 4 lateral femoral condyle, 7 trochlea	12, 18, and 24 months	At 12 months, Tegner = 2.4 to 6 ($p = 0.009$), IKDC = 37.7 to 65.1 ($p = 0.004$), KOOS = improved 46 points ($p < 0.001$) At 24 months no significant difference	[115]
			–	Clinical case	2 cases: 43-year-old male three plugs (two 7 mm and one 11 mm) 34-year-old female two 7 mm plugs	Failure at 9 and 20 months	Adverse foreign-body giant cell reaction that was clinically correlated with progressive symptomatic failure at extended follow-up periods of 20 months and 9 months	[116]

–	Clinical	20 patients, 15 patients were followed up during 12 months	6 and 12 months	3 failed due to foreign body giant cells indicated by histology at 12 months; good filling but fibrous vascularised tissue, disorganised ECM and a lot of fibroblasts KOOS = 214.16 ± 92.47 pre-op to 263.94 ± 139.09 ($P < 0.151$) at 6 months to 358.69 ± 112.37 ($P < 0.019$) at 12 months Stable MOCART score over time Modest short-term clinical and MRI outcome	[117]
–	Clinical	1 case: semi-professional footballer, 3 plugs	6 months and 2 years follow-up	Delayed incorporation of plug: at 6 months inflammation and synovitis; at 9 months pain and an effusion; at 12 months improvement, but mild synovitis and lack of subchondral bone filling; at 24 months return to impact sport	[118]
–	Clinical	10 patients, grades III and IV, patella defects mean size of 2.64 cm2	6, 12, and 24 months	Short-time improvement in symptoms specially small lesions, but after 2 years failure in restoring subchondral bone: some degree of plug incorporation failure (replacement of subchondral bone with cyst), although predominantly hyaline but lesions (collapse and fissuring) appeared → MRI 70% reoperating	[119]

(continued)

Table 12.2 (continued)

Trade name	Martials	Mechanical properties	Fixation	Clinical	No. of samples	Duration	Findings	References
			–	Clinical	13 patients, 1.9 ± 0.7 cm, 2 in femoral condyle	12 ± 4 months	After 1 year the bony part of plug is still visible in MRI. The plug did not damage the surrounding surface, and neo-tissue had cartilage-like appearance in MRI	[120]
			–	Clinical	6 patients, cystic subchondral bone in ankle: medial talar dome and distal tibia	12 months	Fibrous rather than hyaline cartilage. Improvement in AOFAS and AOS. MRI showed resolution of bony cyst, but bony plug remained. Authors believe much of the improvement is due to a stable filling of the cystic lesion	[114]
	Platelet-rich fibrin matrix (PRFM) + cartilage fragments with fibrin glue+ hyaluronic acid felt		Press fit	Pilot in vivo	18 goats (36 defects) unilateral trochlear defects (7 mm × 3.5 mm)	1, 3 and 6 months		[121]
ChondraColl	Microfibrillar collagen type I (bovine)/HAp/ collagen II (porcine)/HyA			Commercial	Separate scaffolds in 90 rats for 4 and 8 weeks			[122, 123]
	Collagen I on 3D printed PLA, the interface charged with hyaluronic acid and rhBMP-2		Press fit	Animal	70 rabbits, femoral condyle, defect size of 3 mm × 3 mm	6, 12, 24 weeks	Using morphogen led to better integration of the regenerated tissue with the host cartilage. Bone reconstitution and attachment of subchondral bone to cartilage repair tissue were greater with morphogens; with implant only the repair is good but lacks GAGs deposition	[124]

| Agili-C | Hyaluronic acid+aragonite (cartilage) – aragonite (bone) | Press fit | Clinical | 1 case: 47 years old, 2 cm^2 defect size, 2 scaffolds | 24 months post-op | Articular surface appeared to restore (MRI), patient returned to sports IKDC: 50 to 78 Tegner: 5 to 8 | [92, 125] |
| | | | Animal | 14 goats (28 lesions), 6 mm × 8 mm defect size | 6 months | Combined histological evaluation according to the ICRS II (2010) and O'Driscoll et al. 34 ± 4 n = 7 in channelled scaffolds; integration of newly formed hyaline (due to proteoglycans, col. II and absence of col. I) with surrounding tissue | |

AE adverse events (in KOOS and VAS evaluation systems), *IKDC* International Knee Documentation Committee, *MOCART* Magnetic Resonance Observation of Cartilage Repair Tissue, *ICRS* International Cartilage Repair Society, *VAS* Visual Analogue Scale did not increase new bone formation. No information regarding the quality of neo-cartilage was given in this study [95, 96].

Deng et al. [97] have designed a scaffold comprising gelatin–chondroitin sulfate–sodium hyaluronate (GCH) for cartilage and gelatin–ceramic bovine bone (GCBB) for bone compartments. The mechanical properties (modulus) of this scaffold were superior to the scaffold made from collagen and TCP as reported by Gotterbarm [96]. The bone section showed a modulus of 13.4 MPa and a modulus of 5.7 MPa for cartilage section. The scaffolds were implanted in large patella OCDs of rabbits with and without addition of cells. It was shown that the scaffolds containing cells encouraged formation of hyaline cartilage at 6, 12 and 24 weeks. In comparison with cell-containing scaffolds, the scaffolds alone prompted formation of a fibrous tissue. Although favourable results were observed in terms of the quality of the tissue, the size of the animal model as well as the short duration of the study did not give any indications of their long-term performance in clinical settings (i.e. human).

In another study [100], a composite scaffold of different types of silk with and without incorporation of growth factors was studied in rats for 8 weeks. A satisfactory biointegration of scaffold with the surrounding tissue was observed. The researchers also observed formation of a neo-osteochondral tissue, with mulberry silk favouring osseous differentiation (type I collagen) and non-mulberry directing chondrogenic differentiation (proteoglycan with both type I collagen and type II collagen). This was very encouraging in terms of achieving the correct tissue type; however, again the animal model and duration of study were the limiting factors.

A polylactic acid poly-ε-caprolactone (PLCL) support structure was fabricated using laser micromachining technology and thermal crimping to create a functionally graded open pore network scaffold [126]. This scaffold was evaluated in a rabbit model with and without MSCs. The authors observed no evidence of inflammation or giant cells and concluded that the acellular constructs performed better than cell-seeded constructs with endogenous progenitor cells homing through microtunnels. However, the duration of study was very short (4 weeks), and although the scaffold is branded as an osteochondral scaffold, no information about the subchondral bone regeneration was given.

Recently, a multilayered biomimetic scaffold comprising a bone layer of type I collagen/HAp, an intermediate layer of type I/type II/HAp and a superficial layer of type I/type II collagen/hyaluronic acid was developed and tested in femoral condyles of eight rabbits. After 12 weeks, it was shown that the International Cartilage Repair Society (ICRS) score of joints treated with the scaffold was higher than the non-treated joints, showing a grade II cartilage (nearly normal) compared to grade III (abnormal) in non-treated groups. The level of bone formation was also significantly higher in groups treated with the scaffold [94]. However, looking closely at micro-CT images (Fig. 12.6a), it can be seen that there are areas of incomplete bone regeneration after 12 weeks. The scaffolds were further evaluated in caprine model and compared to a commercial scaffold – TruFit [127]. Complete bone regeneration took place after 12 months (Fig. 12.6).

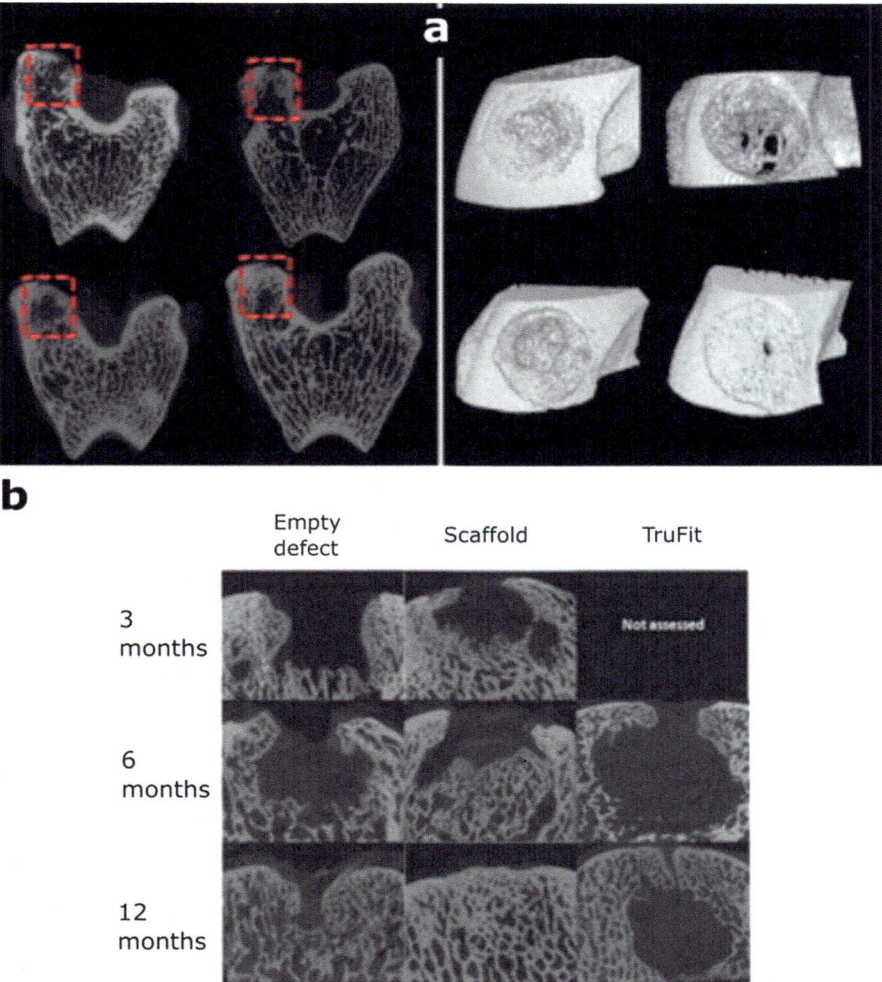

Fig. 12.6 (**a**) Micro-CT analysis at 12 weeks post-surgery showed greater levels of bone repair in the multilayered scaffold group than the empty defect group [94] (rabbit model); (**b**) micro-CT analysis showed improved subchondral bone repair in the multilayered scaffold group in medial femoral condyle [127] (caprine model)

12.3.2   Clinical Performance of Typical Existing OC Scaffolds

A search in Clinicaltrials.gov with the key phrase "osteochondral scaffold" resulted in nine entries, four of which with high relevance to the current review are discussed below.

BiPhasic Cartilage Repair Implants (Exactech, Taiwan) Biphasic osteochondral scaffolds [128] or BiPhasic Cartilage Repair Implants (BiCRI, Exactech

Taiwan) are used in a matrix-assisted autologous chondrocyte implantation (MACI). The biphasic cylindrical plugs (8.5 × 8.5 mm) are made from polylactic-co-glycolic acid (PLGA) and PLGA plus b-tricalcium phosphate (TCP) by particulate leaching method. PLGA, which comprises 1.5 mm of the cylinder height, serves as the cartilage layer, and PLGA+TCP serves as the osseous phase. A reservoir between these two layers is created for double-minced chondrocytes harvested from the graft. The structure of the BiPhasic Cartilage Repair Implant is illustrated in Fig. 12.7. Ten patients with grade III and IV lesions (size less than 20 mm) of the knee femoral condyle were treated with this osteochondral scaffold and assessed using Knee Injury and Osteoarthritis Outcome *Score* (KOOS) and Visual Analogue Scale (VAS) scores at 6 weeks and 3, 6, 12, and 24 months intervals. Magnetic resonance imaging (MRI) scans were performed at 12 months post-operation to assess cartilage formation. No serious adverse events (AE) were reported but KOOS score increased significantly only after 2 years. It was reported that the interface between the graft and the neighbouring native bone was distinguishable by MRI. Second-look arthroscopy after 12 months showed that in 70% of patients the neo-cartilage was well-integrated, while 30% showed incomplete integration and presence of fibrous tissue. Histology of the biopsy samples stained positive with col. II and Alcian blue showed mostly a hyaline cartilage with viable columnar cells. The study later moved into a phase III clinical trial. However, no results have been published yet. As it was observed from MRI scans of patients after 12 months, the boundary of the bony pit was still distinguishable [128].

Chondromimetic (TiGenix, Belgium) Chondromimetic implant is intended to serve as a scaffold for cellular and tissue ingrowth in small osteochondral defect. The plug consists of a chondral layer with collagen and glycosaminoglycans (GAG) and an osseous layer with collagen, GAG and calcium phosphate. The clinical trial was terminated due to slow patient recruitment rate. A preclinical study in goats is available [129] with implantation of scaffolds in nine goats for 26 weeks. It was shown that hyaline-like cartilage (50% in medial femoral condyle (MFC) defects and 83% in lateral femoral condyle (LFC) defects) with mechanical properties close to that of native cartilage was formed. The efficacy was described as a viable alternative to marrow simulation. Getgood and colleagues compared the performance of Chondromimetic with that of TruFit in critical sized defects of MFC and lateral trochlear sulcus [130]. The scaffold was also combined with BMP-7 and rhFGF18 and evaluated in ovine model. Statistical analysis demonstrated significant improvements in gross repair, with respect to the mechanical properties and histological score, over empty defects when Chondromimetic was combined with rhFGF18 [131]. However, subchondral bone regeneration was incomplete, and subchondral cysts were formed in the case of combining with BMP-7, as revealed by histological examination shown in Fig. 12.8.

MaioRegen (Finceramica, Italy) MaioRegen is one of the most investigated scaffolds for OC tissue engineering. It was developed as a tri-layered composite with layers of different compositions representing cartilage (100% type I colla-

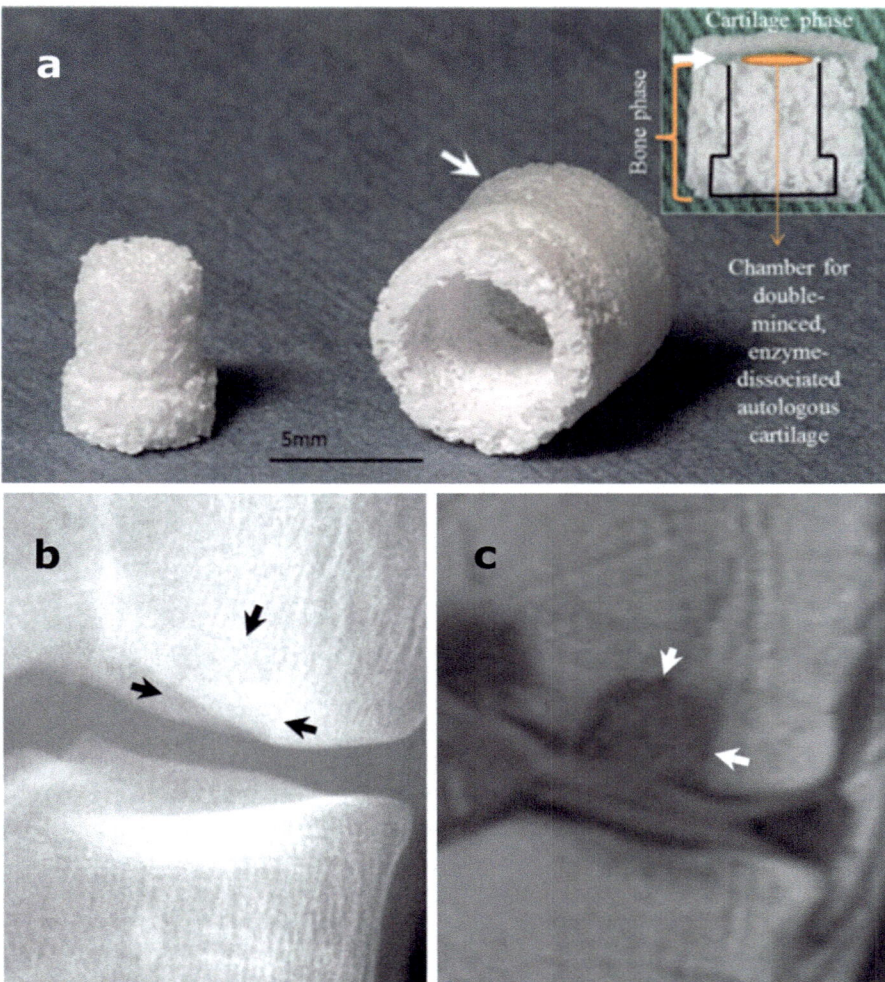

Fig. 12.7 Biphasic cylindrical scaffold. (**a**) The *white arrow* indicates the 1.5 mm PLGA layer for chondral phase, (**b**) at 12 months post-operation, the surgically created pit at the grafted site was filled with radiopaque trabecule. The *black arrows* indicate contour of the bony pit; (**c**) MRI (coronal acquisition with T1 sequence) at 12 months still clearly demarcates the bony pit, which is depicted by a signal different from the native cancellous bone. The *white arrows* indicate interface between graft and host bone. (Adapted from Ref. [128], with permission from Elsevier)

gen), calcified cartilage (60% collagen and 40% HAp) and subchondral bone regions (30% collagen and 70% HAp) [132], as illustrated in Fig. 12.4a. MaioRegen is reported to have a porosity of 45–65% and porosity-dependent Young's modulus in the region of 1.50–6.85 GPa [132]. It has been evaluated in animal studies, as well as several clinical trials cases (see Table 12.2), and it is currently in its phase 4 clinical trials.

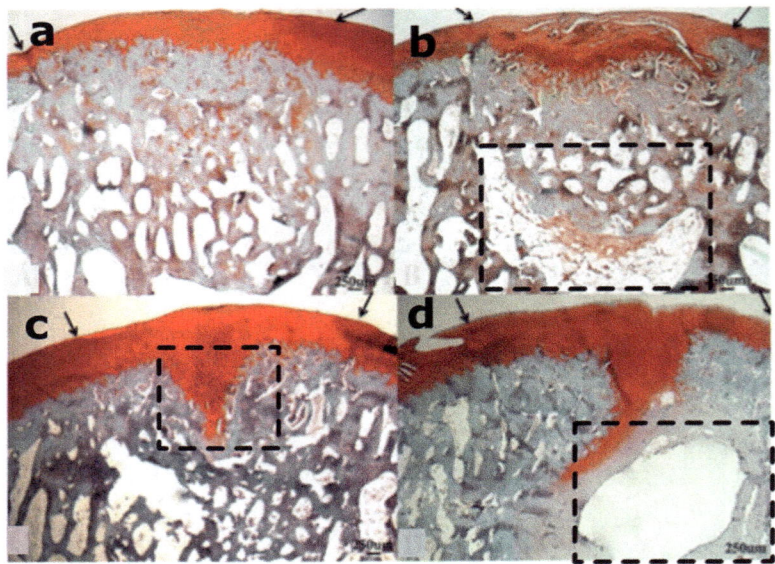

Fig. 12.8 *Histological sections stained with Safranin O of Chondromimetic in a goat model.* (**a**) Empty defect, (**b**) scaffold alone, (**c**) scaffold + rhFGF-18 and (**d**) scaffold + BMP-7. In (**c**) and (**d**), there appears to be a cartilage cleft, with significant proteoglycan staining extending down into the subchondral bone. In (**d**) there is a large subchondral cyst (*black rectangle*). The arrows denote the margins of the defect. (Adapted with permission from Ref. [131] from Springer)

The animal study was conducted on two horses, with tri-layered scaffolds implanted in MFC lesions (10 mm × 8–10 mm). It is reported that newly formed bone and cartilage-like tissues integrated well with the surrounding host tissue after a 6-month period, and formation of tidemark line was described [104]. However, the newly formed cartilage was fibrocartilage, not integrated hyaline cartilage as expected in the articulating joints. The first clinical trial was completed on 13 patients (15 lesions) on MFC, LFC, patellae and trochlear lesions of grades IV and V with a mean defect size of 2.8 cm^2 (range, 1.5–5.9 cm^2). After a 6-month follow-up period, a complete graft attachment in 13 sites and 2 partial detachments were reported. The examination of biopsies, taken from two patients who needed additional surgery, revealed that the newly formed tissue in cartilage compartment was fibrocartilage containing mostly type I collagen and no proteoglycans [110].

Subsequently, the scaffolds were used in another clinical trial on 27 patients (32 defects) with MFC, LFC lesions, patellae, trochleae and tibial plateaus with a lesion size of 1.5–6 cm^2. The 2–5-year follow-up study demonstrated a significant improvement in the International Knee Documentation Committee (IKDC) score. The IKDC score increased to 76.5 ± 14.5 (2 years) and 77.1 ± 18.0 (5 years), respectively, from its initial scores of 40.0 ± 15.0. Similarly, the Tegner score increased to 4.0 ± 1.8 (2 years) and 4.1 ± 1.9 (5 years) from its initial score of 1.6 ± 1.1. MRI examination showed significant improvement in Magnetic Resonance Observation of Cartilage Repair Tissue (MOCART) score and subchondral bone status from 2 to 5 years. At

5 years post-operation, complete filling of the cartilage was observed in 78.3% of the patients, complete integration of the graft in 69.6% of the patients, intact repair tissue surface in 60.9% of the patients, and a homogeneous repair tissue structure was observed in 60.9% of the cases. However, the study did not include negative controls, and due to complex nature of some lesions, combined surgeries took place in some cases [104, 105]. Consequently, 23 patients with osteochondritis dissecans (OCD) grade III or IV with an average lesion size of 3.5 ± 1.43 cm^3 were treated with this scaffold, which resulted in an increase in ICRS and EuroQol Visual Analogue Scale (EQ-VAS) score; Tegner score increased significantly after a 2-year follow-up. However, it did not reach the pre-op levels. MRI data showed that 80% of the lesion was filled and that scaffold at the bone interface was still detectable after 2 years [112]. Furthermore, the scaffolds were used for treatment of large OC knee lesions (size 4.35 ± 1.26 cm^2) in 49 patients, and the results showed a significant increase in IKDC, VAS and Tegner scores after 2 years compared to pre-op. Five biopsies were taken, four of which were from the failed grafts. Macroscopic examination indicated a well integration of the grafts in all specimens. Histological examinations demonstrated that cartilage region was stained positively for type II collagen and MRI scans showed no signs of edema. It was found that the age of the patients, type of defects (OCD better outcome) and level of previous activity affected the clinical outcomes [113]. Similar results were observed in 79 patients with medial, femoral and trochlea grade III and IV lesions; IKDC score showed that 82.2% of patients had improved symptoms at a 2-year follow-up, with Tegner and MOCART scores increasing significantly at 1 and 2 years compared to pre-operation levels. It was again shown that a better clinical result was obtained in OCD lesions compared to degenerative lesions [103].

The MaioRegen scaffold has also been reported with favourable outcome in a series of complex cases, including a 46-year-old athletic patient, a 31-year-old Olympic level athlete and a 50-year-old woman. Due to complexity of lesions in each case, different reconstructive approaches (e.g. alignment correction, autologous OC transplant and microfracturing) were combined. The results after 1- and 2-year follow-up were good in terms of MRI showing signals for hyaline-like cartilage and little/no edema of the subchondral bone. The patients had lower levels of pain and were able to resume their previous level of activity [107–109].

Most recently however, a clinical study on ten patients with OC defects using MaioRegen scaffolds showed a poor OC repair in 1- and 2.5-year postoperative assessments [133]. In this study, two patients were reoperated due to treatment failure and were excluded from the results. None of the patients showed a complete regeneration of the subchondral bone observed from 2.5 years of follow-up CT examinations [133]. This was in contrary to previous studies that reported 62–72% of patients had complete subchondral bone regeneration after 2 years [103]. It was further shown that none of the patients had an intact articular surface or complete integration with the surrounding host tissue, as shown in Fig. 12.9. Since this report the clinical trials have been discontinued in Denmark, and it is advised that MaioRegen be used with caution.

Yan et al. have examined the outcomes of clinical trials on eight patients and reported that subchondral edema, sclerosis and cyst were observed in most of the

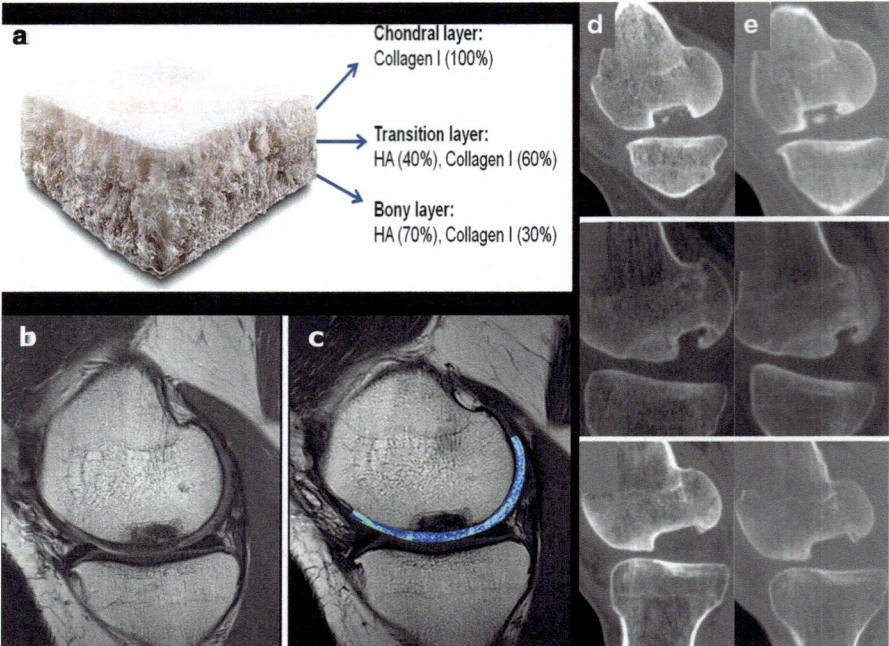

Fig. 12.9 MaioRegen scaffold. (**a**) shows the three layers in MaioRegen scaffold; (**b**) and (**c**) MRI scan of OC lesion repair after 18 months, and bone cysts are observed; (**d**) and (**e**) CT scans of three different patients at 1 and 2.5 years after implantation; a clear cylindrical bone cavity is seen in all of the cases. (Images adapted from Refs. [134–136], with permission from ACS (**a**) and Springer (**b–e**))

cases. A complete integration of the scaffold into the border zone was described; however, T2 mapping data and the zonal T2 index significantly differed in the repair tissue compared to the healthy control cartilage ($P < 0.001$) which indicates a limited quality of the repair cartilage obtained using this scaffold [134] as observed in Fig. 12.9.

TruFit (Smith & Nephew) TruFit is another well-explored scaffold for clinical OC treatment, which has produced controversial results. This biphasic plug was originally developed as a backfill for donor sites after autologous OC transplantation; however, it has been used as a scaffold for OCDs as well. Structurally, the osseous phase of the scaffold consists of calcium sulfate and polyglycolide (PGA) fibres with the chondral phase made of PLGA. Implantation in MFC and lateral trochlear groove showed good histological results of cartilage regeneration after 12 months, followed by encouraging clinical outcomes in small case series. However, later it was observed that in large OCDs the bone incorporation is delayed leading to instability of the graft and failure. This affects the cartilage, both surrounding and opposing the plug, which can be damaged because of the direct articulation or increased contact pressure by this instability, and has led to failure or only

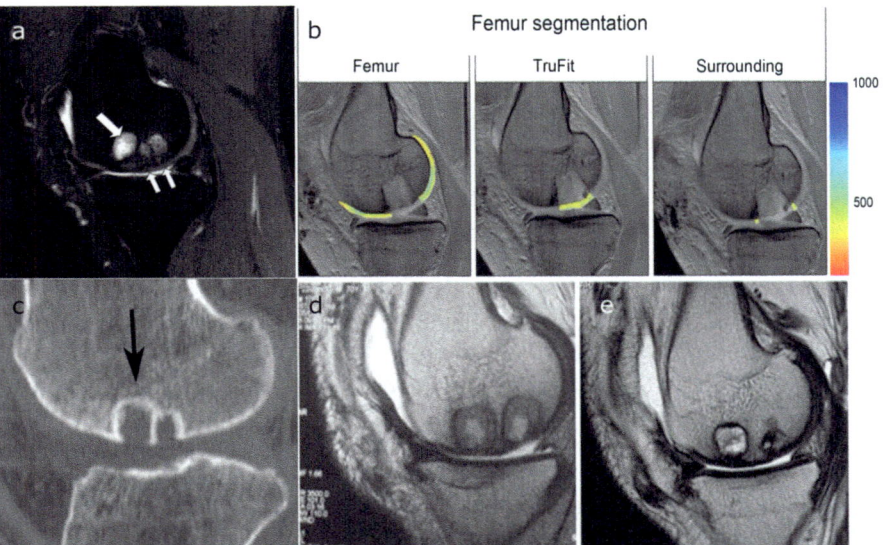

Fig. 12.10 A TruFit plug. (**a**) MRI (1 T) of an OC lesion of the medial femoral condyle treated with a TruFit plug after 2 years. Despite good scaffold integration and partial filling of the chondral layer (*double arrow*), subchondral bone changes were clearly seen (*arrow*) in this MRI sequence [138]; (**b**) sagittal magnetic resonance images of postoperative situation when an osteochondritis dissecans lesion of approximately 1.5 cm was treated using two TruFit BGS plugs. The bone-plug interface is clearly visible and could be used to define the regions of interests. The colour bar represents the calculated dGEMRIC index (T1gd), where a high T1gd (1000 ms) is depicted as blue and a low T1gd as red [120]; (**c**) CT scan failed to show bone ingrowth (*arrow*) [139]; (**d, e**) MRI at 12 and 48 months showing failure of integration [140]

modest results in the large defects [120]. The results of studies using TruFit have been contradictory in short and long terms: up to 12 months, improvements [119, 120] have been reported (although Dhollander [117] reports 20% failure), while after 12 months worsening and 70% failure was observed by Joshi. This coincidence with Verhaegen's study reported in 2015 [137] that no bone ingrowth was observed and instead bone edema and cyst sclerosis were detected ([114, 117, 119]) by MRI and CT examinations, as shown in Fig. 12.10. Further to questionable results, the scaffold was withdrawn from the global market in 2013.

12.3.3   What Have We Learnt from the Clinical Study of OC Scaffolds?

The significance of subchondral bone integration in maintaining a healthy articular cartilage is well established [133, 141], and it was discussed earlier (Sect. 12.2.1.2) from biomechanical and nutritive perspectives.

In general, during physiological loading, a range of mechanical forces are exerted on cartilage such as compressive and shear stress. These external stresses induce hydrostatic pressure in the cartilage and biofluid flow in and out of the cartilage. The function of subchondral bone is to support the overlying cartilage and protect the underlying cancellous bone from high stresses. Changes in the properties of the subchondral bone lead to increased strain generated in the cartilage layer, thereby initiating/maintaining matrix degradation, which can contribute to initiation/progression of OA [29]. Delivery of oxygen and nutrition to different zones of articular cartilage occurs either through diffusion from synovial fluid or through diffusion from micro-blood vessels within subchondral bone depending on the zone of cartilage. Both diffusions are needed to maintain a healthy articular cartilage. Therefore, degeneration of cartilage in the long run is expected if the support from subchondral bone is compromised, pointing to a possible reason for failure of healthy regeneration of cartilage as reported in the clinical studies.

To better understand the relationship between cartilage defect and subchondral bone changes, we conducted a study on osteoarthritic femoral heads collected from total hip replacement operations. The cartilage on the femoral head was graded using Outerbridge classification system. A typical femoral head with cartilage grade is shown in Fig. 12.11a. The specimens were scanned with a micro-CT and peripheral quantitative CT (pQCT) system to determine the subchondral bone structural changes and volumetric bone mineral density (vBMD, mg/cm^3) distribution within the femoral head. Typical micro-CT and pQCT images of OA femoral head are shown in Fig. 12.11b, c, respectively. It was revealed that subchondral bone cysts with varied sizes existed in the subchondral bone, and these cysts are normally observed at regions of greatest cartilage loss. The cysts formation leads to the changes of loading condition in the joint. As a result, the vBMD increased in the subchondral bone that was dependent on the degree of cartilage degeneration. This is in line with what was observed in other studies regarding the advanced stages of OA [1].

The cavities in subchondral bone, which are usually referred to as "subchondral bone cysts", are normally reported in patients with OA. Usually, cysts observed in OA joints are in the range of 0.1–2.5 cm in diameter and appear in multiple. While smaller cysts are detected in the subchondral bone closer to the joint surface, larger cysts typically extend more deeply [142]. There are two main hypotheses about the origin of subchondral bone cysts in OA. The "synovial fluid intrusion" theory suggests that due to cracks in the OC region occurred by repetitive overloading, synovial fluid enters into subchondral bone and leads to formation of these cysts [142], while the "bone contusion" theory suggests that the necrotic lesions in subchondral bone, induced by abnormal mechanical stress and subsequent microcracks, edema and focal bone resorption, are responsible for the cyst formations [143]. Subchondral bone cysts are recognisable in MRI images as areas of fluid signal and in radiographic images as lucent areas with sclerotic rims [143, 144]. The cysts observed in the terminal osteoarthritic cases in our study resembled those of "unfilled bone voids" observed in TruFit [117, 119, 139], MaioRegen and

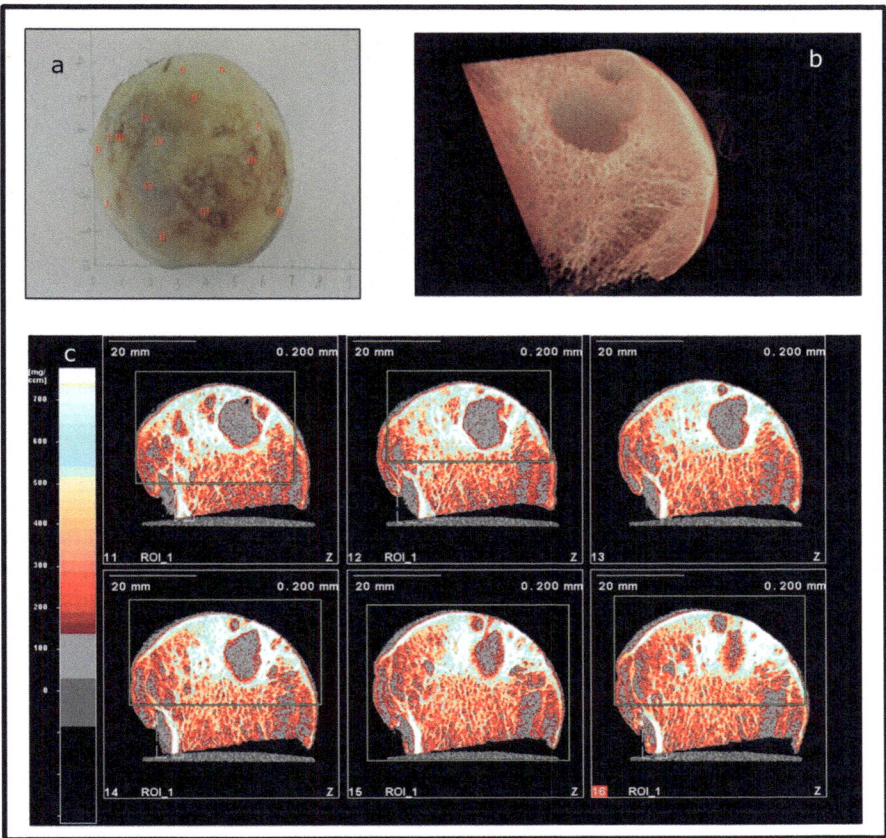

Fig. 12.11 Osteoarthritic femoral head. (**a**) Visual inspection of OA femoral head showing cartilage and subchondral bone in different stages of disease progression, (**b**) micro-CT scan showing subchondral bone cyst and (**c**) pQCT scans showing subchondral bone cysts and bone mineral density distribution surrounding the cysts

Chondromimetic (see Figs. 12.8, 12.9, and 12.10). The "cyst-like" cavities in scaffold developed by seem to resolve after 12 months [94, 127] (see Fig. 12.6).

Based on these results, a paradigm was developed for cartilage defect progression (Fig. 12.12): damage in the articular cartilage changes the loading pattern on the subchondral bone, which leads to bone remodelling (increase in vBMD and formation of cysts). This affects the physical environment supporting the overlying cartilage and hence enhances the progression of cartilage degeneration. The existing hypothetical model for OA pathogenesis looks at the repetitive joint loading, which causes an initial increase in bone remodelling activity, perhaps as to repair the damage caused by the loading. This increased remodelling is associated with increased vascular invasion of the deep layers of the cartilage [145], which allows access to the cartilage by chondrolytic enzymes. This process has several effects,

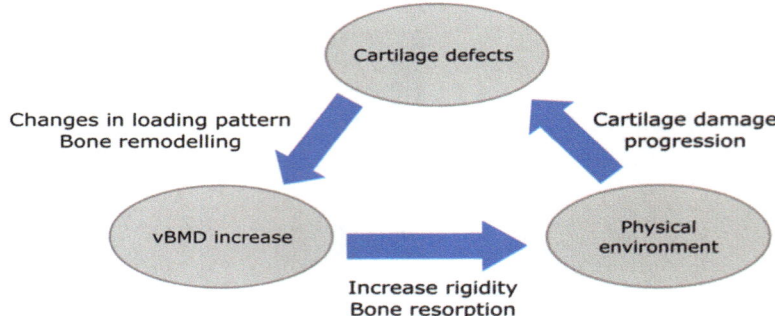

Paradigm for cartilage defect progression

Fig. 12.12 Paradigm for cartilage defect progression

which include secondary synovial thickening with or without secondary inflammation and loss of B cells from the synovial lining with subsequent additional impairment of enzymatic inhibition. The loss of cartilage integrity caused by the loss of aggrecan, which normally maintains cartilage matrix compressive stiffness, will increase the overload of the joint feeding back to an elevation of bone formation as the joint attempts to adapt to the greater loads. Ultimately, this positive feedback loop will promote the continued loss of cartilage integrity, allowing deterioration to progress to clinically evident OA [27].

Although the primacy of the onset of articular cartilage degeneration and OA is still debatable [146], there is no doubt that the subchondral bone plays an important role in the progression of the cartilage degeneration. In fact, there is evidence of communication, biomechanically and biochemically, between cartilage and subchondral bone. Where a healthy homeostatic crosstalk leads to regulated bone remodelling and joint maintenance, a catabolic unhealthy crosstalk leads to dysregulated bone remodelling and progressive damage [36].

This paradigm of OA progression can also be applied to OC tissue regeneration in TE, as it emphasises the importance of subchondral bone in cartilage regeneration and maintenance. If the scaffold for OC defect repair does not provide a mechanically stable compartment in the region of subchondral bone, the forces on the joint will not be transferred to the walls of the defect to stimulate the cells. When bone marrow cells at the periphery of the defect are unable to regenerate bone, the osseous walls of the defect are resorbed which leads to formation of large cavities and collapse of the surrounding cartilage and subchondral bone [101]. These are the cysts that were observed in OA bones and in the cases where the bones were treated with the above-mentioned scaffolds. The variation in subchondral bone stiffness (e.g. due to existence of cysts) results in interfacial stresses with subchondral bone, as well as varied stress in overlying cartilage when subjected to the dynamic loading. As such, without the subchondral bone healing, the elevated contact stress gradients in normal cartilage near the defects may inhibit normal repair [147].

12.3.4 Improved Biomechanical Fixation Enhances Cartilage Fill

As empirically observed in the commercial scaffolds, the dominant factor in scaffold failing to support healthy cartilage regeneration and restore the joint function satisfactorily seems to be the insufficient bone ingrowth and integration with the host tissues. Without a stable biomechanical support, the newly formed cartilage would "collapse". The "collapsed" cartilage would not be subjected to mechanical stimulation, which is a critical factor for healthy hyaline cartilage formation. As a result, poor cartilage fill and associated fibrocartilaginous repair rather than the hyaline cartilage, as well as poor OC repair, are often observed in the clinical trials of a few commercially available OC scaffolds as previously reported.

The authors believed that providing an appropriate physical environment (that includes the generation of an appropriate biomechanical environment and hydrostatic pressure) to support cartilage healing is critical for cartilage fill and hyaline cartilage formation. The researchers at UCL have recently developed a novel biomimetic OC scaffold based on a "sandwich" composite system comprising titanium, PLGA and collagen matrices. The titanium and PLGA supporting frameworks are fabricated by a 3D rapid prototyping technique, and the porous matrix is filled with cross-linked type I collagen which is spatially graded to form a structural and compositional scaffold.

The biomimetic OC scaffold has been evaluated in sheep condyle model. The in vivo sheep study demonstrated that the new bone growth into the titanium matrix at the bone section provided a strong mechanical fixation at 12 weeks post-operation. This provides a strong support to the overlying cartilage layer leading to the improved cartilage fill compared to a commercially available collagen-/hydroxyapatite-based OC scaffold, as demonstrated in Fig. 12.13.

The scaffold has also been tested in a clinical dog shoulder model where an OCD had occurred due to natural development of OA in the dog. A 10 mm × 10 mm biomimetic scaffold was implanted in the shoulder defect. The 3-month follow-up arthroscopic examination revealed the cartilage had regenerated well, matching the curvature of the joint perfectly. Recent reports from the dog owner suggested the dog's shoulder function has recovered completely. A glimpse of how this scaffold will perform has been given, with promising results, by Professor Noel Fitzpatrick of the Channel 4 TV series *The Supervet*.

This biomimetic OC scaffold has the strength needed to bear the physical load of the joints, and its biomechanical structure encourages consistent cartilage fill and a smooth articular surface. It has the potential to address the unmet clinical need for repair of large OCDs. This functional biomimetic OC scaffold bridges the gap between small OCD treatment and joint replacement. It is hoped that it will provide clinicians with a viable treatment option in situations where the disease has progressed beyond a small defect but where a full joint replacement could still be avoided. This would lead to tangible and clinically relevant results in a one-step surgical procedure for the treatment of large cartilage and OCDs, relieving pain and improving quality of life by keeping people active.

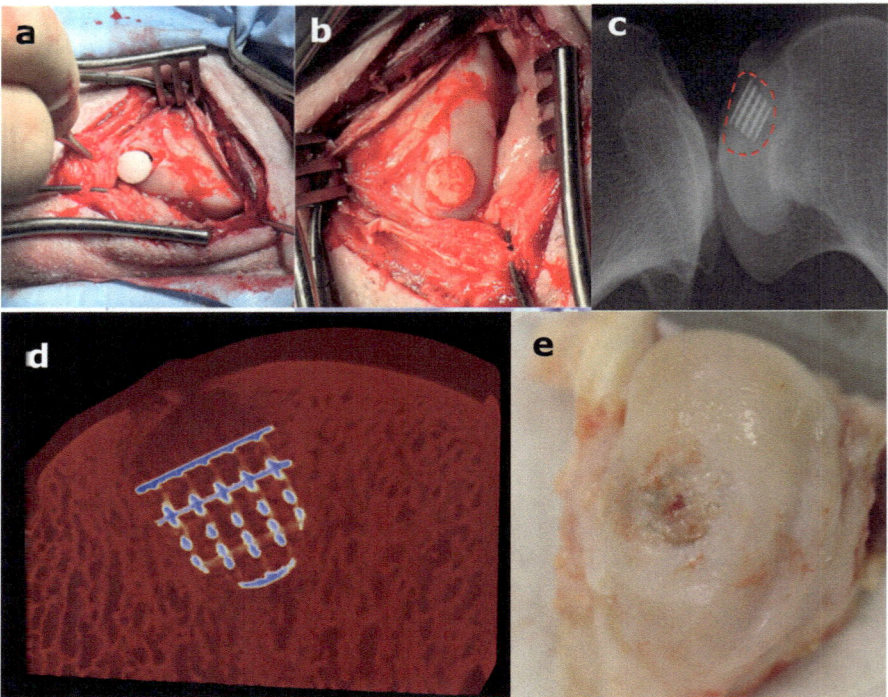

Fig. 12.13 Sheep condyle model was used to evaluate the in vivo performance of the scaffold. (**a**) and (**b**) show the scaffold during and after implantation, respectively; (**c**) shows the X-ray image of the joint after surgery; (**d**) shows the micro-CT image of the scaffold (*red line*) inside the condyle after 12 weeks; and (**e**) shows the regenerated cartilage after 12 weeks

12.4 Perspective Summary

OCDs, typically derived by traumatic injuries or OA, involve articular cartilage and associated subchondral bone. These defects are characterised by unbalanced degeneration and regeneration of the articular cartilage and bone where the intrinsic repair mechanisms are insufficient. Stopping or delaying progression of OCDs would have significant impact in health care.

The treatment of cartilage and OC defects remains a challenge because treatments to date have failed to achieve a complete restoration of the joint cartilage surface and its properties. Many new technologies, such as OC tissue engineering and stem cell therapies, have been studied and applied to the repair of OC defects. The goal of a tissue engineering approach is to repair the defect in the joint and restore its function in order to delay or remove the need for a joint replacement.

Numerous OC scaffolds have been developed by different research groups around the world, and there are many commercially available products. However, few of these products promote satisfactory durable regeneration of large OC defects.

The authors believe that the subchondral bone and adjacent cartilage form a functional unit. OC scaffold that simultaneously support the regeneration of cartilage and subchondral bone is critical for the successful repair of cartilage and OC defects. Lessons learnt from the clinical trials and animal studies suggest that an improved biomechanical fixation of the OC scaffold would provide an appropriate physical environment for healthy growth of the overlying cartilage.

Development of a functionally biomimetic OC scaffold which will bridge the gap between small OC defect treatment and joint replacement is still a grand challenge. However, with the advancing of OC scaffold biotechnology, it is hoped that, in the near future, a novel OC scaffold with improved capability for biomechanical and biological fixation would lead to tangible and clinically relevant results in a one-step surgical procedure for the treatment of large OCDs, relieving pain and improving quality of life by keeping people active.

References

1. Goldring SR, Goldring MB (2016) Changes in the osteochondral unit during osteoarthritis: structure, function and cartilage-bone crosstalk. Nat Rev Rheumatol 12(11):632
2. Temenoff JS, Mikos AG (2000) Review: tissue engineering for regeneration of articular cartilage. Biomaterials 21:431–440
3. Yousefi A-M, Hoque ME, Prasad RGSV, Uth N (2015) Current strategies in multiphasic scaffold design for osteochondral tissue engineering: a review. J Biomed Mater Res A 103:2460–2481
4. Das SK, Farooqi A (2008) Osteoarthritis. Best Pract Res Clin Rheumatol 22:657–675
5. Madry H, Van Dijk CN, Mueller-Gerbl M (2010) The basic science of the subchondral bone. Knee Surg Sports Traumatol Arthrosc 18:419–433
6. Martel-Pelletier J, Boileau C, Pelletier J-P, Roughley PJ (2008) Cartilage in normal and osteoarthritis conditions. Best Pract Res Clin Rheumatol 22:351–384
7. National Joint Registry for England, W. A. N. I (2014) 11th annual report
8. Zhang W, Moskowitz RW, Nuki G, Abramson S, Altman RD, Arden N, Bierma-Zeinstra S, Brandt KD, Croft P, Doherty M, Dougados M, Hochberg M, Hunter DJ, Kwoh K, Lohmander LS, Tugwell P (2007) OARSI recommendations for the management of hip and knee osteoarthritis, Part I: Critical appraisal of existing treatment guidelines and systematic review of current research evidence. Osteoarthr Cartil 15:981–1000
9. Lories RJ, Luyten FP (2011) The bone-cartilage unit in osteoarthritis. Nat Rev Rheumatol 7:43–49
10. Bijlsma JWJ, Berenbaum F, Lafeber FPJG (2011) Osteoarthritis: an update with relevance for clinical practice. Lancet 377:2115–2126
11. Walker-Bone K, Javaid K, Arden N, Cooper C (2000) Medical management of osteoarthritis. BMJ 321:936–940
12. Lee CR, Breinan HA, Nehrer S, Spector M (2000) Articular cartilage chondrocytes in type I and type II collagen-GAG matrices exhibit contractile behavior in vitro. Tissue Eng 6:555–565
13. Melton JT, Wilson AJ, Chapman-Sheath P, Cossey AJ (2010) TruFit CB® bone plug: chondral repair, scaffold design, surgical technique and early experiences. Expert Rev Med Devices 7:333–341
14. Angele P, Niemeyer P, Steinwachs M, Filardo G, Gomoll AH, Kon E, Zellner J, Madry H (2016) Chondral and osteochondral operative treatment in early osteoarthritis. Knee Surg Sports Traumatol Arthrosc 24:1743–1752

15. Gomoll AH, Madry H, Knutsen G, Van Dijk N, Seil R, Brittberg M, Kon E (2010) The subchondral bone in articular cartilage repair: current problems in the surgical management. Knee Surg Sports Traumatol Arthrosc: Off J of the ESSKA 18:434–447

15. Hangody L, Dobos J, Baló E, Pánics G, Hangody LR, Berkes I (2010) Clinical experiences with autologous osteochondral mosaicplasty in an athletic population. Am J Sports Med 38:1125–1133

17. Gomoll AH, Filardo G, Almqvist FK, Bugbee WD, Jelic M, Monllau JC, Puddu G, Rodkey WG, Verdonk P, Verdonk R, Zaffagnini S, Marcacci M (2012) Surgical treatment for early osteoarthritis. Part II: allografts and concurrent procedures. Knee Surg Sports Traumatol Arthrosc 20:468–486

13. O'shea TM, Miao X (2008) Bilayered scaffolds for osteochondral tissue engineering. Tissue Eng Part B Rev 14:447–464

19. Martin I, Wendt D, Heberer M (2004) The role of bioreactors in tissue engineering. Trends Biotechnol 22:80–86

20. Vunjak-Novakovic G, Meinel L, Altman G, Kaplan D (2005) Bioreactor cultivation of osteochondral grafts. Orthod Craniofac Res 8:209–218

21. Schaefer D, Martin I, Shastri P, Padera RF, Langer R, Freed LE, Vunjak-Novakovic G (2000) In vitro generation of osteochondral composites. Biomaterials 21:2599–2606

22. Mikos AG, Herring SW, Ochareon P, Elisseeff J, Lu HH, Kandel R, Schoen FJ, Toner M, Mooney D, Atala A, Van Dyke ME, Kaplan D, Vunjak-Novakovic G (2006) Engineering complex tissues. Tissue Eng 12:3307

23. Nukavarapu SP, Dorcemus DL (2013) Osteochondral tissue engineering: current strategies and challenges. Biotechnol Adv 31:706–721

24. Buckwalter JA, Mankin HJ (1997) Articular cartilage. Part II: Degeneration and osteoarthrosis, repair, regeneration and transplantation. J Bone Joint Surg (Am Vol) 79A:612–632

25. Newman AP (1998) Articular cartilage repair. Am J Sports Med 26:309–324

26. Lu HH, Subramony SD, Boushell MK, Zhang X (2010) Tissue engineering strategies for the regeneration of orthopedic interfaces. Ann Biomed Eng 38:2142–2154

27. Burr DB, Gallant MA (2012) Bone remodelling in osteoarthritis. Nat Rev Rheumatol 8:665–673

28. Flachsmann ER, Broom ND, Oloyede A (1995) A biomechanical investigation of unconstrained shear failure of the osteochondral region under impact loading. Clin Biomech 10:156–165

29. McMahon LA, O'Brien FJ, Prendergast PJ (2008) Biomechanics and mechanobiology in osteochondral tissues. Regen Med 3:743–759

30. Martin I, Miot S, Barbero A, Jakob M, Wendt D (2007) Osteochondral tissue engineering. J Biomech 40:750–765

31. Layton MW, Goldstein SA, Goulet RW, Feldkamp LA, Kubinski DJ, Bole GG (1988) Examination of subchondral bone architecture in experimental osteoarthritis by microscopic computed axial tomography. Arthritis Rheum 31:1400–1405

32. Eckstein F, Muller-Gerbl M, Putz R (1992) Distribution of subchondral bone density and cartilage thickness in the human patella. J Anat 180(Pt 3):425–433

33. Shepherd DET, Seedhom BB (1999) Thickness of human articular cartilage in joints of the lower limb. Ann Rheum Dis 58:27–34

34. Arkill KP, Winlove CP (2008) Solute transport in the deep and calcified zones of articular cartilage. Osteoarthr Cartil 16:708–714

35. Berry JL, Thaeler-Oberdoerster DA, Greenwald AS (1986) Subchondral pathways to the superior surface of the human talus. Foot Ankle 7:2–9

36. Findlay DM, Kuliwaba JS (2016) Bone-cartilage crosstalk: a conversation for understanding osteoarthritis. Bone Res 4:16028

37. Imhof H, Sulzbacher I, Grampp S, Czerny C, Youssefzadeh S, Kainberger F (2000) Subchondral bone and cartilage disease: a rediscovered functional unit. Investig Radiol 35:581–588

38. Malinin T, Ouellette EA (2000) Articular cartilage nutrition is mediated by subchondral bone: a long-term autograft study in baboons. Osteoarthr Cartil 8:483–491

39. Freed LE, Vunjak-Novakovic G (1998) Culture of organized cell communities. Adv Drug Deliv Rev 33:15–30
40. Sachlos E, Czernuszka JT (2003) Making tissue engineering scaffolds work: review on the application of solid freeform fabrication technology to the production of tissue engineering scaffolds. Eur Cell Mater 5:29–40
41. Tamaddon M, Czernuszka JT (2013) The need for hierarchical scaffolds in bone tissue engineering. Hard tissue 2:4–37
42. Yahyouche A, Xia Z, Triffitt JT, Czernuszka JT, Clover AJP (2013) Improved angiogenic cell penetration in vitro and in vivo in collagen scaffolds with internal channels. J Mater Sci Mater Med 24:1571–1580
43. Guilak F, Cohen DM, Estes BT, Gimble JM, Liedtke W, Chen CS (2009) Control of stem cell fate by physical interactions with the extracellular matrix. Cell Stem Cell 5:17–26
44. Jingjing Z, Yingnan W, Tanushree T, Eng Hin L, Zigang G, Zheng Y (2014) The influence of scaffold microstructure on chondrogenic differentiation of mesenchymal stem cells. Biomed Mater 9:035011
45. Singh M, Berkland C, Detamore MS (2008) Strategies and applications for incorporating physical and chemical signal gradients in tissue engineering. Tissue Eng Part B Rev 14:341–366
46. Anselme K (2000) Osteoblast adhesion on biomaterials. Biomaterials 21:667–681
47. Bacakova L, Filova E, Rypacek F, Svorcik V, Stary V (2004) Cell adhesion on artificial materials for tissue engineering. Physiol Res 53:S35–S45
48. Martino S, D'angelo F, Armentano I, Kenny JM, Orlacchio A (2012) Stem cell-biomaterial interactions for regenerative medicine. Syst Biol Biomed Innov 30:338–351
49. Zhao C, Tan A, Pastorin G, Ho HK (2013) Nanomaterial scaffolds for stem cell proliferation and differentiation in tissue engineering. Biotechnol Adv 31:654–668
50. Bacakova L, Filova E, Parizek M, Ruml T, Svorcik V (2011) Modulation of cell adhesion, proliferation and differentiation on materials designed for body implants. Biotechnol Adv 29:739–767
51. Boyan BD, Hummert TW, Dean DD, Schwartz Z (1996) Role of material surfaces in regulating bone and cartilage cell response. Biomaterials 17:137–146
52. Chen M, Le DQS, Baatrup A, Nygaard JV, Hein S, Bjerre L, Kassem M, Zou X, Bünger C (2011b) Self-assembled composite matrix in a hierarchical 3-D scaffold for bone tissue engineering. Acta Biomater 7:2244–2255
53. Malafaya PB, Silva GA, Reis RL (2007) Natural–origin polymers as carriers and scaffolds for biomolecules and cell delivery in tissue engineering applications. Adv Drug Deliv Rev 59:207–233
54. O'Brien FJ (2011) Biomaterials & scaffolds for tissue engineering. Mater Today 14:88–95
55. Hutmacher DW (2000) Scaffolds in tissue engineering bone and cartilage. Biomaterials 21:2529–2543
56. Amjad Z, Koutsoukos P, Tomson MB, Nancollas GH (1978) Growth of hydroxyapatite from solution – new constant composition method. J Dent Res 57:909–909
57. Porter JR, Ruckh TT, Popat KC (2009) Bone tissue engineering: a review in bone biomimetics and drug delivery strategies. Biotechnol Prog 25:1539–1560
58. Leng P, Wang Y, Zhang H (2013) Repair of large osteochondral defects with mix-mosaicplasty in a Goat Model. Orthopedics 36:e331
59. Livingston TL, Gordon S, Archambault M (2003) Mesenchymal stem cells combined with biphasic calcium phosphate ceramics promote bone regeneration. J Mater Sci Mater Med 14:211-218
60. Murphy CM, O'brien FJ, Little DG, Schindeler A (2013) Cell-scaffold interactions in the bone tissue engineering triad. Eur Cell Mater 26:120–132
61. Liu Y, Lim J, Teoh S-H (2013) Review: development of clinically relevant scaffolds for vascularised bone tissue engineering. Biotechnol Adv 31:688–705

62. Kuboki Y, Takita H, Kobayashi D, Tsuruga E, Inoue M, Murata M, Nagai N, Dohi Y, Ohgushi H (1998) BMP-Induced osteogenesis on the surface of hydroxyapatite with geometrically feasible and nonfeasible structures: Topology of osteogenesis. J Biomed Mater Res 39:190–199
63. Delise AM, Fischer L, Tuan RS (2000) Cellular interactions and signaling in cartilage development. Osteoarthr Cartil 8:309–334
64. Goldring MB, Tsuchimochi K, Ijiri K (2006) The control of chondrogenesis. J Cell Biochem 97:33–44
65. Folkman J, Moscona A (1978) Role of cell shape in growth control. Nature 273:345–349
66. Paluch E, Heisenberg C-P (2009) Biology and physics of cell shape changes in development. Curr Biol 19:R790–R799
67. Zanetti NC, Solursh M (1984) Induction of chondrogenesis in limb mesenchymal cultures by disruption of the actin cytoskeleton. J Cell Biol 99:115–123
68. Mcbride SH, Falls T, Knothe Tate ML (2008) Modulation of stem cell shape and fate B: mechanical modulation of cell shape and gene expression. Tissue Eng A 14:1573–1580
69. Chen CS, Alonso JL, Ostuni E, Whitesides GM, Ingber DE (2003) Cell shape provides global control of focal adhesion assembly. Biochem Biophys Res Commun 307:355–361
70. Hutmacher DW (2001) Scaffold design and fabrication technologies for engineering tissues – state of the art and future perspectives. J Biomater Sci Polym Ed 12:107–124
71. Davidenko N, Gibb T, Schuster C, Best SM, Campbell JJ, Watson CJ, Cameron RE (2012) Biomimetic collagen scaffolds with anisotropic pore architecture. Acta Biomater 8:667–676
72. Nowicki MA, Castro NJ, Plesniak MW, Zhang LG (2016) 3D printing of novel osteochondral scaffolds with graded microstructure. Nanotechnology 27:414001
73. Do AV, Khorsand B, Geary SM, Salem AK (2015) 3D printing of scaffolds for tissue regeneration applications. Adv Healthc Mater 4:1742–1762
74. O'brien CM, Holmes B, Faucett S, Zhang LG (2015) Three-dimensional printing of nanomaterial scaffolds for complex tissue regeneration. Tissue Eng Part B Rev 21:103–114
75. Liu CZ, Xia ZD, Han ZW, Hulley PA, Triffitt JT, Czernuszka JT (2008) Novel 3D collagen scaffolds fabricated by indirect printing technique for tissue engineering. J Biomed Mater Res B Appl Biomater 85B:519–528
76. Armstrong JPK, Burke M, Carter BM, Davis SA, Perriman AW (2016) 3D bioprinting using a templated porous bioink. Adv Healthc Mater 5:1724–1730
77. Gao G, Yonezawa T, Hubbell K, Dai G, Cui X (2015) Inkjet-bioprinted acrylated peptides and PEG hydrogel with human mesenchymal stem cells promote robust bone and cartilage formation with minimal printhead clogging. Biotechnol J 10:1568–1577
78. Cao T, Ho KH, Teoh SH (2003) Scaffold design and in vitro study of osteochondral coculture in a three-dimensional porous polycaprolactone scaffold fabricated by fused deposition modeling. Tissue Eng 9(Suppl 1):S103–S112
79. Shishkovsky IV, Volova LT, Kuznetsov MV, Morozov YG, Parkin IP (2008) Porous biocompatible implants and tissue scaffolds synthesized by selective laser sintering from Ti and NiTi. J Mater Chem 18:1309–1317
80. Discher DE, Janmey P, Wang YL (2005) Tissue cells feel and respond to the stiffness of their substrate. Science 310:1139–1143
81. Engler AJ, Sen S, Sweeney HL, Discher DE (2006) Matrix elasticity directs stem cell lineage specification. Cell 126:677–689
82. Freyman TM, Yannas IV, Pek YS, Yokoo R, Gibson LJ (2001) Micromechanics of fibroblast contraction of a collagen-GAG matrix. Exp Cell Res 269:140–153
83. Reinhart-King CA, Dembo M, Hammer DA (2008) Cell-cell mechanical communication through compliant substrates. Biophys J 95:6044–6051
84. Ashby MF (2006) The properties of foams and lattices. Philos Trans R Soc A Math Phys Eng Sci 364:15–30
85. Harley BA, Leung JH, Silva ECCM, Gibson LJ (2007) Mechanical characterization of collagen-glycosaminoglycan scaffolds. Acta Biomater 3:463–474

86. Murphy CM, Matsiko A, Haugh MG, Gleeson JP, O'brien FJ (2012) Mesenchymal stem cell fate is regulated by the composition and mechanical properties of collagen-glycosaminogly-can scaffolds. Special Issue Tissue Eng 11:53–62

87. Wendt D, Jakob M, Martin I (2005) Bioreactor-based engineering of osteochondral grafts: from model systems to tissue manufacturing. J Biosci Bioeng 100:489–494

88. Braccini A, Wendt D, Jaquiery C, Jakob M, Heberer M, Kenins L, Wodnar-Filipowicz A, Quarto R, Martin I (2005) Three-dimensional perfusion culture of human bone marrow cells and generation of osteoinductive grafts. Stem Cells 23:1066–1072

89. Davisson T, Sah RL, Ratcliffe A (2002) Perfusion increases cell content and matrix synthesis in chondrocyte three-dimensional cultures. Tissue Eng 8:807–816

90. Chang C-H, Lin F-H, Lin C-C, Chou C-H, Liu H-C (2004) Cartilage tissue engineering on the surface of a novel gelatin–calcium-phosphate biphasic scaffold in a double-chamber bio-reactor. J Biomed Mater Res B Appl Biomater 71B:313–321

91. Shimomura K, Moriguchi Y, Murawski CD, Yoshikawa H, Nakamura N (2014) Osteochondral tissue engineering with biphasic scaffold: current strategies and techniques. Tissue Eng Part B Rev 20(5):468–476

92. Kon E, Filardo G, Robinson D, Eisman JA, Levy A, Zaslav K, Shani J, Altschuler N (2014d) Osteochondral regeneration using a novel aragonite-hyaluronate bi-phasic scaffold in a goat model. Knee Surg Sports Traumatol Arthrosc 22:1452–1464

93. Levingstone TJ, Matsiko A, Dickson GR, O'brien FJ, Gleeson JP (2014b) A biomi-metic multi-layered collagen-based scaffold for osteochondral repair. Acta Biomater 10:1996–2004

94. Levingstone TJ, Thompson E, Matsiko A, Schepens A, Gleeson JP, O'brien FJ (2016b) Multi-layered collagen-based scaffolds for osteochondral defect repair in rabbits. Acta Biomater 32:149–160

95. Gotterbarm T, Breusch SJ, Jung M, Streich N, Wiltfang J, Berardi Vilei S, Richter W, Nitsche T (2014) Complete subchondral bone defect regeneration with a tricalcium phosphate col-lagen implant and osteoinductive growth factors: A randomized controlled study in Göttingen minipigs. J Biomed Mater Res B Appl Biomater 102:933–942

96. Gotterbarm T, Richter W, Jung M, Berardi Vilei S, Mainil-Varlet P, Yamashita T, Breusch SJ (2006) An in vivo study of a growth-factor enhanced, cell free, two-layered collagen–trical-cium phosphate in deep osteochondral defects. Biomaterials 27:3387–3395

97. Deng T, Lv J, Pang J, Liu B, Ke J (2014) Construction of tissue-engineered osteochondral composites and repair of large joint defects in rabbit. J Tissue Eng Regen Med 8:546–556

98. Chen J, Chen H, Li P, Diao H, Zhu S, Dong L, Wang R, Guo T, Zhao J, Zhang J (2011a) Simultaneous regeneration of articular cartilage and subchondral bone in vivo using MSCs induced by a spatially controlled gene delivery system in bilayered integrated scaffolds. Biomaterials 32:4793–4805

99. Petersen JP, Ueblacker P, Goepfert C, Adamietz P, Baumbach K, Stork A, Rueger JM, Poertner R, Amling M, Meenen NM (2008) Long term results after implantation of tissue engineered cartilage for the treatment of osteochondral lesions in a minipig model. J Mater Sci Mater Med 19:2029–2038

100. Saha S, Kundu B, Kirkham J, Wood D, Kundu SC, Yang XB (2013) Osteochondral tissue engineering in vivo: a comparative study using layered silk fibroin scaffolds from Mulberry and Nonmulberry silkworms. PLoS One 8:e80004

101. Schlichting K, Schell H, Kleemann RU, Schill A, Weiler A, Duda GN, Epari DR (2008) Influence of scaffold stiffness on subchondral bone and subsequent cartilage regeneration in an Ovine model of osteochondral defect healing. Am J Sports Med 36:2379–2391

102. Lyons FG, Al-Munajjed AA, Kieran SM, Toner ME, Murphy CM, Duffy GP, O'Brien FJ (2010) The healing of bony defects by cell-free collagen-based scaffolds compared to stem cell-seeded tissue engineered constructs. Biomaterials 31:9232–9243

103. Kon E, Filardo G, Perdisa F, Di Martino A, Busacca M, Balboni F, Sessa A, Marcacci M (2014c) A one-step treatment for chondral and osteochondral knee defects: clinical results

of a biomimetic scaffold implantation at 2 years of follow-up. J Mater Sci Mater Med 25:2437–2444

104. Kon E, Filardo G, Di Martino A, Busacca M, Moio A, Perdisa F, Marcacci M (2014b) Clinical results and MRI evolution of a nano-composite multilayered biomaterial for osteochondral regeneration at 5 years. Am J Sports Med 42:158–165

105. Filardo G, Kon E, Di Martino A, Busacca M, Altadonna G, Marcacci M (2013) Treatment of knee osteochondritis dissecans with a cell-free biomimetic osteochondral scaffold: clinical and imaging evaluation at 2-year follow-up. Am J Sports Med 41:1786–1793

106. Kon E, Mutini A, Arcangeli E, Delcogliano M, Filardo G, Nicoli Aldini N, Pressato D, Quarto R, Zaffagnini S, Marcacci M (2010b) Novel nanostructured scaffold for osteochondral regeneration: pilot study in horses. J Tissue Eng Regen Med 4:300–308

107. Filardo G, Di Martino A, Kon E, Delcogliano M, Marcacci M (2012) Midterm results of a combined biological and mechanical approach for the treatment of a complex knee lesion. Cartilage 3:288–292

108. Kon E, Delcogliano M, Filardo G, Altadonna G, Marcacci M (2009) Novel nano-composite multi-layered biomaterial for the treatment of multifocal degenerative cartilage lesions. Knee Surg Sports Traumatol Arthrosc 17:1312–1315

109. Perdisa F, Filardo G, Di Matteo B, Di Martino A, Marcacci M (2014) Biological knee reconstruction: a case report of an Olympic athlete. Eur Rev Med Pharmacol Sci 18:76–80

110. Kon E, Delcogliano M, Filardo G, Pressato D, Busacca M, Grigolo B, Desando G, Marcacci M (2010a) A novel nano-composite multi-layered biomaterial for treatment of osteochondral lesions: technique note and an early stability pilot clinical trial. Injury 41:693–701

111. Filardo G, Drobnic M, Perdisa F, Kon E, Hribernik M, Marcacci M (2014) Fibrin glue improves osteochondral scaffold fixation: study on the human cadaveric knee exposed to continuous passive motion. Osteoarthr Cartil 22:557–565

112. Delcogliano M, Menghi A, Placella G, Speziali A, Cerulli G, Carimati G, Pasqualotto S, Berruto M (2014) Treatment of osteochondritis dissecans of the knee with a biomimetic scaffold. A prospective multicenter study. Joints 2:102–108

113. Berruto M, Delcogliano M, De Caro F, Carimati G, Uboldi F, Ferrua P, Ziveri G, De Biase CF (2014) Treatment of large knee osteochondral lesions with a biomimetic scaffold: results of a multicenter study of 49 patients at 2-year follow-up. Am J Sports Med 42:1607–1617

114. Pearce CJ, Gartner LE, Mitchell A, Calder JD (2012) Synthetic osteochondral grafting of ankle osteochondral lesions. Foot Ankle Surg 18:114–118

115. Spalding T, Carey-Smith R, Carmont M, Dunn K (2009) TruFit plugs for articular cartilage repair in the knee: 2 year experience, results and MRI appearances (SS-59). Arthroscopy: J Arthrosc Relat Surg 25:e32–e33

116. Sgaglione NA, Florence AS (2009) Bone graft substitute plug failure with giant cell reaction in the treatment of osteochondral lesions of the distal femur: a report of 2 cases with operative revision. Arthroscopy: J Arthrosc Relat Surg 25:815–819

117. Dhollander AAM, Liekens K, Almqvist KF, Verdonk R, Lambrecht S, Elewaut D, Verbruggen G, Verdonk PCM (2012) A pilot study of the use of an osteochondral scaffold plug for cartilage repair in the knee and how to deal with early clinical failures. Arthroscopy: J Arthrosc Relat Surg 28:225–233

118. Carmont MR, Carey-Smith R, Saithna A, Dhillon M, Thompson P, Spalding T (2009) Delayed incorporation of a TruFit plug: perseverance is recommended. Arthroscopy: J Arthrosc Relat Surg 25:810–814

119. Joshi N, Reverte-Vinaixa M, Díaz-Ferreiro EW, Domínguez-Oronoz R (2012) Synthetic resorbable scaffolds for the treatment of isolated patellofemoral cartilage defects in young patients: magnetic resonance imaging and clinical evaluation. Am J Sports Med 40:1289–1295

120. Bekkers JE, Bartels LW, Vincken KL, Dhert WJ, Creemers LB, Saris DB (2013) Articular cartilage evaluation after TruFit plug implantation analyzed by delayed gadolinium-enhanced MRI of cartilage (dGEMRIC). Am J Sports Med 41:1290–1295

121. Marmotti A, Bruzzone M, Bonasia DE, Castoldi F, Von Degerfeld MM, Bignardi C, Mattia S, Maiello A, Rossi R, Peretti GM (2013) Autologous cartilage fragments in a composite scaf-

fold for one stage osteochondral repair in a goat model. Eur Cell Mater 26:15–31. discussion 31-2

122. Levingstone TJ, Matsiko A, Dickson GR, O'brien FJ, Gleeson JP (2014a) A biomimetic multi-layered collagen-based scaffold for osteochondral repair. Acta Biomater 10:1996–2004

123. Farrell E, Both SK, Odorfer KI, Koevoet W, Kops N, O'brien FJ, Baatenburg De Jong RJ, Verhaar JA, Cuijpers V, Jansen J, Erben RG, Van Osch GJ (2011) In-vivo generation of bone via endochondral ossification by in-vitro chondrogenic priming of adult human and rat mesenchymal stem cells. BMC Musculoskelet Disord 12:31

124. Frenkel SR, Bradica G, Goldman SM, Kronengold RT, Iesaka K, Issack P, Bong MR, Hua Tian JG, Coutts RD, Brekke JH (2003) A multiphasic device engineered to regenerate articular cartilage. *49th Annual Meeting of the Orthopaedic Research Society*

125. Kon E, Drobnic M, Davidson PA, Levy A, Zaslav KR, Robinson D (2014a) Chronic post-traumatic cartilage lesion of the knee treated with an acellular osteochondral-regenerating implant: case history with rehabilitation guidelines. J Sport Rehabil 23:270–275

126. Barron V, Neary M, Mohamed KMS, Ansboro S, Shaw G, O'malley G, Rooney N, Barry F, Murphy M (2016) Evaluation of the early in vivo response of a functionally graded macroporous scaffold in an osteochondral defect in a Rabbit Model. Ann Biomed Eng 44:1832–1844

127. Levingstone TJ, Ramesh A, Brady RT, Brama PAJ, Kearney C, Gleeson JP, O'brien FJ (2016a) Cell-free multi-layered collagen-based scaffolds demonstrate layer specific regeneration of functional osteochondral tissue in caprine joints. Biomaterials 87:69–81

128. Chiang H, Liao CJ, Hsieh CH, Shen CY, Huang YY, Jiang CC (2013) Clinical feasibility of a novel biphasic osteochondral composite for matrix-associated autologous chondrocyte implantation. Osteoarthr Cartil 21:589–598

129. Getgood A, Brooks R, Lynn A, Simon T, Aberman H, Rushton N (2010) Preclinical articular cartilage repair using a regionally specific collagen/glycosaminoglycan osteochondral scaffold. J Bone Joint Surg Br Vol 93-B:412

130. Getgood AMJ, Kew SJ, Brooks R, Aberman H, Simon T, Lynn AK, Rushton N (2012) Evaluation of early-stage osteochondral defect repair using a biphasic scaffold based on a collagen–glycosaminoglycan biopolymer in a caprine model. Knee 19:422–430

131. Getgood A, Henson F, Skelton C, Brooks R, Guehring H, Fortier LA, Rushton N (2014) Osteochondral tissue engineering using a biphasic collagen/GAG scaffold containing rhFGF18 or BMP-7 in an ovine model. J Exp Orthop 1:1–11

132. Tampieri A, Sandri M, Landi E, Pressato D, Francioli S, Quarto R, Martin I (2008) Design of graded biomimetic osteochondral composite scaffolds. Biomaterials 29:3539–3546

133. Christensen BB, Foldager CB, Jensen J, Jensen NC, Lind M (2016) Poor osteochondral repair by a biomimetic collagen scaffold: 1- to 3-year clinical and radiological follow-up. Knee Surg Sports Traumatol Arthrosc 24:2380–2387

134. Brix M, Kaipel M, Kellner R, Schreiner M, Apprich S, Boszotta H, Windhager R, Domayer S, Trattnig S (2016) Successful osteoconduction but limited cartilage tissue quality following osteochondral repair by a cell-free multilayered nano-composite scaffold at the knee. Int Orthop 40:625–632

135. Christensen BB, Foldager CB, Jensen J, Jensen NC, Lind M (2015) Poor osteochondral repair by a biomimetic collagen scaffold: 1- to 3-year clinical and radiological follow-up. Knee Surg Sports Traumatol Arthrosc 24:2380–2387

136. Yan L-P, Oliveira JM, Oliveira AL, Reis RL (2015) Current concepts and challenges in osteochondral tissue engineering and regenerative medicine. ACS Biomater Sci Eng 1:183–200

137. Verhaegen J, Clockaerts S, Van Osch GJVM, Somville J, Verdonk P, Mertens P (2015) TruFit plug for repair of osteochondral defects—where is the evidence? Systematic review of literature. Cartilage 6:12–19

138. Gelber PE, Batista J, Millan-Billi A, Patthauer L, Vera S, Gomez-Masdeu M, Monllau JC (2014) Magnetic resonance evaluation of TruFit(R) plugs for the treatment of osteochondral lesions of the knee shows the poor characteristics of the repair tissue. Knee 21:827–832

139. Barber FA, Dockery WD (2011) A computed tomography scan assessment of synthetic multiphase polymer scaffolds used for osteochondral defect repair. Arthroscopy: J Arthrosc Relat Surg 27:60–64

140. Dell'osso G, Bottai V, Bugelli G, Manisco T, Cazzella N, Celli F, Guido G, Giannotti S (2016) The biphasic bioresorbable scaffold (Trufit®) in the osteochondral knee lesions: long-term clinical and MRI assessment in 30 patients. Musculoskelet Surg 100:1–4

141. Radin EL, Rose RM (1986) Role of subchondral bone in the initiation and progression of cartilage damage. Clin Orthop Relat Res 213:34–40

142. Landells JW (1953) The bone cysts of osteoarthritis. J Bone Joint Surg Br 35-b:643–649

143. Li G, Yin J, Gao J, Cheng TS, Pavlos NJ, Zhang C, Zheng MH (2013) Subchondral bone in osteoarthritis: insight into risk factors and microstructural changes. Arthritis Res Ther 15:223

144. Carrino JA, Blum J, Parellada JA, Schweitzer ME, Morrison WB (2006) MRI of bone marrow edema-like signal in the pathogenesis of subchondral cysts. Osteoarthr Cartil 14:1081–1085

145. Mapp PI, Walsh DA (2012) Mechanisms and targets of angiogenesis and nerve growth in osteoarthritis. Nat Rev Rheumatol 8:390–398

146. Madry H, Orth P, Cucchiarini M (2016) Role of the subchondral bone in articular cartilage degeneration and repair. J Am Acad Orthop Surg 24:e45–e46

147. Brown TD, Pope DF, Hale JE, Buckwalter JA, Brand RA (1991) Effects of osteochondral defect size on cartilage contact stress. J Orthop Res 9:559–567

Part IV
Osteochondral Tissue Engineering Approaches

Chapter 13
Combination of Polymeric Supports and Drug Delivery Systems for Osteochondral Regeneration

Luis Rojo

Abstract Musculoskeletal conditions have been defined by European National Health systems as one of the key themes which should be featured during the present decade as a consequence of the significant healthcare and social support costs. Among others, articular cartilage degeneration due to traumatic and degenerative lesion injury or other pathologies commonly results in the development of musculoskeletal disorders such as osteoarthritis and arthritis rheumatoid, eventually leading to progressive articular cartilage and joint destruction especially at osteochondral interphase, that account for more disability among the elderly than any other diseases constituting a global social challenge that needs a multidisciplinary response from the scientific community. Current treatments for damaged osteoarthritic joint cartilage include the use of disease-modifying drugs and ultimately joint arthroplasty as unavoidable surgical intervention due to the limited ability of articular cartilage to self-regenerate. However, potential future regenerative therapies based on tissue engineering strategies are likely to become more important to facilitate the recruitment of repairing cells and improve musculoskeletal metabolism. In addition, emerging bioprinting technologies in combination with implemented manufacturing techniques such electrospinning or cryogelation processes have permitted the development of new tissue substitutes with precise control of sizes and shapes to recreate the complex physiological, biomechanical and hierarchical microstructure of osteochondral interphases. Thus, this chapter will provide an upgrade on the state of the art focusing the most relevant developments on polymer scaffolds and drug delivery systems for osteochondral regeneration.

Keywords Biomimetic scaffolds · Osteochondral regeneration · Cartilage · Drug delivery

L. Rojo (✉)
Instituto de Ciencia y Tecnología de Polímeros CSIC, Madrid, Spain

Consorcio CIBER-BBN, Instituto de Salud Carlos III, Madrid, Spain
e-mail: rojodelolmo@ictp.csic.es

© Springer International Publishing AG, part of Springer Nature 2018
J. M. Oliveira et al. (eds.), *Osteochondral Tissue Engineering*,
Advances in Experimental Medicine and Biology 1059,
https://doi.org/10.1007/978-3-319-76735-2_13

13.1   Introduction

Musculoskeletal conditions including spinal, joint and bone disorders have been considered the most prevalent occupational diseases and lead to significant health-care and social support costs with an economic burden estimated at 37 billion euros to the European National Health systems [1, 2]. Consequently these diseases have been defined by H2020 priorities as one of the key themes which should be featured during the present decade [3]. Among others, articular cartilage degeneration due to traumatic and degenerative lesion injury or other pathologies commonly results in the development of musculoskeletal disorders such as osteoarthritis and arthritis rheumatoid, eventually leading to progressive articular cartilage and joint destruction [4] especially at osteochondral interphase, that account for more disability among the elderly than any other diseases constituting a global social challenge that needs a multidisciplinary response from the scientific community [5] (Fig. 13.1).

Current treatments of damaged or osteoarthritic joint cartilage include the use of disease-modifying drugs and ultimately joint arthroplasty as unavoidable surgical intervention due to the limited ability of articular cartilage to self-regenerate. However, potential future regenerative therapies based on tissue engineering strategies are likely to become more important to facilitate the recruitment of repairing cells and improve musculoskeletal metabolism. These new therapies include the administration of different stem cell lineages such as human mesenchymal stem cells (hMSCs), together with acellular scaffolds based on hydrogels and soft materials fabricated from synthetic polymers such as PLGA and naturally occurring extracellular matrix (ECM) components derived from collagen, glycosaminoglycans (CAG) like hyaluronic acid (HA) or chondroitin sulphate (CS) and signalling molecules such as transforming growth factors (TGFβ1 and TGFβ3), insulin-like growth factor (IGF-1), (growth/differentiation factor GDF-5) and bone morphogenetic proteins (BMP-2) [6–10]. In addition to this, novel polymer therapeutics for the controlled delivery of disease-modifying drugs, anti-inflammatory conjugates or inhibitors of matrix metalloproteinases (MMPs) and immunomodulators such as strontium folate [11, 12], ibuprofen [13], imidazole [14] or dapsone [15],

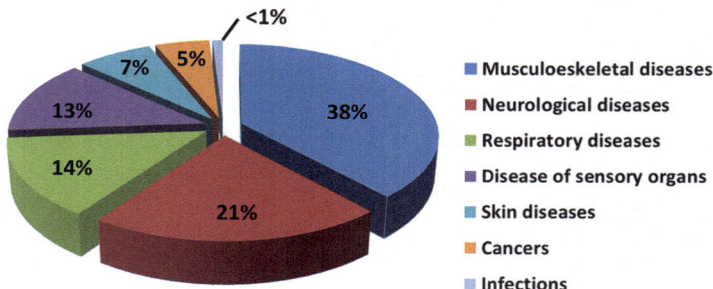

Fig. 13.1 Proportion of occupational diseases in EU according to the European Musculoskeletal Conditions Surveillance and Information Network (www.eumusc.net)

respectively, constitute effective treatments for musculoskeletal disorders, especially in osteoarthritis and rheumatoid arthritis in which the effectiveness of tissue regeneration is limited by joint inflammation [16, 17].

In this chapter, it will be considered the most recent advances on osteochondral regeneration from a polymer materials science perspective expanding the examples mentioned above and providing an overview of the most promising system currently used in preclinical or clinical studies.

13.2 Osteochondral Tissue Complexity: A Challenge for Materials Science

Articular cartilage at osteochondral interface is a well-organized and specialized tissue with hierarchical structure bridging subchondral bone and articular cartilage. The organization and composition of cartilage extracellular matrix change dramatically though a gradient transition from deep to middle and superficial zones, in which highly calcified phase changes towards a dense cell-populated articular surface, rich in orthogonally oriented collagen. In addition to this, a complex cocktail of proteoglycan, protein and cell population coexists within a highly hydrated avascular three-dimensional niche that plays a key role in maintaining the cartilage integration for load-bearing stress and friction-free locomotion. Thus, the osteochondral regenerative processes play a central role in maintaining joint movement and lubrication of this hierarchically structure. Unlike other tissues, damaged cartilage regeneration is very limited as a consequence of its low cell density and lack of cell recruitment through blood supply, and therefore clinical regenerative procedures based on classical resurfacing strategies have achieve only partial advances [18]. Contrary, modern musculoskeletal tissue engineering therapies based on local delivery of morphogens, cells and scaffolds have achieved significant progress towards the successful translation of regenerative medicine solutions to the clinic (Fig. 13.2).

The creation of the adequate environment to promote osteochondral cartilage regeneration can be achieved by the implantation of three-dimensional porous scaffolds that mimic the physiological properties, microstructure and functionality of native extracellular matrix as closely as possible. Thus, the ideal scaffolds do not only have to maintain its physical properties at the implanted environment but may also provide support to the populating cells and drive their proliferation by means of the orchestration of a controlled delivery of bioactive molecules to aid the differentiation of chondroprogenitor cells. In addition, the scaffolds should mimic the biomechanics, porosity and hierarchical microstructure of osteochondral interface with a synchronized degradation time with the new tissue forming rate. In order to overcome all this challenges, macromolecular chemistry, among different disciplines, plays a key role in the design and preparation of synthetic extracellular matrices with specific structures, morphologies and properties, using molecules that give not only high molecular weight polymers but also functionality, structural order and self-organization of microdomains with modulated biodegradability (Fig. 13.3).

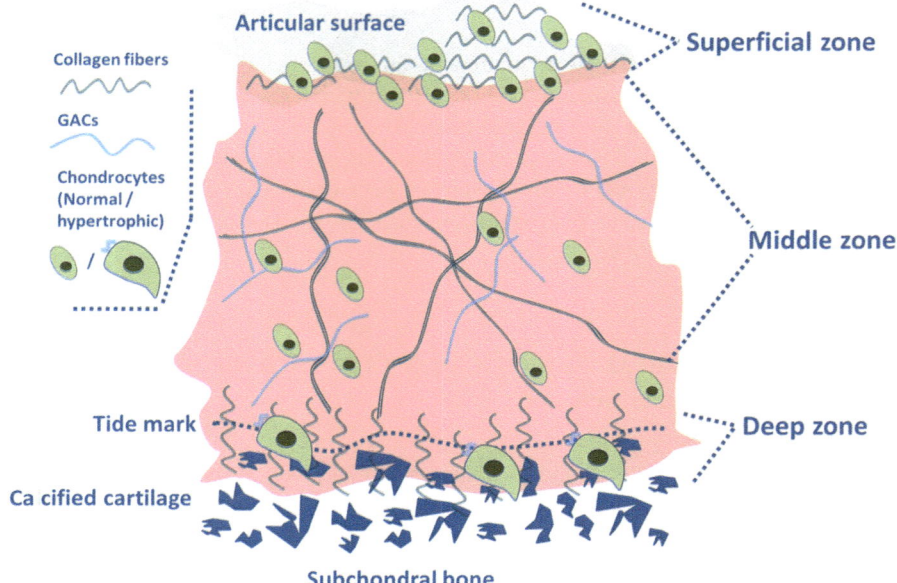

Fig. 13.2 Schematic representation of native hyaline cartilage showing the stratified microstructure at osteochondral region

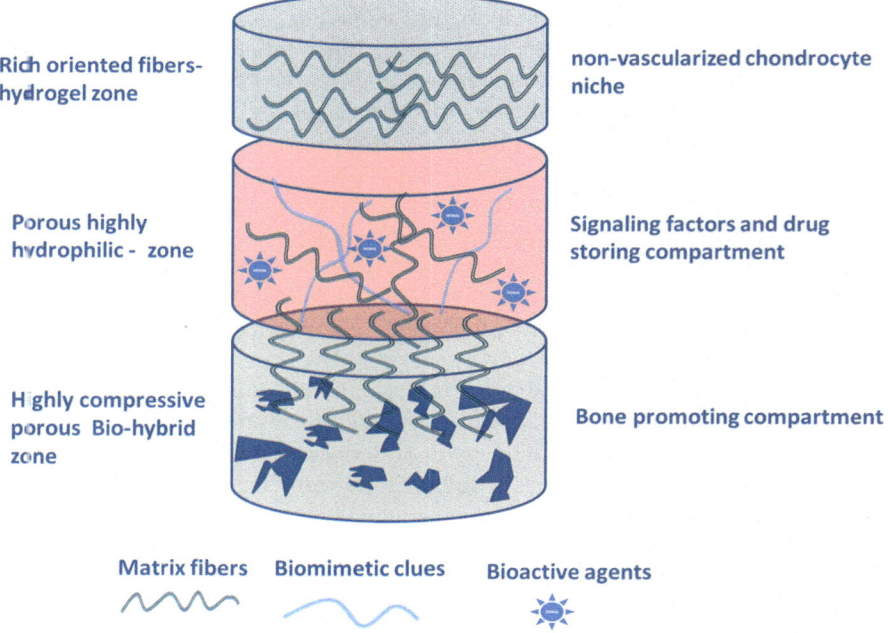

Fig. 13.3 Diagram of the ideal composition and microstructure of multiphasic scaffold for the regeneration of osteochondral lesson

Naturally derived and synthetic polymers with the specific functionality offer uncountable possibilities for design and development of advanced cell supportive materials in which the combination of microstructural design and preparation of molecular polymer architectures, synthetic pathways and technological methodologies provides a powerful tool for the preparation of engineering scaffolds and drug delivery devices for regenerative medicine. However, due to the intrinsic complexity of the osteochondral tissue, only few systems have been considered for clinical trials where hydrogels have shown the most promising results among other polymeric systems due to their capability to form highly hydrated three-dimensional matrices from hydrophilic polymers through chemical or physical cross-linking reactions while keeping the incorporated cells or biomolecules undamaged [19, 20].

13.3   Polymer Scaffolds Used in Osteochondral Regeneration

Biocompatible hydrogels based on natural macromolecules such as hyaluronic acid, chondroitin sulphate, fibrin, gelatine or collagen sponge-like matrices contain specific molecular domains which can stimulate cells to differentiate and maintain their chondrocyte phenotype [21–23], while in combination with hydrophilic networks based on polysaccharides like alginate and chitosan, these have been extensively used to deliver loaded morphogens, drugs and bioactive ions [24–28]. Compared to natural polymers, hydrophilic synthetic polymers including poly(lactic-co-glycolic acid) (PLGA), poly(ethylene glycol) (PEG) and poly(vinyl alcohol) (PVA) can be mass produced with specific structures and molecular weights offering improved control of the chemical composition and matrix architecture facilitating the controlled delivery of bioactive molecules with therapeutic properties and the formation of multiphasic scaffolds mimicking the native osteochondral microstructure [29–33]. An ideal scaffold for osteochondral regeneration does not exist. Instead, different polymeric materials based on naturally occurring extracellular matrix components in combination with synthetic polymer systems and biphasic ceramic composites have been developed to fit the desired biological, physical, chemical and biomechanical properties of the osteochondral interface tissue to be repaired. Among other fabrication methods, electrospinning [34], cryogelation-lyophilisation [35] and additive manufacturing [36] techniques have resulted most promising for the preparation of stratified multiphasic composite cartilage substitutes. However, these methods usually imply manufacturing conditions and processes that can easily affect to biological agents rendering to their partial degradation or biological inactivation and thus requiring the incorporation of carriers and/or conjugates that permit their optimal stabilization and spatial-temporal deliver.

13.4 Latest Developments on Polymer Scaffolds and Drug Delivery Osteochondral Regeneration

Few extensive reviews have been recently published providing a vast overview of the most relevant advances on the preparation of scaffolds for musculoskeletal regenerative applications [37–42]. Many of these developments have led into innovative technologies and patented products that are currently under clinical evaluation for the preparation of cartilage substitutes to be used in the treatment of osteochondral diseases [37]. However, emerging bioprinting technologies in combination with implemented manufacturing techniques such as electrospinning or cryogelation processes have permitted the development of new tissue substitutes with precise control of sizes and shapes to recreate the complex physiological, biomechanical and hierarchical microstructure of osteochondral interphases. Thus, this chapter will provide an upgrade on the state of the art focussing on the most relevant developments on polymer scaffolds and drug delivery systems for osteochondral regeneration (Table 13.1).

Table 13.1 Main vehiculization strategies of bioactive molecules in current scaffold-based regenerative approaches of osteochondral defects

Vehiculization type	Incorporated bioactive molecule	References
Direct loading	IGF-1 and TGF-β1	Gugjoo et al. [43]
	MGF and TGF-β3	Luo et al. [44]
	IGF-1 and BMP-2	Lu et al. [45]
	Curcumin	Kim et al. [46]
	Chondroitin sulphate	Zhou et al. [47]
	Chondroitin sulphate and SDF-1	Hen et al. [48]
	rhAC	Frohbergh et al. [49]
	TGFβ3 and Y27632	Hung et al. [50]
Micro-nano-encapsulation	Chitosan/Nell-1	Wang et al. [51]
	Zein/TCH	Fereshteh et al. [52]
	Poly-L-arginine/dextran/BSA	Mercato et al. [53]
	CaP/cBMP2/cTGF-β3	Lee et al. [54]
Polymer drug conjugation	Resveratrol	Wang et al. [55]
	Ibuprofen/imidazole	Suarez et al. [13]
	Dapsone	Rojo et al. [15]
	PEI/HA/cSOX/cTrio/cRUNX2	Needham et al. [56]

BMP, bone morphogenetic protein; BSA, ovine serum albumin; CaP, calcium phosphate; cBMP, bone morphogenetic protein coding gene; cRUNX2, runt-related transcription factor coding gene; cSOX, sox-related transcription factor coding gene; cTGF, transforming growth factor coding gene; cTrio, trio rho guanine nucleotide exchange factor coding gene; HA, hyaluronic acid; IGF, insulin-like growth factor; Nell, protein kinase C-binding protein; PEI, polyethylenimine; rhAC, recombinant acid ceramidase; SDF, stromal cell-derived factor; TCH, tetracycline hydrochloride; TGF, transforming growth factor; Y27632, 1R,4r-4-((R)-1-aminoethyl)-N-(pyridin-4-yl) cyclohexanecarboxamide

The load of growth factors as bioactive agent into drug delivery scaffolds constitutes one of the most reproduced systems in the literature for tissue strategies [30]. Among others, the family of transforming growth factors TGFβs, insulin growth factor IGFs and bone morphogenetic growth factors BMPs have been used more extensively and allowed for significant advances both in vitro and in vivo for osteochondral regeneration [57]. On this direction, Gugjoo et al. studied the application of MSCs in laminin scaffolds in combination with IGF-1 and TGF-β1 and their enhancing capacity to induce healing of osteochondral defects in a rabbit model by protection of synovium and decreased chronic inflammation as a consequence of the anabolic effect of IGF-1 on cartilage matrix synthesis and its bioactivity stimulating stem cell proliferation and downregulating inflammatory marker gene expression. In addition, the combination of TGF-β1 stimulated new matrix synthesis by hMSCs via initial cell–cell interactions impairing the formation of hyaline cartilage instead of fibrocartilage in comparison with growth factor-free laminin-based scaffolds [43]. Similar findings were reporting by Luo and colleagues where mechano growth factor (MGF), an analogue to IGF-1, together with TGF-β3 were loaded into sponge-like silk fibroin scaffolds to enhance cartilage repair, synergically activated by directed differentiation of endogenous stem cells and inhibited fibrosis, indicating the facilitation of cartilage regeneration [44]. This dual growth factor controlled release strategy was studied recently by Lu et al. to deliver IGF-1 and BMP-2 using a bilayered oligo(poly(ethylene glycol) fumarate) (OPF) hydrogel for the evaluation of degree of subchondral bone and osteochondral tissue repair. Interestingly, this system led to an increase in bone mineral density, percent bone volume and intersection surface and a decrease in bone-specific surface. In addition, delivery of both growth factors in separate layers had a synergistic effect on subchondral bone repair but not on cartilage repair, in an osteochondral defect [45].

The identification of new signalling pathways for cartilage formation has permitted the isolation and application of alternative growth factors with synergic bioactivity. It is important to highlight that, in order to preserve the biostability and function of this new morphogens during the scaffold fabrication and implantation time it is very often necessary, if not always, to load these molecules in combination with delivery vehicles, protective shells or stabilizing polymer conjugates. In this sense, Wang and col. have recently developed oriented and large-sized scaffolds by two-phase electrospinning nanofibers of PLLA incorporating chitosan nanoparticles loaded with Nell-1 growth factors for mimicking the oriented structure of native articular cartilage as well as protect Nell-1's bioactivity. Released Nell-1 from the scaffolds showed bioactivity protection effects by chitosan nanoparticles in the process of electrospinning and their function as two-skin layers enhancing hMSCs chondrogenic differentiation, extracellular matrix production and cartilage regeneration with fewer side effects [51].

From a regenerative point of view, damaged osteochondral interphase can be difficult to heal once the inflammatory body reactions have been triggered around the affected zone. Indeed, the control of inflammation by local delivery of disease-modifying osteoarthritis drugs such as matrix metalloproteinase inhibitors, immunomodulators, ROS scavenger species and conventional anti-inflammatory drugs is considered as one of the first actions to take for slow progress of the cartilage degradation and facilitate the recruitment of chondroprogenitors and matrix deposition on the affected area. However, only few scaffolds for cartilage repair reported anti-inflammation properties. Kim and collaborators reported the fabrication of an efficient bioengineered porous silk-based scaffolds with different concentrations of curcumin as anti-inflammatory bioactive molecule with the capacity to maintain cell phenotype of chondrocytes and promote the formation of cartilaginous matrix [46]. In a similar way, Zhou and collaborators have developed silk-based scaffold loaded with chondroitin sulphate with anti-inflammatory effects both in vitro and in vivo, promoting the inherent cartilage regeneration in an animal model [47]. Hen et al. also demonstrated the healing of osteochondral defects capacity from radially oriented channel collagen scaffold laden with anti-inflammatory and bioadhesive components such as chondroitin sulphate, in combination with SDF-1 growth factor promoting the formation of normally functional cartilage by inhibition expression of matrix hydrolysing enzymes related with inflammation such as Col1 and MMP13 [48]. Likewise, Wang and collaborators fabricated anti-inflammatory resveratrol-loaded scaffold that eventually offered a protective effect on chondrocytes and BMSCs in a traumatic ROS environment and high efficiency for repairing osteochondral defects in rabbits being absent of cells and growth factors. The drug was conjugated into a macromolecular system in order to improve its water solubility and compatibility with collagen scaffold showing a compressive strength comparable to normal cartilage and a slow degradation rate promoting the sustained release capacity of Res without promoting foreign body reaction after implantation into the osteochondral defect created on rabbit models [55].

Other authors have included specific chondrocyte target proteins into delivery scaffolds to test the efficacy and wound-healing capabilities. Frohbergh and collaborators used commercial Bio-Gide collagen scaffolds loaded with recombinant acid ceramidase (rhAC) into osteochondral defects created in Sprague-Dawley rats. The results showed that these scaffolds provided a better quality repair in vivo and demonstrate the positive effects of rhAC treatment on chondrocyte growth and phenotype *revealing* the in vivo effects of the treated cells with rhAC on cartilage repair [49]. Encapsulating proteins like zein, a class of the main storage proteins with pharmaceutical applications, have been also used for vehiculization and controlled release of active additives or drugs. Fereshteh et al. developed a novel type of drug delivery scaffold based on poly(ε-caprolactone) (PCL) and zein blends with oriented microtubule structure via improved unidirectional freeze-drying method and loaded with tetracycline hydrochloride (TCH). The effectiveness of the scaffolds for osteochondral defect regeneration was evaluated by means of tuning the composition and tubule orientation in column-like pores, and it was found that the protein release from the aligned scaffold occurs in a more controlled manner and TCH released in a more sustained and controlled manner as compared to the PCL scaffold without zein. Thus, due to the especial pore architecture of the PCL/zein scaffold, these scaffolds were found suitable for bone and cartilage tissue engineering applications, including osteochondral defect regeneration [52].

Mercato et al. have evaluated the regenerative capacity of multifunctional microcapsule-integrated scaffolds for controlled delivery of bioactive molecules.

Three different polyelectrolyte multilayer microcapsules made from poly-L-arginine were prepared by a simple method based on freeze-drying and layer-by-layer encapsulating cationic (bovine serum albumin labelled with fluorescein) or anionic (dextran-rhodamine and dextran-fluorescein) microcapsules into their cavities. The applied procedure allowed engineering porous 3D collagen scaffolds with protein-loaded microcapsules while preserving both the scaffold and microcapsule structures and their suitability for soft tissue engineering applications [53]. Furthermore, Hung et al. developed water-based polyurethane 3D-printed scaffolds with controlled release of encapsulated growth factor TGFβ3 and drug Y27632 (Y27632: an inhibitor for rho-associated coiled-coil-containing protein kinase (ROCK) 1R,4r)-4-((R)-1-aminoethyl)-N-(pyridin-4-yl) cyclohexanecarboxamide, $C_{14}H_{21}N_3O$) for customized cartilage tissue engineering. The multicomponent bioactive compounds showed timely release from the scaffolds without requiring pre-wetting for cell seeding, thus avoiding the need of giving any exogenous inductive medium being very effective in regenerating rabbit cartilage defect [50].

Other tissue engineering strategies aimed to drive the regenerative process by including delivery systems of DNA encoding transcription factors which can be transported within polymeric vehicles to the desired site of action [14]. Needham and collaborators demonstrated that the regenerated cartilage tissue can be enhanced by local delivery of DNA vectors for gene encoding transcription factors SOX trio and RUNX2. These vectors were complexed with the branched poly(ethylenimine)-hyaluronic acid and loaded within porous bilayered hydrogel scaffolds made from oligo[poly(ethylene glycol) fumarate] for implantation directly in an osteochondral defect. This combination has been found very promising and resulted in greater tissue generation and quality, especially in the bone layer, demonstrating the ability to apply polymeric gene therapy and transcription factors in vivo [56]. Furthermore, Lee and collaborators demonstrated the capacity of 3D enzymatic-cross-linked bilayer composite scaffolds to promote prolonged transgene expression and stimulation of hMSCs differentiation into the osteogenic and chondrogenic lineages and their effect to accelerate healing process for osteochondral tissue regeneration by spatial control release of TGF-β3 and BMP-2 coding plasmids loaded into a non-viral gene carrier nanoparticles [54].

13.5 Conclusions

Despite the vast advances achieved in regenerative medicine, the consistent repair of articular cartilage at osteochondral interphases still remains a significant scientific and clinical challenge. However, recent developments based on multidisciplinary approaches with significance in tissue engineering therapies for musculoskeletal disorders are beginning to address many of the major clinical limitations. Along this chapter, it has been discussed the progress in the last 3-year period giving special attention to innovative developments on stratified multiphasic, drug delivery scaffolds with the capacity of mimicking the hierarchical microstructure and composition of the native osteochondral tissue. Indeed, it is clear that the future trends on osteochondral regenerative strategies will be driven by developments on materials sciences by means of engineering biomimetic systems based on synthetic polymers and natural existing biomacromolecules such us collagen, gelatine and hyaluronic acid in combination with the spatiotemporally delivery of multiple bioactive cell signalling molecules, morphogens and disease-modifying drugs aimed to provide in a single treatment the appropriate 3D milieu with tuned cascade of biochemical signals and controlled anti-inflammatory and inmunomodulative pharmacological properties to promote mesenchymal chondroprogenitors to orchestrate a complete articular cartilage and osteochondral tissue regeneration.

References

1 Smith E, Hoy D, Cross M, Merriman TR, Vos T, Buchbinder R, Woolf A, March L (2014) The global burden of gout: estimates from the Global Burden of Disease 2010 study. Ann Rheum Dis 73(8):1470–1476. https://doi.org/10.1136/annrheumdis-2013-204647

2. Storheim K, Zwart J-A (2014) Musculoskeletal disorders and the Global Burden of Disease study. Ann Rheum Dis 73(6):949–950. https://doi.org/10.1136/annrheumdis-2014-205327

3. Vieira S, Vial S, Maia F, Carvalho M, Reis R, Granja P, Oliveira J (2015) Gellan gum-coated gold nanorods: an intracellular nanosystem for bone tissue engineering. RSC Adv 5:77996–78005

4. Buchbinder R, Maher C, Harris IA (2015) Setting the research agenda for improving health care in musculoskeletal disorders. Nat Rev Rheumatol 11(10):597–605. https://doi.org/10.1038/nrrheum.2015.81

5. Hasani-Sadrabadi MM, Pour Hajrezaei S, Hojjati Emami S, Bahlakeh G, Daneshmandi L, Dashtimoghadam E, Seyedjafari E, Jacob KI, Tayebi L (2015) Enhanced osteogenic differentiation of stem cells via microfluidics synthesized nanoparticles. Nanomedicine 11(7):1809–1819. https://doi.org/10.1016/j.nano.2015.04.005

6. Ringe J, Burmester GR, Sittinger M (2012) Regenerative medicine in rheumatic disease—progress in tissue engineering. Nat Rev Rheumatol 8(8):493–498

7. Ye K, Felimban R, Moulton S, Wallace G, Di Bella C, Traianedes K, Choong PFM, Myers D, Choong P (2013) Bioengineering of articular cartilage: past, present and future. Regen Med 8(3):333–349

8. Ringe J, Sittinger M (2014) Regenerative medicine: selecting the right biological scaffold for tissue engineering. Nat Rev Rheumatol 10(7):388–389. https://doi.org/10.1038/nrrheum.2014.79

9. Gothard D, Smith EL, Kanczler JM, Black CR, Wells JA, Roberts CA, White LJ, Qutachi O, Peto H, Rashidi H, Rojo L, Stevens MM, El Haj AJ, Rose F, Shakesheff KM, Oreffo ROC (2015) In vivo assessment of bone regeneration in alginate/bone ECM hydrogels with incorporated skeletal stem cells and single growth factors. PLoS One 10(12):e0145080. https://doi.org/10.1371/journal.pone.0145080

10. Fabbri M, Soccio M, Costa M, Lotti N, Gazzano M, Siracusa V, Gamberini R, Rimini B, Munari A, García-Fernández L, Vázquez-Lasa B, San Román J (2016) New fully bio-based PLLA triblock copoly(ester urethane)s as potential candidates for soft tissue engineering. Polym Degrad Stab 132(Supplement C):169–180. https://doi.org/10.1016/j.polymdegradstab.2016.02.024

11. Rojo L, Radley-Searle S, Fernandez-Gutierrez M, Rodriguez-Lorenzo LM, Abradelo C, Deb S, Roman JS (2015) The synthesis and characterisation of strontium and calcium folates with

potential osteogenic activity. J Mater Chem B 3(13):2708–2713. https://doi.org/10.1039/c4tb01969e

12. Del Campo MM, Alvarado-Estrada K, Rojo L, Sampedro JG, Rosales-Ibáñez R, Román JS (2015) Effect and application of 3D-Scaffolds in restoration of bone defects. Dent Mater 31(Supplement 1):e65. https://doi.org/10.1016/j.dental.2015.08.142

13. Suarez P, Rojo L, Gonzalez-Gomez A, San Roman J (2013) Self-assembling gradient copolymers of vinylimidazol and (acrylic)ibuprofen with anti-inflammatory and zinc chelating properties. Macromol Biosci 13(9):1174–1184. https://doi.org/10.1002/mabi.201300141

14. Velasco D, Réthoré G, Newland B, Parra J, Elvira C, Pandit A, Rojo L, San Román J (2012) Low polydispersity (N-ethyl pyrrolidine methacrylamide-co-1-vinylimidazole) linear oligomers for gene therapy applications. Eur J Pharm Biopharm 82(3):465–474. https://doi.org/10.1016/j.ejpb.2012.08.002

15. Rojo L, Fernandez-Gutierrez M, Deb S, Stevens MM, San Roman J (2015) Designing dapsone polymer conjugates for controlled drug delivery. Acta Biomater 27:32–41. https://doi.org/10.1016/j.actbio.2015.08.047

16. Ndlovu M, Bedson J, Jones PW, Jordan KP (2014) Pain medication management of musculoskeletal conditions at first presentation in primary care: analysis of routinely collected medical record data. BMC Musculoskelet Disord 15:418. https://doi.org/10.1186/1471-2474-15-418

17. Chang D, Lamothe M, Stevens R, Sigal L (1996) Dapsone in rheumatoid arthritis. Semin Arthritis Rheum 25(6):390–403

18. Sakata R, Iwakura T, Reddi AH (2015) Regeneration of articular cartilage surface: morphogens, cells, and extracellular matrix scaffolds. Tissue Eng Part B Rev 21(5):461–473. https://doi.org/10.1089/ten.TEB.2014.0661

19. Kim BS, Park IK, Hoshiba T, Jiang HL, Choi YJ, Akaike T, Cho CS (2011) Design of artificial extracellular matrices for tissue engineering. Prog Polym Sci 36(2):238–268. https://doi.org/10.1016/j.progpolymsci.2010.10.001

20. Ulery BD, Nair LS, Laurencin CT (2011) Biomedical applications of biodegradable polymers. J Polym Sci B Polym Phys 49(12):832–864. https://doi.org/10.1002/polb.22259

21. Basad E, Wissing FR, Fehrenbach P, Rickert M, Steinmeyer J, Ishaque B (2015) Matrix-induced autologous chondrocyte implantation (MACI) in the knee: clinical outcomes and challenges. Knee Surg Sports Traumatol Arthrosc 23(12):3729–3735. https://doi.org/10.1007/s00167-014-3295-8

22. Muzzarelli RAA, Greco F, Busilacchi A, Sollazzo V, Gigante A (2012) Chitosan, hyaluronan and chondroitin sulfate in tissue engineering for cartilage regeneration: a review. Carbohydr Polym 89(3):723–739. https://doi.org/10.1016/j.carbpol.2012.04.057

23. Mora-Boza A, Puertas-Bartolomé M, Vázquez-Lasa B, San Román J, Pérez-Caballer A, Olmeda-Lozano M (2017) Contribution of bioactive hyaluronic acid and gelatin to regenerative medicine. Methodologies of gels preparation and advanced applications. Eur Polym J 95(Supplement C):11–26. https://doi.org/10.1016/j.eurpolymj.2017.07.039

24. Man Z, Hu X, Liu Z, Huang H, Meng Q, Zhang X, Dai L, Zhang J, Fu X, Duan X, Zhou C, Ao Y (2016) Transplantation of allogenic chondrocytes with chitosan hydrogel-demineralized bone matrix hybrid scaffold to repair rabbit cartilage injury. Biomaterials 108(Supplement C):157–167. https://doi.org/10.1016/j.biomaterials.2016.09.002

25. Park H, Lee HJ, An H, Lee KY (2017) Alginate hydrogels modified with low molecular weight hyaluronate for cartilage regeneration. Carbohydr Polym 162(Supplement C):100–107. https://doi.org/10.1016/j.carbpol.2017.01.045

26. Place ES, Rojo L, Gentleman E, Sardinha JP, Stevens MM (2011) Strontium- and zinc-alginate hydrogels for bone tissue engineering. Tissue Eng Part A 17(21–22):2713–2722. https://doi.org/10.1089/ten.tea.2011.0059

27. Sartori M, Pagani S, Ferrari A, Costa V, Carina V, Figallo E, Maltarello MC, Martini L, Fini M, Giavaresi G (2017) A new bi-layered scaffold for osteochondral tissue regeneration: in vitro and in vivo preclinical investigations. Mater Sci Eng C 70(Part 1):101–111. https://doi.org/10.1016/j.msec.2016.08.027

28. Peniche H, Reyes-Ortega F, Aguilar MR, Rodriguez G, Abradelo C, Garcia-Fernandez L, Peniche C, San Roman J (2013) Thermosensitive macroporous cryogels functionalized with bioactive chitosan/bemiparin nanoparticles. Macromol Biosci 13(11):1556–1567. https://doi.org/10.1002/mabi.201300184

29. Chang N-J, Lin C-C, Shie M-Y, Yeh M-L, Li C-F, Liang P-I, Lee K-W, Shen P-H, Chu C-J (2015) Positive effects of cell-free porous PLGA implants and early loading exercise on hyaline cartilage regeneration in rabbits. Acta Biomater 28(Supplement C):128–137. https://doi.org/10.1016/j.actbio.2015.09.026

30. Rojo L, Vazquez B, San Roman J (2014) Synthetic polymers for tissue engineering scaffolds: biological design, materials, and fabrication. In: Migliaresi C, Motta A (eds) Scaffolds for tissue engineering: biological design, materials and fabrication. Pan Stanford Publishing, Singapore, pp 263–300

31. Wang J, Zhang F, Tsang WP, Wan C, Wu C (2017) Fabrication of injectable high strength hydrogel based on 4-arm star PEG for cartilage tissue engineering. Biomaterials 120(Supplement C):11–21. https://doi.org/10.1016/j.biomaterials.2016.12.015

32. Radhakrishnan J, Subramanian A, Sethuraman S (2017) Injectable glycosaminoglycan–protein nano-complex in semi-interpenetrating networks: a biphasic hydrogel for hyaline cartilage regeneration. Carbohydr Polym 175(Supplement C):63–74. https://doi.org/10.1016/j.carbpol.2017.07.063

33. Moeinzadeh S, Pajoum Shariati SR, Jabbari E (2016) Comparative effect of physicomechanical and biomolecular cues on zone-specific chondrogenic differentiation of mesenchymal stem cells. Biomaterials 92:57–70. https://doi.org/10.1016/j.biomaterials.2016.03.034

34. Jiang T, Carbone EJ, Lo KWH, Laurencin CT (2015) Electrospinning of polymer nanofibers for tissue regeneration. Prog Polym Sci 46(Supplement C):1–24. https://doi.org/10.1016/j.progpolymsci.2014.12.001

35. Hixon KR, Lu T, Sell SA (2017) A comprehensive review of cryogels and their roles in tissue engineering applications. Acta Biomater 62(Supplement C):29–41. https://doi.org/10.1016/j.actbio.2017.08.033

36. Bracaglia LG, Smith BT, Watson E, Arumugasaamy N, Mikos AG, Fisher JP (2017) 3D printing for the design and fabrication of polymer-based gradient scaffolds. Acta Biomater 56(Supplement C):3–13. https://doi.org/10.1016/j.actbio.2017.03.030

37. Lopez-Ruiz E, Jimenez G, Garcia MA, Antich C, Boulaiz H, Marchal JA, Peran M (2016) Polymers, scaffolds and bioactive molecules with therapeutic properties in osteochondral pathologies: what's new? Expert Opin Ther Pat 26(8):877–890. https://doi.org/10.1080/13543776.2016.1203903

38. Smith BD, Grande DA (2015) The current state of scaffolds for musculoskeletal regenerative applications. Nat Rev Rheumatol 11(4):213–222. https://doi.org/10.1038/nrrheum.2015.27

39. Moreira Teixeira LS, Patterson J, Luyten FP (2014) Skeletal tissue regeneration: where can hydrogels play a role? Int Orthop 38(9):1861–1876. https://doi.org/10.1007/s00264-014-2402-2

40. Yang J, Zhang YS, Yue K, Khademhosseini A (2017) Cell-laden hydrogels for osteochondral and cartilage tissue engineering. Acta Biomater 57(Supplement C):1–25. https://doi.org/10.1016/j.actbio.2017.01.036

41. Lam J, Lu S, Kasper FK, Mikos AG (2015) Strategies for controlled delivery of biologics for cartilage repair. Adv Drug Deliv Rev 84(Supplement C):123–134. https://doi.org/10.1016/j.addr.2014.06.006

42. Rojo L, Deb S (2015) Polymer therapeutics in relation to dentistry. Biomaterials for oral and craniomaxillofacial applications. Front Oral Biol 17:13–21. https://doi.org/10.1159/000381688

43. Gugjoo MB, Amarpal, Abdelbaset-Ismail A, Aithal HP, Kinjavdekar P, Pawde AM, Kumar GS, Sharma GT (2017) Mesenchymal stem cells with IGF-1 and TGF-β1 in laminin gel for osteochondral defects in rabbits. Biomed Pharmacother 93(Supplement C):1165–1174. https://doi.org/10.1016/j.biopha.2017.07.032

44. Luo Z, Jiang L, Xu Y, Li H, Xu W, Wu S, Wang Y, Tang Z, Lv Y, Yang L (2015) Mechano growth factor (MGF) and transforming growth factor (TGF)-β3 functionalized silk scaffolds

enhance articular hyaline cartilage regeneration in rabbit model. Biomaterials 52(Supplement C):463–475. https://doi.org/10.1016/j.biomaterials.2015.01.001

45. Lu S, Lam J, Trachtenberg JE, Lee EJ, Seyednejad H, van den JJJP B, Tabata Y, Wong ME, Jansen JA, Mikos AG, Kasper FK (2014) Dual growth factor delivery from bilayered, biodegradable hydrogel composites for spatially-guided osteochondral tissue repair. Biomaterials 35(31):8829–8839. https://doi.org/10.1016/j.biomaterials.2014.07.006

46. Kim DK, Kim JI, Sim BR, Khang G (2017) Bioengineered porous composite curcumin/silk scaffolds for cartilage regeneration. Mater Sci Eng C 78(Supplement C):571–578. https://doi.org/10.1016/j.msec.2017.02.067

47. Zhou F, Zhang X, Cai D, Li J, Mu Q, Zhang W, Zhu S, Jiang Y, Shen W, Zhang S, Ouyang HW (2017) Silk fibroin-chondroitin sulfate scaffold with immuno-inhibition property for articular cartilage repair. Acta Biomater 63:64–75. https://doi.org/10.1016/j.actbio.2017.09.005

48. Chen P, Tao J, Zhu S, Cai Y, Mao Q, Yu D, Dai J, Ouyang H (2015) Radially oriented collagen scaffold with SDF-1 promotes osteochondral repair by facilitating cell homing. Biomaterials 39(Supplement C):114–123. https://doi.org/10.1016/j.biomaterials.2014.10.049

49. Frohbergh ME, Guevara JM, Grelsamer RP, Barbe MF, He X, Simonaro CM, Schuchman EH (2016) Acid ceramidase treatment enhances the outcome of autologous chondrocyte implantation in a rat osteochondral defect model. Osteoarthr Cartil 24(4):752–762. https://doi.org/10.1016/j.joca.2015.10.016

50. Hung K-C, Tseng C-S, Dai L-G, Hsu S-h (2016) Water-based polyurethane 3D printed scaffolds with controlled release function for customized cartilage tissue engineering. Biomaterials 83(Supplement C):156–168. https://doi.org/10.1016/j.biomaterials.2016.01.019

51. Wang C, Hou W, Guo X, Li J, Hu T, Qiu M, Liu S, Mo X, Liu X (2017) Two-phase electrospinning to incorporate growth factors loaded chitosan nanoparticles into electrospun fibrous scaffolds for bioactivity retention and cartilage regeneration. Mater Sci Eng C 79(Supplement C):507–515. https://doi.org/10.1016/j.msec.2017.05.075

52. Fereshteh Z, Fathi M, Bagri A, Boccaccini AR (2016) Preparation and characterization of aligned porous PCL/zein scaffolds as drug delivery systems via improved unidirectional freeze-drying method. Mater Sci Eng C 68(Supplement C):613–622. https://doi.org/10.1016/j.msec.2016.06.009

53. del Mercato LL, Passione LG, Izzo D, Rinaldi R, Sannino A, Gervaso F (2016) Design and characterization of microcapsules-integrated collagen matrixes as multifunctional three-dimensional scaffolds for soft tissue engineering. J Mech Behav Biomed Mater 62(Supplement C):209–221. https://doi.org/10.1016/j.jmbbm.2016.05.009

54. Lee Y-H, Wu H-C, Yeh C-W, Kuan C-H, Liao H-T, Hsu H-C, Tsai J-C, Sun J-S, Wang T-W (2017) Enzyme-crosslinked gene-activated matrix for the induction of mesenchymal stem cells in osteochondral tissue regeneration. Acta Biomater 63:210–226. https://doi.org/10.1016/j.actbio.2017.09.008

55. Wang W, Sun L, Zhang P, Song J, Liu W (2014) An anti-inflammatory cell-free collagen/resveratrol scaffold for repairing osteochondral defects in rabbits. Acta Biomater 10(12):4983–4995. https://doi.org/10.1016/j.actbio.2014.08.022

56. Needham CJ, Shah SR, Dahlin RL, Kinard LA, Lam J, Watson BM, Lu S, Kasper FK, Mikos AG (2014) Osteochondral tissue regeneration through polymeric delivery of DNA encoding for the SOX trio and RUNX2. Acta Biomater 10(10):4103–4112. https://doi.org/10.1016/j.actbio.2014.05.011

57. Karimi T, Moeinzadeh S, Jabbari E (2015) 3-Growth factors for musculoskeletal tissue engineering. In: Nukavarapu SP, Freeman JW, Laurencin CT (eds) Regenerative engineering of musculoskeletal tissues and interfaces. Woodhead Publishing, Amsterdam, pp 43–76. https://doi.org/10.1016/B978-1-78242-301-0.00003-3

Chapter 14
Osteochondral Angiogenesis and Promoted Vascularization: New Therapeutic Target

Luis García-Fernández

Abstract The control of the different angiogenic process is an important point in osteochondral regeneration. Angiogenesis is a prerequisite for osteogenesis in vivo; insufficient neovascularization of bone constructs after scaffold implantation resulted in hypoxia and cellular necrosis. Otherwise, angiogenesis must be avoided in chondrogenesis; vascularization of the cartilage contributes to structural damage and pain. Finding a balance between these processes is important to design a successful treatment for osteochondral regeneration. This chapter shows the most important advances in the control of angiogenic process for the treatment of osteochondral diseases focused on the administration of pro- or anti-angiogenic factor and the design of the scaffold.

Keywords Angiogenic factors · Neovascularization · Osteochondral angiogenesis · Biomaterials · Osteochondral regeneration

14.1 Introduction

Osteochondral tissue presents a very complex morphology that result in the development of complex treatments. On one side, articular cartilage is an avascular tissue that lines the ends of bones in diarthrodial joints, serves to support and acts as a shock absorber, and facilitates joint's motion in low friction [1]. On the other side, the bony layer is comprised of a variety of cell population, inorganic compounds, extracellular matrix (ECM), and other components designed to provide mechanical support to the body [2]. The osteochondral tissue comprises the interface between the bone and cartilage and is a gradual transition in which the key constituents of

L. García-Fernández (✉)
Institute of Polymer Science and Technology, Spanish National Research Council
(ICTP-CSIC), Madrid, Spain
e-mail: luis.garcia@csic.es

© Springer International Publishing AG, part of Springer Nature 2018
J. M. Oliveira et al. (eds.), *Osteochondral Tissue Engineering*,
Advances in Experimental Medicine and Biology 1059,
https://doi.org/10.1007/978-3-319-76735-2_14

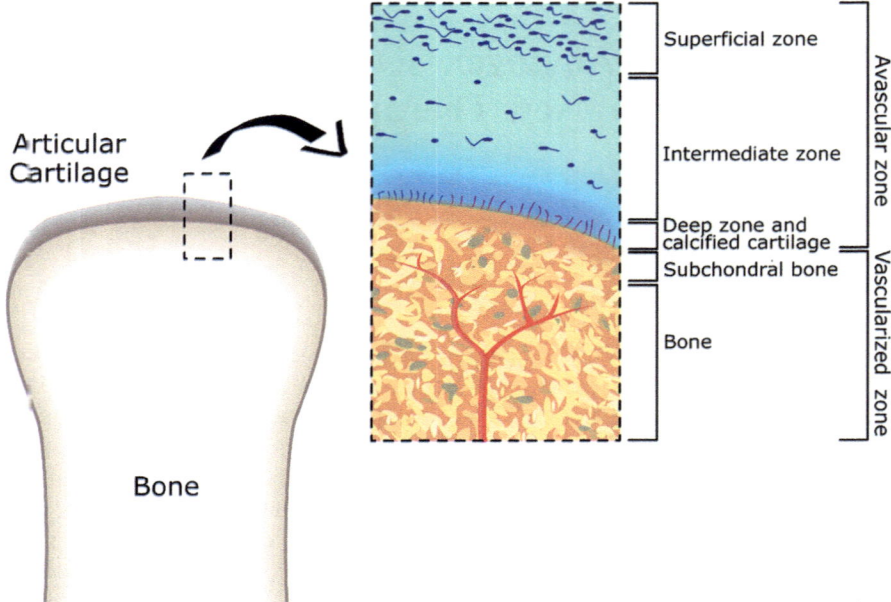

Fig. 14.1 Osteochondral tissue structure

each tissue undergo an exchange in predominance. The structure of the osteochondral joint could be defined in four different regions (Fig. 14.1):

- Superficial zone: This is incompressible tissue with a high-water content. The collagen fiber is oriented parallel to the joint surface and presents few chondrocytes with low metabolic activity [3].
- Intermediated zone: This zone is mainly composed of proteoglycans. The chondrocytes in this region assume a more spheroidal morphology, suggesting a more dominant role in the ECM synthesis [4, 3].
- Deep zone and calcified cartilage: This area is where the transition from soft to stiff occurs. In the deep zone, the collagen fibers align perpendicular to the joint surface. The calcified cartilage is responsible for attaching to the subchondral bone [3].
- Subchondral bone: This is composed of the bony lamella and the trabeculae. The trabecula **is the only vascularized area** in the osteochondral joint and contains nutrient for both itself and the adjacent articular cartilage [5, 6].

The promotion of angiogenesis (the growth of new blood vessels from the preexisting ones [7]) in the subchondral area is an important point in the nutrition of the entire osteochondral tissue. Oxygen consumption and nutrient diffusion occur within the cartilage via perfusion at the articular surface [8]. Continual nutrition of the cartilage is essential to remain functional and to avoid degeneration. One of the principal problems in the osteochondral angiogenesis is the invasion of new blood vessels from the subchondral bone to the usually avascular cartilage. The vascularization of the articular cartilage may potentiate the ossification and innervation of the deep and intermediate zone implying a joint damage and increasing pain [9]. The interface located between the hyaline cartilage and the subchondral bone acts as a barrier preventing the invasion of vessels from the bone. For these reasons, a good design of the scaffold and a good control of the angiogenesis are the key points in osteochondral regeneration. In this chapter, we are going to describe the latest advanced in osteochondral control of angiogenesis.

14.2 Angiogenesis in Osteochondral Environment

Oftentimes, the scaffolds used in osteochondral regeneration do not offer enough support to regenerate the bone and cartilage. In addition, the cartilage-subchondral bone interface presents a fragile balance between anti-angiogenesis and angiogenesis. It is possible to differentiate two different areas [10]:

- *The cartilage area* (superficial zone, intermediated zone, and deep zone): This is an anti-angiogenic area. The expression of anti-angiogenic protease inhibitors and chondromodulin-I (ChM-I) makes the cartilage avascular and resistant to vascular invasion [9, 11]. Low levels of ChM-I can derive in vascular invasion at the cartilage-subchondral bone and, in consequence, in the development of osteoarthritis [12].
- *The bone area* (subchondral bone and bone): The bone is a living tissue, and the presence of angiogenesis is critical for bone reconstruction. Different growth factors, such as vascular endothelial growth factor (VEGF), basic fibroblast growth factor (bFGF), endothelin-1 (ET-1), and others, play an essential role in the promotion of angiogenesis [13].

Our body naturally presents the different growth factors needed for osteochondral repair; however, the natural concentration is too low to regenerate a damage; for this reason, in tissue engineering, it must be used with a higher concentration to achieve their goal and the correct factors to avoid undesirable results [14, 15].

14.2.1 The Control of Angiogenesis and Differentiation in the Cartilage Area

The modulation of the angiogenesis in this area is an important point but also the modulation of cell differentiation by the use of growth factors.

14.2.1.1 Gene Therapy

Different authors have investigated cell-based anti-angiogenic gene therapies for inducing overexpression of different anti-angiogenic factors. Some of these factors are described below.

Chondromodulin (ChM) ChMs are cartilage-derived growth factors formed by 120 amino acids, including ChM-I, ChM-II, and ChM-III. The most important is ChM-I that can act as an angiogenesis inhibitor and stimulate the chondrogenesis [16]. Recent studies showed that the use of mesenchymal stem cells transfected to overexpress ChM-I seeded on scaffolds of PLLA/chitosan can promote chondrocytes growth and inhibit vascular endothelial cell invasion [16, 17].

Endostatin Endostatin is a 20 kDa proteolytic fragment of type XVIII collagen that inhibits endothelial cell proliferation and migration and promoted anabolic activity in the cartilage [18, 19]. Jeng et al. transfected MSC to overexpress endostatin and seeded in different covalently cross-linked collagen hydrogels. Endostatin expressed in the cells has the capacity to interact with the glycosaminoglycan (GAG) [20]. The inhibition of GAG induces a decrease in the angiogenic process because they are necessary to activate VEGF-induced angiogenesis.

Soluble fms-like tyrosine kinase 1 (sFlt1) receptor sFlt1 is a soluble receptor of VEGF. The overexpression of soluble VEGF receptor enhanced cartilage regeneration in osteochondral defect models [21]. Marsano and coworkers used bone marrow-derived mesenchymal stromal/stem cells (MSCs) transduced with a retroviral vector to express sFlt1. The overexpression of VEGF receptors improved the persistence of the repaired articular cartilage by preventing vascularization and bone invasion into the repaired articular cartilage [22].

14.2.1.2 Anti-angiogenic Drugs

The angiogenic process involves the infiltration of mononuclear cells and several growth factors, such as VEGF, which are powerful chemoattractant for this kind of cells. This aspect underlines the importance to control the signaling of VEGF in the cartilage area. The use of gene therapy (like in the case of ChM or sFlt1) has several limitations in terms of direct clinical translation [23]. To avoid this problem, the use of a biomaterial-based anti-angiogenic drug-release system could provide an appropriate environment for the control of angiogenesis [7, 24]. Centola et al. developed a hyaluronan-/fibrin-based porous scaffold that was functionalized by the incorporation of the anti-angiogenic drug bevacizumab (currently used as an anti-angiogenic therapeutic drug in the treatment of metastatic colorectal cancer, metastatic kidney cancer, and glioblastoma) [25]. Bevacizumab is a humanized monoclonal anti-VEGF antibody that binds to human VEGF. The results obtained by Centola and coworkers suggested that blocking angiogenesis supports the formation of a long-term stable engineered cartilage, as it effectively preserves its avascular nature and prevents its resorption [25]. Another typical anticancer drug that can be used as anti-angiogenic drug for osteochondral defects is suramin [26]. Hunziker et al. incorporated suramin in different ways into fibrin scaffolds. Some scaffolds were loaded with free suramin, and others were loaded with liposome-encapsulated suramin to obtain a delayed-release system. The scaffolds were partially effective in suppressing bone tissue upgrowth into the cartilaginous defect space when present in a free form alone, but for the liposome-encapsulated suramin, the cartilaginous compartment of most defects (69%) contained no bone due to the inhibition of the neovascularization [27].

14.2.2   The Control of Angiogenesis in the Bone Area

The neovascularization of this area is a critical point for a correct regeneration of the bone. In the last years, physical architecture (microchannel, interconnected pores, etc.) was used as initiator for cellular infiltration, but this alone did not induce angiogenesis [28]. The presence of proangiogenic stimulus is necessary for a correct and complete vascularization of the bone area. There are different ways to control the angiogenic process: the classical method implies the use of growth factors (i.e., VEGF, bFGF) to promote the angiogenesis, but recent studies also showed that the use of inorganic cations can stimulate the angiogenic process.

14.2.2.1   Vascular Endothelial Growth Factor (VEGF)

It is a potent mitogen and angiogenic factor for endothelial cells. Normally, VEGF is highly expressed on chondrocytes, during osteoclast differentiation, and in hematopoietic and endothelial cells [29–31]. VEGF activity is not only focused on the activation of angiogenic process; recent studies suggested that VEGF is also involved in the regulation of osteoblast differentiation at early stages of bone development [32, 33]. VEGF has a short half-life of 6–8 h; the preparation of VEGF-loaded scaffolds to control the delivery of VEGF is required to ensure the sustained activity. The use of natural polymers to prepare scaffold for VEGF delivery, such as PLGA, is a common way to prepare scaffold for bone treatment [34]. García and coworkers have developed new integrin-specific hydrogels functionalized with VEGF for vascularization and bone regeneration. Maleimide-functionalized four-arm PEG macromer was functionalized with recombinant human VEGF and cross-linked using the bi-cysteine peptide. In this case, VEGF-modified hydrogels increased vascularization compared to VEGF-free hydrogels [35]. However, the use of VEGF alone does not yield optimal bone regeneration; the limitation of monotherapy with VEGF therapy has been verified in different studies. Kastern et al. studied angiogenesis and bone formation on platelet-rich plasma and VEGF-transfected mesenchymal stem cells. Platelet-rich plasma is a mixture of different growth factors (VEGF, platelet-derived growth factor (PDGF), TGF-β1, TGF-β2, insulin-like growth factor (IGF), epidermal growth factor, and endothelial cell growth factor), and in this research, Kastern concluded that while VEGF was sufficient to improve revascularization, a combination of growth factors led to better bone repair compared to VEGF alone [36].

14.2.2.2 Fibroblast Growth Factor Basic (bFGF)

bFGF is a member of FGF family. It is produced by fibroblasts, endothelial cells, and osteoblasts [37] and plays an important role in the induction of angiogenesis and mitogenesis of osteoblasts [38]. Like VEGF, the effectiveness of bFGF is associated to its dosage and release kinetics. The encapsulation of the growth factors in microspheres to control the delivery is a common way to control the dosage. Perets et al. design a scaffold of sodium alginate embedded with bFGF-loaded PLGA microspheres [39]. In these conditions, the release rate of bFGF from the alginate composite scaffolds was constant and promoted angiogenesis in vivo.

14.2.2.3 Combination of Growth Factors

The use of two or more growth factors is due to the complexity of angiogenic signaling pathways and the difficulty to find an appropriate release rate and dosage of single growth factor to achieve a correct angiogenic process. VEGF in particular showed their potential to promote therapeutic angiogenesis, but the vessels induced frequently display morphological and functional abnormalities [40, 41]. Therefore, the combination of different growth factors, like PDGF, with VEGF improves the neovascularization process [42]. Table 14.1 presents a description of some relevant studies for osteochondral regeneration.

14.2.2.4 Inorganic Cations

Many trace elements such as Mg^{2+}, Si^{4+}, Sr^{2+}, Zn^{2+}, and Na^+ are also present in the bone mineral [50, 51]. In many cases, the use of these cations in the scaffold structure revealed high mechanical properties and improved biological responses [52–54]. In human physiology, magnesium and silicon play a vital role in stabilizing protein structures, modulating cell proliferation, and improving bone formation [53, 55]. In recent studies, Mg^{2+} and Si^{4+} revealed their capacity to induce neovascularization in combination with other cations [54, 56–58]. Tarafder et al. studied the effect of Sr^{2+}- and Mg^{2+}-doped tricalcium phosphate scaffolds on the promotion of angiogenesis in vivo. The presence of Sr^{2+} and Mg^{2+} in combination with a porous structure increases the neovascularization of the scaffold in comparison with the scaffolds without the cations [54]. Fielding and Bose investigated SiO_2 and ZnO as dopants in three-dimensionally printed

Table 14.1 Dual or multiple growth factor delivery systems to promote angiogenesis in osteochondral regeneration

Principal growth factor	Secondary growth factor	Scaffold composition	Biological effect	Reference (year)
VEGF	PDGF	PDLG-loaded PLGA microspheres and VEGF-dispersed alginate	Elevated vessel density and maturity	[42–44] (2001, 2007, 2010)
VEGF	KGF PDGF	Hyaluronic acid hydrogels	Largest vascularization response	[45] (2010)
VEGF	PDGF TGF-β1	Alginate scaffold	Elevated vessel density and maturity	[46] (2009)
bFGF	VEGF	Freeze-dried collagen sponge	Elevated vessel density and maturity	[47] (2007)
		PEG hydrogels	HUVEC migration and vessel formation	[48] (2011)
BMP-2	VEGF	Silk fibroin	Synergistic effect on bone formation and angiogenesis	[49] (2016)

tricalcium phosphate scaffolds. Samples containing SiO_2/ZnO presented higher blood vessel formation with larger vessels and more complex vessel formation [56].

Due to the substantial evidence from literature on the positive effect of Mg^{2+} and Si^{4+}, Bose et al. examined the influence of both cations in 3D-printed TCP scaffolds (3DP). Figure 14.2 shows von Willebrand factor (vWF)-stained tissue sections (an endothelial cell marker for angiogenesis). TCP scaffolds induced new blood vessel formation, but an increase in blood vessel formation, between 3 and 6%, is observed in Mg-Si-TCP compared to pure TCP scaffolds. Histomorphometric analysis revealed significantly higher blood vessel area formation at early time points for Mg-Si-TCP [57]. The effect of these cations over the angiogenic process is due to the upregulated expression of receptors of angiogenic cytokines in stem cells when they found a high concentration of Mg^{2+} or Si^{4+} [59]. Other cations, like Co^{2+}, also show proangiogenic activity [60–62]. Co^{2+} have the capacity to upregulate the hypoxia-inducible factor in cells and activate the angiogenic process by creating a hypoxia-mimicking condition [62]. Recent studies combined fluoride and cobalt into calcium phosphates to obtain a system of antibacterial (fluoride), angiogenic (cobalt), and osteogenic (calcium) activities [61, 60].

14.3 Scaffold Designs for Angiogenic Control in Osteochondral Tissue

Tissue engineering tries to provide different methods for osteochondral tissue regeneration, which has been applied in the reconstruction of many tissues and organs. Traditionally, most of the scaffolds used in tissue engineering come from

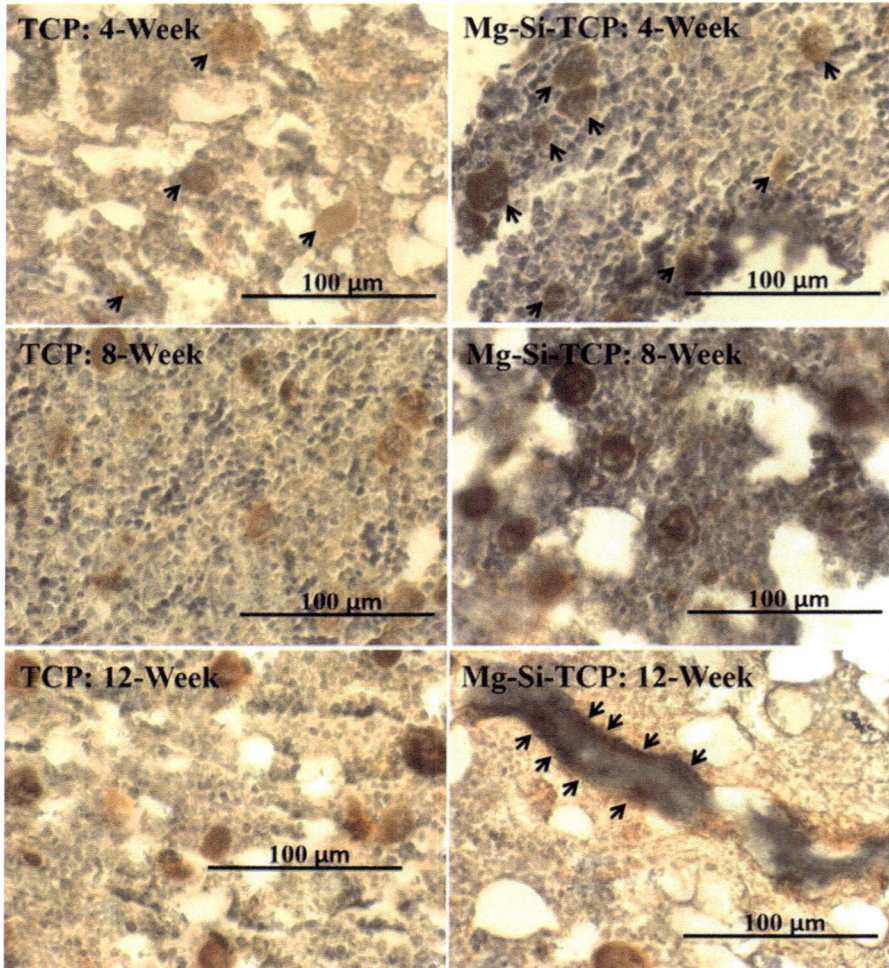

Fig. 14.2 Photomicrograph of vWF-stained tissue sections showing blood vessel formation after 4, 8, and 12 weeks in 3DP pure TCP implants and Mg-Si-doped TCP implants. Arrows indicate newly formed blood vessels inside scaffolds. vWF-positive signals are brown with hematoxylin counterstaining. (Reprinted from [57] with permission of Springer)

natural or bioinspired polymers and are homogeneous in composition [63]. However, the use of traditional homogeneous scaffolds in osteochondral regeneration is limited due to the different properties and requirements of the different strata. Therefore, complex scaffolds with different angiogenic properties (and different mechanical properties) are required to meet the demands (Fig. 14.3).

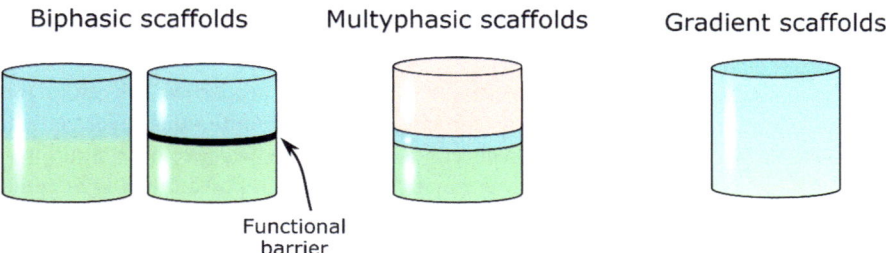

Fig. 14.3 Different kinds of scaffold designs for osteochondral regeneration

14.3.1 Biphasic Scaffolds

The most common strategy to mimic the osteochondral tissue is through biphasic materials composed of natural or synthetic polymers and loaded with the needed growth factors or drugs to control the angiogenic process [64]. Natural hydrogels are the most used materials for the chondral layer and degradable ceramics materials, such as hydroxyapatite or TCP for the subchondral bone phase. Recent studies try to design biphasic scaffolds to improve the regeneration of both the bone and cartilage. In this way, Sartori et al. designed a bilayered scaffold based on type I collagen. Chondral layer was synthesized with type I collagen, and in bone layer, magnesium-doped hydroxyapatite (Mg-HA) was coprecipitated into the collagen. The presence of magnesium ions promotes the angiogenic process, and the low release of this cation did not interfere on the cartilage regeneration [65]. One of the problems of promoting angiogenesis is the vascularization and the subsequent calcification of the cartilage area [9]. After stimulation by proangiogenic compounds, the blood vessels can grow from the bony area and invade the cartilaginous compartment. One of the options to avoid the blood vessel invasion is the use of anti-angiogenic compounds as we saw previously [16–25]. Another option that is being investigated is the use of a functional barrier at the cartilage-bone interface [66]. This functional barrier consists in a cell-excluding membrane that prevents the cartilaginous compartment from angiogenic activities. Hunziker et al. studied the use of Millipore and Goretex membranes to avoid blood vessel invasion (Fig. 14.3). These structural barriers were effective in obstructing the upgrowth of blood vessels into the cartilage compartment [66]. However, the appropriate placement of the membrane cannot be easily achieved, and vascularization occurs through the space between the membrane and defect walls.

An alternative therapeutic approach would be to apply a functional barrier, such as an anti-angiogenic substance, which would chemically impede rather than physically block vascular upgrowth into the cartilaginous compartment [27]. Hunziker et al. used in further researches this principle to avoid blood vessel invasion by adding suramin to fibrin-gelatin matrix used in osteochondral regeneration [27].

14.3.2 Multiphasic Scaffolds

Osteochondral tissue is a multiphasic tissue with different properties. Therefore, the development of multiphasic scaffolds is a research field in continuous development. From the point of view of angiogenesis control, recent studies developed multiphasic scaffolds following the idea of the functional barrier to prevent the angiogenic process on the chondral layer [67–69]. Levingstone et al. developed a multiphasic scaffold based on collagen mixed with other compounds depending on the layer (hydroxyapatite for the bone layer, different kinds of collagen and hydroxyapatite for the intermediate layer, and hyaluronic acid for the cartilage layer). The inclusion of the intermediate layer promoted the formation of a tidemark that restricted the formation of bone and vessel to the subchondral region [67, 70, 68].

The functional barrier effect could be enhanced by the addition of proangiogenic factors to the bony layer. Kon et al. developed a triphasic scaffold, mimicking the whole osteochondral anatomy. The mineral phase consists of a mineralized blend of type I collagen and Mg-doped hydroxyapatite (MG-HA). The cartilaginous layer, consisting of type I collagen and the intermediate layer, consists of a combination of type I collagen and Mg-HA [71]. Currently, this scaffold is commercially available with the name of MaioRegen®.

14.3.3 Gradient Scaffolds

Continuous gradient scaffold is the principal alternative to bilayer and multilayer scaffold. The possibility to display systematic gradients in distribution of angiogenic or anti-angiogenic factors provides additional means for mimicking the variations observed in native tissues. A scaffold with a gradient of angiogenic factors can be designed by immobilization of the molecules to the polymer network or designing delivery systems to provide spatially distinct cues [72].

Mohan et al. design a scaffold with simultaneous gradient of both bioactive signals (growth factor) and material composition [73]. In this study, different growth factors were encapsulated in PLGA, PLGA-HA, and HA microspheres. The scaffold was prepared pumping suspensions of different particles at controlled flow rates over a filter. In this way, the particles remain on the filter until a height of 3 mm [73, 74].

Another way to obtain spatial regulation of controlled bioactive factor is by nano-patterning using irradiation [75]. In this way, Wylie et al. designed a system in which FGF-2 was photochemically patterned in three-dimensional scaffolds. Agarose hydrogels were modified with two-photon labile 6-bromo-7-hydroxycoumarin-protected thiols. After two-photon irradiation, FGF-2 was immobilized through disulfide bonds between free cysteine groups present on FGF-2 and agarose-thiol groups (Fig. 14.4) [76].

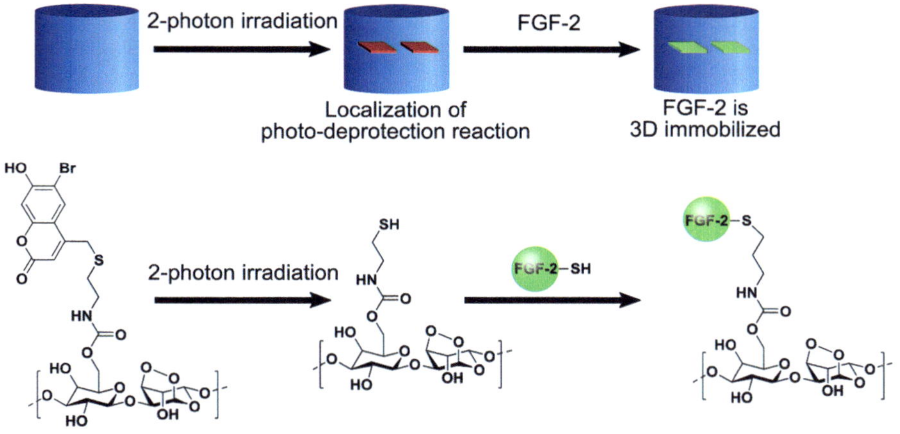

Fig. 14.4 3D immobilization of FGF-2 to agarose through disulfide bonds. (**A**) Schematic diagram demonstrating the 3D photodeprotection of thiols in agarose-thiol-Bhc gels for the coupling of FGF-2. (**B**) FGF-2 was immobilized to agarose-thiol through disulfide bonds. Thiols are deprotected by two-photon excitation of coumarin (740 nm), which subsequently forms disulfide bonds with free cysteines on FGF-2. (Reprinted and adapted with permission from [76]. Copyright 2011 American Chemical Society)

The amount of immobilized FGF-2 was dependent on the number of laser scans; this allows the creation of a spatial concentration gradient of growth factors along the scaffold.

14.4 Conclusion and Future Trends

The control of the different angiogenic process through osteochondral regeneration will generate a better integration and reinforcement of the scaffold, but an uncontrolled angiogenic process can produce some problems like the vascularization and calcification of the cartilage. From the point of view of the angiogenesis control, it is possible to differentiate two zones: the cartilage area, which is an avascular area (in consequence, neovascularization should be avoided), and the bone area, where the promotion of angiogenesis improves the integration and the production of bone tissue.

The complexity of the osteochondral tissue makes the scaffold design for angiogenic control really complicated. The research on biphasic, multiphasic, or gradient scaffolds and the spatiotemporal release of activator or inhibitor angiogenic compounds is under continuous development.

The future of osteochondral angiogenesis will depend on the research development in different directions. The combination of different proangiogenic compounds or pro- and anti-angiogenic compounds in the same scaffold should be the starting point for future research, trying to find the best combination and the optimal spatiotemporal distribution. The application of new methodologies, like 3D print, opens a wide range of opportunities due to the possibility to prepare scaffolds on demand with the desired physical properties and distribution of pro- or anti-angiogenic factors. All of these fields of research will grow further in the future and will expand the application of these systems to unforeseen fields.

References

1 Moutos FT, Freed LE, Guilak F (2007) A biomimetic three-dimensional woven composite scaffold for functional tissue engineering of cartilage. 6:162. https://doi.org/10.1038/nmat1822

2 O'Keefe RJ, Mao J (2011) Bone tissue engineering and regeneration: from discovery to the clinic—an overview. Tissue Eng Part B Rev 17(6):389–392. https://doi.org/10.1089/ten.teb.2011.0475

3 Meachim G, Stockwell RA (1979). The matrix. In M.A.R. Freeman (Ed.), Adult Articular Cartilage (Second edition, pp.1–67) Pitman medical

4 Franzen A, Inerot S, Hejderup SO, Heinegard D (1981) Variations in the composition of bovine hip articular cartilage with distance from the articular surface. Biochem J 195(3):535–543

5 Kawcak CE, CW MI, Norrdin RW, Park RD, James SP (2001) The role of subchondral bone in joint disease: a review. Equine Vet J 33(2):120–126

6 Madry H, van Dijk CN, Mueller-Gerbl M (2010) The basic science of the subchondral bone. Knee Surg Sports Traumatol Arthrosc 18(4):419–433. https://doi.org/10.1007/s00167-010-1054-z

7 García-Fernández L, Halstenberg S, Unger RE, Aguilar MR, Kirkpatrick CJ, San Román J (2010) Anti-angiogenic activity of heparin-like polysulfonated polymeric drugs in 3D human cell culture. Biomaterials 31(31):7863–7872. https://doi.org/10.1016/j.biomaterials.2010.07.022

8 Keeney M, Pandit A (2009) The osteochondral junction and its repair via bi-phasic tissue engineering scaffolds. Tissue Eng Part B Rev 15(1):55–73. https://doi.org/10.1089/ten.teb.2008.0388

9 Franses RE, McWilliams DF, Mapp PI, Walsh DA (2010) Osteochondral angiogenesis and increased protease inhibitor expression in OA. Osteoarthr Cartil 18(4):563–571. https://doi.org/10.1016/j.joca.2009.11.015

10. Nukavarapu SP, Dorcemus DL (2013) Osteochondral tissue engineering: current strategies and challenges. Biotechnol Adv 31(5):706–721. https://doi.org/10.1016/j.biotechadv.2012.11.004

11. Hayami T, Funaki H, Yaoeda K, Mitui K, Yamagiwa H, Tokunaga K, Hatano H, Kondo J, Hiraki Y, Yamamoto T, Duong LT, Endo N (2003) Expression of the cartilage derived anti-angiogenic factor chondromodulin-I decreases in the early stage of experimental osteoarthritis. J Rheumatol 30(10):2207–2217

12. Deng B, Chen C, Gong X, Guo L, Chen H, Yin L, Yang L, Wang F (2017) Chondromodulin-I expression and correlation with angiogenesis in human osteoarthritic cartilage. Mol Med Rep 16(2):2142–2148. https://doi.org/10.3892/mmr.2017.6775

13. Chim SM, Tickner J, Chow ST, Kuek V, Guo B, Zhang G, Rosen V, Erber W, Xu J (2013) Angiogenic factors in bone local environment. Cytokine Growth Factor Rev 24(3):297–310. https://doi.org/10.1016/j.cytogfr.2013.03.008

14. Athanasiou KA, Darling EM, Hu JC (2009) Articular cartilage tissue engineering. Synth Lect Tissue Eng 1(1):1–182. https://doi.org/10.2200/S00212ED1V01Y200910TIS003

15. Guo X, Liao J, Park H, Saraf A, Raphael RM, Tabata Y, Kasper FK, Mikos AG (2010) Effects of TGF-beta3 and preculture period of osteogenic cells on the chondrogenic differentiation of rabbit marrow mesenchymal stem cells encapsulated in a bilayered hydrogel composite. Acta Biomater 6(8):2920–2931. https://doi.org/10.1016/j.actbio.2010.02.046

16. Xing S-C, Liu Y, Feng Y, Jiang C, Hu Y-Q, Sun W, Wang X-H, Wei Z-Y, Qi M, Liu J, Zhai L-J, Wang Z-Q (2015) Chondrogenic differentiation of ChM-I gene transfected rat bone marrow-derived mesenchymal stem cells on 3-dimensional poly (L-lactic acid) scaffold for cartilage engineering. Cell Biol Int 39(3):300–309. https://doi.org/10.1002/cbin.10393

17. Zhang X, Prasadam I, Fang W, Crawford R, Xiao Y (2016) Chondromodulin-1 ameliorates osteoarthritis progression by inhibiting HIF-2α activity. Osteoarthr Cartil 24(11):1970–1980. https://doi.org/10.1016/j.joca.2016.06.005

18. Feng Y, Wu YP, Zhu XD, Zhang YH, Ma QJ (2005) Endostatin promotes the anabolic program of rabbit chondrocyte. Cell Res 15(3):201–206. https://doi.org/10.1038/sj.cr.7290287

19. O'Reilly MS, Boehm T, Shing Y, Fukai N, Vasios G, Lane WS, Flynn E, Birkhead JR, Olsen BR, Folkman J (1997) Endostatin: an endogenous inhibitor of angiogenesis and tumor growth. Cell 88(2):277–285. https://doi.org/10.1016/S0092-8674(00)81848-6

20. Jeng L, Olsen BR, Spector M (2012) Engineering endostatin-expressing cartilaginous constructs using injectable biopolymer hydrogels. Acta Biomater 8(6):2203–2212. https://doi.org/10.1016/j.actbio.2012.02.015

21. Kubo S, Cooper GM, Matsumoto T, Phillippi JA, Corsi KA, Usas A, Li G, Fu FH, Huard J (2009) Blocking VEGF with sFlt1 improves the chondrogenic potential of mouse skeletal muscle-derived stem cells. Arthritis Rheum 60(1):155–165. https://doi.org/10.1002/art.24153

22. Marsano A, Medeiros da Cunha CM, Ghanaati S, Gueven S, Centola M, Tsaryk R, Barbeck M, Stuedle C, Barbero A, Helmrich U, Schaeren S, Kirkpatrick JC, Banfi A, Martin I (2016) Spontaneous in vivo Chondrogenesis of bone marrow-derived mesenchymal progenitor cells by blocking vascular endothelial growth factor signaling. Stem Cells Transl Med 5(12):1730–1738. https://doi.org/10.5966/sctm.2015-0321

23. Mulligan RC (1993) The basic science of gene therapy. Science (New York, NY) 260(5110):926–932

24. Peniche H, Reyes-Ortega F, Aguilar MR, Rodríguez G, Abradelo C, García-Fernández L, Peniche C, Román JS (2013) Thermosensitive macroporous cryogels functionalized with bioactive chitosan/bemiparin nanoparticles. Macromol Biosci 13(11):1556–1567. https://doi.org/10.1002/mabi.201300184

25. Centola M, Abbruzzese F, Scotti C, Barbero A, Vadalà G, Denaro V, Martin I, Trombetta M, Rainer A, Marsano A (2013) Scaffold-based delivery of a clinically relevant anti-angiogenic drug promotes the formation of in vivo stable cartilage. Tissue Eng A 19(17–18):1960–1971. https://doi.org/10.1089/ten.tea.2012.0455

26. Firsching-Hauck A, Nickel P, Yahya C, Wandt C, Kulik R, Simon N, Zink M, Nehls V, Allolio B (2000) Angiostatic effects of suramin analogs in vitro. Anti-Cancer Drugs 11(2):69–77

27. Hunziker EB, Driesang IMK (2003) Functional barrier principle for growth-factor-based articular cartilage repair. Osteoarthr Cartil 11(5):320–327. https://doi.org/10.1016/S1063-4584(03)00031-1

28. Yousefi A-M, Hoque ME, Prasad RGSV, Uth N (2015) Current strategies in multiphasic scaffold design for osteochondral tissue engineering: a review. J Biomed Mater Res A 103(7):2460–2481. https://doi.org/10.1002/jbm.a.35356

29. Deckers MML, Karperien M, Van Der Bent C, Yamashita T, Papapoulos SE, Löwik CWGM (2000) Expression of vascular endothelial growth factors and their receptors during osteoblast differentiation. Endocrinology 141(5):1667–1674. https://doi.org/10.1210/en.141.5.1667

30. Mayr-wohlfart U, Waltenberger J, Hausser H, Kessler S, Günther KP, Dehio C, Puhl W, Brenner RE (2002) Vascular endothelial growth factor stimulates chemotactic migration of primary human osteoblasts. Bone 30(3):472–477. https://doi.org/10.1016/S8756-3282(01)00690-1

31. Midy V, Plouet J (1994) Vasculotropin/vascular endothelial growth factor induces differentiation in cultured osteoblasts. Biochem Biophys Res Commun 199(1):380–386. https://doi.org/10.1006/bbrc.1994.1240

32. Duan X, Murata Y, Liu Y, Nicolae C, Olsen BR, Berendsen AD (2015) Vegfa regulates perichondrial vascularity and osteoblast differentiation in bone development. Development 142(11):1984

33. Zavan B, Ferroni L, Gardin C, Sivolella S, Piattelli A, Mijiritsky E (2017) Release of VEGF from dental implant improves Osteogenetic process: preliminary in vitro tests. Materials 10(9). https://doi.org/10.3390/ma10091052

34. Kaigler D, Wang Z, Horger K, Mooney DJ, Krebsbach PH (2006) VEGF scaffolds enhance angiogenesis and bone regeneration in irradiated osseous defects. J Bone Miner Res 21(5):735–744. https://doi.org/10.1359/jbmr.060120

35. García JR, Clark AY, García AJ (2016) Integrin-specific hydrogels functionalized with VEGF for vascularization and bone regeneration of critical-size bone defects. J Biomed Mater Res A 104(4):889–900. https://doi.org/10.1002/jbm.a.35626

36. Kasten P, Beverungen M, Lorenz H, Wieland J, Fehr M, Geiger F (2012) Comparison of platelet-rich plasma and VEGF-transfected mesenchymal stem cells on vascularization and bone formation in a critical-size bone defect. Cells Tissues Organs 196(6):523–533. https://doi.org/10.1159/000337490

37. Sato Y, Shimada T, Takaki R (1991) Autocrinological role of basic fibroblast growth factor on tube formation of vascular endothelial cells in vitro. Biochem Biophys Res Commun 180(2):1098–1102. https://doi.org/10.1016/S0006-291X(05)81179-9

38. Globus RK, Patterson-Buckendahl P, Gospodarowicz D (1988) Regulation of bovine bone cell proliferation by fibroblast growth factor and transforming growth factor beta. Endocrinology 123(1):98–105. https://doi.org/10.1210/endo-123-1-98

39. Perets A, Baruch Y, Weisbuch F, Shoshany G, Neufeld G, Cohen S (2003) Enhancing the vascularization of three-dimensional porous alginate scaffolds by incorporating controlled release basic fibroblast growth factor microspheres. J Biomed Mater Res A 65A(4):489–497. https://doi.org/10.1002/jbm.a.10542

40. Ozawa CR, Banfi A, Glazer NL, Thurston G, Springer ML, Kraft PE, McDonald DM, Blau HM (2004) Microenvironmental VEGF concentration, not total dose, determines a threshold between normal and aberrant angiogenesis. J Clin Invest 113(4):516–527. https://doi.org/10.1172/jci18420

41. von Degenfeld G, Banfi A, Springer ML, Wagner RA, Jacobi J, Ozawa CR, Merchant MJ, Cooke JP, Blau HM (2006) Microenvironmental VEGF distribution is critical for stable and functional vessel growth in ischemia. FASEB J: Off Publ Fed Am Soc Exp Biol 20(14):2657–2659. https://doi.org/10.1096/fj.06-6568fje

42. Richardson TP, Peters MC, Ennett AB, Mooney DJ (2001) Polymeric system for dual growth factor delivery. Nat Biotech 19(11):1029–1034

43. Chen RR, Silva EA, Yuen WW, Mooney DJ (2007) Spatio-temporal VEGF and PDGF delivery patterns blood vessel formation and maturation. Pharm Res 24(2):258–264. https://doi.org/10.1007/s11095-006-9173-4

44. Sun Q, Silva EA, Wang A, Fritton JC, Mooney DJ, Schaffler MB, Grossman PM, Rajagopalan S (2010) Sustained release of multiple growth factors from injectable polymeric system as a novel therapeutic approach towards angiogenesis. Pharm Res 27(2):264–271. https://doi.org/10.1007/s11095-009-0014-0

45. Elia R, Fuegy PW, VanDelden A, Firpo MA, Prestwich GD, Peattie RA (2010) Stimulation of in vivo angiogenesis by in situ crosslinked, dual growth factor-loaded, glycosaminoglycan hydrogels. Biomaterials 31(17):4630–4638. https://doi.org/10.1016/j.biomaterials.2010.02.043

46. Freeman I, Cohen S (2009) The influence of the sequential delivery of angiogenic factors from affinity-binding alginate scaffolds on vascularization. Biomaterials 30(11):2122–2131. https://doi.org/10.1016/j.biomaterials.2008.12.057

47. Nillesen ST, Geutjes PJ, Wismans R, Schalkwijk J, Daamen WF, van Kuppevelt TH (2007) Increased angiogenesis and blood vessel maturation in acellular collagen-heparin scaffolds containing both FGF2 and VEGF. Biomaterials 28(6):1123–1131. https://doi.org/10.1016/j.biomaterials.2006.10.029

48. Zieris A, Prokoph S, Levental KR, Welzel PB, Grimmer M, Freudenberg U, Werner C (2010) FGF-2 and VEGF functionalization of starPEG-heparin hydrogels to modulate biomolecular

and physical cues of angiogenesis. Biomaterials 31(31):7985–7994. https://doi.org/10.1016/j.biomaterials.2010.07.021

49. Li B, Wang H, Qiu G, Su X, Wu Z (2016) Synergistic effects of vascular endothelial growth factor on bone morphogenetic proteins induced bone formation in vivo: influencing factors and future research directions. Biomed Res Int 2016:2869572. https://doi.org/10.1155/2016/2869572

50. Banerjee SS, Tarafder S, Davies NM, Bandyopadhyay A, Bose S (2010) Understanding the influence of MgO and SrO binary doping on the mechanical and biological properties of β-TCP ceramics. Acta Biomater 6(10):4167–4174. https://doi.org/10.1016/j.actbio.2010.05.012

51. Bandyopadhyay A, Bernard S, Xue W, Bose S (2006) Calcium phosphate-based resorbable ceramics: influence of MgO, ZnO, and SiO2 dopants. J Am Ceram Soc 89(9):2675–2688. https://doi.org/10.1111/j.1551-2916.2006.01207.x

52. Rojo L, Radley-Searle S, Fernandez-Gutierrez M, Rodriguez-Lorenzo LM, Abradelo C, Deb S, San Roman J (2015) The synthesis and characterisation of strontium and calcium folates with potential osteogenic activity. J Mater Chem B 3(13):2708–2713. https://doi.org/10.1039/C4TB01969E

53. Jugdaohsingh R (2007) Silicon and bone health. J Nutr Health Aging 11(2):99–110

54. Tarafder S, Dernell WS, Bandyopadhyay A, Bose S (2015) SrO- and MgO-doped microwave sintered 3D printed tricalcium phosphate scaffolds: mechanical properties and in vivo osteogenesis in a rabbit model. J Biomed Mater Res B Appl Biomater 103(3):679–690. https://doi.org/10.1002/jbm.b.33239

55. Wu F, Su J, Wei J, Guo H, Liu C (2008) Injectable bioactive calcium-magnesium phosphate cement for bone regeneration. Biomed Mater (Bristol, England) 3(4):044105. https://doi.org/10.1088/1748-6041/3/4/044105

56. Fielding G, Bose S (2013) SiO2 and ZnO dopants in three-dimensionally printed tricalcium phosphate bone tissue engineering scaffolds enhance osteogenesis and angiogenesis in vivo. Acta Biomater 9(11):9137–9148. https://doi.org/10.1016/j.actbio.2013.07.009

57. Bose S, Tarafder S, Bandyopadhyay A (2017) Effect of chemistry on osteogenesis and angiogenesis towards bone tissue engineering using 3D printed scaffolds. Ann Biomed Eng 45(1):261–272. https://doi.org/10.1007/s10439-016-1646-y

58. Dashnyam K, El-Fiqi A, Buitrago JO, Perez RA, Knowles JC, Kim H-W (2017) A mini review focused on the proangiogenic role of silicate ions released from silicon-containing biomaterials. J Tissue Eng 8:2041731417707339. https://doi.org/10.1177/2041731417707339

59. Zhai W, Lu H, Wu C, Chen L, Lin X, Naoki K, Chen G, Chang J (2013) Stimulatory effects of the ionic products from Ca–Mg–Si bioceramics on both osteogenesis and angiogenesis in vitro. Acta Biomater 9(8):8004–8014. https://doi.org/10.1016/j.actbio.2013.04.024

60. Birgani ZT, Gharraee N, Malhotra A, van Blitterswijk CA, Habibovic P (2016) Combinatorial incorporation of fluoride and cobalt ions into calcium phosphates to stimulate osteogenesis and angiogenesis. Biomed Mater (Bristol, England) 11 (1):015020. doi:https://doi.org/10.1088/1748-6041/11/1/015020

61. Zhou J, Zhao L (2016) Multifunction Sr, Co and F co-doped microporous coating on titanium of antibacterial, angiogenic and osteogenic activities. Sci Rep 6:29069. https://doi.org/10.1038/srep29069

62. Perez RA, Kim JH, Buitrago JO, Wall IB, Kim HW (2015) Novel therapeutic core-shell hydrogel scaffolds with sequential delivery of cobalt and bone morphogenetic protein-2 for synergistic bone regeneration. Acta Biomater 23:295–308. https://doi.org/10.1016/j.actbio.2015.06.002

63. Raeisdasteh Hokmabad V, Davaran S, Ramazani A, Salehi R (2017) Design and fabrication of porous biodegradable scaffolds: a strategy for tissue engineering. J Biomater Sci Polym Ed 28(16):1797–1825. https://doi.org/10.1080/09205063.2017.1354674

64. Lopa S, Madry H (2014) Bioinspired scaffolds for osteochondral regeneration. Tissue Eng Part A 20(15–16):2052–2076. https://doi.org/10.1089/ten.tea.2013.0356

65. Sartori M, Pagani S, Ferrari A, Costa V, Carina V, Figallo E, Maltarello MC, Martini L, Fini M, Giavaresi G (2017) A new bi-layered scaffold for osteochondral tissue regeneration:

in vitro and in vivo preclinical investigations. Mater Sci Eng C 70(Part 1):101–111. https://doi.org/10.1016/j.msec.2016.08.027

66. Hunziker EB, Driesang IM, Saager C (2001) Structural barrier principle for growth factor-based articular cartilage repair. Clin Orthop Relat Res 391 Suppl:S182–S189

67. Levingstone TJ, Matsiko A, Dickson GR, O'Brien FJ, Gleeson JP (2014) A biomimetic multi-layered collagen-based scaffold for osteochondral repair. Acta Biomater 10(5):1996–2004. https://doi.org/10.1016/j.actbio.2014.01.005

68. Levingstone TJ, Thompson E, Matsiko A, Schepens A, Gleeson JP, O'Brien FJ (2016) Multi-layered collagen-based scaffolds for osteochondral defect repair in rabbits. Acta Biomater 32(Supplement C):149–160. https://doi.org/10.1016/j.actbio.2015.12.034

69. Frenkel SR, Bradica G, Brekke JH, Goldman SM, Ieska K, Issack P, Bong MR, Tian H, Gokhale J, Coutts RD, Kronengold RT (2005) Regeneration of articular cartilage – evaluation of osteochondral defect repair in the rabbit using multiphasic implants. Osteoarthr Cartil 13(9):798–807. https://doi.org/10.1016/j.joca.2005.04.018

70. Levingstone TJ, Ramesh A, Brady RT, Brama PAJ, Kearney C, Gleeson JP, O'Brien FJ (2016) Cell-free multi-layered collagen-based scaffolds demonstrate layer specific regeneration of functional osteochondral tissue in caprine joints. Biomaterials 87(Supplement C):69–81. https://doi.org/10.1016/j.biomaterials.2016.02.006

71. Kon E, Delcogliano M, Filardo G, Fini M, Giavaresi G, Francioli S, Martin I, Pressato D, Arcangeli E, Quarto R, Sandri M, Marcacci M (2010) Orderly osteochondral regeneration in a sheep model using a novel nano-composite multilayered biomaterial. J Orthop Res 28(1):116–124. https://doi.org/10.1002/jor.20958

72. Cao L, Mooney DJ (2007) Spatiotemporal control over growth factor signaling for therapeutic neovascularization. Adv Drug Deliv Rev 59(13):1340–1350. https://doi.org/10.1016/j.addr.2007.08.012

73. Mohan N, Dormer NH, Caldwell KL, Key VH, Berkland CJ, Detamore MS (2011) Continuous gradients of material composition and growth factors for effective regeneration of the osteochondral interface. Tissue Eng Part A 17(21–22):2845–2855. https://doi.org/10.1089/ten.tea.2011.0135

74. Mohan N, Gupta V, Sridharan BP, Mellott AJ, Easley JT, Palmer RH, Galbraith RA, Key VH, Berkland CJ, Detamore MS (2015) Microsphere-based gradient implants for osteochondral regeneration: a long-term study in sheep. Regen Med 10(6):709–728. https://doi.org/10.2217/rme.15.38

75. Samorezov JE, Alsberg E (2015) Spatial regulation of controlled bioactive factor delivery for bone tissue engineering. Adv Drug Deliv Rev 84:45–67. https://doi.org/10.1016/j.addr.2014.11.018

76. Wylie RG, Shoichet MS (2011) Three-dimensional spatial patterning of proteins in hydrogels. Biomacromolecules 12(10):3789–3796. https://doi.org/10.1021/bm201037j

Chapter 15
Models of Disease

Gema Jiménez, Elena López-Ruiz, Cristina Antich, Carlos Chocarro-Wrona, and Juan Antonio Marchal

Abstract Osteochondral (OC) lesions are a major cause of chronic musculoskeletal pain and functional disability, which reduces the quality of life of the patients and entails high costs to the society. Currently, there are no effective treatments, so in vitro and in vivo disease models are critically important to obtain knowledge about the causes and to develop effective treatments for OC injuries. In vitro models are essential to clarify the causes of the disease and the subsequent design of the first barrier to test potential therapeutics. On the other hand, in vivo models are anatomically more similar to humans allowing to reproduce the pattern and progression of

G. Jiménez (✉) · C. Antich · J. A. Marchal (✉)
Biopathology and Regenerative Medicine Institute (IBIMER), Centre for Biomedical Research (CIBM), University of Granada, Granada, Spain

Biosanitary Research Institute of Granada (ibs.GRANADA), University Hospitals of Granada-University of Granada, Granada, Spain

Department of Human Anatomy and Embryology, Faculty of Medicine, University of Granada, Granada, Spain

Excellence Research Unit "Modeling Nature" (MNat), University of Granada, Granada, Spain
e-mail: gemajg@ugr.es; cantich@ugr.es; jmarchal@ugr.es

E. López-Ruiz

Biopathology and Regenerative Medicine Institute (IBIMER), Centre for Biomedical Research (CIBM), University of Granada, Granada, Spain

Biosanitary Research Institute of Granada (ibs.GRANADA), University Hospitals of Granada-University of Granada, Granada, Spain

Excellence Research Unit "Modeling Nature" (MNat), University of Granada, Granada, Spain

Department of Health Science, University of Jaén, Jaén, Spain
e-mail: elruiz@ujaen.es

C. Chocarro-Wrona
Biosanitary Research Institute of Granada (ibs.GRANADA), University Hospitals of Granada-University of Granada, Granada, Spain

Excellence Research Unit "Modeling Nature" (MNat), University of Granada, Granada, Spain
e-mail: cchocarro@ugr.es

© Springer International Publishing AG, part of Springer Nature 2018
J. M. Oliveira et al. (eds.), *Osteochondral Tissue Engineering*,
Advances in Experimental Medicine and Biology 1059,
https://doi.org/10.1007/978-3-319-76735-2_15

the lesion in a controlled scene and offering the opportunity to study the symptoms and responses to new treatments. Moreover, in vivo models are the most suitable preclinical model, being a fundamental and a mandatory step to ensure the successful transfer to clinical trials. Both in vitro and in vitro models have a number of advantages and limitation, and the choice of the most appropriate model for each study depends on many factors, such as the purpose of the study, handling or the ease to obtain, and cost, among others. In this chapter, we present the main in vitro and in vivo OC disease models that have been used over the years in the study of origin, progress, and treatment approaches of OC defects.

Keywords Osteoarthritis · Osteochondral defects · Disease models · In vitro models · In vivo models

Highlights
- Disease models are essential to understand the pathology of OC defects and to guide the development of new therapeutic approaches.
- In vitro disease models are easy handling and least expensive of OC lesions. These models allow to reproduce the disease conditions in controlled conditions, suitable to elucidate the molecular mechanisms and pathways involved in the pathology.
- In vivo disease models give the possibility to obtain a more accurate idea of the naturally occurring OC defects. These models are the previous step before the clinical trials of new therapeutic interventions. It is important to select the in vivo model most suitable to the purpose of the study in order to transfer the data obtained to the human.

15.1 Introduction

Osteochondral (OC) lesions can originate from diverse causes, varying from trauma-related lesions to naturally occurring degradation that involve the degeneration of articular cartilage, subchondral bone, and the OC tissue complex [103]. Osteoarthritis (OA) is the major cause of natural OC lesions. It is a chronic and degenerative disease of the articular joint involving the bone, synovium, cartilage, ligaments, tendon, meniscus, and periarticular muscle [34]. Moreover, OA is a disease that decreases the patient's quality of life and that entails important socio-economic impact [35, 54]. The proportions of people suffering symptomatic knee OA are likely to increase due to the aging of the population and the high rate of overweight, expecting that OA incidence will rise in the near future and at least 40% of patients aged above 65 years will be affected [74]. Depending on the severity of the lesion, there are different OA treatment options: microfractures, mosaicplasty, autologous chondrocyte implantation (ACI), matrix-induced autologous chondrocyte implantation (MACI), and, ultimately, prosthesis.

Although these current treatments increase the patient's quality of life and reduce pain, the lack of regenerative potential restricts the full healing of articular cartilage [87]. In addition, it has not yet developed disease-modifying anti-OA drugs (DMOADs) with enough safety properties and efficacy to its use in the clinical ambit [9, 62]. The main cause for this shortcoming is in part due to the poor knowledge of the mechanisms involved in the initiation and progression of the pathology. To overcome these limitations, experimental models of the disease can be very useful to reproduce such complex processes and to gather insights about tissue development and regeneration of OC lesions. In this sense, several approaches have been addressed to study the development of OC tissue, the evolution of OC lesions, and potential treatments.

In order to study cartilage or bone tissue development, and also strategies to induce chondrogenesis or osteogenesis differentiation, setting up the proper 3D cell culture model could help to elucidate the underlying mechanisms that influence cellular interactions, morphogenetic events, and the deposition of the extracellular matrix (ECM). Among these models, three main types have been developed: high-density micromass cultures, cell pellet cultures, and 3D cell cultures on biomaterial scaffolds. High-density micromass culture method and pellet culture system have been widely applied to analyze the regulation of cellular condensation and differentiation [110], the maturation of cartilage analogue [112], as well as cartilage hypertrophy [13] and calcification [66]. On the other hand, many natural and synthetic biomaterials such as sponges, fibrils, or hydrogels have been tested to generate scaffolds for the support of cartilage and bone formation. Among the most widely used synthetic polymers, poly(glycolic acid) (PGA), poly(lactic acid) (PLA), and poly(ε-caprolactone) (PCL) [107] present better properties. Regarding natural polymers, collagen type I and chitosan are the most commonly used to form rigid 3D structures [84]. Further, the addition of growth factors allows to create the most appropriate and close to natural molecular environments that enables to develop the experimental models of OC disease as suitable as possible [39, 96].

Although the development of in vitro models is essential, to deal with the OC lesions or defects, it is necessary to use in vivo models of disease to understand how the disease develops and testing potential treatment approaches. In this chapter, we present the main in vitro and in vivo OC disease models that have been used over the years in the origin, progress, and new treatment approaches (Fig. 15.1). We address the different options offered by both models, making a review of the advantages and limitations of each model and the possibilities offered according to the study to be carry out. In addition, this review also covers the role of potential OC disease models as preclinical studies and the requirements that the regulatory rules establish.

15.2 In Vitro Models

In vitro models are based on simulation of the OC intra-articular environment in cellular cultures through the effect of diverse factors, such as the action of pro-inflammatory molecules or mechanical loads among others. These models are characterized by being easier to manipulate and less expensive because they require less time to be carried out. Currently, there are a great number of in vitro OC disease models that can be broadly grouped according to the specimen, either cell cultures, 3D cultures, or ex vivo tissues. In this context, it is possible to find descriptive models to detect specific pathological hallmarks and functional models to elucidate the molecular mechanisms and pathways involved in pathophysiology [41].

In this chapter, in vitro disease models have been classified into 2D and 3D depending on how the culture is performed (Fig. 15.1).

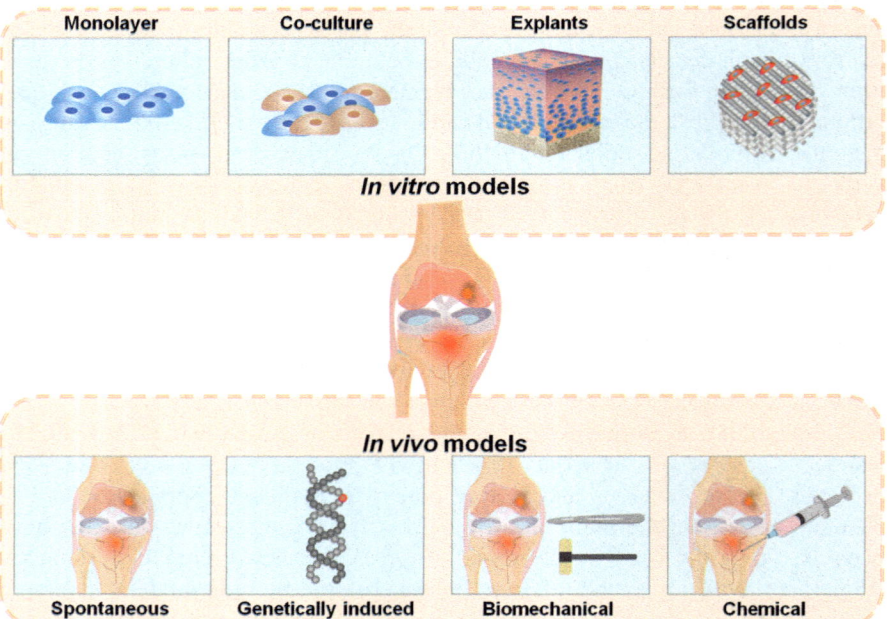

Fig. 15.1 Osteochondral disease models. The schematic shows the different options of in vitro and in vivo models available for studying OC defects

15.2.1 2D Models

One of the classic in vitro models to study OC disease is monolayer cell culture. This model can be developed from primary or immortalized cell lines, and together with its easy availability, management and expansion capacity make it the most widely used [17, 77, 106]. Given that cartilage destruction is one of the most common characteristics of OC diseases such OA, the first model option is chondrocyte culture. It has been used to perform comparative studies between cells derived from healthy and pathological tissue, allowing the identification of new factors that could participate in the disease development [3, 5, 33, 92]. Moreover, chondrocyte culture has enabled the investigation of the mechanisms that underlie OC-related diseases and the role of associated factors such as mechanical stress, inflammation, or susceptibility genes. In order to induce OC monolayer model and exploration of subsequent effect, mechanical loading [101], cytokine stimulation [77, 119], or candidate genes alterations (via loss- or gain-of-function mutations) are used [88, 121].

Although monolayer cultures are the main way to research on this field, chondrocyte culture instability offers the possibility to study them without inductive stimuli requirement, since the changes that occur during dedifferentiation are very similar to those occurring naturally in the disease. Otherwise, deregulation of bone homeostasis can also lead to several OC diseases [59, 81]. Hence, in OA there is a debate about whether it begins in the cartilage or in the subchondral bone component. Therefore, in vitro model based on osteocyte culture has also been used to study bone remodeling in such diseases. As chondrocyte cultures, comparative studies have been performed between osteocytes derived from healthy and pathological tissue to identify disease hallmarks and new factors involved in bone remodeling [93]. Also, it has been analyzed the roles of mechanical load and cytokines in such process [82, 111] .

Recognizing that OC disease affects and involves the interaction between multiple tissues [9], models have been developed based on cellular coculture in monolayer. This model has enabled to contemplate in vitro molecular and cellular interactions that could mean key mechanisms to disease development. First attempts have used chondrocytes-osteoblasts coculture exposed to certain factors (cytokines or mechanical stress) or co-cultives in which some of its cellular components derived from pathological tissue [55, 94]. Currently, synovium-cartilage cocultures are being employed as models to elucidate the role of the synovium in some OC disease or, alternatively, as a method to reproduce in vitro complexity of the pro-inflammatory cascade, because synovium is the primary source of these mediators [6].

Although the 2D culture models previously described have been suitable for exploration of particular aspects of diseases, it is well known that cell behavior in 2D is often not a representative of what occurs in physiologically relevant 3D contexts. Therefore, it has been required that other models be able to reproduce the complexity of pathological in vivo process.

15.2.2 3D Models

Explant-based models are the first approach to study in vitro OC disease in physiologically 3D contexts. This model is based on isolated OC tissue that will allow to assess the response of cells embedded in their natural ECM and elucidate the contribution of different tissues in the joint. On one hand, explants derived from pathological tissue can provide not only analytical data about disease features but also visually, as well as data progression (spatial and temporal changes) [92, 99]. On the other hand, explants derived from healthy tissue can be used to develop functional studies. As in monolayer model, in 3D models the pathogenesis has been easily induced by experimentally controlled conditions, such as enzymatic cleavage, mechanical load, and cytokine stimuli [16, 21, 41]. However, explant-based models also have several drawbacks, such as a limited availability, ethical considerations referred to tissue obtaining, and high inter-experimental variability.

Recently, tissue engineering has developed a novel in vitro model that combines cell culture and explant-based approaches. This 3D culture system involves both scaffolds and cell component organized in a pattern that mimics OC complex. Generally, engineered constructs are multiphasic structures that include a stiff, porous section containing osteoblasts, corresponding to subchondral bone, and a hydrated, viscoelastic section containing chondrocytes referring to cartilage region [7]. In some cases, this structure also shows a transitional zone, but this is almost always an unexplored phenomenon. In OC disease research, they have been used as a model for functional studies that allows to elucidate the pathogenesis and sequence of signaling events during disease development and to study potential OA therapeutic molecules. However, their use is not yet widespread, due to the difficulties that OC unit culture entails. Until now, tissue engineering has attempted to develop an in vitro 3D disease model for OA based on a single-phase construct. In such approach, cells from one of OC regions, generally chondrocytes, were cultured within hydrogel or scaffolds under disease conditions (inflammation, overload, or physical injury) [7, 14, 75], resulting in an increased expression of chemokines, cytokines, growth factors controlling ECM remodeling, ossification, and apoptosis.

15.3 In Vivo Models

Animal models are useful to gain knowledge about the causes and progression of OC diseases, as well as to develop and evaluate new diagnostic tools and therapeutic approaches. The ideal animal model must be as close as possible to humans with respect to cartilage and bone characteristics, progression of joint damage and pain, diagnosis, and tissue engineering [105]. When animal model is necessary for a research, it is recommended to apply the "3R" principles: reduce the number of animals used, refine to enhance efficiency of the experimental protocol, and replace animal studies with an alternative approach whenever possible [32, 58].

OC disease animal models are classified in small and large animals. Small animal models include mouse, rat, rabbit, guinea pig, and zebrafish, and large animal models include dog, goat/sheep, and horse. The choice of the animal depends on several factors such as the type of study, length of time, costs, easy of handling, ability to use routine diagnostic imaging, and outcome measurements (Table 15.1). In general, small animal models are faster, cheaper, and easier to implement. These models are used to study the pathogenesis and pathophysiology as the first screening model. On the other hand, the anatomy of large animal models is more similar to humans and is easy for diagnostic imaging. Large animal models are used to study the disease process and treatment, generating more clinically relevant data, and are required as preclinical information before approval of therapeutic interventions [25, 51].

Nowadays, there are different forms to develop or induce OC defects, which include spontaneous models, genetically modified models, and experimentally induced models (Fig. 15.1). A major challenge consists in choosing the model that is best suited to the study requirements, taking into account that there is not a single model that accurately reflects all human aspects [83].

15.3.1 Spontaneous Models

Certain breeds of animals are genetically predisposed to develop naturally OC diseases, for which it has constituted a very suitable spontaneous model. The spontaneous OA model is the most represented and studied, since it allows the characterization of the OA process at different levels: development, grossly, histological, and biochemical. Moreover, it presents an important advantage with respect to other disease models, because it permits addressing the changes associated to the disease [122] and studying the therapies that prevent the progression [90, 104].

OA spontaneous models include small (mouse, rabbit, and guinea pig) and large animal models (dog, nonhuman primates, and horse). Mice models include STR/ORT, C57Bl/6, and BALB/c strains, which allow to study the genetic subjacent to OA development [38]. Moreover, these mice models provide information on diverse processes associated with cartilage degeneration, such as the role of inflammatory markers, collagen type II degradation, loss of glycosaminoglycans, disturbed protein transport, or sugar synthesis [52, 102, 117]. Nowadays, the Dunkin Hartley guinea pig has been positioned as the most widely used animal to study spontaneous OA model [40]. The studies performed with guinea pig evidenced the role of the subchondral bone in the development of the OA, showing that subchondral bone is fragile and its ultrastructure changes before the onset of cartilage degeneration [76]. Guinea pig spontaneous model has provided other contributions to OA knowledge, like alteration of TGF-β signaling [124], collagen type II deposition in cruciate ligament before OA [120], the increment of collagen and proteoglycan content prior to OA development, and its consequent decrease once OA is evident [114]. Further, the potential of new therapies such as RNA interference [95], gene therapy [36], and human mesenchymal stem cells [97] has also been evaluated in guinea pig models.

Table 15.1 Characteristics of animal models for OC defects

Specie	Skeletal maturity	Advantages	Disadvantages
Mouse	10 weeks	Low cost Easy handling Transgenic and knockout models available Lesions develop rapidly Genome sequenced	Thin cartilage Size limits: (i) tissue and fluid collection and (ii) technical and image procedures
Rat	12 weeks	Low cost Easy handling Genome sequenced Lesions develop rapidly Zonal structure of cartilage comparable to human	Size limits: (i) tissue and fluid collection and (ii) technical and image procedures
Rabbit	8–9 months	Easy handling Size facilitates (i) tissue and fluid collection and (ii) technical and image procedures Lesions develop rapidly	More expensive Spontaneous healing of cartilage lesions Gait generates different load characteristics that impedes functional studies
Guinea pig	6 months	Histopathology and biochemistry close to human Relative easy handling Size facilitates (i) tissue and fluid collection and (ii) technical and image procedures	More expensive Sedentary lifestyle
Dog	9–18 months	Relative easy handling Share with humans: Treatments and rehabilitation regime Size facilitates (i) tissue and fluid collection and (ii) technical and image procedures	High cost Ethical considerations
Goat/ sheep	2 years	Relative easy handling Large joint Biomechanically similar to human Size facilitates (i) tissue and fluid collection and (ii) technical and image procedures	High cost Limitations in oral treatments due to ruminant digestive
Horse	2 years	Large joint Large defects similar to human Biomechanically similar to human Active lifestyle similar to athletes Size facilitates (i) tissue and fluid collection and (ii) technical and image procedures	High cost Difficult handling

Table adapted from Refs. [22, 25, 70, 106]

In general, the anatomy of large animal models is more similar to that of humans, so that studies of cartilage degeneration and OC defects are much more useful. For example, it has been proven that changes in cartilage and subchondral bone in the development of spontaneous OA in dogs and nonhuman primates, such as baboons, rhesus macaque, and cynomolgus macaque, are very similar to human OA [12, 57, 85]. However, ethical considerations and the proximity with human have also been given as reasons for their exclusion for research. On the other hand, horse articular cartilage is the most similar animal model to the human cartilage with similar thickness, cellular structure, and biochemical-biomechanical characteristics, making it the most suitable and useful translational model [19, 60]. In addition, as it happens in humans, the OA in horse is related with the activity and lifestyle [65], so the use of horse as spontaneous animal model has contributed to OA knowledge, for example, with the development of advanced imaging modalities [43], identification of fluid-based biomarkers [18, 109], and bone cysts and osteophyte formation as a consequent of bone remodeling [78] in OC defects. Moreover, horse model has contributed to find interesting treatments for joint pathology, such as cell [47] and gene therapy [20].

15.3.2 Genetically Induced Models

Genetically induced models (or transgenic models) are obtained as a consequence of the modification of the genome that can involve the addition or deletion of genetic material. This model is useful for studying the function of a specific gene and its interaction with others [56]. Genetically modified mice are the most useful and best tool for studying the functional role of a particular gene or specific molecule. Moreover, they have played a crucial role in understanding specific genetic contributions in the development of OC pathologies such as OA [56], chondrodysplasia [113], and bone abnormalities [26, 28].

Knockout and/or overexpressed target genes are related with cartilage matrix components and matrix degradation, chondrocyte differentiation or apoptosis, bone turnover, and/or inflammation [56]. The components of the cartilage ECM are the target of principal genetic modifications in the creation of mice models, such as mutation or inactivation of collagen type II (Col2a1) [30], collagen type IX (Col9a1) [4], aggrecan [30], and COMP [10]. Also, genetically modified models allow to study another very important aspect of OC pathogenesis such as the role of matrix-related enzymes [68]. The creation of matrix metalloproteinases (MMPs) knockout mice permits to investigate the potential in vivo effect of these enzymes, which participate in the normal physiology of connective tissues during development, morphogenesis, bone remodeling, and wound healing [80]. Moreover, thanks to knockout mice for these enzymes inhibitors, called tissue inhibitors of metalloproteinases (TIMPs), it is known that TIMPs lead to develop joint pathologies [91]. It has also been successfully demonstrated that ablation of a disintegrin and metalloproteinase with thrombospondin motifs (ADAMTS) leads to cartilage degeneration similar to that in OA patients [37, 89]. On the other hand, overexpression of cathepsin K in mice models has made known the role of this enzyme in the development of osteopenia in the trabecular bone and in the increment of porosity in the cortical bone, which leads to OC defects [73].

Recently, zebrafish has been proposed as a potential genetically modified model to study skeletogenesis [24, 45] and OA [50]. For instance, a stable transgenic zebrafish for collagen type 10a1 promoter that drives GFP reporter expression in bone structures provides a useful tool for identifying osteoblast-specific molecular pathways [46]. Mitchell et al. describe the expression pattern of six OA susceptibility genes in zebrafish larvae (Mcf2l, Gdf5, PthrP/Pthlh, Col9a2, and Col10a1) and the generation of two new transgenic lines for collagen type 2a1 and 10a1 [69]. Although zebrafish represents a great model to study skeletal development, it presents some limitations that mean it will not replace the mammalian models but may complement the gap between in vitro models and mice models.

The advantage of genetically modified models is that it is possible to control the effect of specific genetic contributions to the pathogenesis; however, these models simplify the process by not taking into account other contributing genes in a polygenic pathology.

15.3.3   Experimentally Induced Models

Experimentally induced models that involve creation of OC lesions include biomechanical and chemical models. These models present the advantage that do not require a long waiting time and are less variable compared to genetically induced or spontaneous models. Moreover, they are more representative of severe alterations than slower or spontaneous changes that occur during the development of early stages of OC damage and in OA.

15.3.3.1   Biomechanical Models

Biomechanical models include surgical and nonsurgical models that induce structural damage and/or instability. As is necessary an adequate number of cell layers and complex zonal cartilage structure to mimic the OC lesion, ideal surgical models include rodent and non-rodent animal models [8]. A number of different surgical procedures are available, among the most used included medial meniscal tear, partial meniscectomy, complete meniscectomy, or anterior cruciate ligament transection (ACLT). Meniscal injuries have a poor self-healing potential and lead to degenerative changes in the knee. Meniscal tears have been commonly performed to evaluate the use of cell-load scaffolds to improve the healing and progress of the lesion and in order to find methods to decrease the need for partial meniscectomy [79, 116]. Medial meniscal tear model can be created after a lateral parapatellar arthrotomy, which is performed to identify the medial meniscus where an incision at the middle segment is made [64]. Size and characteristics of meniscal tears performed vary depending on the animal used [42, 123]. Partial meniscectomy of the injured portion of the meniscus is the most common surgical treatment for meniscal tears; however, removal of the meniscus increases contact pressures on the cartilage and alters the knee kinematics. Consequently, partial meniscectomy has been used as a model to induce OC lesions. Differences in load bearing on the menisce exist among each animal model, and, therefore, the site for the surgical procedure also varies. Thus, rabbits load their lateral meniscus more than their medial, while human and pigs often load the medial site of the knee [51]. Partial meniscectomy has been carried out to the medial meniscus of several animal models such as rabbit and sheep, in order to analyze histological and/or biochemical changes of OC tissue [27]. A similar mechanism of injury is performed in total meniscectomy; however, compared to partial meniscectomy, this model results in severe OC lesions. Large animals like sheep are often used. For example, Kon et al. performed total meniscectomies in six sheep, and after 12 months cartilage damage and exposed subchondral bone were observed in all the animals [48].

Anterior cruciate ligament (ACL) injury is common in athletic and youth populations [100] and constitutes one of the most popular surgery for induced biochemical models. Compared to meniscectomy, the ACLT model induces OC lesions that progress more slowly. Researchers have used animal models of ACLT to obtain information about the early stages of OA to examine the relationship between subchondral bone changes and cartilage degradation and for evaluating pharmacotherapies. ACLT has been extensively described in many species, such as rat, rabbit, dog, and sheep, being the last the most indicated as it has the closest anatomy to the human knee.

Otherwise, chondral and OC defects have been surgically created on articular surface of femoral condyles for studying the clinical applicability of scaffolds for cartilage and subchondral bone repair [63]. For example, Yan et al. have used the rabbit knee critical size as OC defect models to test the OC defect regeneration using a bilayered scaffold composed by a silk fibroin layer and another layer of silk-nano calcium phosphate [118]. He et al. [29], in order to study the implantation of engineered cartilage for tissue-specific repair of OC defects, created two defects with a cylindrical shape (10-mm diameter, 4-mm depth) into the subchondral bone at the weight-bearing area of the femoral medial and lateral condyles in a swine model [29].

Extra-articular procedures could also induce OC lesions; among these procedures ovariectomy (OVX) models are useful to study OC changes associated with postmenopausal women. Indeed, it has been demonstrated that OVX induced deleterious effects on rat cartilage, leading to an early stage of OA. For this reason, OVX models are often used for the study of different antiosteoporotic agents and antiresorptive therapies to prevent bone loss and, therefore, cartilage damage [67]. On the other hand, physical alteration models have been proposed as induced models that do not need any chemical or surgical intervention. The procedure consists in the creation of an injury with precision and without causing skin lesions through a mechanical impact. As a result, subchondral bone injury, edema, microfractures, and OA changes are induced [53].

15.3.3.2 Chemical Factors

As another approach to induce OC lesions, experimental structural models have used chemical factors by intra-articular administration of chemical agents, which induce alteration of the structural components of the joints. One chemically established model consists in the use of collagenase. Intra-articular injection of collagenase causes the digestion of collagen from the ECM, resulting in degeneration of the articular cartilage and changes in other articular structures such as the synovial membrane and the subchondral bone [49]. This model has been used for the study of histopathological alterations of the knee joint after pharmacologic treatments, particularly in mice, rats, and rabbits [1]. For example, in a mouse model, histomorphometric analysis and in vivo imaging revealed a high degree of cartilage proteoglycan loss after 8 weeks of collagenase injection [31].

Other compounds have been used to create OC injuries at the joint structures. Among them, proteolytic enzyme papain, which has been used to degrade proteoglycans of joint cartilage, has triggered the digestion of the noncalcified cartilage layer [44]. Also, monosodium iodoacetate has been intra-articularly injected to induce cell death through inhibition of the enzyme glyceraldehyde-3-phosphate dehydrogenase, which disrupts the cellular aerobic glycolysis pathway in chondrocytes [11]. Quinolone antibiotics also cause irreversible loss of cartilage components such as proteoglycans and degenerative lesions in the upper layers from the beginning of the treatment. Therefore, quinolones should be avoided for studying the effect of other animal models of OA [98]. In addition, factors including fibroblast growth factor-2, bone morphogenetic protein-4, or inhibitors of transforming growth factor beta have also been used to create OC injuries at the joint [53].

15.4 Preclinical Considerations of Osteochondral Disease Models

Animal models provide a useful model to study the pathogenesis of the disease, the therapeutic security, and the efficacy of the new treatments and significantly contribute to the translatability of new treatments toward clinical realization [115]. However, there is not just one model that represents all features of the disease, and this translation of the results varies depending on the model chosen [61].

In 1999, the US Food and Drug Administration (FDA) provided draft guidelines in an attempt to make OA animal models properly used. In this guidance, the FDA proposes five questions that should be considered when an OA animal model is used and that have not been modified to this day:

1. How accurately does the model replicate human OA?
2. What are the structural determinants of pain and loss of function?
3. Do structural changes (identified with MRI or X-ray) correlate with clinical (pain, motion, weight distribution, gait) or biochemical (cartilage composition, enzymatic activity, pain mediators, receptor expression) markers?
4. Is the model useful for studying prophylactic strategies or for studying structural arrest or reversal?
5. Can the model be used to assess long-term toxicity [15]?

Moreover, Osteoarthritis Research Society International (OARSI) group was created in order to standardize preclinical OA studies, particularly those that use animal models. As a result, it was provided a guideline with recommendation that address some of the deficiencies observed in preclinical studies such as lack of defining clear distinction of OA subsets, established clinical trial endpoints, evaluation of biomarkers, histopathology, and exclusion of other arthritis types. In addition, OARSI develops a new grading system that would meet the guiding principles of simplicity, utility, scalability, extendibility, and comparability [86], as well as the design of a guideline for each animal used in OC disease models and the creation of a tool to facilitate comparison between species [23].

While small animal models are commonly used as proof of concept or degradation and safety profile, large animal studies are required as preclinical data before approval of therapeutic interventions. Large animals more appropriated for preclinical evaluation include dogs, which have been beneficial as natural models in preclinical trials of therapeutic intervention [71, 72], and sheep/goats due to ease of handling, low cost (compared to other large animal species), and similar anatomy with human [64]. However, equine model is the most adequate preclinical model due to the anatomy and morphology, the development of spontaneous or traumatic OA defects [2], and its suitability for magnetic resonance imaging (MRI) evaluation, as recommended by OARSI [108].

15.5   Conclusions

Current treatments for OC defects are not as effective as required, so disease models are a very important source of knowledge to the whole pathological process. It is fundamental to guide the identification of new therapeutic targets that, coupled with the development of novel strategies, could prevent, reduce, or stop the progression of OC diseases or, alternatively, resolve the existing damage of the joint. In vitro models are important tools to elucidate the molecular mechanisms and pathways that are involved in OC tissue physiology under normal and pathological conditions. Also, these models have the advantage of making it possible to obtain high amount of samples that are cheaper and easy to handle in a short time and, however, presents important limitations when it comes to transfer to the clinic. In contrast, in vivo models are more complex, with numerous species and ways of obtaining OC damage. Animal models provide a very close idea of (i) lesions that OC causes, (ii) the development and responses to treatment in the whole animal and (iii) they are closer to what happens in humans, making it the most suitable preclinical model. However, these models are very expensive and ethically debated. Cost, anatomy, maturity, joint biomechanics, length of the experiment, as well as postsurgical protocol must be taken into account when an animal model is used to make the best possible choice for the study. For preclinical approaches, large animal models will better replicate the human clinical scenario, but just one model does not represent all feature of the disease or the treatment, which has forced the OARSI to create a guideline for each animal model in order to ensure the success of the transfer to clinical trials.

Acknowledgements This work was supported by the Ministerio de Economía, Industria y Competitividad (FEDER funds, project RTC-2016-5451-1). G. J. acknowledges the Junta de Andalucía for providing a post-doctoral fellowship. Also, C. A. acknowledges the predoctoral fellowship from the Spanish Ministry of Education, Culture and Sports (BOE-A-2014-13539).

References

1. Adaes S et al (2014) Intra-articular injection of collagenase in the knee of rats as an alternative model to study nociception associated with osteoarthritis. Arthritis Res Ther 16(1):R10
2. Ahern BJ et al (2009) Preclinical animal models in single site cartilage defect testing: a systematic review. Osteoarthr Cartil 17(6):705–713
3. Aigner T, McKenna L (2002) Molecular pathology and pathobiology of osteoarthritic cartilage. Cell Mol Life Sci 59(1):5–18
4. Allen KD et al (2009) Decreased physical function and increased pain sensitivity in mice deficient for type IX collagen. Arthritis Rheum 60(9):2684–2693
5. Asahara H (2016) Current status and strategy of microRNA research for cartilage development and osteoarthritis pathogenesis. J Bone Metab 23(3):121–127
6. Beekhuizen M et al (2011) Osteoarthritic synovial tissue inhibition of proteoglycan production in human osteoarthritic knee cartilage: establishment and characterization of a long-term cartilage-synovium coculture. Arthritis Rheum 63(7):1918–1927
7. Bhattacharjee M et al (2015) Tissue engineering strategies to study cartilage development, degeneration and regeneration. Adv Drug Deliv Rev 84:107–122. Available at: https://doi.org/10.1016/j.addr.2014.08.010
8. Bian Q et al (2012) Osteoarthritis: genetic factors, animal models, mechanisms, and therapies. Front Biosci (Elite Ed) 4:74–100
9. Bijlsma JWJ, Berenbaum F, Lafeber FPJG (2011) Osteoarthritis: an update with relevance for clinical practice. Lancet 377(9783):2115–2126
10. Blumbach K et al (2008) Ablation of collagen IX and COMP disrupts epiphyseal cartilage architecture. Matrix Biol: J Int So Matrix Biol 27(4):306–318
11. Bove SE et al (2003) Weight bearing as a measure of disease progression and efficacy of anti-inflammatory compounds in a model of monosodium iodoacetate-induced osteoarthritis. Osteoarthr Cartil 11(11):821–830. Available at: http://www.ncbi.nlm.nih.gov/pubmed/14609535. Accessed 26 May 2017

12. Carlson CS et al (1994) Osteoarthritis in cynomolgus macaques: a primate model of naturally occurring disease. J Orthop Res: Off Publ Orthop Res Soc 12(3):331–339

13. Coleman CM, Tuan RS (2003) Functional role of growth/differentiation factor 5 in chondrogenesis of limb mesenchymal cells. Mech Dev 120(7):823–836

14. De Croos JNA et al (2006) Cyclic compressive mechanical stimulation induces sequential catabolic and anabolic gene changes in chondrocytes resulting in increased extracellular matrix accumulation. Matrix Biol 25(6):323–331

15. FDA (1999) Guidance for industry: clinical development programs for drugs, devices, and biological products intended for the treatment of osteoarthritis (OA). www.FDA.gov (July)

16. Fehrenbacher A et al (2003) Rapid regulation of collagen but not metalloproteinase 1, 3, 13, 14 and tissue inhibitor of metalloproteinase 1, 2, 3 expression in response to mechanical loading of cartilage explants in vitro. Arch Biochem Biophys 410(1):39–47

17. Finger F et al (2003) Molecular phenotyping of human chondrocyte cell lines T/C-28a2, T/C-28a4, and C-28/I2. Arthritis Rheum 48(12):3395–3403

18. Frisbie DD et al (2008) Changes in synovial fluid and serum biomarkers with exercise and early osteoarthritis in horses. Osteoarthr Cartil 16(10):1196–1204

19. Frisbie DD, Cross MW, McIlwraith CW (2006) A comparative study of articular cartilage thickness in the stifle of animal species used in human pre-clinical studies compared to articular cartilage thickness in the human knee. Vet Comp Orthop Traumatol: VCOT 19(3):142–146

20. Frisbie DD, McIlwraith CW (2000) Evaluation of gene therapy as a treatment for equine traumatic arthritis and osteoarthritis. Clin Orthop Relat Res 379 Suppl:S273–S287

21. Gabriel N et al (2010) Development of an in vitro model of feline cartilage degradation. J Feline Med Surg 12(8):614–620. Available at: https://doi.org/10.1016/j.jfms.2010.03.007

22. Garner BC et al (2011) Using animal models in osteoarthritis biomarker research. J Knee Surg 24(4):251–264. Available at: http://www.ncbi.nlm.nih.gov/pubmed/22303754. Accessed 26 May 2017

23. Gerwin N et al (2010) The OARSI histopathology initiative - recommendations for histological assessments of osteoarthritis in the rat. Osteoarthr Cartil 18(Suppl 3):S24–S34

24. Gouttenoire J et al (2007) Knockdown of the intraflagellar transport protein IFT46 stimulates selective gene expression in mouse chondrocytes and affects early development in zebrafish. J Biol Chem 282(42):30960–30973

25. Gregory MH et al (2012) A review of translational animal models for knee osteoarthritis. Arthritis 2012:764621

26. Groma G et al (2012) Abnormal bone quality in cartilage oligomeric matrix protein and matrilin 3 double-deficient mice caused by increased tissue inhibitor of metalloproteinases 3 deposition and delayed aggrecan degradation. Arthritis Rheum 64(8):2644–2654

27. Gruchenberg K et al (2015) In vivo performance of a novel silk fibroin scaffold for partial meniscal replacement in a sheep model. Knee Surg Sports Traumatol Arthrosc: Off J ESSKA 23(8):2218–2229

28. Hafez A et al (2015) Col1a1 regulates bone microarchitecture during embryonic development. J Dev Biol 3(4):158–176

29. He A et al (2017) Repair of osteochondral defects with in vitro engineered cartilage based on autologous bone marrow stromal cells in a swine model. Sci Rep 7:40489

30. Hering TM et al (2014) Changes in type II procollagen isoform expression during chondrogenesis by disruption of an alternative 5′ splice site within Col2a1 exon 2. Matrix Biol: J Int Soc Matrix Biol 36:51–63

31. Hillen J et al (2017) Structural cartilage damage attracts circulating rheumatoid arthritis synovial fibroblasts into affected joints. Arthritis Res Ther 19(1):40

32. Hooijmans CR, Leenaars M, Ritskes-Hoitinga M (2010) A gold standard publication checklist to improve the quality of animal studies, to fully integrate the three Rs, and to make systematic reviews more feasible. Altern Lab Anim: ATLA 38(2):167–182

33. Hsueh MF, Önnerfjord P, Kraus VB (2014) Biomarkers and proteomic analysis of osteoarthritis. Matrix Biol 39:56–66. Available at: https://doi.org/10.1016/j.matbio.2014.08.012
34. Hunter DJ, Felson DT (2006) Osteoarthritis. BMJ (Clinical research ed) 332(7542):639–642
35. Hunter DJ, Schofield D, Callander E (2014) The individual and socioeconomic impact of osteoarthritis. Nat Rev Rheumatol 10(7):437–441. Available at: https://doi.org/10.1038/nrrheum.2014.44
36. Ikeda T et al (1998) Adenovirus mediated gene delivery to the joints of Guinea pigs. J Rheumatol 25(9):1666–1673
37. Ilic MZ et al (2007) Distinguishing aggrecan loss from aggrecan proteolysis in ADAMTS-4 and ADAMTS-5 single and double deficient mice. J Biol Chem 282(52):37420–37428
38. Jaeger K et al (2008) The genetics of osteoarthritis in STR/ort mice. Osteoarthr Cartil 16(5):607–614
39. Jiménez G et al (2015) Activin a/BMP2 chimera AB235 drives efficient redifferentiation of long term cultured autologous chondrocytes. Sci Rep 5(1):16400. Available at: http://www.ncbi.nlm.nih.gov/pubmed/26563344. Accessed 26 May 2017
40. Jimenez PA et al (1997) Spontaneous osteoarthritis in Dunkin Hartley Guinea pigs: histologic, radiologic, and biochemical changes. Lab Anim Sci 47(6):598–601
41. Johnson CI, Argyle DJ, Clements DN (2016) In vitro models for the study of osteoarthritis. Vet J 209:40–49
42. Kawanishi Y et al (2014) Intra-articular injection of synthetic microRNA-210 accelerates avascular meniscal healing in rat medial meniscal injured model. Arthritis Res Ther 16(6):488
43. Kawcak CE et al (2008) Effects of exercise vs experimental osteoarthritis on imaging outcomes. Osteoarthr Cartil 16(12):1519–1525
44. Kerckhofs G et al (2013) Contrast-enhanced nanofocus computed tomography images the cartilage subtissue architecture in three dimensions. Eur Cell Mater 25:179–189. Available at: http://www.ncbi.nlm.nih.gov/pubmed/23389752. Accessed 26 May 2017
45. Kim Y-I et al (2015) Cartilage development requires the function of estrogen-related receptor alpha that directly regulates sox9 expression in zebrafish. Sci Rep 5:18011
46. Kim Y-I et al (2013) Establishment of a bone-specific col10a1:GFP transgenic zebrafish. Mol Cells 36(2):145–150
47. Koch TG, Betts DH (2007) Stem cell therapy for joint problems using the horse as a clinically relevant animal model. Expert Opin Biol Ther 7(11):1621–1626
48. Kon E et al (2008) Tissue engineering for total meniscal substitution: animal study in sheep model. Tissue Eng A 14(6):1067–1080
49. van der Kraan PM et al (1990) Degenerative knee joint lesions in mice after a single intra-articular collagenase injection. A new model of osteoarthritis. J Exp Pathol (Oxford) 71(1):19–31
50. van der Kraan PM (2013) Relevance of zebrafish as an OA research model. Osteoarthr Cartil 21(2):261–262
51. Kuyinu EL et al (2016) Animal models of osteoarthritis: classification, update, and measurement of outcomes. J Orthop Surg Res 11:19
52. Kyostio-Moore S et al (2011) STR/ort mice, a model for spontaneous osteoarthritis, exhibit elevated levels of both local and systemic inflammatory markers. Comp Med 61(4):346–355
53. Lampropoulou-Adamidou K et al (2014) Useful animal models for the research of osteoarthritis. Eur J Orthop Surg Traumatol: Orthop Traumatol 24(3):263–271
54. Lawrence RC et al (2008) Estimates of the prevalence of arthritis and other rheumatic conditions in the United States: part II. Arthritis Rheum 58(1):26–35. Available at: https://doi.org/10.1002/art.23176
55. Lin YY et al (2010) Applying an excessive mechanical stress alters the effect of subchondral osteoblasts on chondrocytes in a co-culture system. Eur J Oral Sci 118(2):151–158
56. Little CB, Zaki S (2012) What constitutes an "animal model of osteoarthritis"--the need for consensus? Osteoarthr Cartil 20(4):261–267

57. Liu W et al (2003) Spontaneous and experimental osteoarthritis in dog: similarities and differences in proteoglycan levels. J Orthop Res: Off Publ Orthop Res Soc 21(4):730–737
58. Madden JC et al (2012) Strategies for the optimisation of in vivo experiments in accordance with the 3Rs philosophy. Regul Toxicol Pharmacol 63(1):140–154. Available at: http://www.sciencedirect.com/science/article/pii/S0273230012000578
59. Mahjoub M, Berenbaum F, Houard X (2012) Why subchondral bone in osteoarthritis? The importance of the cartilage bone interface in osteoarthritis. Osteoporos Int 23(8 SUPPL):841–846
60. Malda J et al (2012) Comparative study of depth-dependent characteristics of equine and human osteochondral tissue from the medial and lateral femoral condyles. Osteoarthr Cartil 20(10):1147–1151
61. Malfait A-M, Little CB (2015) On the predictive utility of animal models of osteoarthritis. Arthritis Res Ther 17:225
62. Martel-Pelletier J, Wildi LM, Pelletier J-P (2012) Future therapeutics for osteoarthritis. Bone 51(2):297–311
63. Martinez-Diaz S et al (2010) In vivo evaluation of 3-dimensional polycaprolactone scaffolds for cartilage repair in rabbits. Am J Sports Med 38(3):509–519
64. McCoy AM (2015) Animal models of osteoarthritis: comparisons and key considerations. Vet Pathol 52(5):803–818
65. McIlwraith CW, Frisbie DD, Kawcak CE (2012) The horse as a model of naturally occurring osteoarthritis. Bone Joint Res 1(11):297–309
66. Mello MA, Tuan RS (2006) Effects of TGF-beta1 and triiodothyronine on cartilage maturation: in vitro analysis using long-term high-density micromass cultures of chick embryonic limb mesenchymal cells. J Orthop Res: Off Publ Orthop Res Soc 24(11):2095–2105
67. Mierzwa AGH et al (2017) Different doses of strontium ranelate and mechanical vibration modulate distinct responses in the articular cartilage of ovariectomized rats. Osteoarthr Cartil 25(7):1179–1188
68. Miller RE et al (2013) Genetically engineered mouse models reveal the importance of proteases as osteoarthritis drug targets. Curr Rheumatol Rep 15(8):350
69. Mitchell RE et al (2013) New tools for studying osteoarthritis genetics in zebrafish. Osteoarthr Cartil 21(2):269–278
70. Moran CJ et al (2016) The benefits and limitations of animal models for translational research in cartilage repair. J Exp Orthop 3(1):1. Available at: http://www.jeo-esska.com/content/3/1/1. Accessed 26 May 2017
71. Moreau M et al (2014) A medicinal herb-based natural health product improves the condition of a canine natural osteoarthritis model: a randomized placebo-controlled trial. Res Vet Sci 97(3):574–581. Available at: http://linkinghub.elsevier.com/retrieve/pii/S0034528814002483. Accessed 26 May 2017
72. Moreau M et al (2013) A posteriori comparison of natural and surgical destabilization models of canine osteoarthritis. Biomed Res Int 2013:180453. Available at: http://www.hindawi.com/journals/bmri/2013/180453/. Accessed 26 May 2017
73. Morko J et al (2005) Spontaneous development of synovitis and cartilage degeneration in transgenic mice overexpressing cathepsin K. Arthritis Rheum 52(12):3713–3717
74. Moyer RF et al (2017) Osteoarthritis year in review 2014: mechanics – basic and clinical studies in osteoarthritis. Osteoarthr Cartil 22(12):1989–2002. Available at: https://doi.org/10.1016/j.joca.2014.06.034
75. Murab S et al (2013) Matrix-embedded cytokines to simulate osteoarthritis-like cartilage microenvironments. Tissue Eng A 19(15–16):1733–1753. Available at: http://online.liebertpub.com/doi/abs/10.1089/ten.tea.2012.0385
76. Muraoka T et al (2007) Role of subchondral bone in osteoarthritis development: a comparative study of two strains of Guinea pigs with and without spontaneously occurring osteoarthritis. Arthritis Rheum 56(10):3366–3374

77. Novakofski KD, Torre CJ, Fortier LA (2012) Interleukin-1??, −6, and −8 decrease Cdc42 activity resulting in loss of articular chondrocyte phenotype. J Orthop Res 30(2):246–251
78. Olive J et al (2009) Imaging and histological features of central subchondral osteophytes in racehorses with metacarpophalangeal joint osteoarthritis. Equine Vet J 41(9):859–864
79. Osawa A et al (2013) The use of blood vessel-derived stem cells for meniscal regeneration and repair. Med Sci Sports Exerc 45(5):813–823
80. Page-McCaw A, Ewald AJ, Werb Z (2007) Matrix metalloproteinases and the regulation of tissue remodelling. Nat Rev Mol Cell Biol 8(3):221–233
81. Pape D et al (2010) Disease-specific clinical problems associated with the subchondral bone. Knee Surg Sports Traumatol Arthrosc 18(4):448–462
82. Pecchi E et al (2012) A potential role of chondroitin sulfate on bone in osteoarthritis: inhibition of prostaglandin E 2 and matrix metalloproteinases synthesis in interleukin-1β- stimulated osteoblasts. Osteoarthr Cartil 20(2):127–135. Available at: https://doi.org/10.1016/j.joca.2011.12.002
83. Percie du Sert N (2012) Maximising the output of osteoarthritis research: the ARRIVE guidelines. Osteoarthr Cartil 20(4):253–255
84. Pina S, Oliveira JM, Reis RL (2015) Natural-based nanocomposites for bone tissue engineering and regenerative medicine: a review. Adv Mater (Deerfield Beach, Fla) 27(7):1143–1169
85. Pritzker KP et al (1989) Rhesus macaques as an experimental model for degenerative arthritis. P R Health Sci J 8(1):99–102
85. Pritzker KPH et al (2006) Osteoarthritis cartilage histopathology: grading and staging. Osteoarthr Cartil 14(1):13–29
87. Redman SN, Oldfield SF, Archer CW (2005) Current strategies for articular cartilage repair. Eur Cell Mater 9:23–32
83. Reynard LN, Loughlin J (2013) Insights from human genetic studies into the pathways involved in osteoarthritis. Nat Rev Rheumatol 9(10):573–583. Available at: http://www.nature.com/doifinder/10.1038/nrrheum.2013.121
83. Rogerson FM et al (2008) Evidence of a novel aggrecan-degrading activity in cartilage: studies of mice deficient in both ADAMTS-4 and ADAMTS-5. Arthritis Rheum 58(6):1664–1673
93. Sabatini M et al (2005) Effect of inhibition of matrix metalloproteinases on cartilage loss in vitro and in a Guinea pig model of osteoarthritis. Arthritis Rheum 52(1):171–180
91. Sahebjam S, Khokha R, Mort JS (2007) Increased collagen and aggrecan degradation with age in the joints of Timp3(−/−) mice. Arthritis Rheum 56(3):905–909
92. Sanchez C et al (2016) Chondrocyte secretome: a source of novel insights and exploratory biomarkers of osteoarthritis. Osteoarthr Cartil. Available at: https://doi.org/10.1016/j.joca.2017.02.797
93. Sanchez C et al (2008) Phenotypic characterization of osteoblasts from the sclerotic zones of osteoarthritic subchondral bone. Arthritis Rheum 58(2):442–455
94. Sanchez C et al (2005) Subchondral bone osteoblasts induce phenotypic changes in human osteoarthritic chondrocytes. Osteoarthr Cartil 13(11):988–997
95. Santangelo KS, Bertone AL (2011) Effective reduction of the interleukin-1beta transcript in osteoarthritis-prone Guinea pig chondrocytes via short hairpin RNA mediated RNA interference influences gene expression of mediators implicated in disease pathogenesis. Osteoarthr Cartil 19(12):1449–1457
96. Santo VE et al (2013) Controlled release strategies for bone, cartilage, and osteochondral engineering–part II: challenges on the evolution from single to multiple bioactive factor delivery. Tissue Eng Part B Rev 19(4):327–352. Available at: http://www.ncbi.nlm.nih.gov/pubmed/23249320. Accessed 26 May 2017
97. Sato M et al (2012) Direct transplantation of mesenchymal stem cells into the knee joints of Hartley strain Guinea pigs with spontaneous osteoarthritis. Arthritis Res Ther 14(1):R31
98. Sendzik J, Lode H, Stahlmann R (2009) Quinolone-induced arthropathy: an update focusing on new mechanistic and clinical data. Int J Antimicrob Agents 33(3):194–200. Available at: http://linkinghub.elsevier.com/retrieve/pii/S0924857908003531. Accessed 26 May 2017

99. Da Silva MA et al (2009) Cellular and epigenetic features of a young healthy and a young osteoarthritic cartilage compared with aged control and OA cartilage. J Orthop Res 27(5):593–601

100. Simon D et al (2015) The relationship between anterior cruciate ligament injury and osteoarthritis of the knee. Adv Orthop 2015:928301

101. Smith RL et al (1995) Effects of fluid-induced shear on articular chondrocyte morphology and metabolism in vitro. J Orthop Res 13(6):824–831. Available at: http://doi.wiley.com/10.1002/jor.1100130604. Accessed 19 May 2017

102. Stoop R et al (1999) Type II collagen degradation in spontaneous osteoarthritis in C57Bl/6 and BALB/c mice. Arthritis Rheum 42(11):2381–2389

103. Swieszkowski W et al (2007) Repair and regeneration of osteochondral defects in the articular joints. Biomol Eng 24(5):489–495. Available at: http://www.sciencedirect.com/science/article/pii/S1389034407000834

104. Taniguchi S et al (2012) Long-term oral administration of glucosamine or chondroitin sulfate reduces destruction of cartilage and up-regulation of MMP-3 mRNA in a model of spontaneous osteoarthritis in Hartley Guinea pigs. J Orthop Res: Off Publ Orthop Res Soc 30(5):673–678

105. Teeple E et al (2013) Animal models of osteoarthritis: challenges of model selection and analysis. AAPS J 15(2):438–446

106. Thysen S, Luyten FP, Lories RJU (2015) Targets, models and challenges in osteoarthritis research. Dis Model Mech 8(1):17–30. Available at: http://dmm.biologists.org/cgi/doi/10.1242/dmm.016881

107. Tortelli F, Cancedda R (2009) Three-dimensional cultures of osteogenic and chondrogenic cells: a tissue engineering approach to mimic bone and cartilage in vitro. Eur Cell Mater 17:1–14. Available at: http://www.ncbi.nlm.nih.gov/pubmed/19579210. Accessed 26 May 2017

108. Tremoleda JL et al (2011) Imaging technologies for preclinical models of bone and joint disorders. EJNMMI Res 1(1):11

109. Trumble TN et al (2008) Joint dependent concentrations of bone alkaline phosphatase in serum and synovial fluids of horses with osteochondral injury: an analytical and clinical validation. Osteoarthr Cartil 16(7):779–786

110. Tufan AC et al (2002) AP-1 transcription factor complex is a target of signals from both WnT-7a and N-cadherin-dependent cell-cell adhesion complex during the regulation of limb mesenchymal chondrogenesis. Exp Cell Res 273(2):197–203

111. Vazquez M et al (2014) A new method to investigate how mechanical loading of osteocytes controls osteoblasts. Front Endocrinol 5(DEC):1–19

112. Wang W-G et al (2003) In vitro chondrogenesis of human bone marrow-derived mesenchymal progenitor cells in monolayer culture: activation by transfection with TGF-beta2. Tissue Cell 35(1):69–77

113. Watanabe H, Yamada Y (2002) Chondrodysplasia of gene knockout mice for aggrecan and link protein. Glycoconj J 19(4–5):269–273

114. Wei L, Svensson O, Hjerpe A (1997) Correlation of morphologic and biochemical changes in the natural history of spontaneous osteoarthrosis in Guinea pigs. Arthritis Rheum 40(11):2075–2083

115. Wendler A, Wehling M (2010) The translatability of animal models for clinical development: biomarkers and disease models. Curr Opin Pharmacol 10(5):601–606

116. Whitehouse MR et al (2017) Repair of torn avascular meniscal cartilage using undifferentiated autologous mesenchymal stem cells: from in vitro optimization to a first-in-human study. Stem Cells Transl Med 6(4):1237–1248

117. Yamamoto K et al (2005) Morphological studies on the ageing and osteoarthritis of the articular cartilage in C57 black mice. J Orthop Surg (Hong Kong) 13(1):8–18

118. Yan L-P et al (2015) Bilayered silk/silk-nanoCaP scaffolds for osteochondral tissue engineering: in vitro and in vivo assessment of biological performance. Acta Biomater 12:227–241. Available at: http://www.ncbi.nlm.nih.gov/pubmed/25449920. Accessed 26 May 2017

119. Yang B et al (2014) Effect of microRNA-145 on IL-1β-induced cartilage degradation in human chondrocytes. FEBS Lett 588(14):2344–2352. Available at: https://doi.org/10.1016/j.febslet.2014.05.033

120. Young RD et al (2002) Type II collagen deposition in cruciate ligament precedes osteoarthritis in the Guinea pig knee. Osteoarthr Cartil 10(5):420–428

121. Yu XM et al (2015) MicroRNAs' involvement in osteoarthritis and the prospects for treatments. Evid Based Complement Alternat Med 2015:236179

122. Zamli Z et al (2013) Increased chondrocyte apoptosis is associated with progression of osteoarthritis in spontaneous Guinea pig models of the disease. Int J Mol Sci 14(9):17729–17743

123. Zellner J et al (2013) Stem cell-based tissue-engineering for treatment of meniscal tears in the avascular zone. J Biomed Mater Res B Appl Biomater 101(7):1133–1142

124. Zhao W et al (2016) Cartilage degeneration and excessive subchondral bone formation in spontaneous osteoarthritis involves altered TGF-beta signaling. J Orthop Res: Off Publ Orthop Res Soc 34(5):763–770

Part V
In Vitro Models for Osteochondral Regeneration

Chapter 16
Tissue Engineering Strategies
for Osteochondral Repair

F. Raquel Maia, Mariana R. Carvalho, J. Miguel Oliveira, and Rui L. Reis

Abstract Tissue engineering strategies have been pushing forward several fields in the range of biomedical research. The musculoskeletal field is not an exception. In fact, tissue engineering has been a great asset in the development of new treatments for osteochondral lesions. Herein, we overview the recent developments in osteochondral tissue engineering. Currently, the treatments applied in a clinical scenario have shown some drawbacks given the difficulty in regenerating a fully functional hyaline cartilage. Among the different strategies designed for osteochondral regeneration, it is possible to identify cell-free strategies, scaffold-free strategies, and advanced strategies, where different materials are combined with cells. Cell-free strategies consist in the development of scaffolds in the attempt to better fulfill the requirements of the cartilage regeneration process. For that, different structures have been designed, from monolayers to multilayered structures, with the intent to mimic the osteochondral architecture. In the case of scaffold-free strategies, they took advantage on the extracellular matrix produced by cells. The last strategy relies in the development of new biomaterials capable of mimicking the extracellular matrix. This way, the cell growth, proliferation, and differentiation at the lesion site are expedited, exploiting the self-regenerative potential of cells and its interaction

F. Raquel Maia and Mariana R. Carvalho contributed equally to this work.

F. R. Maia (✉) · M. R. Carvalho (✉)
3B's Research Group – Biomaterials, Biodegradables and Biomimetics, University of Minho, Headquarters of the European Institute of Excellence on Tissue Engineering and Regenerative Medicine, Barco, Guimarães, Portugal

ICVS/3B's – PT Government Associate Laboratory, Braga/Guimarães, Portugal
e-mail: raquel.maia@dep.uminho.pt; mariana.carvalho@dep.uminho.pt

J. M. Oliveira · R. L. Reis
3B's Research Group – Biomaterials, Biodegradables and Biomimetics, University of Minho, Headquarters of the European Institute of Excellence on Tissue Engineering and Regenerative Medicine, Barco, Guimarães, Portugal

ICVS/3B's – PT Government Associate Laboratory, Braga/Guimarães, Portugal

The Discoveries Centre for Regenerative and Precision Medicine, Headquarters at University of Minho, Guimarães, Portugal

© Springer International Publishing AG, part of Springer Nature 2018   353
J. M. Oliveira et al. (eds.), *Osteochondral Tissue Engineering*,
Advances in Experimental Medicine and Biology 1059,
https://doi.org/10.1007/978-3-319-76735-2_16

with biomolecules. Overall, despite the difficulties associated with each approach, tissue engineering has been proven a valuable tool in the regeneration of osteochondral lesions and together with the latest advances in the field, promises to revolutionize personalized therapies.

Keywords Osteochondral regeneration · Cell-free strategies · Scaffold-free strategies · Biomaterials · Extracellular matrix mimicry

16.1 Introduction

Tissue engineering approaches are still a promising and powerful toolbox in a wide range of biomedical research areas and are applied in different areas such as cardiology, neurosciences, cancer, immunology, virology, and musculoskeletal.

A lot of attention is focused on osteochondral lesions since most of these lesions do not heal naturally [1]. It is accepted in the field that for successful long-term reconstruction or regeneration, the therapy should not only address the cartilage but also the underlying bone in order to reestablish joint homeostasis. In fact, the cartilage and adjacent subchondral bone has a limited potential for healing and if not properly treated can lead to the development of osteoarthritis. Additionally, nowadays, more young people suffer from these lesions, demanding faster and highly functional treatments. Several regenerative procedures have been used to treat such lesions. For example, microfracturing or subchondral drilling has been used to treat damaged cartilage due to the recruitment of mesenchymal stem cells [2]. With this approach, bone marrow is stimulated releasing progenitor cells from the bone marrow into the defect. These progenitor cells have the capacity to differentiate into chondrocytes, which, ultimately, produce a fibrocartilaginous cartilage. Nevertheless, this result tissue is not fully functional and many times weakens over time. For so, autologous chondrocytes or mesenchymal stem cells have been implanted as an alternative method. Even so, it is associated with some drawbacks, as the requirement of several surgical procedures and the fact that it is limited to application in small defects. To overcome some of the drawback, different sources have been studied [3]. Overall, the current therapies fail in regenerate hyaline cartilage, which hinders the motion and promotes progressive degeneration. Another possible clinical treatment is mosaicplasty. That method consists on the application of autologous osteochondral grafts (Fig. 16.1), which are obtained by the removal of cylindrical plugs of hyaline cartilage and the underlying subchondral bone from a healthy area [4].

To address the limitations, osteochondral tissue engineering has been pursued to develop more effective treatments, based on cell-free strategies, scaffold-free strategies, or even advanced strategies (Fig. 16.2) [5]. Within the strategies that involve the use of scaffolds, it should work as a provisional matrix for tissue regeneration, mimicking some features of the native tissue. By combining materials

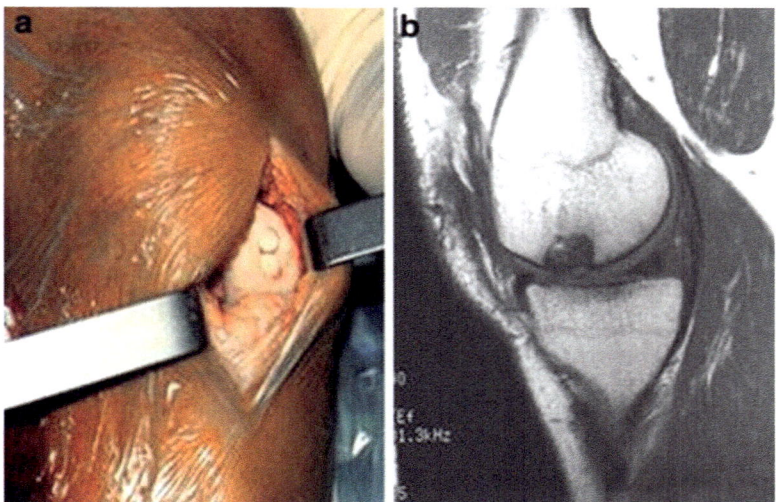

Fig. 16.1 (**A**) Image of osteochondral grafts implanted in injury area. (**B**) MRI depicting the defect site after 24 months [4]. Copyright © 2012, Springer-Verlag

and advanced technologies, it is possible to develop a new generation of three-dimensional scaffolds with similar features to the articular cartilage and bone. Scaffolds can be implanted with or without cells, allowing to address larger defects. But, in the case where scaffolds are implanted without cells (cell-free approaches), microfracture is used to promote chondrogenesis. Still, some issues need to be addressed as the mimicry of unique biofunctioning of articular cartilage and osteochondral interface tissues. Trying to answer this, new approaches appeared, such as the use of new biomaterials that are able to mimic the extracellular matrix [6]. These biomaterials are hydrogels. Many critical properties of hydrogels are advantageous in osteochondral regeneration, such as mechanical stiffness, elasticity, water content, bioactivity, and degradation profiles. Besides, hydrogels can be tuned for both soft and hard tissue regeneration. This way, cell growth, proliferation, and differentiation at the sites of defects are facilitated, and the regeneration of tissues becomes dependent on the interaction of cells and biomolecules.

Another advanced strategy for osteochondral regeneration is three-dimensional bioprinting [7]. This technology allows the fabrication of constructs with high control over spatial resolution, shape, and mechanical properties. Also, the main advantage of this technique is the accurate positioning of biomaterials, cells, and biological cues in a layer-by-layer fashion. However, bioprinting is at its early steps, and significant milestones have to be achieved before this technology can be translated to widespread clinical applications. Currently, the most promising materials, or "bioinks," for cell-based three-dimensional bioprinting are based on hydrogels. Hydrogels enable homogeneous cell encapsulation in a hydrated and mechanically supportive three-dimensional environment.

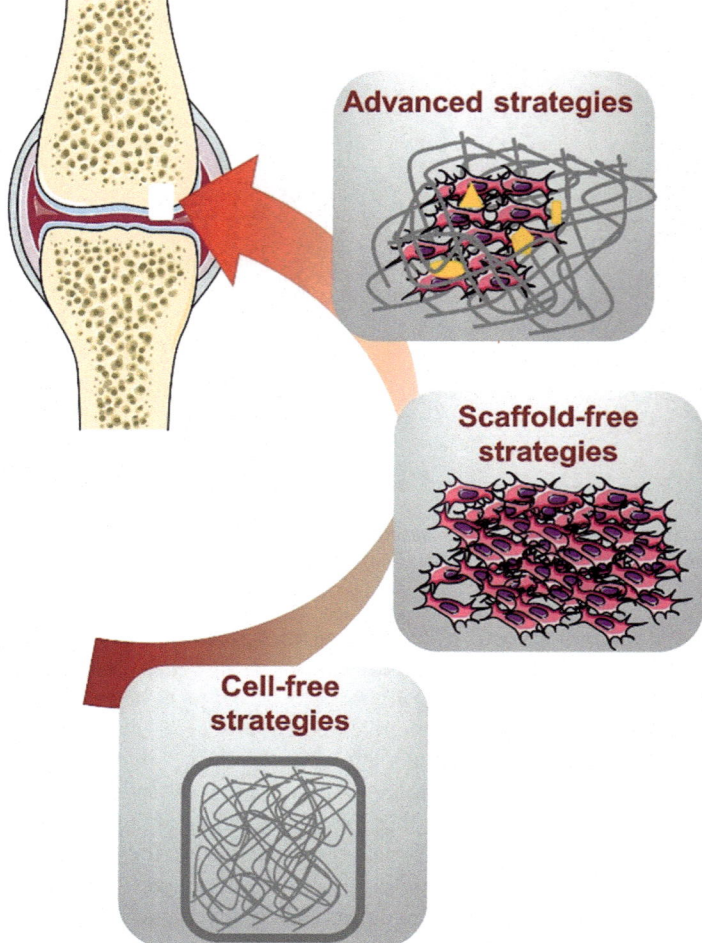

Fig. 16.2 Different strategies for osteochondral defect repair: Cell-free strategies, scaffold-free strategies and advanced strategies

16.2   Cell-Free Strategies for Osteochondral Regeneration

The use of cell-free strategies, i.e., scaffolds without cells, has been pursued to overcome some limitations of the current procedures, such as the fact that current strategies can only be applied into injuries with small size. Cell-free strategies rely on the implantation of scaffolds and creation of microfractures to promote the recruitment of cells present within the articular area. In this sense, mesenchymal stem cells, or other potential cells, are recruited to colonize the scaffold, which ultimately promotes the regeneration of the damage tissue. Noteworthy, for an appropriate tissue regeneration, scaffolds should be biocompatible and biodegradable. This way, it should allow the adhesion and proliferation of recruited cells and provide physical and biochemical cues to direct cells along differentiation pathways. Finally, it should degrade as the new tissue is formed.

Among the materials used for scaffolds production, natural biomaterials, such as collagen, hyaluronan, or other polymer-based materials, have shown improved results. In the case of collagen, since it can be found in high percentages in the cartilage and bone, it is easy to understand why it has been chosen for many studies. For example, Wang et al. [8] used it to produce a scaffold with anti-inflammatory properties and improved mechanical strength for cartilage regeneration. For that, they incorporated into collagen a macromolecular drug composed of resveratrol (natural polyphenolic compound) and polyacrylic acid. Then, the authors implanted it into rabbit osteochondral defects and after 12 weeks observed that the defects were completely repaired. Additionally, the neocartilage was well integrated with its surrounding tissue and subchondral bone. However, osteochondral defects most often involved not only damage cartilage but also damage the subchondral bone, which can be caused by direct trauma or due to chronic overload and degenerative changes. With this in mind, scaffolds composed of two different layers have been studied with the purpose to regenerate cartilage and bone [9]. For example, bilayer scaffolds composed of collagen, for cartilage regeneration, and hydroxyapatite, for bone regeneration, have been studied for knee chondral or osteochondral lesions [10]. These scaffolds were completely integrated and showed to enable the repair of the tissue surface. However, some abnormal bone overgrowth was also observed.

Despite this bilayer approach being more physiologically relevant due to the similarities with native structure of articular tissue, there is still a layer that is not taken into consideration: the calcified cartilage layer. The calcified cartilage layer functions as a barrier inhibiting the vascularization of the cartilage layer and the spread of subchondral bone into the cartilage. With this in mind, Levingstone et al. [11] mixed collagen with hydroxyapatite and hyaluronic acid to develop a multi-layered structure. This strategy relies on the production of a scaffold that emulates the different regions of the osteochondral interface. One layer was composed of collagen type I and hydroxyapatite to promote bone repair, another layer was composed of collagen type I and hyaluronic acid to promote the development of calcified cartilage, and, finally, another layer was composed of collagen type I, collagen type II, and hyaluronic acid to promote cartilage repair. This structure was tested in critical size defect produced in a caprine model. Upon 12 months post implantation, it was observed a complete regeneration of subchondral bone, with formation of well-structured subchondral trabecular bone and hyaline-like cartilage tissue through the recruitment of host cells.

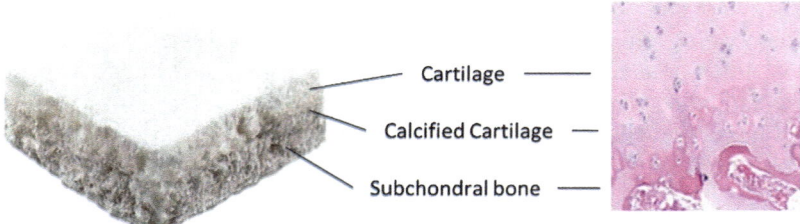

Fig. 16.3 MaioRegen®, a commercialized multilayer scaffold with a structure similar with native articular cartilage. Copyright © Kon et al. [16]; licensee Springer 2014

In fact, the study of multilayer scaffolds resulted in the development of a commercial product named MaioRegen® (JRI Orthopaedics Ltd). This product is composed of a layer of collagen type I, to mimic the cartilage; a layer of magnesium-enriched hydroxyapatite and collagen, to mimic the calcified cartilage; and a layer of magnesium-enriched hydroxyapatite, to mimic the subchondral bone (Fig. 16.3). The evaluation of regenerated tissue quality and scaffold integration showed successful regeneration of the defect. Nonetheless, the quality of the superficial cartilage repair tissue was limited [12]. Additionally, it was observed subchondral laminar and bone changes [13]. Noteworthy, despite the apprehension related with biological repair, the associated risk showed to be acceptable [14]. It is important to mention that another scaffold named TruFit™ bone graft substitute (BGS) is commercialized. It is composed of poly (d,l-lactide co-glycolide) and mimics the porous structure of bone. But reports indicate that established techniques, such as autologous osteochondral mosaicplasty, result in better outcomes than the use of these scaffolds [15].

Despite the majority of the scaffolds developed for osteochondral lesions' treatments have collagen in their composition, other materials has shown promising results. For example, the case of porous poly(lactic-co-glycolic acid) scaffolds. These scaffolds were implanted in the rabbit knee joint and subjected to early loading exercise. The exercise played a key role as a mechanical stimuli, envisioning the self-renew of the neighboring stem cells and benefiting the chondrogenesis and osteogenesis [17]. After 12 weeks, these scaffolds stimulated the full-thickness osteochondral regeneration, with formation of hyaline-like cartilage and development of columnar rounded chondrocytes and functionalized trabecular bone with osteocytes. In a different study, the use of a hyaluronic acid-based scaffold revealed encouraging results with complete repair of the defect after 24 months [18]. But the clinical significance is still not known.

More recently, decellularized tissues have been studied. In fact, they present the same structure and function as the original tissues, resulting in a bioactive and natural material for osteochondral lesions repair [19]. Furthermore, decellularized matrices have been used in conjugation with synthetic materials in order to improve them [20]. These enhanced scaffolds showed to support the growth of endogenous cells and to promote good cartilage repair upon implantation into an osteochondral defect.

16.3 Scaffold-Free Strategies for Osteochondral Regeneration

Different approaches were pursued with the intent to improve osteochondral treatments. The use of scaffolds has shown significant improvements comparing with the golden standards applied nowadays, such as drilling, debridement, microfracture, mosaicplasty, and autologous chondrocyte implantation. However, the implantation of such materials could promote an acute rejection and foreign body reactions. Furthermore, the preparation of constructs composed of cells and scaffolds still needs to be optimized, namely, in terms of cell seeding and cell differentiation. To overcome these issues, the use of a scaffold-free strategy became more relevant. In fact, strategies that rely only in the use of cells at high densities can promote cell-to-cell interactions and mimic the embryonic development, being the native extracellular matrix produced by these cells [21]. In fact, in a study where they produced cell sheets of mesenchymal stem cells and transplanted into osteochondral defects, their contribution to the regeneration of osteochondral defects was clearly shown [22].

Among the different types of cells studied to develop a cartilage construct, chondrocytes have evident advantages over other types. However, implantation of chondrocyte constructs showed better results in the regeneration of chondral defects than osteochondral defects [23]. Additionally, chondrocytes have shown other drawbacks. In fact, only a few cells are obtained upon isolation, and these cells are known to dedifferentiate during proliferation, which results in the loss of their appealing cartilage character. For so, mesenchymal stem cells have gained popularity for osteochondral treatments. These cells have the ability of self-renewal and differentiation into multiple lineages, including the bone and cartilage. Additionally, they can be isolated from different tissues, such as bone marrow, adipose tissue, and umbilical cord blood, among others. Although mesenchymal stem cells of different origins had similar characteristics, they do not have the same phenotype, nor differentiation potential. Despite that, most of the time, the selection between each type is focused on the accessibility and isolation yield. Bone marrow mesenchymal stem cells are the best characterized, since they were the first being isolated by Friedenstein et al. [24], and the most used for skeletal tissue repair. Noteworthy, adipose-derived mesenchymal stem cells have become more relevant, once they require less invasive methods for isolation and require easier harvesting procedures.

The use of monolayers or sheets of mesenchymal stem cells in osteochondral defects showed to be mechanically instable and to require an anchorage system in order to retain the cells into the defect site, as the use of fibrin [25]. To overcome these issues, cellular agglomerates have been used. Additionally, it is known that

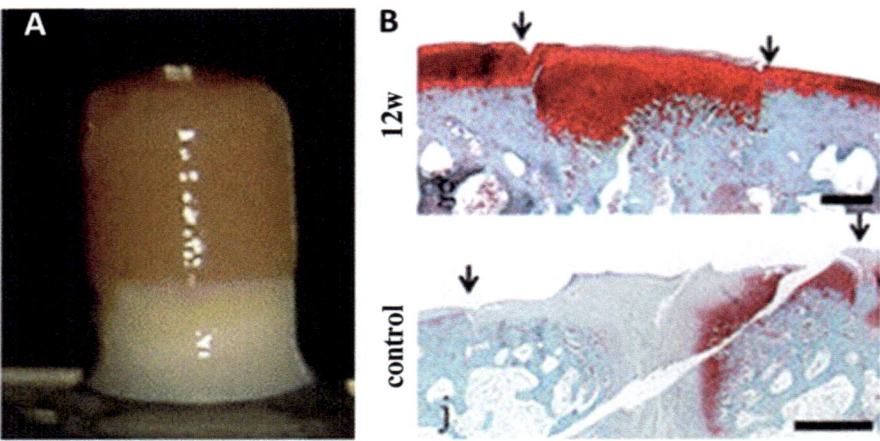

Fig. 16.4 (**A**) Image of a columnar structure composed of fused spheroids of bone marrow mesenchymal stem cells. (**B**) Histological analysis of regeneration of osteochondral defect using a columnar structure. Safranin O staining of (g) columnar structure implantation at 12 weeks and (j) control (without any implantation) (arrows indicate the borders of the defect). Scale bar = 1 mm. Copyright © Ishihara et al. [27]; licensee BioMed Central Ltd. 2014

cellular agglomerates promote preferentially chondrogenesis, which can induce the formation of regenerated cartilage. For example, it was demonstrated that a scaffold-free composed of mesenchymal stem cells and its extracellular matrix was capable of facilitating cartilage repair [26]. Furthermore it showed to adhere to the adjacent cartilage, resulting into a secure integration. In a different strategy, with the intent to create enough thickness to fill in a defect, spheroids of bone marrow mesenchymal stem cells were fused [27]. This way, a columnar structure was developed without using scaffolds, which successfully regenerated both the bone and cartilage (Fig. 16.4). Additionally, promising results were observed when adipose-derived mesenchymal stem cells were used to obtain the columnar structure [28]. In fact, these cells have several advantages over bone marrow mesenchymal stem cells, such as their proliferation rate which doesn't decrease with the age of the animal, and it is abundantly present into adipose tissue which is easily accessed.

Even so, the attachment of agglomerates to the subchondral bone is still an issue. In fact, it is necessary to confer stability to the implant and to allow a correct transfer of forces from the cartilage to the bone. With this in mind, scaffold-free strategies have been combined with other materials to provide fixation to the bone. For example, cells have been combined with porous tantalum [29]. For that, agglomerates of chondrocytes are produced, which results into neocartilage formation. The attachment of the cells to the tantalum is ensured by cartilage ingrowth into the pores. In a different approach, cells were combined with an interconnected hydroxyapatite-based artificial bone (NEOBONE®, MMT Co. Ltd.) and implanted in an osteochondral defect model [30]. After 6 months, it was observed an accelerated and improved osteochondral repair.

16.4 Advanced Strategies for Osteochondral Regeneration

Tissue engineering strategies have been used to develop a next generation of osteochondral constructs, which promise breakthroughs in clinic practices. There are two main approaches for the regeneration of osteochondral tissues. The first is to develop scaffolds that mimic the architectural features and mechanical properties such as stiffness (and thus biological functions of native osteochondral tissues) [31, 32]. The second tissue engineering approach highlights the regeneration of tissues per se [33]. The critical idea is to deliver appropriate biomaterials as artificial extracellular matrices to induce cell growth, proliferation, and differentiation at the defect sites, being the regeneration of the articular cartilage and subchondral bone in charge of the native biological processes. These processes involve interactions among cells and biomolecules (e.g. growth factors and paracrine signaling). In these cases, the extracellular matrices like materials don't need to be mechanically strong to mimic the native healthy tissues and withstand loadings, as they only serve as short-term three-dimensional microenvironments for the chondrogenic/osteogenic progenitor cells to generate real cartilage and bone tissues.

The most recent tissue engineering approaches for the development of three-dimensional osteochondral models include hydrogels and bioprinted cell-laden hydrogels.

Hydrogels have gained increasing popularity during the past few years. Hydrogels are adaptable and attractive biomaterials for tissue engineering and cell therapy applications, due to their unique properties similar to natural extracellular matrices, such as high water content, biodegradability, porosity, and biocompatibility [34]. Depending on the unique characteristics and/or the presence of specific functional groups, hydrogels made from natural polymers (including their derivatives) and synthetic polymers can be fabricated via different cross-linking mechanisms. In general, there are two main mechanisms, i.e., physical cross-linking and chemical (covalent) cross-linking, which result in hydrogels with distinct structures and properties [35]. Both synthetic and natural-based hydrogels offer the ability to mimic various distinctive requirements of an extracellular matrix-like physico-chemical environment.

Hydrogels based on natural polymers including chitosan, alginate, hyaluronic acid (HA), gellan gum (GG), agarose, collagen, gelatin, and fibroin have been extensively documented [35]. The use of a variety of hydrogels based on synthetic polymers (e.g., poly(ethylene glycol), polymer oligo(poly(ethylene glycol) fumarate), polyvinyl alcohol, poly(N,Ndimethylacrylamide), and methoxy poly(ethylene glycol)-poly(e-caprolactone)) have also been widely reported [36–40].

Alginate hydrogels and their derivatives are one of naturally occurring polysaccharide polymers, typically obtained from brown seaweed and various bacteria. One unique property of alginate is the ability to be physically cross-linked by divalent cations such as Ca^{2+} at room temperature [41]. Alginate gels are proven to maintain growth and proliferation of encapsulated chondrocytes, being useful in this particula⁻ biomedical application [41]. Also, when prepared in microspheres, it was found that the preferable pore size on the developed microcavitary alginate hydrogel is 80–120 μm for better growth and extracellular matrix synthesis [41]. An alternative approach consists of mixing decellularized and demineralized bone extracellular matrix together with alginate hydrogel loaded with skeletal stem cells (SSCs) and microparticles-encapsulated growth factors to investigate bone development and tissue formation in vivo. The bone extracellular matrix component was combined with alginate in an attempt to improve structural stability and exhibited inherent osteoinductive capacity leading to enhanced mineralization. It is worth noticing that mineralization was not further enhanced with additional growth factor concentrations revealing the potential of alginate itself in osteochondral application [42]. Moreover, alginate hydrogels were successfully used to deliver bone progenitor cells, including mesenchymal stem cells for bone regeneration. Encapsulated cells could produce collagenous extracellular matrix that was well integrated with the host tissue [43]. As with any other biomaterial, it can be modified to increase its potential. In this sense, 45S5 bioglass was introduced into alginate hydrogels to generate composite hydrogel beads as cell carriers. Results confirmed that extracts of composite hydrogel beads could simulate proliferation and osteogenic differentiation of mesenchymal stem cells as well as angiogenesis of endothelial cells [44]. However, alginate has some drawbacks such as the low adhesive points.

For so, other materials were investigated, as the case of chitosan. Chitosan is the second most abundant natural biopolymer, and it is known for its good biocompatibility and biodegradability. In this sense it is easy to understand why it is an attractive candidate material for osteochondral tissue engineering applications. Chitosan hydrogels prepared by enzymatic cross-linking can support the proliferation of chondrocytes and mesenchymal stem cells while maintaining the chondrogenic phenotype and morphology and boost the deposition of cartilaginous extracellular matrix in vitro, as depicted in Fig. 16.5 [45]. In this study, collagen type II and chondroitin sulfate, another molecule widely present in cartilage, were incorporated into injectable chitosan hydrogels, and its gelation occurred upon exposure to visible blue light (VBL) in the presence of riboflavin. Results show that unmodified chitosan hydrogel supported proliferation and deposition of cartilaginous extracellular matrix by encapsulated chondrocytes and mesenchymal stem cells. Additionally, the incorporation of native collagen type II and chondroitin sulfate into chitosan hydrogels revealed to be an advantage, as it further increased chondrogenesis [45]. Many chemical derivatizations of chitosan have been used to promote new biological activities and to modify its mechanical properties. The addition of polyol salts such as sodium glycerol phosphate to chitosan creates chi-

→

Fig. 16.5 (continued) Col II (MeGC/Col II-L), 0.4% w/v of Col II (MeGC/Col II-H), 0.5% w/v of CS (MeGC/CS-L), or 1% w/v of CS (MeGC/CS-H). Cell adhesion to the hydrogels. The scale bar is 10 μm. (II) Histological analysis of 3D-cultured chondrocytes in hydrogel systems during 6-week culture. Sections are stained with safranin-O (sGAG: orange red) and immunohistochemical staining of Col II (brown) at 1, 3, and 6 weeks in culture. The scale bar is 200 μm. Reprinted with permission from [45]. Copyright © 2014, American Chemical Society

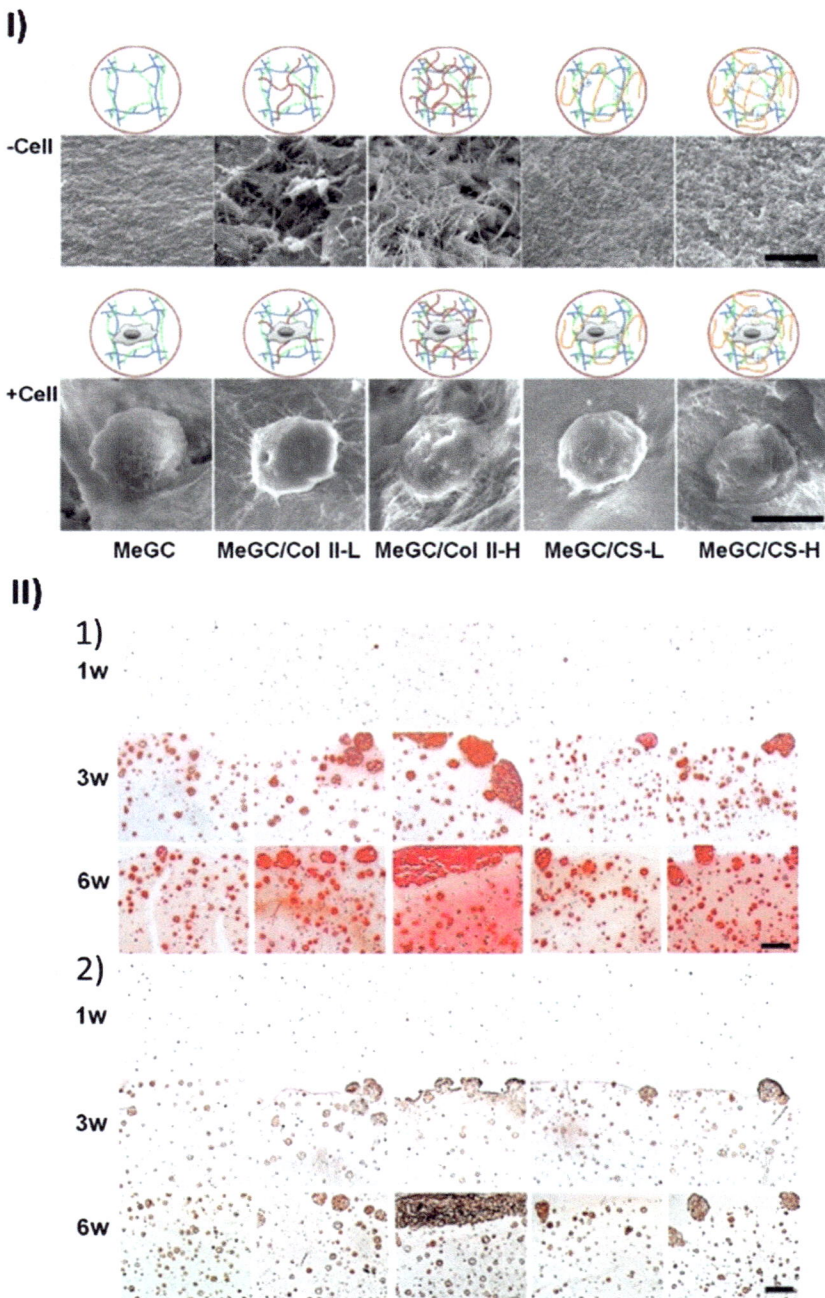

Fig. 16.5 Natural hydrogels for cartilage tissue engineering. (I) SEM images of hydrogels. Interior morphology of MeGC (methacrylated glycol chitosan) hydrogels containing 0.2% w/v of

tosan-glycerol phosphate (GP), a thermosensitive hydrogel [46]. This hydrogel forms a viscous liquid at room temperature or below and converts into a semisolid gel at body temperature. This appealing characteristic leads to its investigation in osteochondral applications. For that, it was tested in horse joints submitted to high mechanical loads. Results confirm that cells were able to synthesize collagen type II and proteoglycans, similar to those synthesized in normal cartilage and in healing tissue without implant [46].

Boyer et al. [47] developed a self-hardening and mechanically reinforced hydrogel (Si-HPCH) composed of silanized (hydroxypropyl)methyl cellulose (Si-HPMC) mixed with chitosan. Subcutaneous explants of human nasal chondrocytes (hNC) with Si-HPMC or Si-HPCH showed the formation of cell clusters surrounded by a cartilage-like extracellular matrix [47]. However, allergenic reactions of chitosan may be an issue for its clinical translation, especially in intolerant patients, and should be furthered studied.

In the case of collagen, it is clearly a natural choice for osteochondral regenerative medicine, since it is the most common protein found on the extracellular matrix, making for around 90% of the dry weight of articular cartilage [6]. Collagen type I hydrogel was constructed and used as a scaffold for cartilage tissue engineering. It was demonstrated to exert immunoisolation effects on the encapsulated chondrocytes and thus enhance in vitro chondrogenesis and cartilage extracellular matrix formation during the culture [48]. Although collagen type I hydrogels alone support growth/adhesion and chondrogenic differentiation [6], combinations are possible, as shown by Mohan et al. [49]. The team developed a chitosan-hyaluronic acid dialdehyde hydrogel for in vivo cartilage regeneration. The hydrogel alone or in combination with encapsulated chondrocytes was applied for cartilage regeneration. Results show that the repair tissue formed in sham rabbits appeared fibrous and opaque, whereas those that received gel had texture similar to the surrounding native cartilage. However, there was no significant enhancement in the quality of regenerated cartilage in the presence of encapsulated chondrocytes [49]. Parmar et al. [50] tried to go further and recapitulate features of the native tissue biochemical microenvironment by adding bioactive and biodegradable peptide motifs. The release of the RGDS peptide from the degradable constructs led to significantly enhanced expression of collagen type II, aggrecan, and SOX9 by human mesenchymal stem cells undergoing chondrogenesis, as well as greater extracellular matrix accumulation, a sign of enhanced osteochondral regeneration [50].

Natural silk, composed of silk fibroin protein core with sericin proteins, is produced from the silkworm cocoons from *Bombyx mori*. Silk fibroin has become a new biomaterial for tissue engineering applications due to its robust mechanical properties, excellent biocompatibility, slow degradability, and abundant supply source [51]. Although it's a relatively new biomaterial to be used in osteochondral applications, studies now report that highly tunable silk hydrogels will be able to mimic the tribological behavior of cartilage, with controlled pore sizes and optimized mechanical properties. Parkes et al. [52] performed exhaustive mechanical tests and showed silk hydrogels with pore sizes between 10 μm and 40 μm had a comparable compressive modulus to cartilage, with stiffness improved by decreasing pore size [52]. Yodmuang et al. [53] developed silk microfiber-reinforced silk hydrogel composites with favorable chondrocyte response, even more when silk fiber reinforcement in the hydrogels was applied, resulting in the development of constructs with properties approaching those of native cartilage [53]. Combinations of materials are also possible, as Wu et al. [54] showed by combining chitosan/silk fibroin /hydroxyapatite/glycerophosphate (chitosan/SF/HA/GP) in the preparation of the hydrogels. Results obtained from in vivo degradation demonstrated that degradation endurance of the optimized chitosan/SF/GP and chitosan/SF/HA/GP gels was significantly enhanced as compared to the chitosan/GP gel, and the degradation rate of the gels could be regulated by the SF component alone or by the combination of SF and HA components [54], which matches one of the most important requirements in osteochondral regeneration: control over degradation rate and its balance with the formation of new tissues.

Regarding the bone, fibroin/sodium alginate hydrogels have been used as template for promoting controlled biomineralization of hydroxyapatite crystals for bone repair [55]. However promising, these last examples still need biological experimentation to determine how bone/cartilage cells respond to these hydrogel constructs.

Although incomparable in terms of biocompatibility for osteochondral cell growth, proliferation, and phenotype maintenance, natural biopolymers present some drawbacks, mainly when it comes to mechanical properties and controlled degradation, which are superior in synthetic polymer-based hydrogels. For instance, PEG-based hydrogels support adhesion and proliferation of chondrocytes, mesenchymal stem cells, and embryonic stem cells [36]. Three-dimensional hydrogels consisting of chondroitin sulfate methacrylate and poly(ethylene) glycol diacrylate were used to study the effects of interactions between adipose-derived stem cells and neonatal chondrocytes and resulted in the formation of large neocartilage nodules. This work shows that hydrogel composition mimics the function of native cartilage extracellular matrix and raises the potential of utilizing stem cells to catalyze tissue formation by neonatal chondrocytes via paracrine signaling [36].

Nowadays, a new cutting-edge technique has been used for the production of three-dimensional osteochondral engineered hydrogels: the bioprinting. The three-dimensional printing is restructuring the biotechnology field and has a transformative effect not only on design but also in the production of materials. Three-dimensional printing was first described by Charles Hull in early 1986. He was the first to develop a system for generating three-dimensional objects by creating a cross-sectional pattern of an object using UV light cross-linking. Then, three-dimensional printing has spread across the field and studied for its advantages, including precise control of the three-dimensional architectures, highly customizable structures, automated manufacturing processes, and relatively cheap production. After the first impact came the development of different materials such as synthetic polymers, cross-linkers, and cells to form three-dimensional integral constructs. More recently, bioprinting has begun to show

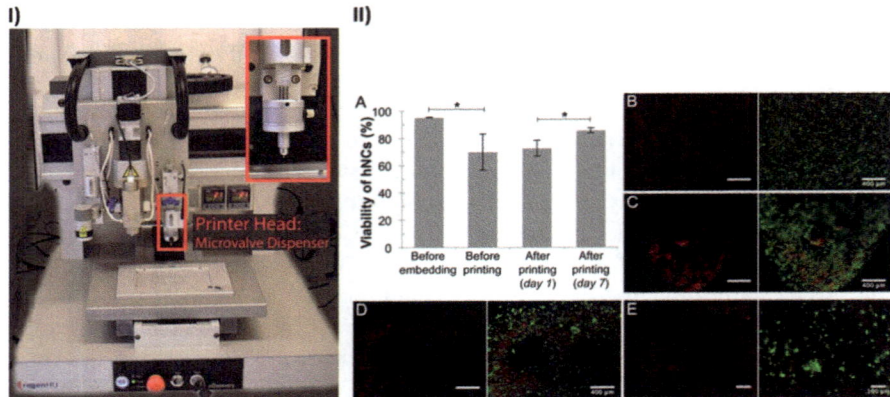

Fig. 16.6 Three-dimensional bioprinting. (I) Three-dimensional bioprinter 3D discovery (Switzerland). (II) (**A**) Viability of human nasoseptal chondrocytes (hNC) before and after three-dimensional bioprinting. Representative images showing dead (red) and live (green) cells (**B**) before and (**C**) after bioprinting hNC in Ink8020 and three-dimensional culture for 1 day. (**D** and **E**) Representative images (at 4× and 10× magnifications) showing dead and live cells in three-dimensional bioprinted constructs after 7 days of culture. Reprinted with permission from [57]. Copyright © 2015, American Chemical Society

great promise in advancing the development of functional tissue/organ replacements, likely to generate well-organized tissue constructs with complex shapes and gradient composition/structure by a multilayered deposition process of bioinks and cells [35]. Nevertheless, progress in this field is currently slow due to limited choices of bioink for cell encapsulation and cytocompatible cross-linking mechanisms. For every new complex technology that arises, it faces some hurdles, such as the lack appropriate mechanical strengths and structural integrity that are required for maintaining their specific shapes and enduring external stress after implantation. Low resolution and long cell viability of cells printed in biomaterials make it hard to fabricate biomimetic constructs with fine structures at the nanoscale [56]. Still, there are several promising examples, such as the one developed by Markstedt et al. and depicted in Fig. 16.6 [57], which bioprinted alginate-based cell-laden bioinks. The embedded chondrocytes in bioprinted hydrogels exhibited good viability after 7 days [57]. A mixture of silk fibroin and gelatin cross-linked by tyrosinase revealed to be a suitable bioink and allowed not only the viability of mesenchymal stem cells but also their differentiation into chondrocytes/osteoblasts when cultured in chondrogenic/osteogenic medium [57].

The idea of vascularized constructs has been a goal in tissue engineering since its early steps. However, engineered tissues are yet to be fully applied in clinical therapies due to lack of biological functions. One major obstacle regarding biologically functioned tissue constructs is the need of vascular network in engineered tissues, since it is proven essential in engineered tissue with thickness superior to 100–200 μm for transport of nutrients and oxygen to cells. However, the complexity of blood vessel networks makes the regeneration process complicated. In an approach that tried to tackle the microvascular issue, Holmes et al. [58] printed three-dimensional bone scaffolds with enhanced osteogenic bone regeneration and vascular cell growth by playing with anisotropic pore patterns. Scaffolds were printed with polylactic acid and further functionalized with no crystalline hydroxyapatite [58]. Promising results show enhanced human mesenchymal stem cell adhesion and growth on scaffolds with small microchannels and hydroxyapatite modification. On the other hand, scaffolds with large channels increased human umbilical vein endothelial cell growth [58].

Undertaking the issue of chondrocyte density gradient and its importance within the articular cartilage as a feature in cartilage tissue engineering, Ren et al. [59] developed a bioprinted collagen type II hydrogel/chondrocyte construct comprising cell density gradient. Results show that the cell density gradient distribution resulted in a correspondent gradient distribution of extracellular matrix and that the chondrocytes' biosynthetic ability was affected by both the total cell density and the cell distribution pattern. However, this study was conducted in static conditions, not having in consideration the mechanical environment in articular cartilage, being further studies required to confirm these findings [59].

A different approach was used by Cui et al. [37], which used poly(ethylene glycol) dimethacrylate in a bioprinting system with simultaneous photopolymerization. Human chondrocytes were printed to repair defects in osteochondral plugs in layer-by-layer assembly, achieving a similar native human articular cartilage compressive modulus. Also, the presence of native cartilage when implanted in vivo promoted extracellular matrix production by the encapsulated human chondrocytes, and enhanced proteoglycan deposition was observed at the interface between printed biomaterial and native cartilage [37].

16.5 Conclusions

The fields of developmental biology and tissue engineering both pursue the understanding and achievement of physiological cues needed to stimulate cells to reconstruct tissues and organs. In particular, osteochondral defect repair is still a major challenge for orthopedic surgeons and tissue engineers.

The main problem is that articular cartilage has a highly complex functional area, with different cell densities, organizations, and even cell types, with variable parameters and modulators from bone and cartilage microenvironment. Also, the extracellular matrix composition/distribution varies from zone to zone, as well as the biomechanical functions for each zone. Given such complexity, there are many key challenges to be overcome in the regeneration of osteochondral tissues. Since it has been established that three-dimensional models best recapitula e the complexity of tissues, perhaps the greatest challenge is to match the degradation of these three-dimensional scaffolds with the growth of cartilage and osteochondral tissues. The degradation rate of the scaffolds has an important effect on the tissue regeneration in vitro and in vivo. Other critical parameters to be studied are the cell source, cell differentiation, and release of growth factors, which have to be fine-tuned to obtain the best results on the macroscale as well as or micro- and nanoscales.

Conveniently, progresses have been made in this area, especially in the last decade, in order to tackle these issues. Advanced manufacturing techniques including microfluidic biofabrication and bioprinting have been developed to generate cell-laden hydrogel constructs with gradient composition and organized zonal architecture to mimic native osteochondral and cartilage tissues, promising next-generation personalized therapeutics. However, it remains a great challenge to regenerate cartilage and osteochondral tissues with fully restored zonal composition, structure, and functions. Furthermore, the translation to the clinic requires a meticulous choice of the most suitable techniques and materials with robust features such as mechanical properties, compatibility, degradability, and regenerative potential assessed in appropriate preclinical trials.

Acknowledgments The authors thank the funds obtained through the Nanotech4als (ENMed/0008/2015), HierarchiTech (M-ERA-NET/0001/2014), and FROnTHERA (NORTE-01-0145-FEDER-0000232) projects. FRM acknowledges the Portuguese Foundation for Science and Technology (FCT) for her postdoc grant (SFRH/BPD/117492/2016); MRC acknowledges the Doctoral Program financed by Programa Operacional Regional do Norte, Fundo Social Europeu, Norte 2020 for her PhD grant (NORTE-08-5369-FSE-000044 TERM&SC); and JMO thanks FCT for the distinction attributed under the Investigator FCT program (IF/01285/2015).

References

1. Yan L-P et al (2015) Current concepts and challenges in osteochondral tissue engineering and regenerative medicine. ACS Biomater Sci Eng 1(4):183–200
2. Panseri S et al (2012) Osteochondral tissue engineering approaches for articular cartilage and subchondral bone regeneration. Knee Surg Sports Traumatol Arthrosc 20(6):1182–1191
3. Correia SI et al (2017) Posterior talar process as a suitable cell source for treatment of cartilage and osteochondral defects of the talus. J Tissue Eng Regen Med 11(7):1949–1962
4. Espregueira-Mendes J et al (2012) Osteochondral transplantation using autografts from the upper tibio-fibular joint for the treatment of knee cartilage lesions. Knee Surg Sports Traumatol Arthrosc 20(6):1136–1142
5. Kon E et al (2015) Scaffold-based cartilage treatments: with or without cells? A systematic review of preclinical and clinical evidence. Arthroscopy 31(4):767–775
6. Yang J et al (2017) Cell-laden hydrogels for osteochondral and cartilage tissue engineering. Acta Biomater 57:1–25
7. Jeong CG, Atala A (2015) 3D printing and biofabrication for load bearing tissue engineering. Adv Exp Med Biol 881:3–14
8. Wang W et al (2014) An anti-inflammatory cell-free collagen/resveratrol scaffold for repairing osteochondral defects in rabbits. Acta Biomater 10(12):4983–4995

9. Yan LP et al (2014) Silk fibroin/Nano-CaP bilayered scaffolds for osteochondral tissue engineering. Key Eng Mater 587:245–248
10. Perdisa F et al (2017) One-step treatment for patellar cartilage defects with a cell-free osteochondral scaffold: a prospective clinical and MRI evaluation. Am J Sports Med 45(7):1581–1588
11. Levingstone TJ et al (2016) Cell-free multi-layered collagen-based scaffolds demonstrate layer specific regeneration of functional osteochondral tissue in caprine joints. Biomaterials 87:69–81
12. Brix M et al (2016) Successful osteoconduction but limited cartilage tissue quality following osteochondral repair by a cell-free multilayered nano-composite scaffold at the knee. Int Orthop 40(3):625–632
13. Verdonk P et al (2015) Treatment of osteochondral lesions in the knee using a cell-free scaffold. Bone Joint J 97-b(3):318–323
14. Stadler N, Trieb K (2016) Osteochondritis dissecans of the medial femoral condyle : new cell-free scaffold as a treatment option. Orthopade 45(8):701–705
15. Hindle P et al (2014) Autologous osteochondral mosaicplasty or TruFit™ plugs for cartilage repair. Knee Surg Sports Traumatol Arthrosc 22(6):1235–1240
16. Kon E et al (2014) Clinical results of multilayered biomaterials for osteochondral regeneration. J Exp Orthop 1:10
17. Chang NJ et al (2015) Positive effects of cell-free porous PLGA implants and early loading exercise on hyaline cartilage regeneration in rabbits. Acta Biomater 28:128–137
18. Sofu H et al (2017) Results of hyaluronic acid-based cell-free scaffold application in combination with microfracture for the treatment of osteochondral lesions of the knee: 2-year comparative study. Arthroscopy 33(1):209–216
19. Fermor HL et al (2015) Development and characterisation of a decellularised bovine osteochondral biomaterial for cartilage repair. J Mater Sci Mater Med 26(5):186
20. Nogami M et al (2016) A human Amnion-derived extracellular matrix-coated cell-free scaffold for cartilage repair: in vitro and in vivo studies. Tissue Eng Part A 22(7–8):680–688
21. Li H et al (2016) Osteochondral repair with synovial membrane-derived mesenchymal stem cells. Mol Med Rep 13(3):2071–2077
22. Itokazu M et al (2016) Transplantation of scaffold-free cartilage-like cell-sheets made from human bone marrow mesenchymal stem cells for cartilage repair: a preclinical study. Cartilage 7(4):361–372
23. Oda K et al (2014) Comparison of repair between cartilage and osteocartilage defects in rabbits using similarly manipulated scaffold-free cartilage-like constructs. J Orthop Sci 19(4):637–645
24. Friedenstein AJ et al (1968) Heterotopic of bone marrow. Analysis of precursor cells for osteogenic and hematopoietic tissues. Transplantation 6(2):230–247
25. Sridharan B et al (2017) In vivo evaluation of stem cell aggregates on osteochondral regeneration. J Orthop Res 35:1606–16
26. Nansai R et al (2011) Surface morphology and stiffness of cartilage-like tissue repaired with a scaffold-free tissue engineered construct. J Biomed Sci Eng 6(1):40–48
27. Ishihara K et al (2014) Simultaneous regeneration of full-thickness cartilage and subchondral bone defects in vivo using a three-dimensional scaffold-free autologous construct derived from high-density bone marrow-derived mesenchymal stem cells. J Orthop Surg Res 9:98
28. Murata D et al (2015) A preliminary study of osteochondral regeneration using a scaffold-free three-dimensional construct of porcine adipose tissue-derived mesenchymal stem cells. J Orthop Surg Res 10:35
29. Whitney GA et al (2012) Methods for producing scaffold-free engineered cartilage sheets from auricular and articular chondrocyte cell sources and attachment to porous tantalum. Biores Open Access 1(4):157–165

30 Shimomura K et al (2014) Osteochondral repair using a scaffold-free tissue-engineered construct derived from synovial mesenchymal stem cells and a hydroxyapatite-based artificial bone. Tissue Eng Part A 20(17–18):2291–2304

31 Oliveira JM et al (2006) Novel hydroxyapatite/chitosan bilayered scaffold for osteochondral tissue-engineering applications: scaffold design and its performance when seeded with goat bone marrow stromal cells. Biomaterials 27(36):6123–6137

32 Yan LP et al (2015) Bilayered silk/silk-nanoCaP scaffolds for osteochondral tissue engineering: in vitro and in vivo assessment of biological performance. Acta Biomater 12:227–241

33 Grassel S, Lorenz J (2014) Tissue-engineering strategies to repair chondral and osteochondral tissue in osteoarthritis: use of mesenchymal stem cells. Curr Rheumatol Rep 16(10):452

34 Caló E, Khutoryanskiy VV (2015) Biomedical applications of hydrogels: a review of patents and commercial products. Eur Polym J 65:252–267

35 Radhakrishnan J et al (2017) Injectable and 3D bioprinted polysaccharide hydrogels: from cartilage to osteochondral tissue engineering. Biomacromolecules 18(1):1–26

36 Lai JH et al (2013) Stem cells catalyze cartilage formation by neonatal articular chondrocytes in 3D biomimetic hydrogels. Sci Rep 3:3553

37 Cui X et al (2012) Direct human cartilage repair using three-dimensional bioprinting technology. Tissue Eng Part A 18(11–12):1304–1312

38 Bichara DA et al (2014) Osteochondral defect repair using a polyvinyl alcohol-polyacrylic acid (PVA-PAAc) hydrogel. Biomed Mater 9(4):045012

39 de Girolamo L et al (2015) Repair of osteochondral defects in the minipig model by OPF hydrogel loaded with adipose-derived mesenchymal stem cells. Regen Med 10(2):135–151

40 Inagaki Y et al (2014) Effects of culture on PAMPS/PDMAAm double-network gel on chondrogenic differentiation of mouse C3H10T1/2 cells: in vitro experimental study. BMC Musculoskelet Disord 15:320

41 Zeng L et al (2014) Effect of microcavitary alginate hydrogel with different pore sizes on chondrocyte culture for cartilage tissue engineering. Mater Sci Eng C 34:168–175

42 Gothard D et al (2015) In vivo assessment of bone regeneration in alginate/bone ECM hydrogels with incorporated skeletal stem cells and single growth factors. PLoS One 10(12):e0145080

43 Fonseca KB et al (2014) Injectable MMP-sensitive alginate hydrogels as hMSC delivery systems. Biomacromolecules 15(1):380–390

44 Zeng Q et al (2014) Bioglass/alginate composite hydrogel beads as cell carriers for bone regeneration. J Biomed Mater Res B Appl Biomater 102(1):42–51

45 Choi B et al (2014) Cartilaginous extracellular matrix-modified chitosan hydrogels for cartilage tissue engineering. ACS Appl Mater Interfaces 6(22):20110–20121

46 Martins EA et al (2014) Evaluation of chitosan-GP hydrogel biocompatibility in osteochondral defects: an experimental approach. BMC Vet Res 10:197

47 Boyer C et al (2017) Si-HPMC/Si-Chitosan hybrid hydrogel for cartilage regenerative medicine: from in vitro to in vivo assessments in nude mice and canine model of osteochondral defects. Osteoarthr Cartil 25(Supplement 1):S77

48 Yuan T et al (2014) Collagen hydrogel as an immunomodulatory scaffold in cartilage tissue engineering. J Biomed Mater Res B Appl Biomater 102(2):337–344

49 Mohan N et al (2017) Chitosan-hyaluronic acid hydrogel for cartilage repair. Int J Biol Macromol 104:1936–1945

50 Parmar PA et al (2015) Collagen-mimetic peptide-modifiable hydrogels for articular cartilage regeneration. Biomaterials 54:213–225

51 Chao P-HG et al (2010) Silk hydrogel for cartilage tissue engineering. J Biomed Mater Res B Appl Biomater 95(1):84–90

52 Parkes M et al (2015) Tribology-optimised silk protein hydrogels for articular cartilage repair. Tribol Int 89:9–18

53. Yodmuang S et al (2015) Silk microfiber-reinforced silk hydrogel composites for functional cartilage tissue repair. Acta Biomater 11:27–36
54. Wu J et al (2016) Rheological, mechanical and degradable properties of injectable chitosan/silk fibroin/hydroxyapatite/glycerophosphate hydrogels. J Mech Behav Biomed Mater 64:161–172
55. Ming J et al (2015) Silk fibroin/sodium alginate fibrous hydrogels regulated hydroxyapatite crystal growth. Mater Sci Eng C 51:287–293
56. Fedorovich NE et al (2012) Biofabrication of osteochondral tissue equivalents by printing topologically defined, cell-laden hydrogel scaffolds. Tissue Eng Part C Methods 18(1):33–44
57. Markstedt K et al (2015) 3D bioprinting human chondrocytes with Nanocellulose-alginate bioink for cartilage tissue engineering applications. Biomacromolecules 16(5):1489–1496
58. Holmes B et al (2016) A synergistic approach to the design, fabrication and evaluation of 3D printed micro and nano featured scaffolds for vascularized bone tissue repair. Nanotechnology 27(6):064001–064001
59. Ren X et al (2016) Engineering zonal cartilage through bioprinting collagen type II hydrogel constructs with biomimetic chondrocyte density gradient. BMC Musculoskelet Disord 17:301

Chapter 17
In Vitro Mimetic Models for the Bone-Cartilage Interface Regeneration

Diana Bicho, Sandra Pina, J. Miguel Oliveira, and Rui L. Reis

Abstract In embryonic development, pure cartilage structures are in the basis of bone-cartilage interfaces. Despite this fact, the mature bone and cartilage structures can vary greatly in composition and function. Nevertheless, they collaborate in the osteochondral region to create a smooth transition zone that supports the movements and forces resulting from the daily activities. In this sense, all the hierarchical organization is involved in the maintenance and reestablishment of the equilibrium in case of damage. Therefore, this interface has attracted a great deal of interest in order to understand the mechanisms of regeneration or disease progression in osteoarthritis. With that purpose, in vitro tissue models (either static or dynamic) have been studied. Static in vitro tissue models include monocultures, co-cultures, 3D cultures, and ex vivo cultures, mostly cultivated in flat surfaces, while dynamic models involve the use of bioreactors and microfluidic systems. The latter have emerged as alternatives to study the cellular interactions in a more authentic manner over some disadvantages of the static models. The current alternatives of in vitro mimetic models for bone-cartilage interface regeneration are overviewed and discussed herein.

Keywords Static models · Dynamic models · Monocultures · Co-cultures · Ex vivo cultures

D. Bicho (✉)
3B's Research Group – Biomaterials, Biodegradables and Biomimetics, Headquarters of the European Institute of Excellence on Tissue Engineering and Regenerative Medicine, University of Minho, Barco, Guimarães, Portugal

ICVS/3B's - PT Government Associate Laboratory, Braga, Guimarães, Portugal
e-mail: dianabicho@dep.uminho.pt

S. Pina
ICVS/3B's - PT Government Associate Laboratory, Braga, Guimarães, Portugal

J. M. Oliveira · R. L. Reis
The Discoveries Centre for Regenerative and Precision Medicine, Headquarters at University of Minho, Guimarães, Portugal

ICVS/3B's - PT Government Associate Laboratory, Braga, Guimarães, Portugal

© Springer International Publishing AG, part of Springer Nature 2018
J. M. Oliveira et al. (eds.), *Osteochondral Tissue Engineering*,
Advances in Experimental Medicine and Biology 1059,
https://doi.org/10.1007/978-3-319-76735-2_17

17.1 Introduction

In the embryonic stage, after the formation of long bones from blastema (embryonic mesenchymal stem cells), a cartilaginous structure (or cartilage anlage) highly rich in collagen type II is formed. This structure will continue to grow until a thin vascularized calcified layer within the bone-cartilage interface is also formed [1]. This layer is on the basis of the articular cartilage, but the mechanisms that lead to its formation are still not completely understood [2]. During the postnatal period, the thick neonatal articular cartilage starts thinning due to its load-bearing activity, and the bone develops layer by layer through an oppositional expansion around the blood vessels and subjacent to the calcified cartilage [3, 4].

The bone-cartilage interface helps to maintain the structural functionality of the osteochondral (OC) entity. This structure possesses a thickness of approximately 100–200 µm and is composed of four interconnected structures, namely, hyaline cartilage, a thin layer of calcified cartilage, and subchondral and cancellous bone [5]. Different types of cells are involved in the structured composition of these uninterrupted compartments, namely, chondrocytes, hypertrophic chondrocytes, and osteoblasts [6, 7]. These cells are important in daily routine activities, in load-bearing applications, and allow a smooth motion of the joint. Moreover, during the correct performance of such activities, some types of forces, including tensile, compressive, and shear forces, are transmitted from the articular cartilage to the subchondral bone [1, 8]. Therefore, it is understandable that any change in the arrangement and composition of the bone-cartilage interface results in disturbances of the joint integrity and in the consequent loss of function, thus causing OC defects and lesions. This type of defects refers to significant injuries that affect both articular cartilage and subchondral bone and present limited regenerative capacity usually leading to Osteoarthritis (OA), one of the most incapacitating degenerative diseases [9].

Current strategies to treat large OC defects include noninvasive treatments such as the immobilization and the administration of nonsteroidal anti-inflammatory drugs, which only can provide symptomatic relief. Alternative methods that start with debridement of the lesion are also applied to remove the degenerated calcified cartilage or other joint components [10]. Another technique applied in the treatment of OC defects can involve drilling the underlying subchondral bone, which allows the formation of a blood clot rich in mesenchymal stem cells (MSCs) aiming their differentiation into the cell types present in the damaged tissues [11]. The use of autologous tissue transplantation is specially indicated for the implantation of hyaline-like cartilage. This technique makes use of cartilage that is removed by arthroscopy, from minor load-bearing area, and after in vitro growth, the chondrocytes are applied in the lesion site. However, this technique have short number of studies, and only one commercial ACI technique has the Food and Drug Administration (FDA) approval [12]. Finally, the implantation of artificial joints also represents an alternative to treat OC defects. Nevertheless, the prosthesis usually lasts around 20 years, can be loose after some time, or even cause infections,

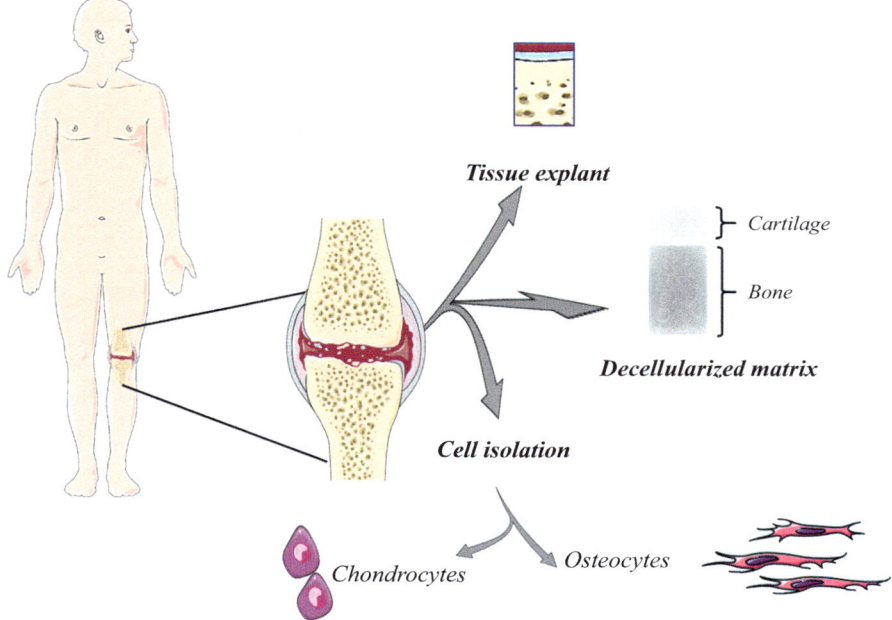

Fig. 17.1 Schematic representation of the sources extracted from bone and cartilage tissues in order to study the crosstalk in the inflammation, vascularization, and ossification processes during OA, as well as the interface regeneration mechanism

years after the surgery [13]. In this sense, new findings have led to improvements in an attempt to regenerate hyaline cartilage and subchondral bone [14, 15]. For instance, mosaicplasty represents a technique that focuses on OC plugs to fill articular cartilage defects. Though, it is limited by the area of the injury and the local site morbidity [16]. For that reason, the best solution for the complete regeneration of the bone-cartilage interface needs further research efforts. Therefore, different sources of tissues, cells, and matrices (Fig. 17.1) have been used to study the healing process of OC defects through in vitro models. However, some extra care should be taken, due to the hurdles related with in vitro models for OC studies [17]. First, it is important to bear in mind that different conditions are needed to mimic bone and cartilage considering their specific structure and composition. Bone is in contact with vasculature and bone marrow in a normoxic environment [18]. On the other hand, cartilage is imbibed in a hydrated viscoelastic matrix, and as deeper as we go in the tissue, less oxygen is needed to its homeostasis (2–4%) [19]. In fact, it is known that hypoxia is a strong promoter of matrix deposition by the chondrocytes [20]. In order to replicate these environments, different conditions are required, for example, growth media and 3D structures in case of using 3D cell culture. The second hurdle is related to the mechanical properties of the interface, a feature determined by matrix composition, water content, and structure of each tissue [21]. Finally, the third challenge when envisioning the development of an

Table 17.1 Advantages and limitations of the available in vitro mimetic models for the regeneration of bone-cartilage interface

Model	Advantages	Limitations	Ref
Static vitro models			
Monoculture	Cheap Easy to handle and optimize Fast expansion of cellular samples	Do not mimic in vivo conditions: Cells lose their histological organization and biochemical behavior The phenotype is changed without their natural ECM	[25, 26]
Co-culture	Easy to handle with fast expansion Crosstalk between cells can be studied	Different medium culture conditions must be considered for each type of cell	[33, 35]
3D culture	Important to simulate in vivo conditions Gives 3D support to the cells Mechanomodulation	The structures are not resistant enough to support load bearing forces After seeding, the isolation and expansion of cells is necessary	[63, 78]
Ex vivo culture	Preservation of the histological structure Maintenance of the in vivo conditions for a determined period of time	Cell death at cut edge of tissue Necrosis of the central zone of the explant after a few days: Insufficient oxygen access Physical features can change in vitro	[80, 81]
Dynamic in vitro *models*			
Bioreactors	Minimize the risk of contamination Reduce the costs associated with in vitro cell culture	Insufficient mechanical stress Difficult to monitor the accurate physiological state	[85, 86, 93]
Microfluidics	Automation and real-time, on-chip analysis and data acquisition Controlled cultures with perfusion Low consumption of reagents Reduced contamination risk	Expensive and complex chip design and operational control Limited by dimensions, which makes it difficult to achieve large tissues and organs	[96, 102, 103]

in vitro bone-cartilage model is related with mechanobiological stimulation, which refers to the cellular responses triggered by mechanical forces. In the construction of this type of models, the force that influences that has the biggest impact in bone and cartilage cells is generally controlled compression [22].

This chapter is focused on the most commonly used in vitro mimetic models, both in static and dynamic conditions, for the repair and regeneration of bone-cartilage interface. Table 17.1 summarizes the advantages and limitations of the available in vitro bone-cartilage interface models in order to develop suitable therapeutic approaches.

17.2   Static In Vitro Models

Static in vitro models refer to assays where the cells are allowed to settle on 2D or 3D adhesive substrates cultured with no flow and no mechanical stimulation [23]. These models are highly used because they are easy to handle and provide high throughput screening. In addition, their maintenance is cheap and no specialized laboratory equipment is needed. However, they can fail to mimic dynamic in vivo conditions of the human body, providing sometimes ambiguous results with regard to specific receptors, growth factor release, gene and protein expression, and even cellular proliferation, viability, and differentiation [24]. In this section, it is discussed two-dimensional (2D) and three-dimensional (3D) mono- and co-cultures and ex vivo cultures.

17.2.1   Monocultures

The majority of the studies using cell lines are developed in 2D monocultures. Conventional 2D cultures are preferably used by culturing cells in flat surfaces like Petri dishes, or in culture flasks, granting easy environment manipulation and similar access to nutrients and stimulus to the cell population [25]. Cell viability can be maintained by a monoculture of low density of cells that are in continuous contact with culture medium. The main problem arises when these conventional cultures need to be scaled up, since they need several changes of culture medium, increasing the risk of infection [26]. Nevertheless, the results obtained provide important conclusions that can set the course of a research line.

The initial in vitro techniques used to study chondrocyte differentiation required high density of confluent monocultures. These studies used embryonic limb bud mesenchymal cells from chick or mouse [27]. More recently, it was studied the monoculture of sclerotic osteoblasts from OA patients in comparison with non-sclerotic osteoblasts [28]. The data showed that sclerotic cells presented higher gene expression levels of matrix metalloproteinase 13 (MMP13), COL1A1 and COL1A2, osteopontin (OPN), alkaline phosphatase (ALP), osteocalcin (OCN), and vascular endothelial growth factor (VEGF). These parameters are indicative of matrix degradation in the bone-cartilage interface, which is evidenced by the osteoclast activation (showed by the high values of MMP13, OCN, and ALP). In addition, OPN and OCN are potent inhibitors of hydroxyapatite crystal formation, reinforcing the bone matrix degradation phenomenon. In the particular case of OA, these observations elucidate the reason why the extracellular matrix (ECM) in the sclerotic areas presents low mineralization [29]. Another study using monoculture in comparison to micro-masses not only described the relevance of using chondrocyte cell lines (chondrosarcoma cell lines, JJO12 and H-EMC-SS, and the immortalized chondrocytes, C-28/I2) and primary adult human chondrocytes to achieve anabolic and catabolic responses but also investigated the properties of chondroprotective ard anti-inflammatory drug effect [30]. However, because C-28/I2 had chondrocyte-like phenotype, they were chosen to produce micromasses to test several stimuli in order to study matrix anabolism and catabolism in chondrocytic cells. The results showed that C-28/I2 micromasses allowed the accumulation of glycosaminoglycan (GAG) which is an important factor of chondrocytic phenotype. Likewise, the retrospective validation of bioactive agents already tested in vivo showed comparable results using the system of C-28/I2 micromasses, giving confidence for this protocol as an alternative to assays based on primary cells. Finally, it was reported that high-density micromass cultures of these cell lines represent a reliable tool to study chondrocytes processes showing its advantages over monocultures. In fact, since the development of micromass cultures by Solursh and co-workers, that technique was regarded as a convenient way to analyze some biological processes [31]. Osteoblasts cultured in monocultures represents the most frequent method to investigate the effect of growth, differentiation, disease progression, and angiogenesis in vitro [32].

The works herein described refer to monocultures of bone and cartilage. However, their interface should be studied in co-cultures preferably under 3D conditions, mostly because monocultures are not able to mimic in vivo conditions, leading to possible changes in phenotype.

17.2.2   Co-Cultures

The surrounding environment of living organisms is highly dynamic, and it involves other cell types and compounds that usually influence cell response and function and gene expression. Therefore, when trying to mimic multilayered tissues, or tissue interfaces such as bone-cartilage interface, 2D or 3D cell co-cultures are very useful to evaluate the mechanism of vascularization [33], bone [34, 35], cartilage [36], and OC tissue regeneration [37].

In co-culture systems, critical parameters like the type of cells, culture media, and the order by which cells are cultured, or even the system used for the culture, will greatly influence the maintenance of the physiological patterns and phenotypic characteristics of the cells. Even though these technical details still need some optimization, the approach of co-culturing cells is advantageous to study interfaces and to reduce the time and production costs of engineered technologies, allowing an easier translation to the clinic [38]. For instance, some works have proved that in vitro co-culture has the benefit of allowing the differentiation of MSCs into chondrocytes, as long as they are in co-culture with chondrocytes, with specific cocktails of growth factors [39, 40]. This feature is due to the trophic effects and direct cell-to-cell contact employed by both MSCs and chondrocyte cell types [41]. Some of the factors that are secreted by mature cartilage are transforming growth factor-β (TGF-β) and insulin-like growth factor-1 (IGF-1), which induce chondrocyte differentiation of BMSCs in co-cultures [42]. The chondroinduction is influenced via the integrins of the ECM and via gap junction in the cell-to-cell contact or

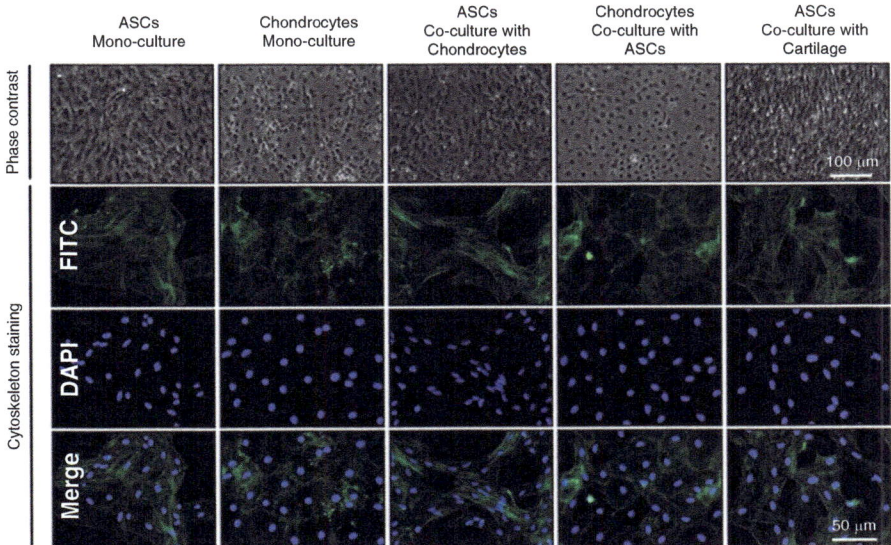

Fig. 17.2 Immunofluorescent staining of the 2D cultures of ASCs and chondrocytes after 7 days of mono- and co-culture. Cell morphological features are shown in phase-contrast (first line) and with immunofluorescent staining for the cytoskeleton (second line) and nucleus (third line). These show that the cells cultured with low-density serum in the media maintained their normal cellular viability and complete cellular integrity. (Adapted with permission from [45]. Copyright © 2016 Sichuan University)

transmembrane protein such as cadherins [43]. In another study, cadaveric primary chondrocytes were used in co-culture with cells having different passages, to explore the possibility of using cells, from allogenic cartilage in a sufficient number that allowed the repair of injured zones. Collagen membrane (Geistlich Bio-Gide®), with and without co-cultured chondrocytes, were implanted in the subcutaneous tissue of athymic mice and left to grow for 3 months. The data showed no significant difference between the numbers of chondrocytes isolated from live and cadaveric donors (48 h post-death), in terms of its ability to form cartilage-like tissue [44]. A different approach was applied by Zhong et al. [45] who explored the paracrine action of adipose-derived stem cells (ASCs) on chondrocytes by screening the presence of specific growth factors (Fig. 17.2). The co-culture showed an increase of bone morphogenetic protein-2 (BMP-2), vascular endothelial growth factor β (VEGF- β), hypoxia-inducible factor-1α (HIF-1α), fibroblast growth factor-2 (FGF-2), and TGF-β1 suggesting that the crosstalk between ASCs and chondrocytes is beneficial for the regeneration of injured cartilage. Data from a preclinical study that used a murine model with induced OA treated with a single injection of ASCs showed the inhibition of synovial activation and formation of chondrophyte/osteophyte, as well as cartilage destruction probably due to the macrophage suppression [46]. In general terms, this suggests that the secretome of ASCs possesses several biological factors that influence pathways important for the cellular activation and molecular processes of crosstalk, namely, the protective effect of ASCs exerted in endogenous cartilage.

In the context of bone tissue engineering, co-culture systems have also been exploited to study differentiation processes. Schulze and co-workers have developed an elegant bone cell co-culture with human bone marrow stromal cells (hBMSC) and human peripheral blood monocytes (hPBMC) as osteoblast/osteoclast precursor cells, respectively [47]. The purpose was to avoid external supplements for the induction of differentiation preventing possible adverse effects. The direct co-culture proved itself efficient in bone cell crosstalk while showing osteoclastogenesis effect of hBMSC in hPBMC. In contrast, indirect co-culture exhibited reduced matrix resorption pointing out the importance of the cell-to-cell contact. Likewise, other co-culture platforms have been used to study the ossification process using MSCs. Specifically, in endochondral ossification it is possible to observe the gene expression gradation in chondrocytes, mesenchymal cells, and osteoblasts [48]. For example, hypertrophic chondrocytes resident in the OC interface show upregulation of *Osterix* and *Runx2* essential in bone formation but also collagen type X that is expressed by hypertrophic chondrocytes during endochondral ossification. In addition, the matrix around the late hypertrophic chondrocytes is mineralized through deposition of hydroxyapatite. The angiogenic factors VEGF and FGF-2 are also increased due to the vascularized matrix in the bone-cartilage interface [49–51]. It has been also reported that OC defects of the femoral patellar groove can be regenerated even with low doses of BMP-2 produced by *Escherichia coli* [52].

In vascularization, the uses of co-cultures have been widely employed to understand the mechanisms that rule this complex process. Even though, when in monoculture, endothelial cells (ECs) only form tubes characteristic of blood vessels when supplemented with angiogenic factors, in co-culture with fibroblasts or osteoblasts, due to the crosstalk between the cells, the vessel-like structures can be generated without resorting to external supplements [53, 54]. Nonetheless, it should be kept in mind that ECs only form cord-like structures in 2D cultures. It is only under 3D culture that ECs can form tube structures with lumen that enable the blood/fluids to flow [55]. The development of lumen can be achieved if the cells in co-culture secrete enough ECM to encapsulate the ECs [56]. Evensen et al. showed that after 1 week of co-culturing ECs with vascular smooth muscle cells (vSMCs), the cells experienced some phenotypical changes that resulted in an interconnected network that can be maintained in culture for several weeks [57]. Moreover, in the same study, primary human microvascular endothelial cells (HuMVECs), primary human umbilical cord vein endothelial cells (HUVECs), and osteoblasts were co-cultured with vSMCs and were able to develop indistinguishable networks. It was shown that EC proliferation is inhibited in co-culture, instead they become highly migratory during the first 72 h in the presence of MSCs, and form interconnected anastomosed endothelial networks. This can be attributed to the fact that the co-cultured cells can function as supporting cells to ECs and are able to express angiogenic factors that guide tube formation through a paracrine effect [38]. Another example of co-culture with the aim to recover poor vascular supply in bone healing used ASCs and BMSCs. The culture consisted in a trans-well having a monoculture of BMSC on the lower well and ASCs in the upper well. The results showed higher level of VEGF expression, osteogenic differentiation, and greater mineralization when compared with the control (monoculture of BMSC) [58]. Nevertheless, some cells present an antagonist effect. As example, culturing HuMVECs in 3D fibrin and collagen gels led to the formation of capillary-like networks, if cells are supplemented with the proper growth factor and not with chondrocytes or chondrocyte conditioned media [59].

As referred above, co-cultures can be useful to better understand the bone-cartilage interface, which represents a complex synergy presenting probably both molecular and biochemical crosstalk between cells within the structure. Data has shown that the crosstalk is elevated in OA, in which the microfractures of the bone increase vascularization and facilitate the transport of biomolecules between bone and cartilage [60, 61]. These results were accomplished using a photobleaching (FLIP) method to monitor a low molecular weight fluorescent marker (0.12 mg/mL sodium fluorescein) perfused in articular surfaces or in the epiphyses of the subchondral bone without any damage of this structure [62]. In another perspective, the study of the interaction between zonal populations of the articular cartilage was performed. To this end, co-culture models were established using surface, middle, and deep zones of the articular cartilage. It was found that cellular interactions, particularly between surface and deep zones, are critical for the inhibition of the mineralization in the deep zone. Besides, the regulation of chondrocyte mineralization is regulated by local paracrine factors, such as parathyroid hormone-related peptide (PTHrP). Accordingly, the upregulation of PTHrP was only measured in co-culture suggesting that cellular interaction between the surface and deep zone is a prerequisite for its secretion [63]. Some works exploring the co-culture of osteoblasts and chondrocytes have suggested that heterotypic cellular communications may also be relevant for regulating articular chondrocyte mineralization in a condition similar to the in vivo [28, 63].

Although this type of model provides a more representative in vivo-like tissue structure and allows the study of the crosstalk between cells in interfaces, it still needs special attention regarding the culture conditions. This is important because different cells need different conditions to grow, which in turn may lead to phenotypes not present in the primary lines [33].

17.2.3   3D Cultures

The 3D cultures can make use of natural or synthetic-based scaffolds possibly opening up new prospects in the manipulation, stimulation, and recording of OC activity in bone-cartilage interfaces. This type of culture creates artificial 3D environments that allow cells to grow in all directions, providing more physiologically relevant information to mimic in vivo environments. It is particularly important for chondrocytes since they lose their phenotype in monoculture, unlike osteoblasts that are anchorage-dependent. Therefore, optimization of the co-culture seeding led to a sequential culturing protocol where a high density of chondrocytes is initially seeded and then, osteoblasts are incorporated [63]. Under these conditions, the capacity of osteoblasts to maintain their phenotype can vary greatly. However, the seeding of micromasses during osteoblast culture has been shown to be advantageous for osteoblasts maturation [32]. As a result, biphasic and multilayered scaffolds have been developed as a suitable alternative to engineered OC constructs. For example, recently Çakmak et al. designed a 3D scaffold possessing three different layers of silk fibroin (SF) from *Bombyx mori* cocoons and peptide amphiphile (PA) hydrogels, suitable for bone and cartilage tissues, respectively [64]. Firstly, the bone and cartilage layers were prepared separately using 6% (w/v) of SF plus 5% (w/w) of hydroxyapatite (HAp) and 1.5% (w/v) of PA-RGDS. During 2 weeks, hBMSCs were cultured in the SF + HAp layer with osteogenic media, while hACs were encapsulated and cultured inside the PA-RGDS layer, using chondrogenic media. Then, both layers were combined with an interface layer of soft silk sponge (Silk-IS) with 4% (w/v) SF. Finally, the construct was kept in an OC cocktail medium without any growth factor. The analysis of the gene expression allowed speculating that hACs in the PA-RGDS initiated their chondrocytic activity at the beginning of the first week. On the other hand, the mineralization of hBMSCs was observed specially on the pore walls of the silk scaffolds that became thicker and darker, due to the higher calcium deposition in the co-culture system. One of the biggest challenges of this area is to engineer anatomically accurate OC constructs. Therefore, researchers have been taking advantage of BMSC which have a natural tendency to become hypertrophic chondrocytes and also experience endochondral ossification [65–67]. Mesallati et al. developed an OC construct suitable for partial or total resurfacing of diseased joints (Fig. 17.3) [37]. They demonstrated that BSMCSs seeded in cylindrical alginate hydrogels were able to support the in vivo osseous component (bottom layer) of the construct. Then, the chondral layer (top) of the bilayered construct was molded using a cell suspension in a 2% (w/v) agarose gel or using a self-assembly approach. Several cells were tested in this layer, namely, chondrocytes, BMSCs, infrapatellar fat pad-derived stem cells (FPSCs), a co-culture of BMSCs plus chondrocytes (with 4:1 ratio), and FPSC plus chondrocytes (with 4:1 ratio). In vivo results showed that OC constructs were better integrated in the chondral layer when co-cultures were used. Specifically, the co-culture of BMSC and chondrocytes produced a homogenous and phenotypically stable layer of articular cartilage. Data also suggested that endochondral ossification can progress in the osseous region of the implant (collagen type I and type X were detected), despite the presence of chondrocytes in the top chondral region of the implant.

Another technique applied to produce scaffolds is based on the ability to mimic structural features of natural interfaces to support gradual mechanical stress and functional tissue regeneration. Accordingly, aligned scaffolds have been developed in order to facilitate the cell adhesion and to prepare suitable porosity for long-term cell colonization and adequate cell–material interactions [68]. For example, Tellado et al. developed aligned SF constructs to simulate the gradient of the natural collagen molecules present at the tendon/ligament-to-bone interface [69]. These structures had two types of pore alignment, i.e., anisotropic at the tendon/ligament side and

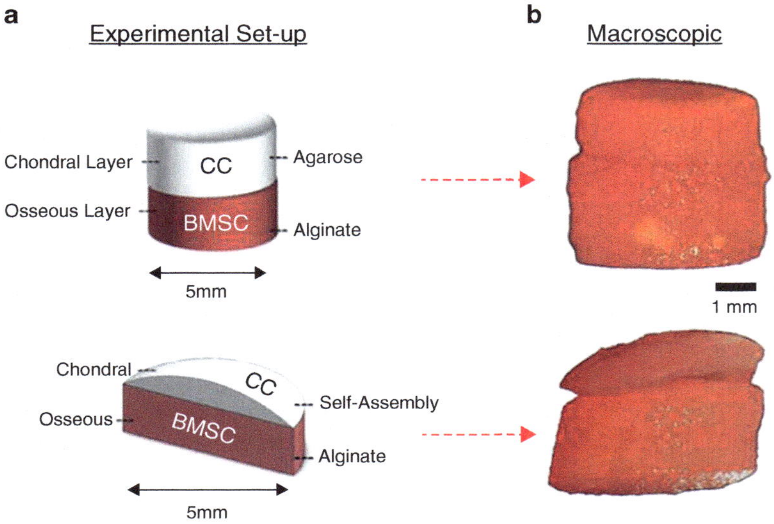

Fig. 17.3 Representation of preimplantation of 3D bilayered constructs constituted by chondrocytes (CCs) in the top layer (chondral) and bone marrow-derived MSCs (BMSCs) in the bottom layer (osseous). (**a**) Experimental setup of bilayered constructs, with the chondral layers formed through agarose encapsulation or self-assembly (SA). (**b**) Macroscopic images of the scaffold after 6 weeks of culture in vitro. (Adapted with permission from [37])

isotropic at the bone side. According to the reported results, isotropic scaffolds were more promising for this purpose given their smooth transition zone in which the lamellar pores penetrated the round pores. This smoother transition zone may have resulted in enhanced mechanical behavior for isotropic scaffold. The data obtained with the seeding of ASCs in isotropic structures showed a decrease over the time in the gene expression of collagen type I and collagen type III, reinforcing the application in tissue engineering tendon/ligament-to-bone.

One of the most used strategies to mimic the organization of articular cartilage and subchondral bone with the intention to treat OC defects has been bilayered scaffolds [70]. Bilayered scaffolds allow to better mimic the native ECM for each tissue type. Naturally, this type of hybrid strategy is more complex than strategies aiming to replace single tissues. However, they exhibit increased importance since the presence of a proper bone-cartilage interface prevents the growth of osseous into the chondral layer. With this in mind, Oliveira et al. [71] prepared a hydroxyapatite−/chitosan-based (HA/CS) bilayered scaffold to be potentially applied in OC defects. Cell culture studies carried out using goat marrow stromal cells showed adequate support for the attachment, proliferation, and differentiation of these cells into osteoblasts and chondrocytes. Additionally, physicochemical tests showed the great potential of these constructs for TE strategies. In another work, Yan et al. [72] prepared silk/nano-CaP bilayered scaffolds having 16% wt. silk in the cartilage-like layer and the osseous layer containing 16% wt. silk with CaP (hydroxyapatite) that was in situ synthesized in the silk solution. The scaffolds presented a continuous porous and interconnected structure in the construct with good integration, proper hydration degree, and superior mechanical properties.

Decellularized matrices have been proposed for allogenic or xenogeneic scaffolds in OC tissue engineering, due to their ability to recruit stem cells without adding any biological factors [73, 74]. However, it is still unclear if a complete decellularization of articular cartilage is needed, but it is well accepted that this process may reduce immune responses [75]. Johnson et al. treated 36 patients with average OC defects of 2.3 cm^2 using decellularized OC allograft plugs [76]. The treated patients had an average follow-up of 15 months with MRI, and ten of the evaluated patients were considered failures because the scaffolds were not retained. Sometime after the surgery, the repaired zone experienced a full-thickness loss of cartilage together with cystic change, irregularity, and depression. In this sense, the authors advised on the cautions when using decellularized OC allograft plugs. As alternative, the matrix components of the decellularized matrix may be conjugated with other biomaterials to improve some characteristics of the constructs. As an example, Vindas Bolaños et al. applied a cartilage-derived matrix (CDM) scaffold with decellularized cartilage from horses and a 3D printed composite scaffold with CDM and calcium phosphate to fill OC defects in the femuropatellar joints of horses [77]. Histological analysis showed predominantly fibrotic repair tissue with limited quantity of GAGs and collagen type II, but high quantity of collagen type I. Biomechanical analysis revealed that the scaffolds were very soft and had less stiffness when compared with normal adjacent tissue. The equine models only showed mild lameness with limited filling of the defect.

Although 3D structures represent a great improvement in cell culture, the diffusion of oxygen, metabolites, and nutrients through the scaffolds represents the greatest disadvantage of these static systems [78].

17.2.4   Ex Vivo Cultures

Ex vivo cultures can involve the extraction of pieces of tissue, or organs, in extreme sterile conditions and their culture in conditions similar to in vivo [79]. The use of ex vivo tissue model for OC interface studies represents an added value in this research field, mostly because it encloses osteoblasts, osteoclasts, and chondrocytes. The mouse femoral head is the explant culture mostly used for bone modeling and cartilage ECM degradation, signaling, and cell communication, as well to evaluate the potential of a given compound before in vivo tests [80]. Madsen et al. utilized murine femoral heads to induce catabolic (using oncostatin M (OSM) + tumor necrosis factor (TNF)-α) and anabolic responses (with insulin-like growth factor (IGF)-I). The results showed that catabolic stimulation decreases the quantity of proteoglycans that can increase the release of collagen and sulfated glycosaminoglycans (GAG) and increases the osteoclast number and bone resorption. Additionally, it was shown that the ex vivo model is able to induce cartilage formation and degradation, by anabolic or catabolic stimulation, respectively [81]. De Vries-van Melle et al. arranged bovine OC biopsies in which different depths of cartilage defects were prepared to study the applicability of its reparation in vitro by seeding cells in these defects [82]. The data demonstrated that this OC model was viable during 28 days showing low levels of lactate dehydrogenase (LDH), mRNA retrievable, and positive expression in subchondral bone genes. Specifically, tartrate-resistant acid phosphatase (TRAP), calcein, and ALP levels showed osteoclast activity, matrix deposition, and bone remodeling during the period of culture. Even though OC biopsies were not equal after the 28 days, it was verified the significance of the subchondral bone to heal the articular cartilage. Therefore, this model provides a more representative system of the in vivo metabolism. However, the viability of these tissues is very short, especially because insufficient oxygen and nutrients diffuse into the central zone of the explant leading to necrosis.

17.3   Dynamic In Vitro Models

Dynamic culture systems aim to provide an environment similar to in vivo conditions with a continuous supply of oxygen and nutrients, as well as mechanobiological conditions. The bioreactors and microfluidic systems used in dynamic in vitro models are presented and discussed herein.

17.3.1   Bioreactors Systems

Most of the findings described until now underline the importance of the bone-cartilage interface communication as a key to understand the histogenesis and biological mechanisms behind the chondral and osseous components. Consequently, in an attempt to replicate in vivo physiological conditions in OC treatments, bioreactors (Fig. 17.4) have emerged as a promising technology to explore [15]. Bioreactors enable the automation of in vitro cell culture, and they were specially designed to overcome the decrease of nutrients in the center of the scaffolds, one of the biggest challenges when dealing with 3D structures.

This technology emerged as a way to allow the dynamic cultivation of cells under controlled conditions of temperature, pH, CO_2, O_2 diffusion, shear rates, and controlled culture media flow [84]. For example, in a study developed by Grayson and co-workers, it was investigated the effect of different factors in cell growth and differentiation in biphasic OC constructs using either statically cultured or a bioreactor with medium perfusion through the bone region. The results showed that static culture of undifferentiated hMSCs in chondrogenic medium simulated the best cartilage properties, but pre-differentiated osteoblasts in perfusion or undifferentiated hMSCs in cocktail medium had the best osteogenic response. The

a. sealing lid
b. removable insert
c. O-ring
d. chondral construct
e. osseous construct

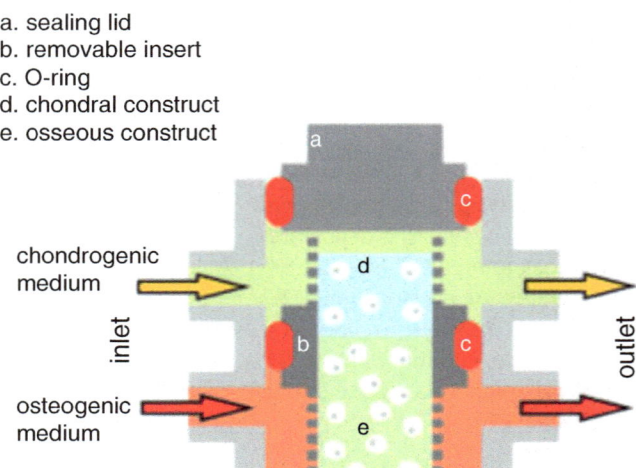

Fig. 17.4 Representation of a dual-chamber bioreactor system that provides tissue-specific compartments, specifically designed for cartilage and bone cell growth into engineered OC scaffolds. This technology allows a separated development of each tissue with different stimulus and culture media. (Adapted with permission from [83])

interface formed in these biphasic constructs is different from what happens in living tissues where the crosstalk between bone and cartilage occurs through mineralized cartilage. These conclusions supported the notion that OC conditions can be simultaneously established in a bioreactor developed with two compartments to enable different stimulus to be given to cells [85]. Thus, it is important that nutrients and growth factors be in a homogeneous distribution within the structure. In this sense, studies have been reported dual-chamber bioreactors, providing specific medias for bone and cartilage [86, 87]. As an example, Canadas et al. developed a dual-chamber bioreactor that allows each chamber to have different conditions to test the dynamics of different interfaces and to test the secretomes of different cell lines in a co-culture [88]. In the same way, Lozito et al. produced an OC scaffold consisting in a layered composite with bone, OC interface, cartilage, and synovium, using a dual-chamber bioreactor with perfusion [89]. The cartilage layer was produced with MSCs seeded in a cross-linked collagen/chitosan gel and TGF-β3. In turn, the synovial lining had MSCs seeded in cross-linked polyethylene glycol (PEG), with or without noninductive medium. The bone construct was peripherally surrounded by ECs to simulate the biological effects of blood vessels and the vasculature on OA. Additionally, physical and chemical stimulation of the cells intended to mimic responses to mechanical, pathological, and inflammatory insults within the micro-tissue system [15].

Bioreactors have been often applied to test different mechanical forces to form, maintain, and change the dynamic and function of 3D living tissues [90, 91]. These types of studies have demonstrated the importance of chondroinductive and chondrodestructive forces. As an example of the biomechanical stimuli as chondroinductive agent in embryonic fibroblasts, the application of an hydrostatic pressure of 5 MPa, 1 Hz, results in twofold increase in collagen synthesis and GAG production [92]. In bone, mechanical stimulation has shown that cells are more responsive to fluid flow than to mechanical strain given that no biochemical responses were detected with mechanical strains of less than 0,5% [83]. In general, there is an improvement in cell proliferation and expression of osteogenic markers in perfusion reactors; however, this is often dependent of the oscillating flow, steady flow, or other mechanisms [93, 94].

As stated above, bioreactors represent great potential for OC constructs to repair defect/injuries in the clinic media. Furthermore, these dynamic cell-culturing platforms enable an extensive monitoring and precise control of specific parameters that impact 3D cell cultures, providing also the possibility to study the function of engineered OC tissue with minimal risk of contamination. In the future, bioreactors can reach the clinic as automated approaches to manufacture OC scaffolds ready to be implanted into the patient where the scaffold would degrade leaving only the regenerated tissue.

17.3.2 Microfluidics

Microfluidics refers to the science and technology that uses small amounts (10^{-9}–10^{-18} liters) of precisely manipulated fluids that are conducted into channels with tens of hundreds of micrometers [95]. Microfluidic devices are able to control several soluble and physical factors simultaneously with high precision and sensitivity [96]. This technology offers a versatile platform and an analytical system in several biomedical applications: genetic assays, protein studies, intracellular signaling, drug discovery and testing, molecule and pathogen detection, cell culture, and tissue engineering [97–99]. Goldman and co-workers utilized a microfluidic hydrogel platform to stimulate OC phenotypes through spatially directed differentiation of bovine MSCs [100]. These researchers designed independent microfluidic networks and evaluated the constructs after 2 weeks of culture for the presence of differential gene expression and matrix composition between the osteogenic and chondrogenic layers. The authors studied the ability of mechanochemical cues, provided by the microfluidic network, to alter the phenotype of the MSCs through gene expression analysis, biochemical composition, and histological staining. Specifically, the chondrogenic region had higher content of GAGs and collagen type II, while the osteogenic region presented significantly higher expression of collagen type I and type X with increase of alizarin red staining. These findings are an important step in the design of an OC in vitro system for the future optimization of scaffolding formulations and bioprocessing parameters. Regarding physical stimulation of stem cells to improve osteogenic differentiation, Kim and co-workers used hMSCs within an osmotic pump-driven microfluidic chip that generated low shear stress. The results showed a significant increase of the nuclear and transcriptional activity of TAZ (transcriptional coactivator with PDZ-binding motif), which means an easier OC differentiation [101].

In brief, microfluidic systems can easily integrate biosensors to noninvasively detect cellular physiological parameters and to analyze external stimuli in situ. This type of devices can also incorporate separated chambers to co-culture cells and allow the study of interfaces even under mechanical changes. It is also possible to control the functionalization of the materials used but also the physical properties such as porosity, stiffness, or roughness to properly mimic in vivo situations. In this sense and even though microfluidic systems provide automated analysis with low consumption of reagents and low risk of contamination, they are very expensive and difficult to design.

17.4 Concluding Remarks

The use of in vitro models to study bone-cartilage interface regeneration is always associated with the specific purpose of the research line. Consequently, the ideal in vitro model must mimic as much as possible the specific in vivo situation, as it may help to better understand the physiological, pathological, and regenerative processes. The literature has been suggesting that even though static systems fail to mimic in vivo conditions and provided ambiguous results, they can serve as an initial screening to set the culture conditions. By its turn, dynamic culturing systems have shown to be the most plausible technologies to implement in the clinic, due to their capacity to produce sterile constructs in highly controlled conditions. Based on this, the use of hierarchical organized scaffolds can provide the necessary cues to the cells to mimic the complexity of the osteochondral unit and facilitate the tissue organization of subchondral bone and articular cartilage but also induce an integrated OC interface. Subsequently, several approaches have been developed regarding the cell source, composition and type of scaffold, and culture media composition, which have produced different outcomes, however, without producing consensus regarding the best overall in vitro strategy.

Acknowledgments The research leading to this work has received funding from the Portuguese Foundation for Science and Technology for the M-ERA.NET/0001/2014 project and for the funds provided under the program Investigador FCT 2012 and 2015 (IF/00423/2012 and IF/01285/2015).

References

1. Hoemann CD, Lafantaisie-Favreau C-H, Lascau-Coman V et al (2012) The cartilage-bone interface. J Knee Surg 25:85–97. https://doi.org/10.1055/s-0032-1319782
2. Haines RW (1975) The histology of epiphyseal union in mammals. J Anat 120:1–25
3. Hunziker EB, Kapfinger E, Geiss J (2007) The structural architecture of adult mammalian articular cartilage evolves by a synchronized process of tissue resorption and neoformation

during postnatal development. Osteoarthr Cartil 15:403–413. https://doi.org/10.1016/j.joca.2006.09.010

4. Hayes AJ, MacPherson S, Morrison H et al (2001) The development of articular cartilage: evidence for an appositional growth mechanism. Anat Embryol (Berl) 203:469–479

5. Khanarian NT, Boushell MK, Spalazzi JP et al (2014) FTIR-I compositional mapping of the cartilage-to-bone Interface as a function of tissue region and age. J Bone Miner Res 29:2643–2652. https://doi.org/10.1002/jbmr.2284

6. Yang L, Tsang KY, Tang HC et al (2014) Hypertrophic chondrocytes can become osteoblasts and osteocytes in endochondral bone formation. Proc Natl Acad Sci U S A 111:12097–12102. https://doi.org/10.1073/pnas.1302703111

7. Atesok K, Doral MN, Karlsson J et al (2016) Multilayer scaffolds in orthopaedic tissue engineering. Knee Surgery, Sport Traumatol Arthrosc 24:2365–2373. https://doi.org/10.1007/s00167-014-3453-z

8. Liu Y, Lian Q, He J et al (2011) Study on the microstructure of human articular cartilage/bone Interface. J Bionic Eng 8:251–262. https://doi.org/10.1016/S1672-6529(11)60037-1

9. Goldring SR, Goldring MB (2016) Changes in the osteochondral unit during osteoarthritis: structure, function and cartilage–bone cross talk. Nat Rev Rheumatol 12:632–644. https://doi.org/10.1038/nrrheum.2016.148

10. Mithoefer K, McAdams TR, Scopp JM, Mandelbaum BR (2009) Emerging options for treatment of articular cartilage injury in the athlete. Clin Sports Med 28:25–40. https://doi.org/10.1016/j.csm.2008.09.001

11. Smith GD, Knutsen G, Richardson JB (2005) A clinical review of cartilage repair techniques. J Bone Jt Surg - Br 87–B:445–449. https://doi.org/10.1302/0301-620X.87B4.15971

12. Iqbal Z, Kumaraswamy V, Srivastava P et al (2013) Role of autologous chondrocyte transplantation in articular cartilage defects: an experimental study. Indian J Orthop 47:129. https://doi.org/10.4103/0019-5413.108878

13. Röhner E, Pfitzner T, Preininger B et al (2016) Temporary arthrodesis using fixator rods in two-stage revision of septic knee prothesis with severe bone and tissue defects. Knee Surgery, Sport Traumatol Arthrosc 24:84–88. https://doi.org/10.1007/s00167-014-3324-7

14. Siclari A, Mascaro G, Gentili C et al (2014) Cartilage repair in the knee with subchondral drilling augmented with a platelet-rich plasma-immersed polymer-based implant. Knee Surgery, Sport Traumatol Arthrosc 22:1225–1234. https://doi.org/10.1007/s00167-013-2484-1

15. Alexander PG, Gottardi R, Lin H et al (2014) Three-dimensional osteogenic and chondrogenic systems to model osteochondral physiology and degenerative joint diseases. Exp Biol Med 239:1080–1095. https://doi.org/10.1177/1535370214539232

16. Worthen J, Waterman BR, Davidson PA, Lubowitz JH (2012) Limitations and sources of bias in clinical knee cartilage research. Arthrosc J Arthrosc Relat Surg 28:1315–1325. https://doi.org/10.1016/j.arthro.2012.02.022

17. Johnson CI, Argyle DJ, Clements DN (2016) In vitro models for the study of osteoarthritis. Vet J 209:40–49. https://doi.org/10.1016/j.tvjl.2015.07.011

18. Gibson JS, Milner PI, White R et al (2007) Oxygen and reactive oxygen species in articular cartilage: modulators of ionic homeostasis. Pflügers Arch - Eur J Physiol 455:563–573. https://doi.org/10.1007/s00424-007-0310-7

19. Grimshaw MJ, Mason RM (2001) Modulation of bovine articular chondrocyte gene expression in vitro by oxygen tension. Osteoarthr Cartil 9:357–364. https://doi.org/10.1053/joca.2000.0396

20. Murphy CL, Sambanis A (2001) Effect of oxygen tension and alginate encapsulation on restoration of the differentiated phenotype of passaged chondrocytes. Tissue Eng 7:791–803. https://doi.org/10.1089/107632701753337735

21. Yang PJ, Temenoff JS (2009) Engineering Orthopedic tissue interfaces. Tissue Eng Part B Rev 15:127–141. https://doi.org/10.1089/ten.teb.2008.0371

22. Madry H, Luyten FP, Facchini A (2012) Biological aspects of early osteoarthritis. Knee Surgery, Sport Traumatol Arthrosc 20:407–422. https://doi.org/10.1007/s00167-011-1705-8
23. Butler LM, McGettrick HM, Nash GB (2016) Static and dynamic assays of cell adhesion relevant to the vasculature. Methods Mol Biol 1430:231–248
24. Elliott NT, Yuan F (2011) A review of three-dimensional in vitro tissue models for drug discovery and transport studies. J Pharm Sci 100:59–74. https://doi.org/10.1002/jps.22257
25. Duval K, Grover H, Han L-H et al (2017) Modeling physiological events in 2D vs. 3D cell culture. Physiology 32:266–277. https://doi.org/10.1152/physiol.00036.2016
26. Sailon AM, Allori AC, Davidson EH et al (2009) A novel flow-perfusion bioreactor supports 3D dynamic cell culture. J Biomed Biotechnol 2009:873816. https://doi.org/10.1155/2009/873816
27. Umansky R (1966) The effect of cell population density on the developmental fate of reaggregating mouse limb bud mesenchyme. Dev Biol 13:31–56. https://doi.org/10.1016/0012-1606(66)90048-0
28. Sanchez C, Deberg MA, Bellahcène A et al (2008) Phenotypic characterization of osteoblasts from the sclerotic zones of osteoarthritic subchondral bone. Arthritis Rheum 58:442–455. https://doi.org/10.1002/art.23159
29. Hunter GK, Hauschka PV, Poole AR et al (1996) Nucleation and inhibition of hydroxyapatite formation by mineralized tissue proteins. Biochem J 317(Pt 1):59–64
30. Greco KV, Iqbal AJ, Rattazzi L et al (2011) High density micromass cultures of a human chondrocyte cell line: a reliable assay system to reveal the modulatory functions of pharmacological agents. Biochem Pharmacol 82:1919–1929. https://doi.org/10.1016/j.bcp.2011.09.009
31. Ahrens PB, Solursh M, Reiter RS (1977) Stage-related capacity for limb chondrogenesis in cell culture. Dev Biol 60:69–82
32. Jähn K, Richards RG, Archer CW, Stoddart MJ (2010) Pellet culture model for human primary osteoblasts. Eur Cell Mater 20:149–161. doi: vol020a13 [pii]
33. Kirkpatrick CJ, Fuchs S, Unger RE (2011) Co-culture systems for vascularization — learning from nature. Adv Drug Deliv Rev 63:291–299. https://doi.org/10.1016/j.addr.2011.01.009
34. Genova T, Munaron L, Carossa S, Mussano F (2016) Overcoming physical constraints in bone engineering: "the importance of being vascularized". J Biomater Appl 30:940–951. https://doi.org/10.1177/0885328215616749
35. Pirraco RP, Marques AP, Reis RL (2010) Cell interactions in bone tissue engineering. J Cell Mol Med 14:93–102. https://doi.org/10.1111/j.1582-4934.2009.01005.x
36. Lee P, Tran K, Zhou G et al (2015) Guided differentiation of bone marrow stromal cells on co-cultured cartilage and bone scaffolds. Soft Matter 11:7648–7655. https://doi.org/10.1039/C5SM01909E
37. Mesallati T, Sheehy EJ, Vinardell T et al (2015) Tissue engineering scaled-up, anatomically shaped osteochondral constructs for joint resurfacing. Eur Cell Mater 30:163–185
38. Baldwin J, Antille M, Bonda U et al (2014) In vitro pre-vascularisation of tissue-engineered constructs a co-culture perspective. Vasc Cell 6:1–16. https://doi.org/10.1186/2045-824X-6-13
39. Hubka KM, Dahlin RL, Meretoja VV et al (2014) Enhancing Chondrogenic phenotype for cartilage tissue engineering: monoculture and co-culture of articular chondrocytes and mesenchymal stem cells. Tissue Eng Part B 20:1–50. https://doi.org/10.1089/ten.TEB.2014.0034
40. Meretoja VV, Dahlin RL, Wright S et al (2013) The effect of hypoxia on the chondrogenic differentiation of co-cultured articular chondrocytes and mesenchymal stem cells in scaffolds. Biomaterials 34:4266–4273. https://doi.org/10.1016/j.biomaterials.2013.02.064
41. Wu L, Leijten JCH, Georgi N et al (2011) Trophic effects of mesenchymal stem cells increase chondrocyte proliferation and matrix formation. Tissue Eng Part A 17:1425–1436. https://doi.org/10.1089/ten.TEA.2010.0517

42. Ahmed N, Gan L, Nagy A et al (2009) Cartilage tissue formation using Redifferentiated passaged chondrocytes in vitro. Tissue Eng Part A 15:665–673. https://doi.org/10.1089/ten.tea.2008.0004

43. Qing C, Wei-ding C, Wei-min F (2011) Co-culture of chondrocytes and bone marrow mesenchymal stem cells in vitro enhances the expression of cartilaginous extracellular matrix components. Brazilian J Med Biol Res = Rev Bras Pesqui medicas e Biol 44:303–310

44. Olivos-Meza A, Velasquillo Martínez C, Olivos Díaz B et al (2017) Co-culture of dedifferentiated and primary human chondrocytes obtained from cadaveric donor enhance the histological quality of repair tissue: an in-vivo animal study. Cell Tissue Bank 18:369–381. https://doi.org/10.1007/s10561-017-9635-4

45. Zhong J, Guo B, Xie J et al (2016) Crosstalk between adipose-derived stem cells and chondrocytes: when growth factors matter. Bone Res 4:15036. https://doi.org/10.1038/boneres.2015.36

46. ter Huurne M, Schelbergen R, Blattes R et al (2012) Antiinflammatory and chondroprotective effects of intraarticular injection of adipose-derived stem cells in experimental osteoarthritis. Arthritis Rheum 64:3604–3613. https://doi.org/10.1002/art.34626

47. Schulze S, Wehrum D, Dieter P, Hempel U (2017) A supplement-free osteoclast-osteoblast co-culture for pre-clinical application. J Cell Physiol. https://doi.org/10.1002/jcp.26076

48. Janardhanan S, Wang MO, Fisher JP (2012) Coculture strategies in bone tissue engineering: the impact of culture conditions on pluripotent stem cell populations. Tissue Eng Part B Rev 18:312–321. https://doi.org/10.1089/ten.TEB.2011.0681

49. Panseri S, Russo A, Cunha C et al (2012) Osteochondral tissue engineering approaches for articular cartilage and subchondral bone regeneration. Knee Surgery, Sport Traumatol Arthrosc 20:1182–1191. https://doi.org/10.1007/s00167-011-1655-1

50. Pesesse L, Sanchez C, Henrotin Y (2011) Osteochondral plate angiogenesis: a new treatment target in osteoarthritis. Jt Bone Spine 78:144–149. https://doi.org/10.1016/j.jbspin.2010.07.001

51. Gawlitta D, Farrell E, Malda J et al (2010) Modulating endochondral ossification of multipotent stromal cells for bone regeneration. Tissue Eng Part B Rev 16:385–395. https://doi.org/10.1089/ten.TEB.2009.0712

52. Tokuhara Y, Wakitani S, Imai Y et al (2010) Repair of experimentally induced large osteochondral defects in rabbit knee with various concentrations of Escherichia coli-derived recombinant human bone morphogenetic protein-2. Int Orthop 34:761–767. https://doi.org/10.1007/s00264-009-0818-x

53. UNGER R, SARTORIS A, PETERS K et al (2007) Tissue-like self-assembly in cocultures of endothelial cells and osteoblasts and the formation of microcapillary-like structures on three-dimensional porous biomaterials. Biomaterials 28:3965–3976. https://doi.org/10.1016/j.biomaterials.2007.05.032

54. Goerke SM, Plaha J, Hager S et al (2012) Human endothelial progenitor cells induce extracellular signal-regulated kinase-dependent differentiation of mesenchymal stem cells into smooth muscle cells upon Cocultivation. Tissue Eng Part A 18:120914061009005. https://doi.org/10.1089/ten.tea.2012.0147

55. Davis GE, Bayless KJ, Mavila A (2002) Molecular basis of endothelial cell morphogenesis in three-dimensional extracellular matrices. Anat Rec 268:252–275. https://doi.org/10.1002/ar.10159

56. Davis GE, Stratman AN, Sacharidou A, Koh W (2011) Molecular basis for endothelial lumen formation and Tubulogenesis during Vasculogenesis and Angiogenic sprouting. Int Rev Cell Mol Biol 288:101–165. https://doi.org/10.1016/B978-0-12-386041-5.00003-0

57. Evensen L, Micklem DR, Blois A et al (2009) Mural cell associated VEGF is required for Organotypic vessel formation. PLoS One 4:e5798. https://doi.org/10.1371/journal.pone.0005798

58. Kim K-I, Park S, Im G-I (2014) Osteogenic differentiation and angiogenesis with cocultured adipose-derived stromal cells and bone marrow stromal cells. Biomaterials 35:4792–4804. https://doi.org/10.1016/j.biomaterials.2014.02.048

59. Pepper MS, Montesano R, Vassalli J-D, Orci L (1991) Chondrocytes inhibit endothelial sprout formation in vitro: evidence for involvement of a transforming growth factor-beta. J Cell Physiol 146:170–179. https://doi.org/10.1002/jcp.1041460122

60. Yuan XL, Meng HY, Wang YC et al (2014) Bone-cartilage interface crosstalk in osteoarthritis: potential pathways and future therapeutic strategies. Osteoarthr Cartil 22:1077–1089. https://doi.org/10.1016/j.joca.2014.05.023

61. Pan J, Wang B, Li W et al (2012) Elevated cross-talk between subchondral bone and cartilage in osteoarthritic joints. Bone 51:212–217. https://doi.org/10.1016/j.bone.2011.11.030

62. Pan J, Zhou X, Li W et al (2009) In situ measurement of transport between subchondral bone and articular cartilage. J Orthop Res 27:1347–1352. https://doi.org/10.1002/jor.20883

63. Jiang J, Leong NL, Mung JC et al (2008) Interaction between zonal populations of articular chondrocytes suppresses chondrocyte mineralization and this process is mediated by PTHrP. Osteoarthr Cartil 16:70–82. https://doi.org/10.1016/j.joca.2007.05.014

64. Çakmak S, Çakmak AS, Kaplan DL, Gümüşderelioğlu M (2016) A silk fibroin and peptide Amphiphile-based co-culture model for osteochondral tissue engineering. Macromol Biosci 16:1212–1226. https://doi.org/10.1002/mabi.201600013

65. Farrell E, Both SK, Odörfer KI et al (2011) In-vivo generation of bone via endochondral ossification by in-vitro chondrogenic priming of adult human and rat mesenchymal stem cells. BMC Musculoskelet Disord 12:31. https://doi.org/10.1186/1471-2474-12-31

66. Janicki P, Kasten P, Kleinschmidt K et al (2010) Chondrogenic pre-induction of human mesenchymal stem cells on β-TCP: enhanced bone quality by endochondral heterotopic bone formation. Acta Biomater 6:3292–3301. https://doi.org/10.1016/j.actbio.2010.01.037

67. Scotti C, Piccinini E, Takizawa H et al (2013) Engineering of a functional bone organ through endochondral ossification. Proc Natl Acad Sci 110:3997–4002. https://doi.org/10.1073/pnas.1220108110

68. Panseri S, Montesi M, Dozio SM et al (2016) Biomimetic scaffold with aligned microporosity designed for dentin regeneration. Front Bioeng Biotechnol 4:48. https://doi.org/10.3389/fbioe.2016.00048

69. Font Tellado S, Bonani W, Balmayor ER et al (2017) Fabrication and characterization of biphasic silk fibroin scaffolds for tendon/ligament-to-bone tissue engineering. Tissue Eng Part A 23:859–872. https://doi.org/10.1089/ten.tea.2016.0460

70. O'Shea TM, Miao X (2008) Bilayered scaffolds for osteochondral tissue engineering. Tissue Eng Part B Rev 14:447–464. https://doi.org/10.1089/ten.teb.2008.0327

71. Oliveira JM, Rodrigues MT, Silva SS et al (2006) Novel hydroxyapatite/chitosan bilayered scaffold for osteochondral tissue-engineering applications: scaffold design and its performance when seeded with goat bone marrow stromal cells. Biomaterials 27:6123–6137. https://doi.org/10.1016/J.BIOMATERIALS.2006.07.034

72. Yan L-P, Silva-Correia J, Oliveira MB et al (2015) Bilayered silk/silk-nanoCaP scaffolds for osteochondral tissue engineering: in vitro and in vivo assessment of biological performance. Acta Biomater 12:227–241. https://doi.org/10.1016/j.actbio.2014.10.021

73. Kang H, Peng J, Lu S et al (2014) In vivo cartilage repair using adipose-derived stem cell-loaded decellularized cartilage ECM scaffolds. J Tissue Eng Regen Med 8:442–453. https://doi.org/10.1002/term.1538

74. Zheng X-F, Lu S-B, Zhang W-G et al (2011) Mesenchymal stem cells on a decellularized cartilage matrix for cartilage tissue engineering. Biotechnol Bioprocess Eng 16:593–602. https://doi.org/10.1007/s12257-010-0348-9

75. Sutherland AJ, Beck EC, Dennis SC et al (2015) Decellularized cartilage may be a Chondroinductive material for osteochondral tissue engineering. PLoS One 10:e0121966. https://doi.org/10.1371/journal.pone.0121966

76. Johnson CC, Johnson DJ, Garcia GH et al (2017) High short-term failure rate associated with Decellularized osteochondral allograft for treatment of knee cartilage lesions. Arthrosc J Arthrosc Relat Surg. https://doi.org/10.1016/j.arthro.2017.07.018

77. Vindas Bolaños RA, Cokelaere SM, Estrada McDermott JM et al (2017) The use of a cartilage decellularized matrix scaffold for the repair of osteochondral defects: the importance of long-term studies in a large animal model. Osteoarthr Cartil 25:413–420. https://doi.org/10.1016/j.joca.2016.08.005

78. Do A-V, Khorsand B, Geary SM, Salem AK (2015) 3D printing of scaffolds for tissue regeneration applications. Adv Healthc Mater 4:1742–1762. https://doi.org/10.1002/adhm.201500168

79. JoVE Science Education Database. Developmental Biology. Explant Culture for Developmental Studies. JoVE, Cambridge, MA, (2018). https://www.jove.com/science-education/5329/explant-culture-fordevelopmental-studies. Accessed 22 Sep 2017

80. Marino S, Staines KA, Brown G et al (2016) Models of ex vivo explant cultures: applications in bone research. Bonekey Rep. https://doi.org/10.1038/bonekey.2016.49

81. Madsen SH, Goettrup AS, Thomsen G et al (2011) Characterization of an ex vivo femoral head model assessed by markers of bone and cartilage turnover. Cartilage 2:265–278. https://doi.org/10.1177/1947603510383855

82. de Vries-van Melle ML, Mandl EW, Kops N et al (2012) An osteochondral culture model to study mechanisms involved in articular cartilage repair. Tissue Eng Part C Methods 18:45–53. https://doi.org/10.1089/ten.TEC.2011.0339

83. You J, Yellowley CE, Donahue HJ et al (2000) Substrate deformation levels associated with routine physical activity are less stimulatory to bone cells relative to loading-induced oscillatory fluid flow. J Biomech Eng 122:387–393

84. Yeatts AB, Fisher JP (2011) Bone tissue engineering bioreactors: dynamic culture and the influence of shear stress. Bone 48:171–181. https://doi.org/10.1016/j.bone.2010.09.138

85. Grayson WL, Bhumiratana S, Grace Chao PH et al (2010) Spatial regulation of human mesenchymal stem cell differentiation in engineered osteochondral constructs: effects of pre-differentiation, soluble factors and medium perfusion. Osteoarthr Cartil 18:714–723. https://doi.org/10.1016/j.joca.2010.01.008

86. Malafaya PB, Reis RL (2009) Bilayered chitosan-based scaffolds for osteochondral tissue engineering: influence of hydroxyapatite on in vitro cytotoxicity and dynamic bioactivity studies in a specific double-chamber bioreactor. Acta Biomater 5:644–660. https://doi.org/10.1016/j.actbio.2008.09.017

87. R, Canadas (2015) Oliveira JM. Marques AP RR, Multi-chambers bioreactor, methods and uses

88. Canadas R, Oliveira JM, Marques AP RR (2014) Rotational dual chamber bioreactor: methods and uses thereof

89. Lozito TP, Alexander PG, Lin H et al (2013) Three-dimensional osteochondral microtissue to model pathogenesis of osteoarthritis. Stem Cell Res Ther 4:S6. https://doi.org/10.1186/scrt367

90. Responte DJ, Lee JK, Hu JC, Athanasiou KA (2012) Biomechanics-driven chondrogenesis: from embryo to adult. FASEB J 26:3614–3624. https://doi.org/10.1096/fj.12-207241

91. Oftadeh R, Perez-Viloria M, Villa-Camacho JC et al (2015) Biomechanics and mechanobiology of trabecular bone: a review. J Biomech Eng 137:108021. https://doi.org/10.1115/1.4029176

92. Elder S, Fulzele K, McCulley W (2005) Cyclic hydrostatic compression stimulates chondroinduction of C3H/10T1/2 cells. Biomech Model Mechanobiol 3:141–146. https://doi.org/10.1007/s10237-004-0058-3

93. Gaspar DA, Gomide V, Monteiro FJ (2012) The role of perfusion bioreactors in bone tissue engineering. Biomatter 2:167–175. https://doi.org/10.4161/biom.22170

94. Amini AR, Laurencin CT, Nukavarapu SP (2012) Bone tissue engineering: recent advances and challenges. Crit Rev Biomed Eng 40:363–408

95. Sackmann EK, Fulton AL, Beebe DJ (2014) The present and future role of microfluidics in biomedical research. Nature 507:181–189. https://doi.org/10.1038/nature13118

96. Zhang J, Wei X, Zeng R et al (2017) Stem cell culture and differentiation in microfluidic devices toward organ-on-a-chip. Futur Sci OA 3:FSO187. https://doi.org/10.4155/fsoa-2016-0091

97. Jun Y, Kang E, Chae S, Lee S-H (2014) Microfluidic spinning of micro- and nano-scale fibers for tissue engineering. Lab Chip 14:2145–2160. https://doi.org/10.1039/C3LC51414E

98. Hasan A, Paul A, Vrana NE et al (2014) Microfluidic techniques for development of 3D vascularized tissue. Biomaterials 35:7308–7325. https://doi.org/10.1016/j.biomaterials.2014.04.091

99. Aroonnual A, Janvilisri T, Ounjai P, Chankhamhaengdecha S (2017) Microfluidics: innovative approaches for rapid diagnosis of antibiotic-resistant bacteria. Essays Biochem 61:91–101. https://doi.org/10.1042/EBC20160059

100. Goldman SM, Barabino GA (2016) Spatial engineering of osteochondral tissue constructs through Microfluidically directed differentiation of mesenchymal stem cells. Biores Open Access 5:109–117. https://doi.org/10.1089/biores.2016.0005

101. Kim KM, Choi YJ, Hwang J-H et al (2014) Shear stress induced by an interstitial level of slow flow increases the osteogenic differentiation of mesenchymal stem cells through TAZ activation. PLoS One 9:e92427. https://doi.org/10.1371/journal.pone.0092427

102. Hasani-Sadrabadi MM, Pour Hajrezaei S, Hojjati Emami S et al (2015) Enhanced osteogenic differentiation of stem cells via microfluidics synthesized nanoparticles. Nanomedicine Nanotechnology, Biol Med 11:1809–1819. https://doi.org/10.1016/j.nano.2015.04.005

103. Whitesides GM (2006) The origins and the future of microfluidics. Nature 442:368–373. https://doi.org/10.1038/nature05058

Chapter 18
Bioreactors and Microfluidics for Osteochondral Interface Maturation

Raphaël F. Canadas, Alexandra P. Marques, Rui L. Reis, and J. Miguel Oliveira

Abstract The cell culture techniques are in the base of any biology-based science. The standard techniques are commonly static platforms as Petri dishes, tissue culture well plates, T-flasks, or well plates designed for spheroids formation. These systems faced a paradigm change from 2D to 3D over the current decade driven by the tissue engineering (TE) field. However, 3D static culture approaches usually suffer from several issues as poor homogenization of the formed tissues and development of a necrotic center which limits the size of in vitro tissues to hundreds of micrometers. Furthermore, for complex tissues as osteochondral (OC), more than recovering a 3D environment, an interface needs to be replicated. Although 3D cell culture is already the reality adopted by a newborn market, a technological revolution on cell culture devices needs a further step from static to dynamic already considering 3D interfaces with dramatic importance for broad fields such as biomedical, TE, and drug development. In this book chapter, we revised the existing approaches for dynamic 3D cell culture, focusing on bioreactors and microfluidic systems, and the future directions and challenges to be faced were discussed. Basic principles, advantages, and challenges of each technology were described. The reported systems for OC 3D TE were focused herein.

R. F. Canadas
3B's Research Group – Biomaterials, Biodegradables and Biomimetics, University of Minho, Headquarters of the European Institute of Excellence on Tissue Engineering and Regenerative Medicine, Barco, Guimarães, Portugal

ICVS/3B's – PT Government Associate Laboratory, Braga/Guimarães, Portugal

A. P. Marques · R. L. Reis (✉) · J. M. Oliveira
3B's Research Group – Biomaterials, Biodegradables and Biomimetics, University of Minho, Headquarters of the European Institute of Excellence on Tissue Engineering and Regenerative Medicine, Barco, Guimarães, Portugal

ICVS/3B's – PT Government Associate Laboratory, Braga/Guimarães, Portugal

The Discoveries Centre for Regenerative and Precision Medicine, Headquarters at University of Minho, Guimarães, Portugal
e-mail: rgreis@dep.uminho.pt

© Springer International Publishing AG, part of Springer Nature 2018   395
J. M. Oliveira et al. (eds.), *Osteochondral Tissue Engineering*,
Advances in Experimental Medicine and Biology 1059,
https://doi.org/10.1007/978-3-319-76735-2_18

Keywords Bioreactors · Microfluidics · Dynamic systems · Osteochondral tissue engineering

18.1   Introduction

Tissue engineering (TE) technologies are based on the biological triad of cells, signaling molecular pathways and ECM. Although TE leads a transition from 2D to 3D cell culture techniques in order to achieve more physiologically relevant tissue substitutes and models, culture techniques are still limited. Typically, current 3D cell culture techniques do not yet allow meeting the multicellular complexity of tissues, do not offer fine control over gradients, and require medium exchange at discrete time points instead of in a continuous manner because they rely on static environments [1]. Additionally, bioreactors and microfluidic devices, as a dynamic stimulus element, may be used as an alternative to or in conjunction with molecular growth factors for the signaling part of the TE triad. A TE bioreactor can be defined as a device to perform maturation of cell-material constructs under a controlled environment, using or not mechanical means to direct biological processes. This generally means that bioreactors are used not only to flow cell culture medium but also to stimulate cells and induce them to produce ECM [2]. Microfluidic devices, as an alternative, were born from the combination of bioreactor principles and microfabrication techniques to reduce either time or costs of cell culture process and diagnostic systems.

The available culture systems, including bioreactors, allow monocultures in 3D scaffolds but are not adapted for co-cultures, which currently is a big challenge of the field [3]. The classic case of the OC interface, which consists of a hyaline cartilage layer and an integrated subchondral bone, is a good example of an interface requiring for more complex scaffold design and a dual environment when cultured in vitro. Interfaced bone-cartilage in vitro models keep being predominantly limited to cell co-culture systems in which bone and cartilage cells are both exposed to the same medium, perhaps a very distant condition from the in vivo environment [4]. Regarding this, when co-culturing stem cells in 3D scaffolds, researchers either need to use costly pre-differentiation operations before cell seeding and keep its phenotype over cell culture or, as a more challenging alternative, simultaneously modulate differentiation down to distinct lineages in a unified culture environment [5]. So, using conventional bioreactor systems based on one single culture chamber, supplementing the culture media with lineage-specific signaling molecules for directed differentiation of stem cells is invalid for biphasic constructs. This limitation has been addressed by using custom made dual-chamber culture systems to spatially direct delivery preventing dominance of one phenotype over the other throughout the construct [5, 6].

Microfluidics, as an alternative to bioreactors at the microscale, allows spatial control over fluids in micrometer-sized channels and has become a valuable tool to further increase the physiological relevance of 3D cell culture by enabling controlled co-cultures, perfusion, and spatial control of signaling gradients [7, 8].

Over the next sections of this chapter, we discuss the current bioreactor and microfluidic systems, the current issues and advantages of using a dynamic cell culture instead of a static one, and the features to be considered when designing the devices for dynamic 3D cell culture approaches. The reported bioreactor and microfluidic studies for OC tissue regeneration and in vitro modeling are overviewed and discussed.

18.2 Dynamic Systems Designs

Bioreactor and microfluidic systems are cell culture environments confined in a reservoir with a shape of a vessel, flask, or even channels, connected to inlets and outlets for continuous flowing of nutrients through the cell culture. A bioreactor can be described as a dynamic device or system for culturing cells or tissues in suspension, 2D or 3D, under controlled biochemical or mechanical conditions. Microfluidic technique was created as an option for the perfusion of cell cultures, since the compartmentalized nature of microfluidic devices interconnected by microchannels allows perfusing media adjacent to or through a population of cells or 3D tissue-like construct.

In general, bioreactors and microfluidic systems are designed to perform at least one of the following five functions:

1. To achieve uniform cell distribution
2. To keep constant and optimized concentration of gasses and nutrients
3. To perform mass transport to the tissue
4. To increase tissue maturation by applying physical stimuli
5. To provide information about the formation of 3D tissue by attaching sensors and designing transparent chambers [9]

A bioreactor or microfluidic device incorporating a flow pump to continuously circulate culture medium is the minimum criteria to define as dynamic a cell culture technique. The pump or motor must be small enough to fit into an incubator and also be usable at 37°C and in a humid environment. The forces needed for cellular stimulation are very small, so it is important to ensure that the pump/motor has the capability to apply small forces accurately. Tissue culture is a continuous, non-steady-state process in which the cultivation and tissue-specific parameters change with time. Furthermore, the scaffold and chamber shapes are of significant importance for controlling the nutrition flow pattern [10].

18.2.1 Mass Transport: Diffusion and Convection

Static culture systems rely primarily on diffusion, and to a lesser extent natural convection, for transport of oxygen to cells. The depth of media in the static conditions limits the supply of oxygen from the gas phase, which is not a big issue when culturing a monolayer of cells [11]. However, in a 3D cell culture approach, improving local perfusion of thick tissue constructs remains a significant challenge for 3D-based devices. Static culture of cell-seeded 3D scaffolds typically results into localized tissue growth in the construct periphery because of lack on mass transport of fresh nutrients through the tissue [12]. Mass transport through the scaffolds can be improved using perfusion methodologies by housing the construct within a flow-through column [13, 14] or by suspending cell spheroids or constructs within rotary culture devices or spinner flasks, generating a dynamic culture [15, 16].

The important role of mass transport in a cell culture device is to keep cell metabolism within a physiological range by providing metabolic substrates and removing toxic degradation products. Understanding how device geometry is related to convective or diffusive transport limitations is therefore a key element of the bioreactor and microfluidic systems design.

The rate of diffusion is proportional to the metabolite concentration gradient, being the constant of proportionality and the diffusion coefficient:

$$D = \frac{kT}{6\pi R}$$

The Stokes-Einstein equation relates the radius of the diffusing particle and temperature to the diffusion coefficient, where k is Boltzmann's constant, T is temperature, and R is the particle radius. The volume of a sphere is proportional to the cube of its radius, $V = 4/3\pi r^3$, so the diffusion coefficient is approximately inversely related to the cube root of molecular weight.

The mass flux is also related to the gradient that can be generated at the interface between cells and medium, according to Fick's first law of diffusion:

$$J = -D\frac{d\varphi}{dx}$$

Fick's equation relates the diffusive flux "J" to the concentration when assuming a steady state (if a system is in a steady state, then the recently observed behavior of the system will continue into the future). The diffusion flux dimension is amount of substance per unit area per unit time, so it is expressed in mol m^{-2} s^{-1} and quantifies the amount of substance that will flow through a unit area during a time interval. D is the diffusion coefficient, and its dimension is area per unit time, so standard units would be m^2/s. φ is the concentration in ideal mixtures, of which the dimension is amount of substance per unit volume as mol/m^3. x is the position, the dimension of which is length; it might thus be expressed in the unit m.

18.2.2 Flow Conditions for Dynamic Cell Cultures

Diffusion is the only driving force to move nutrients and waste in a static culture system. As the size of the scaffold increases, diffusion of nutrients and waste removal to and from the interior of the construct becomes more difficult, leading to necrosis at the core of the scaffold [16]. Defects requiring TE solutions are typically many millimeters in size for which, in static cell culture systems, it can be quite difficult to enable sufficient fresh medium circulation through the engineering construct [11]. Scaffolds in such a size range are easily fabricated. However, problems arise when culturing cells on these scaffolds. The main challenge in preparing constructs larger than few millimeters is to obtain a homogeneous distribution of cells, and hence new tissue, throughout the whole 3D scaffold volume [17]. This problem is increased by the static culture conditions, which result in scaffolds with few cells in the center of the construct. It has been shown that despite uniform cell seeding, over cell culture period, more cells distribute predominantly on the periphery of the constructs [18]. The main reasons for this distribution in scaffolds at millimeter scale or more are cell sedimentation, necrosis, as a consequence of the previous ones, and cell chemotaxis. For instance, mineralized bone matrix reaches a maximum penetration depth of 240 μm when stromal osteoblasts are cultures onto poly(DL-lactic-co-glycolic acid) scaffolds, which is far thinner than an ideal bone graft needed for the clinic [19].

Perfusion flow performed by bioreactors and microfluidic devices provides several benefits such as stable nutrient supply, waste metabolites washed away, and control of oxygen tension distribution. Furthermore, perfusion is one of the key stimuli in vasculature, as it provides shear stress, which affects the cellular morphology and gene expression [20, 21]. So, constant nutrition, oxygen flow supply and metabolic product elimination must be performed when culturing cells up to higher density which is one of the most important tasks to achieve physiologically meaningful functions for TE and sufficient cell number for in vitro construct development [22].

While the first generation of bioreactors for culturing cells was designed simply to pump nutrient through the assembling tissue followed by waste removal [17], the next wave of bioreactors was designed for maturing tissues such as blood vessels [23], cartilage [12], or cardiac muscle [24], subjecting the emerging tissue to mechanical compression, shear stresses, and even culture medium pulsatile flow. Such stresses lead to improved mechanical properties of engineered tissues such as cartilage [25].

18.2.3 Flow Shear Stress and Related Stimuli

Flow-derived shear stress provides a physiologically relevant mechanical stimulation that significantly promotes specific protein expression and elicits paracrine effects by increasing the probability of randomly happen secretion and protein-protein and cell-cell contact [26]. These events are particularly relevant when culturing complex tissues as interfaces requiring co-cultures. Given the central role of several protein pathways in specific disorders, these stimuli represent a significant dynamic, for example, for an in vitro model of a disorder condition as OA.

Mechanical cues are also crucial for tissue specification and maturation, which can be transduced by the bioreactor chamber design [10] or microfluidic channel geometry, as suggested by a study of the effect of shear stress on mature osteoblasts. Kou et al. [27] developed a device capable of applying four different magnitudes of shear stress in parallel channels on one chip. The intracellular calcium intensity peak was proportional to intensity of shear stress, while response time was independent of shear stress for values larger than 0.03 Pa.

While flow rate can be used to modulate media exchange [28], pore size, interconnectivity, and anisotropy can influence different rates of media exchange and shear stress on cells distributed within the construct [29]. Mechanobiological aspects, such as active stretch and tension, are another functional feature that can be added using microfabrication techniques. Although interesting, it has received minor attention in combination with 3D cell culture. Regarding this challenge, sophisticated devices for 3D cell culture are one of the main needs for the current regenerative medicine and TE evolution [30].

18.3 Bioreactor Designs

18.3.1 Bioreactor General Principles

Bioreactors arise as a proposed solution when addressing the reported issues afore. Although the requirements for bioreactor design are application specific, there are a few general principles which have to be considered when developing a bioreactor. Biocompatible and bioinert materials must be selected to fabricate a bioreactor. Although stainless steel can be used if it is treated avoiding chromium ions leaching out into the culture medium, most metals are eliminated by this permit. Furthermore, a bioreactor operates at 37°C in a humid atmosphere, so selected materials cannot change drastically under these conditions, which avoids the use of some plastics. If a material changes volume under this humid condition, it can trigger medium leakage, which can be a problem in any design involving fluids. In most cases, fluid seals are necessary, and a good design should be considered to minimize the number of junctions needing for seals. Gharravi et al. designed a tissue chamber made of stainless steel presenting multiple pores and two inlets and outlets for the media and the gases (O_2 and CO_2) to replicate the perfusion process and oxygen tension in the body. The inside geometry of the central part of culture chamber was designed to mimic ball and socket of the temporomandibular joint for cartilage TE with round shape and as a tissue sized at a large dimension. Alginate was selected for the chondrocytes encapsulation in a sheet configuration [31].

A bioreactor has to operate under aseptic conditions to prevent any contamination of microorganisms. To guarantee the aseptic environment, bioreactor parts can be either sterilized by autoclaving or disinfected by submersion in alcohol or under a sterilizing gas (e.g., ethylene oxide), limiting the range of materials considered for a bioreactor fabrication. Autoclave runs cycles of a severe protocol performing high temperature and pressure, restricting even more the number of materials that can be selected for the bioreactor manufacturing. Alternatively, some non-stabilizable disposable bioreactor parts may be used and replaced after each use. Considering transparent materials for the culture chambers is advantageous for monitoring the construct during culture. For instance, Powers et al. have presented a bioreactor that enables both morphogenesis of 3D constructs under flow perfusion while allowing in situ observation by light microscopy, which is an important feature to take into account when designing a cell culture device [32].

Sensors should be incorporated into the design if parameters such as pH, nutrient (e.g., glucose) concentration, or oxygen levels are to be monitored. However, at the moment, it is not easy to measure all of these variables in "real time," and the existing microsensors are too expensive. Therefore, there is a requirement to develop sensors for "real-time" measurement or alternatively to remove samples for as close as possible a "real-time" analysis. Dissolved oxygen and culture medium pH have been monitored with fluorescent sensors developed to study the metabolic state of cells in culture without removing or damaging cells during cultivation [33].

As a market approach perspective, a prototype bioreactor has to be designed thinking about the scale-up opportunity. This means designing a device that is easy to enlarge without changing its characteristics or projecting a simple and user-friendly device easy to multiply in number. Moreover, thinking about an industrial application, a high-throughput system is the ideal strategy so that numerous scaffolds can be cultured at one time, which currently is not considered in most of the bioreactor systems reported. A bidirectional perfusion bioreactor was developed enabling implementing perfusion flow and flow direction which can be varied along the culturing period. For large-scale applications, the most important feature that this system comprises is its compact and user-friendly design made of autoclavable materials and enabling to culture up to 20 samples simultaneously [34].

A bioreactor can also be designed not to expand cells in number but to obtain de novo tissue with biomechanical properties comparable to the native one, by applying various mechanical or electromagnetic stimuli. Suckosky et al. [35] performed a characterization of the flow field within a spinner flask system working under optimized conditions to produce cartilage. The simulation and experimental collected data have shown that subjecting a scaffold to a dynamic flow provided a uniform cell distribution throughout the 3D construct resulting in a homogenous matrix deposition, whereas Altman et al. [36] have reported that directional strain applied on silicone-based constructs promoted cell differentiation into a ligament-like phenotype instead of bone or cartilage lineages. Electric [37] and magnetic [38] stimuli have also been used experimentally with encouraging results on stimulation of 3D muscle maturation and osteogenesis, respectively.

The new generation of bioreactors should also be able to support the culture of two or more cell types simultaneously, relevant for the regeneration of interfaced tissues. This currently involves first maintaining the different cell types under different static culture conditions to obtain the desired phenotypes and then at appropriate time, switching to a common cultivation protocol in one bioreactor. The formed tissue either presents a lose interface or fused phenotype. The appropriate design should consider chamber compartments (named dual chambers) interfaced by 3D scaffolds [39, 40]. This way the cell adhesion, proliferation, and differentiation phases can be promoted in a whole construct immersed in different environments over all culture time.

18.3.2   Bioreactor Types

The basic principle applied at the first generation of bioreactors for TE was the agitation-based approach. According to this approach, a cell suspension is placed into a container while keeping the suspension in motion. Gentle stirring is used to provide motion to cells. Due to this, cells do not adhere to the walls and form cell-to-cell interactions. Based on this principle, several bioreactor designs were created, which are summarized in Fig. 18.1.

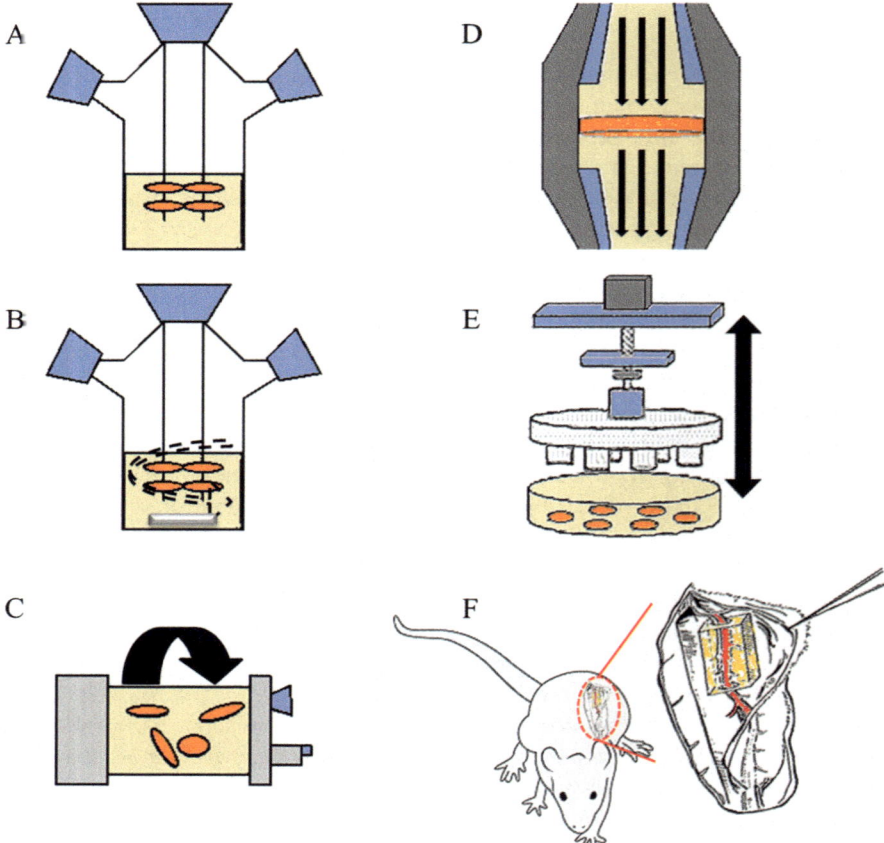

Fig. 18.1 Schematic representation of each bioreactor system for 3D cell culture. (**A**) Basic flask bioreactor, (**B**) spinner flask bioreactors, (**C**) rotating cell culture bioreactor, (**D**) flow perfusion bioreactor, (**E**) compression bioreactor, and (**F**) in vivo bioreactor

18.3.2.1 Spinner Flask Bioreactors

Spinner flasks consist of a container and a stirring element to continuously stir the cell suspension (Fig. 18.1B). These flasks are traditionally used to form cell spheroids but can also be applied to constructs attached to needles from the top cover. The size of the cell spheroids or constructs depends on the volume of the container. Culture medium agitation is performed using a magnetic stir bar placed at the bottom of the flask [41] promoting the nutrients diffusion and removal of waste products keeping the cells fed with fresh culture medium [42].

The main issues associated to the spinner flask bioreactors are related to the shear force of the stirring bar, requiring a larger amount of culture medium and resulting in inconsistency of size of the formed cell spheroids [43]. However, constructs cultured in spinner flasks have a higher cell seeding density and more uniform distribution of cells when compared to a static culture model [44].

The spinner flask bioreactor has been tested to promote osteogenesis. Sikavitsas et al. [45] compared steady flask, spinner flask, and rotating vessel (described below) systems by culturing rat mesenchymal stem cells (MSCs) in 3D scaffolds for a period of 21 days. The author observed that the highest alkaline phosphatase activity and osteocalcin secretion were obtained using the spinner flask system. Moreover, constructs cultured in the agitated systems had higher proliferation rate and calcium content than the steady flask system.

18.3.2.2 Rotating Cell Culture Bioreactors

The rotating bioreactor spins the whole container instead of using a stirrer bar/rod as in the case of the spinner flask bioreactor (Fig. 18.1C). The culture chamber is initially rotated at low speed which is increased when cells start forming large aggregates to maintain the spheroids in suspension. This concept was developed at NASA to simulate microgravity conditions. The systems design consists of two concentric cylindrical containers, within which lies an annular space containing the scaffold [46]. The outside wall rotates, and to obtain a microgravity condition, the gravitational forces must balance the centrifugal forces, subjecting the scaffold to dynamic laminar flow [47]. The main advantage of this system is the reduced shear force applied [48]. Using this approach, Marlovits et al. [49] suspended differentiated chondrocytes in a rotating wall vessel. After 90 days of cultivation under microgravity, cartilage-like neotissue was formed, encapsulated by fibrous tissue that closely resembled the perichondrium, without the use of any scaffolding material.

Saini et al. [50] have also shown the potential of this technique in cartilage TE, but, using porous polylactic acid scaffolds, seeded dynamically in the bioreactor using bovine chondrocytes. Four weeks after cell seeding, constructs from condition seeded at the highest cell densities contained up to 15 M cells, 2 mg glycosaminoglycan, and 3.5 mg collagen per construct and exhibited morphology similar to that of native cartilage. Overall the rotating wall vessel bioreactor has been shown to optimize nutrient transport and promote cartilage growth and differentiation [51].

While this system is simple, allowing easy handling and large-scale and long-term production of spheroids, there is large variability in the size of the spheroids as observed in the spinner flask bioreactors [52].

18.3.2.3 Flow Perfusion Bioreactors

Flow perfusion bioreactors utilize a pump to percolate medium continuously through the scaffold's interconnected pores (Fig. 18.1D). A molecular weight cutoff membrane isolates the chamber inlet and outlet avoiding cells from leaving the container. The enhanced nutrient transfer has been shown to result in improved mass transfer [41], contributing for a homogeneous cell distribution and high seeding efficiency throughout the thickness of the scaffold [53]. Furthermore, the fluid shear forces resulting from flow perfusion have been shown to enhance the expression of the osteoblastic phenotype [54]. Gomes et al. [55] cultured bone marrow MSCs under static and perfusion conditions. A superficial layer of cells was formed when the constructs were cultured statically, while a homogeneous cell distribution filled the scaffolds in the perfusion bioreactor. Although the proliferation rate and alkaline phosphatase activity patterns were similar for both conditions, the constructs cultured under perfused conditions showed a significant increase in calcium deposition.

Goldstein et al. [16] compared the three systems described above: rotating wall vessel, spinner flask, and flow perfusion. Osteoblastic cells were seeded onto PLGA foams and cultured for 2 weeks. Although cell seeding efficiencies and osteocalcin content were similar for the three systems, the spinner flask produced the least uniform cell distribution throughout the foams and the rotating wall vessel system resulted in the lowest levels of alkaline phosphatase activity. Consequently, the flow perfusion system appears to be the most attractive culturing system for bone constructs among those three bioreactors. Moreover, rotary perfusion bioreactor system has the main advantage of eliminating the internal transport limitations of the spinner flask and the rotating wall vessel [41].

18.3.2.4 Compression Bioreactors

Compression bioreactors are designed to exert controllable mechanical forces under physiological environment to reproduce, in vitro, the in vivo mechanical stimuli (Fig. 18.1E). Hydrodynamic shear, hydrostatic pressure, mechanical compression, tension, and friction are some of the mechanical forces applied by this class of bioreactors. One of the main applications of these systems is in cartilage engineering [56]. Correia et al. [57] tested the effect of either a pulsatile or a steady hydrostatic pressure to human adipose-derived stem cells (ASCs) encapsulated in gellan gum hydrogel constructs over a period of 3 weeks. The authors observed that pulsatile hydrostatic pressure regimen led to greater chondrogenic differentiation and matrix deposition, as evidenced by gene expression of aggrecan, collagen type II, and sox-9, metachromatic staining of cartilage extracellular matrix, and immunolocalization of collagens.

Cochis et al. [58] used a bioreactor to mechanically stimulate a hydrogel laden with bone marrow MSCs by simultaneously applying compression and shear forces days using a ceramic hip ball over 21 days. The mechanically stimulated MSCs successfully expressed chondrogenic genes, and the GAG quantification confirmed the higher differentiation of MSCs under compression stimulus. Histological analysis showed the retention of the cells within the polyurethane scaffold pores and the presence of a surrounding matrix of collagen and proteoglycan.

The compression bioreactors have as main advantage the ability to apply very specific and accurate mechanical stimuli which are not possible to promote with any other culture system. On the other side, the specificity of this approach makes these systems not so interesting for large-scale or high-throughput applications.

18.3.2.5 In Vivo Bioreactors

The in vivo bioreactor is a regenerative medicine concept where the bone is grown in vivo (Fig. 18.1F). This bioengineering approach relies on the conductive properties of the implanted scaffold to recruit MSCs from neighboring tissue and takes advantage of the physiological environment to supply the necessary growth factors and nutrients to the construct. Several studies have been made to take advantage from in vivo bioreactors to generate vascularized bone tissue [59, 60].

In the design of an in vivo bioreactor, Holt et al. [59] used a scaffold composed of coralline cylinders supplemented with BMP-2. To recruit MSCs from the blood circulation into the bioreactor by BMP-2, a vascular pedicle channel was incorporated into the scaffold. This closed system isolated by silicone ensured that bone formation would depend on the scaffold and the invading cells. The designed in vivo bioreactor implanted in male rats was harvested after 6 weeks. New bone formation was observed at 11.3% with neovascular ingrowth. In a different approach, Stevens et al. [60] manipulated an artificial space to perform as a bioreactor between the tibia and the periosteum, a layer rich in MSCs, and taking advantage from the body's healing mechanism to leverage neo-bone formation. The authors incorporated alginate gel in New Zealand white rabbits. Bone tissue was formed and showed biomechanical similarities to the native bone. Furthermore, the authors observed enhanced cartilage formation within the bioreactor when angiogenesis was inhibited promoting a more hypoxic environment.

18.4 Microfluidics Designs

Microfluidic devices experienced a fast evolution starting in the 1990s [61] and nowadays contribute with versatile platforms in fields as molecular analysis, laboratory diagnostics, biodefense, and consumer electronics [62]. Applications of microfluidic systems based on cell and tissue culture have been also emerging, and TE is taking advantage from micro- and nanofabrication techniques for the development of sophisticated features on tissue modeling and drug testing as platforms for high-throughput screening [22]. First at two dimensions, but during last 10 years, the third dimension of cell culture has been also revolutionized by the integration of microfluidics. When several integrated chambers culture cells mimicking more than one tissue or organ in a single device, the approach became known as organ-on-a-chip [8].

18.4.1   Microfluidic Systems General Principles

Microfluidic systems typically consist of devices with channel geometries having characteristic length scaled from tens to hundreds of microns [63]. When microfluidics are designed for cell culture, it might encompass from millions to single cells providing a level of flexibility beyond that possible in conventional well plates or even with bioreactors [64, 65]. Structural features in microfluidic devices may be designed to provide spatial control over cell behavior, and interactions between cell populations may be controlled through the use of channels, membranes, and other features incorporated into these systems [66, 67].

Fabrication is typically done by photolithography by applying a standard technique as Radio Corporation of America (RCA) cleaning, thin film deposition, wet hydrofluoric etching, access hole forming, or chip bonding (Fig. 18.4) [61]. Soft lithography and other processing techniques have enabled rapid, simple fabrication of microfluidic devices from a broad range of substrate materials including thermoplastics and thermoset polymers, typically produced in optically transparent formats that may be rigid or elastomeric [68]. Some significant features which make this technology distinguishable are:

1. Microscale resolution and flow conditions match with the cellular structure dimensions and traffic present in the human organism.
2. Spatial control over chemical gradients can mimic the dynamic 3D network existing in vivo.
3. Reduced costs as it requires samples in nanoliter volumes.
4. Design of microfluidic devices is compatible with substrates permeable to oxygen enabling cell culture in 3D.
5. Microfluidics can handle several processes at one time such as culture, replenishment of medium, cell detachment, and subsequent detection. Furthermore, fabrication can use transparent materials allowing microscopic imaging.

18.4.2   Microfluidic Types of Devices

Different types of microfluidic systems have been used to establish and support 3D culture and have been categorized based on the substrates used to fabricate the microdevices, namely, glass-/silicon-based, polymer-based, and paper-based platforms [22].

18.4.2.1 Glass-/Silicon-Based Platforms

Glass-based systems can be reusable and applied for long-term studies because of its stable surface with reproducible and reliable electroosmotic flow. The main advantage of this systems is the enhanced optical properties which are advantageous in high-resolution microscopy [69]. Glass-based channels are impermeable to oxygen, which has been repeatedly utilized to create hypoxic conditions [70]. Khan et al. designed a microfluidic platform able to create gradients of oxygen tension. A glass coating on the inner microfluidic channel prevented multi-directional diffusion of oxygen across Polydimethylsiloxane (PDMS) enabling and keeping the gradient resolution and stability which is monitored by incorporation of sensors [71]. Silicon-based systems, on the contrary, have the disadvantage of being expensive and demand complicated fabrication procedures.

18.4.2.2 Polymer-Based Platforms

Various polymers such as PDMS, polycarbonate, polystyrene, and polymethyl methacrylate (PMMA) have been used as biocompatible substrates for microdevices [22]. Among these polymers, PDMS is the most predominant because it is permeable to oxygen and cost-effective [61]. Microchannels are formed by contacting the PDMS structure with a substrate, and these channels deliver the fluid to restricted areas on the substrate. The microchannels can selectively deliver the materials for cell adhesion or cell suspension to desired areas of a substrate [72, 73].

Natural origin polymers such as agarose, fibrin, and collagens have also been used for 3D cell culture on microfluidic applications [74].

18.4.2.3 Paper-Based Platforms

Paper-based microfluidic systems were born as a relatively simple and cost-effective approach. Moreover, these systems are flexible and can be designed in varied architectures, as demonstrated by Martinez et al. by creating several 3D microfluidic devices fabricated in layered paper and tape [75]. 3D cell culture was recently demonstrated by Derda et al. for the first time on a paper-based microfluidic platform [76]. The authors used chromatographic papers to pattern hydrophobic barriers by wax printing. Cell suspensions were then impregnated on the papers. To mimic the 3D architecture, multiple papers sheets were stacked over each other. These papers can later be detached for layer-by-layer molecular analysis.

18.4.3 Microfluidic Potential for 3D Cell Culture

One of the main advantages of microfluidic approaches is the spatial fine control over fluids at micrometer scale, which can be explored to increase the physiological significance of 3D tissue models. Early examples demonstrate spatial patterning of adhesion molecules [77] and hydrogels [78, 79], which are still used in microfluidic 3D cell culture. Today, the most important drivers for the use of microfluidic techniques in 3D cell culture are:

(i) The integration of perfusion/flow
(ii) The ability of co-culturing cells in a spatially controlled manner
(iii) Generation of and control over (signaling) gradients [1]

Spatial control is essential for the increasing need for more complex tissues in vitro modeling, since our body presents several interfaced tissues and barriers. These are, for example, determinant for testing drug efficacy as drug molecules either have to cross barriers or be effective to treat a tissue without affecting the adjacent one. Microfluidic fabrication allows patterning surfaces for cells and extracellular microenvironment stratified (co-)cultures with basal-apical access, controlling gradient formation and medium perfusion. In classical culture techniques, the spatial control is usually achieved by a membrane to support surface-attached cell growth dividing the culture well in two independent compartments [80]. The recent trend in microfluidic systems is to use hydrogels interfaced by two channels offering a physiologically more relevant environment. By using laminar flow, two or more streams are joined into a single channel flowing parallel to each other without any turbulent mixing, allowing the only mixing by diffusion across the interface. This ability to sustain parallel streams of different solutions in a single microchannel has been applied to pattern cells and their environments [81, 82]. This method can also be used to study subcellular processes by positioning a single cell interfacing two adjacent streams [83]. Furthermore, by patterning a hydrogel between two fluids, stable and predictable linear gradients are formed, which can be controlled by the channel geometry and applied flow rates [84]. To integrate this type of assays into the high-throughput drug-screening pipeline, Trietsch et al. [85] created a microfluidic platform based on a 96-well titer plate format enabling a double flow perfusion to generate a gradient over an hydrogel. Perfusion flow was maintained by passive leveling between two reservoirs, thereby eliminating the need for external pumps. This allows high-throughput migration assays and gradient formation in combination with stratified co-cultures.

An interesting technology to study chondrocytes was introduced by Neve et al. [86]. The developed technique integrates micron-resolution particle image velocimetry with dual optical tweezers that allow for the capture and maintenance of a single chondrocyte in a flow field that can be measured in real time.

From the microfluidic devices was born the organ-on-a-chip concept, which is based on a microfluidic cell culture device created with microchip manufacturing methods that contains continuously perfused chambers inhabited by living cells arranged to simulate tissue- and organ-level physiology. By recapitulating the multicellular architectures, tissue-tissue interfaces, physicochemical microenvironments, and vascular perfusion of the body, these devices could produce levels of tissue and organ functionality not possible with conventional static 2D or 3D culture systems. This concept was born for the creation of tools to enable in vitro analysis of biochemical and metabolic paracrine activities in between different tissues and can have a huge impact in the future after maturation and optimization of the concept (Fig. 18.2).

In addition to the physiological relevance, microfluidic systems can potentially improve reproducibility, cost-effectiveness, and implementation at larger scales for diagnostics and drug screening. For example, the reduced dimensions offer advantages such as reduced consumption of expensive cell material, hydrogels, and screening reagents. Well-defined heights of microfluidic channels improve imaging quality and speed. Precise metering of liquids with microfluidic techniques enables better quantification of assays. However, moving to a microfluidic reality implies changing several exclusive factors to microfluidic from macroscopic cell culture, such as different culture surfaces, reduced media volumes, and vastly different rates of, and methods for, medium exchange. These unique features slowdown the acceptance and adaptation of the current state-of-the-art of cell culture techniques to the dynamic microscale. Furthermore, even though there are reports about 3D cell culture in microfluidic devices, a further push is needed to consolidate this interesting concept for more complex 3D tissues as interfaces and co-culture-based studies.

18.5 Bioreactors vs Microfluidics in OC Tissue Modeling

Over the last decades, strategies to investigate bone-cartilage interactions in vitro were mostly limited to cell co-culture well plates in which bone and cartilage cells are both exposed to the same medium [4], arguably a very distant condition from the in vivo environment. Alternatively, co-cultures imply the use of transwells avoiding direct cell contact which is also crucial, for example, for interfaced tissues. Using TE techniques, two construct pieces can be independently cultured under chondrogenic and osteogenic medium and joined together after tissue maturation, resulting in an interrupted interface.

Traditional bioreactors have not been frequently explored in the development of skeletal tissues interfaces but either applied to bone or cartilage tissue development [90]. When applied to the interfaced OC junction, explants were usually used. Understanding the mechanical properties of the articular surface is the main focus of interest, because the OC junction and the subchondral bone are known to confer significant protective mechanical properties to the overlying cartilage [91]. Specifically, the subchondral bone reduces impact-induced fissuring, chondrocyte cell death, and matrix degradation, all of which are hallmarks of pre-osteoarthritis [92]. Studies of chronic joint disorders have revealed the influence of subchondral bone changes in the etiology of osteoarthritis, but these changes have not been effectively reproduced in vitro [93, 94].

Biological characterization of the OC tissue present in any joint of the human body already revealed the existing and key communication between chondrocytes and osteoblasts across the junction. This interface is characterized by a transition of collagen type I to type II and also of collagen fibers orientation. Moreover, several

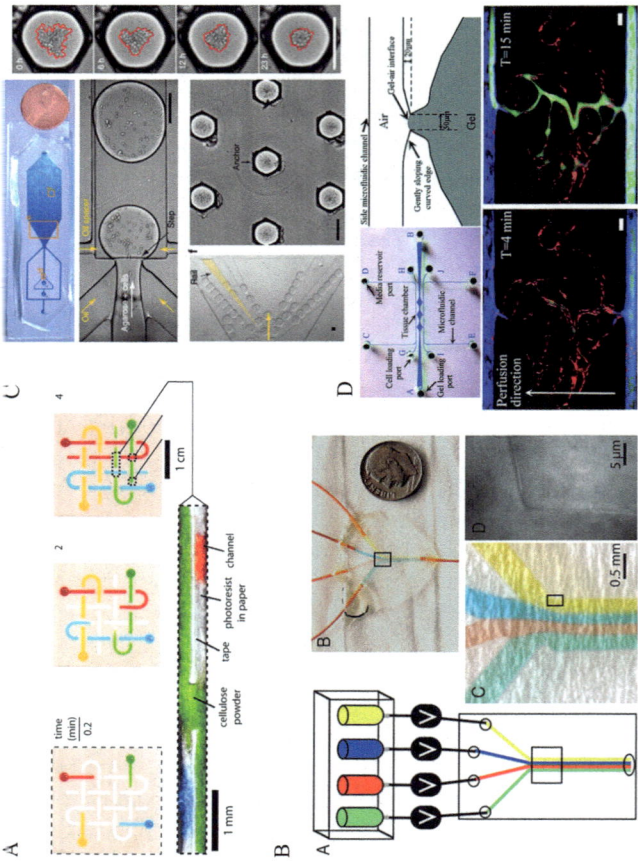

Fig. 18.2 Microfluidic devices designed over several substrates and for broad applications. (**A**) Crossed microfluidic channels creating interfaces in layered paper-tape-based device. (Adapted with permission from [75] Copyright © 2008 National Academy of Sciences). (**B**) Microfluidic concept of parallel channels separated by laminar flow from multiple input that are optically clear on each face, chemically inert, reusable, and inexpensive and can be fabricated on the benchtop in approximately 1 h. (Adapted with permission from [87] Copyright © 2009 Courson, Rock). (**C**) Microfluidic device compartmentalized for 3D cell aggregates analysis at high-throughput scale. (Adapted with permission from [88] Copyright © 2017. Rights Managed by Nature Publishing Group). (**D**) Perfusable vascular network interfacing two microchannels creating an organ-on-a-chip system. (Adapted with permission from [89] Copyright © 2015, Royal Society of Chemistry)

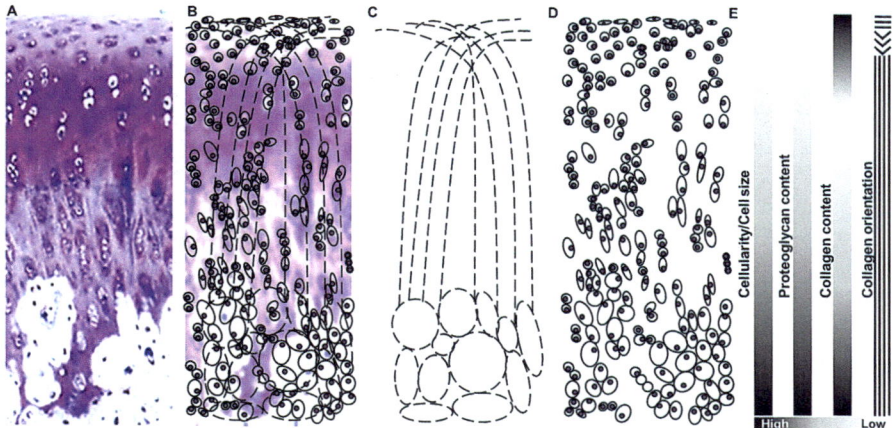

Fig. 18.3 Subchondral bone and articular cartilage interface make the OC tissue which is characterized by specific collagen orientation and several content gradients. (**A**) Histological H&E section of healthy human OC tissue. (**B**) Representation of the organization of collagen and cells. (**C**) Scheme of collagen orientation. (**D**) Scheme of cellular distribution. (**E**) Representation of gradients and collagen orientation

gradients, such as cellularity, cell size, proteoglycans, and collagen content, characterize the OC tissue interface (Fig. 18.3). However, understanding OC phenotype and related disorders through specific communications require an in vitro culture system that supports native or engineered osseous and chondral components of an OC unit.

Bioreactors and microfluidic systems, as mentioned in previous sections, are capable of creating gradients which are of great interest to OC tissue in vitro development. While conventional in vitro culture systems such as static cultures, spinner flasks, rotating wall vessels, and flow perfusion fail to provide physiological conditions capable of reproducing physiological interfaces, the use of in vivo models can give the wrong impression of translation readiness. Indeed, most of the cases fail when translated to clinical trials, mainly because of genotypic and morphological differences. Regarding this, in vitro models, able of mimicking 3D interfaces by using gradient or bilayer scaffolds in general and OC tissue in particular, based on human cells are still urgently needed. Since the traditional culturing systems are not specifically adapted for engineering 3D interfaced tissues, new technologies have been emerging over the last 5 years targeting this gap [5, 40, 95].

Recent developments in the bioreactor field, as the creation of systems adapted for the maturation of interfaces and able to spatially control gradients, will open up new possibilities to foster interfaced TE as it is the case of the OC tissue unit. We described a new bioreactor concept [39, 40], which was designed for the maturation of interfaces by using multi-compartmentalized chambers interconnected by 3D structures (Fig. 18.4). This concept not only allows the co-culture of multiple cell types under different environments but is also designed to avoid cell sedimentation

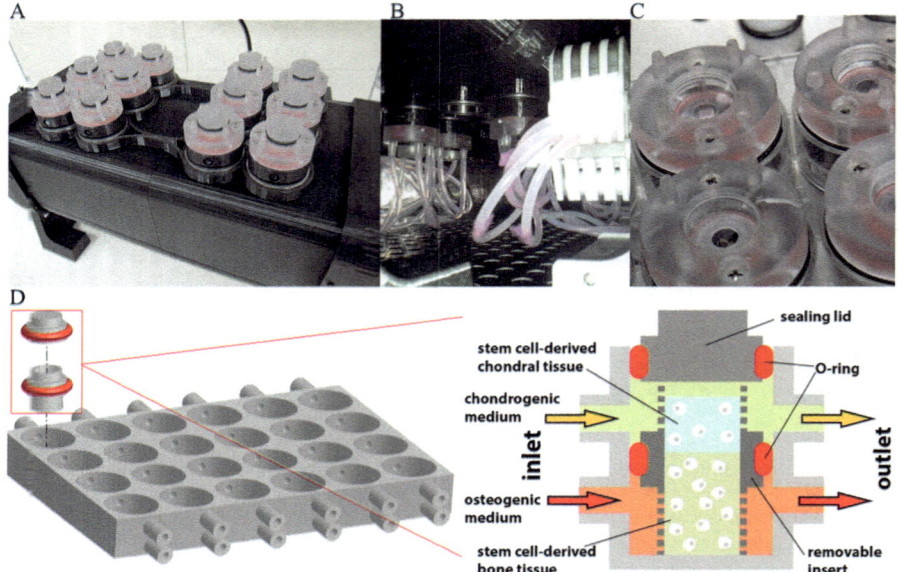

Fig. 18.4 Bioreactor systems adapted for organ-on-a-chip and high-throughput concepts. (**A**) Dynamic platform for 3D cell culture homogenization and increased diffusion by rotational movements. (**B**) The dynamic platform performs inversion of the cell culture chambers avoiding heterogeneous cell distribution in 3D scaffolds. (**C**) Dual-chambers from bioreactor are adaptable to 6-well culture plate as a concept for 3D interfaced in vitro tissues. (**D**) Transition from bioreactors to microfluidics concepts as a high-throughput device for 3D interfaced tissues. (Adapted with permission from ([97], http://pubs.acs.org/doi/abs/10.1021%2Fmp500136b) Copyright © 2014 American Chemical Society)

and to improve culture medium exchange in and out of the tissue constructs by rotational movements and flow perfusion (Fig. 18.4A and B). Moreover, this bioreactor concept allows the connection by flowing medium between several multi-chambers, adapting the concept of organ-on-a-chip microfluidics to the bioreactors reality (Fig. 18.4C).

Concurrently, some dual-chamber bioreactors were tested for OC TE. Chang et al. [96] cultured a gelatin-infused shinbone block to generate OC constructs in a dual-chambered bioreactor for the production of hyaline cartilage within the gelatin portion of the scaffold while the bony portion was acellular. Mahmoudifar and Doran [97] used a similar dual-chambered bioreactor for the maturation of two sutured polyglycolic acid meshes seeded with ASCs. After 2 weeks of culture, both layers were cellularized, but showed statistically undifferentiated GAG content. The main differences in comparison to our system are related to specific dynamic features. While these two reported dual-chambers are under flow perfusion, our bioreactor allows rotational movements to improve medium diffusion and vertical turning movements to increase cell homogenization though the 3D structures.

OC in vitro models based on microfluidic devices were recently reported. Goldman and Babino [98] created a microfluidic dual-chamber device for the control over the osteo- and chondrogenic phenotypes. Bovine MSC were encapsulated in agarose, casted against micromolds of a serpentine network, and stacked to produce tissue constructs containing two independent microfluidic channel networks. Constructs receiving differentiation media showed differential chondrogenic and osteogenic gene expression, which was confirmed at the protein level as collagens I, II, and X. The control group under basal culture medium corroborated the results by showing homogeneous expression of the same biomarkers measured in lower concentrations at both the mRNA and protein level.

Shi et al. [99] compared a strategy based on a gradient generated by a microfluidic device with a conventional approach where two pieces of ASCs-laden hydrogel were cultured under osteogenic and chondrogenic conditions and joined together. The microfluidic system allowed generating a gradient of differentiation mimicking the OC interface. Although the dynamic cell differentiation methods using the microfluidic device consumed more cell culture media than the static method, the microfluidic system continuously supplied new nutrition and transported the wastes produced by cell metabolism out of the device generating a nontoxic environment. In addition, the flowing media could be collected and reused because much of the nutrients in the flowing media are not consumed during the continuous flow of culture media.

Lately, a technological approach using a high-throughput platform was designed, and osteoarthritic condition was already assessed to validate the system (Fig. 18.4D). Lin et al. [100] developed the platform adapted for interfaced OC tissue in a well plate format. This approach represents a transition design in between bioreactors and microfluidics. The system consists in a single bioreactor formed by the inserts and lid in the context of a 24-well plate. The device was designed to accommodate the biphasic nature of an OC plug by creating two separate compartments for the "chondral" and "osseous" microenvironments. The two microenvironments are independently controlled and the medium flow through each row of wells. The authors have shown that MSC-based chondral and osseous tissues respond to IL-1β in a manner that the changes in one tissue compartment are communicated to the other along the OC axis.

While bioreactors are now seeing its concept turned to the field of tissue modeling, microfluidic systems have been tested more often for this goal. However, the main difference of both concepts is that bioreactors were firstly developed for tissue production to be used as grafts for implantation and are now useful platforms for in vitro tissue modeling, while microfluidic devices can only be used for this last purpose as a drug testing or diagnostic tool.

18.6 Future Directions and Conclusions

The combination of culture systems as bioreactors and microfluidics with 3D cell culture has triggered alternatives reporting great potential to provide efficient methods for biomedical applications, TE, and drug screening. However, increased complexity associated with access to cultivated cells in 3D constructs and further sampling for assays is a combination of problematic challenges to be solved. The current systems allow the creation of gradients and the spatial arrangement of cells enabled mainly by microfluidics, but overall lack control of dynamics and spatial presentation of various signals over 3D constructs, which requires meticulous attention.

There is also a strong need of cost-effective and easy-to-use systems. High-throughput systems have been reported to solve these needs, but still require optimization when applied for 3D interfaced tissues as OC. Although organ-on-a-chip has drawn attention for integrated studies of tissue interplay, the integration of microenvironments in a single device still needs further developments. Furthermore, the transition from 2D to 3D adds one more dimension not only in terms of shape and structure but also in terms of data acquisition.

The integration of materials engineering, nanofabrication, and biology already opened up new roots guiding us to the current scenario. However, bringing new tools from bioinformatics, systems biology, and sensors for real-time monitoring may help in overcoming the remaining challenges. In the near future, the development of automated, high-throughput, reproducible, cost-effective, and easy-to-use 3D cell culture systems is expected. Advances of microfluidics and bioreactors cell culturing technologies will trigger a new coming era of developments and discoveries not only about OC-related disorders and therapeutics but in the field of TE and drug discovery in general.

Acknowledgments This work is a result of the project FROnTHERA (NORTE-01-0145-FEDER-000023), supported by Norte Portugal Regional Operational Programme (NORTE 2020), under the Portugal 2020 Partnership Agreement, through the European Regional Development Fund (ERDF). Thanks are also due to the Portuguese Foundation for Science and Technology (FCT) for the project PEst-C/SAU/LA0026/201 and for the distinction attributed to J.M. Oliveira under the Investigator FCT program (IF/00423/2012 and IF/01285/2015). The authors also thank FCT for the Ph.D. scholarship provided to R. F. Canadas (SFRH/BD/92565/2013).

References

1. van Duinen V, Trietsch SJ, Joore J, Vulto P, Hankemeier T (2015) Microfluidic 3D cell culture: from tools to tissue models. Curr Opin Biotechnol [Internet] 35(Supplement C):118–126. Available from: http://www.sciencedirect.com/science/article/pii/S0958166915000713
2. Griffith LG, Swartz MA (2006) Capturing complex 3D tissue physiology in vitro. Nat Rev Mol Cell Biol 7(3):211–224
3. Martin I, Miot S, Barbero A, Jakob M, Wendt D (2007) Osteochondral tissue engineering. J Biomech [Internet] 40(4):750–765. Available from: http://www.sciencedirect.com/science/article/pii/S0021929006000960
4. Lories RJ, Luyten FP (2011) The bone-cartilage unit in osteoarthritis. Nat Rev Rheumatol 7(1):43–49
5. Goldman SM, Barabino GA (2016) Spatial engineering of osteochondral tissue constructs through Microfluidically directed differentiation of Mesenchymal stem cells. Biores Open Access [Internet] 5(1):109–117. Available from: http://www.ncbi.nlm.nih.gov/pmc/articles/PMC4854211/

6. Liu X, Jiang H (2013) Preparation of an osteochondral composite with mesenchymal stem cells as the single-cell source in a double-chamber bioreactor. Biotechnol Lett [Internet] 35(10):1645–1653. Available from: https://doi.org/10.1007/s10529-013-1248-9

7. Khademhosseini A, Langer R, Borenstein J, Vacanti JP (2006) Microscale technologies for tissue engineering and biology. Proc Natl Acad Sci U S A [Internet]. 103(8):2480–2487. Available from: http://www.ncbi.nlm.nih.gov/pmc/articles/PMC1413775/

8. Bhatia SN, Ingber DE (2014) Microfluidic organs-on-chips. Nat Biotech [Internet] 32(8):760–772. Available from: https://doi.org/10.1038/nbt.2989

9. Barron V, Lyons E, Stenson-Cox C, McHugh PE, Pandit A (2003) Bioreactors for cardiovascular cell and tissue growth: a review. Ann Biomed Eng [Internet] 31(9):1017–1030. Available from: https://doi.org/10.1114/1.1603260

10. Costa PF, Vaquette C, Baldwin J, Chhaya M, Gomes ME, Reis RL et al (2014) Biofabrication of customized bone grafts by combination of additive manufacturing and bioreactor know-how. Biofabrication [Internet] 6(3):35006. Available from: http://stacks.iop.org/1758-5090/6/i=3/a=035006

11. Antoni D, Burckel H, Josset E, Noel G (2015) Three-dimensional cell culture: a breakthrough in vivo. Rahaman MN, editor. Int J Mol Sci [Internet] 16(3):5517–5527. Available from: http://www.ncbi.nlm.nih.gov/pmc/articles/PMC4394490/

12. Grayson WL, Bhumiratana S, Cannizzaro C, Chao PH, Lennon DP, Caplan AI et al (2008) Effects of initial seeding density and fluid perfusion rate on formation of tissue-engineered bone. Tissue Eng Part A [Internet] 14(11):1809–1820. Available from: http://www.ncbi.nlm.nih.gov/pmc/articles/PMC2773295/

13. Kleinhans C, Mohan RR, Vacun G, Schwarz T, Haller B, Sun Y et al (2015) A perfusion bioreactor system efficiently generates cell-loaded bone substitute materials for addressing critical size bone defects. Biotechnol J 10(11):1727–1738

14. Gardel LS, Serra LA, Reis RL, Gomes ME (2013) Use of perfusion bioreactors and large animal models for long bone tissue engineering. Tissue Eng Part B Rev [Internet] 20(2):126–146. Available from: https://doi.org/10.1089/ten.teb.2013.0010

15. Massai D, Isu G, Madeddu D, Cerino G, Falco A, Frati C et al (2016) A versatile bioreactor for dynamic suspension cell culture. Application to the culture of cancer cell spheroids. Pesce M, editor. PLoS One [Internet] 11(5):e0154610. Available from: http://www.ncbi.nlm.nih.gov/pmc/articles/PMC4856383/

16. Goldstein AS, Juarez TM, Helmke CD, Gustin MC, Mikos AG (2001) Effect of convection on osteoblastic cell growth and function in biodegradable polymer foam scaffolds. Biomaterials [Internet] 22(11):1279–1288. Available from: http://www.sciencedirect.com/science/article/pii/S0142961200002805

17. Andersson H, van den Berg A (2004) Microfabrication and microfluidics for tissue engineering: state of the art and future opportunities. Lab Chip [Internet] 4(2):98–103. Available from: https://doi.org/10.1039/B314469K

18. Cartmell SH, Porter BD, Garcia AJ, Guldberg RE (2003) Effects of medium perfusion rate on cell-seeded three-dimensional bone constructs in vitro. Tissue Eng 9(6):1197–1203

19. Ishaug SL, Crane GM, Miller MJ, Yasko AW, Yaszemski MJ, Mikos AG (1997) Bone formation by three-dimensional stromal osteoblast culture in biodegradable polymer scaffolds. J Biomed Mater Res 36(1):17–28

20. Polacheck WJ, German AE, Mammoto A, Ingber DE, Kamm RD (2014) Mechanotransduction of fluid stresses governs 3D cell migration. Proc Natl Acad Sci [Internet] 111(7):2447–2452. Available from: http://www.pnas.org/content/111/7/2447.abstract

21. Tarbell JM, Shi Z-D, Dunn J, Jo H (2014) Fluid mechanics, arterial disease, and gene expression. Annu Rev Fluid Mech [Internet] 46:591–614. Available from: http://www.ncbi.nlm.nih.gov/pmc/articles/PMC4211638/

22. Li X (James), Valadez AV, Zuo P, Nie Z (2012) Microfluidic 3D cell culture: potential application for tissue-based bioassays. Bioanalysis [Internet] 4(12):1509–1525. Available from: http://www.ncbi.nlm.nih.gov/pmc/articles/PMC3909686/

23. Song L, Zhou Q, Duan P, Guo P, Li D, Xu Y et al (2012) Successful development of small diameter tissue-engineering vascular vessels by our novel integrally designed pulsatile perfusion-based bioreactor. PLoS One [Internet] 7(8):e42569. Available from: https://doi. org/10.1371/journal.pone.0042569

24. Wang B, Wang G, To F, Butler JR, Claude A, McLaughlin RM et al (2013) Myocardial scaffold-based cardiac tissue engineering: application of coordinated mechanical and electrical stimulations. Langmuir [Internet] 29(35):11109–11117. Available from: https://doi. org/10.1021/la401702w

25. Grodzinsky AJ, Levenston ME, Jin M, Frank EH (2000) Cartilage tissue remodeling in response to mechanical forces. Annu Rev Biomed Eng [Internet] 2(1):691–713. Available from: https://doi.org/10.1146/annurev.bioeng.2.1.691

26. Santoro M, Lamhamedi-Cherradi S-E, Menegaz BA, Ludwig JA, Mikos AG (2015) Flow perfusion effects on three-dimensional culture and drug sensitivity of Ewing sarcoma. Proc Natl Acad Sci U S A [Internet] 112(33):10304–10309. Available from: http://www.ncbi.nlm. nih.gov/pmc/articles/PMC4547215/

27. Kou S, Pan L, van Noort D, Meng G, Wu X, Sun H et al (2011) A multishear microfluidic device for quantitative analysis of calcium dynamics in osteoblasts. Biochem Biophys Res Commun [Internet] 408(2):350–355. Available from: http://www.sciencedirect.com/science/ article/pii/S0006291X11006188

28. Porter B, Zauel R, Stockman H, Guldberg R, Fyhrie D (2005) 3-D computational modeling of media flow through scaffolds in a perfusion bioreactor. J Biomech [Internet] 38(3):543–549. Available from: http://www.sciencedirect.com/science/article/pii/S0021929004001964

29. Godara P, McFarland CD, Nordon RE (2008) Design of bioreactors for mesenchymal stem cell tissue engineering. J Chem Technol Biotechnol [Internet] 83(4):408–420. Available from: https://doi.org/10.1002/jctb.1918

30. Mao AS, Mooney DJ (2015) Regenerative medicine: current therapies and future directions. Proc Natl Acad Sci U S A [Internet] 112(47):14452–14459. Available from: http://www.ncbi. nlm.nih.gov/pmc/articles/PMC4664309/

31. Gharravi AM, Orazizadeh M, Ansari-Asl K, Banoni S, Izadi S, Hashemitabar M (2012) Design and fabrication of anatomical bioreactor systems containing alginate scaffolds for cartilage tissue engineering. Avicenna J Med Biotechnol [Internet] 4(2):65–74. Available from: http://www.ncbi.nlm.nih.gov/pmc/articles/PMC3558208/

32. Powers MJ, Domansky K, Kaazempur-Mofrad MR, Kalezi A, Capitano A, Upadhyaya A et al (2002) A microfabricated array bioreactor for perfused 3D liver culture. Biotechnol Bioeng 78(3):257–269

33. Naciri M, Kuystermans D, Al-Rubeai M (2008) Monitoring pH and dissolved oxygen in mammalian cell culture using optical sensors. Cytotechnology [Internet] 57(3):245–250. Available from: http://www.ncbi.nlm.nih.gov/pmc/articles/PMC2570003/

34. da Costa PF, Martins AMP, Gomes MME, das Neves NJMA, dos Reis RLG (2011) Multichamber bioreactor with bidirectional perfusion integrated in a culture system for tissue engineering strategies [Internet]. Google Patents. Available from: https://www.google.ch/ patents/EP2151491A3?cl=en

35. Sucosky P, Osorio DF, Brown JB, Neitzel GP (2004) Fluid mechanics of a spinner-flask bioreactor. Biotechnol Bioeng [Internet] 85(1):34–46. Available from: https://doi.org/10.1002/ bit.10788

36. Altman GH, Horan RL, Martin I, Farhadi J, Stark PRH, Volloch V et al (2002) Cell differentiation by mechanical stress. FASEB J [Internet] 16(2):270–272. Available from: http://www. fasebj.org/content/early/2002/02/02/fj.01-0656fje.short

37. van der DWJ S, van ACC S, Boonen KJM, Langelaan MLP, Bouten CVC, Baaijens FPT (2013) Engineering skeletal muscle tissues from murine myoblast progenitor cells and application of electrical stimulation. J Vis Exp 73:e4267

38. Tsimbouri PM, Childs PG, Pemberton GD, Yang J, Jayawarna V, Orapiriyakul W et al (2017) Stimulation of 3D osteogenesis by mesenchymal stem cells using a nanovibrational biore-

actor. Nat Biomed Eng [Internet] 1(9):758–770. Available from: https://doi.org/10.1038/s41551-017-0127-4

39. Canadas RF, Marques AP, Oliveira JM, Reis RL (2014) Rotational dual chamber bioreactor: methods and uses thereof [Internet]. WO2014141136 A1. Available from: http://www.google.com/patents/WO2014141136A1?cl=en

40. Canadas RF, Oliveira JM, Marques AP, Reis RL (2016) Multi-chambers bioreactor, methods and uses [Internet]. Association for the Advancement of Tissue Engineering and Cell Based Technologies and Therapies - A4Tec; WO 2016042533 A1. Available from: https://www.google.com/patents/WO2016042533A1?cl=en

41. Bancroft GN, Sikavitsas VI, Mikos AG (2003) Technical note: design of a flow perfusion bioreactor system for bone tissue-engineering applications. Tissue Eng [Internet] 9(3):549–554. Available from: http://www.liebertonline.com/doi/abs/10.1089/107632703322066723

42. Kim JB (2005) Three-dimensional tissue culture models in cancer biology. Semin Cancer Biol [Internet] 15(5):365–377. Available from: http://www.sciencedirect.com/science/article/pii/S1044579X05000301

43. Rodday B, Hirschhaeuser F, Walenta S, Mueller-Klieser W (2011) Semiautomatic growth analysis of multicellular tumor spheroids. J Biomol Screen [Internet] 16(9):1119–1124. Available from: https://doi.org/10.1177/1087057111419501

44. Vunjak-Novakovic G, Obradovic B, Martin I, Bursac PM, Langer R, Freed LE (1998) Dynamic cell seeding of polymer scaffolds for cartilage tissue engineering. Biotechnol Prog [Internet] 14(2):193–202. Available from: https://doi.org/10.1021/bp970120j

45. Sikavitsas VI, Bancroft GN, Mikos AG (2002) Formation of three-dimensional cell/polymer constructs for bone tissue engineering in a spinner flask and a rotating wall vessel bioreactor. J Biomed Mater Res [Internet] 62(1):136–148. Available from: http://doi.wiley.com/10.1002/jbm.10150

46. Vunjak-Novakovic G, Searby N, De Luis J, Freed LE (2002) Microgravity studies of cells and tissues. Ann N Y Acad Sci [Internet] 974(1):504–517. Available from: https://doi.org/10.1111/j.1749-6632.2002.tb05927.x

47. Vunjak-Novakovic G, Martin I, Obradovic B, Treppo S, Grodzinsky AJ, Langer R et al (1999) Bioreactor cultivation conditions modulate the composition and mechanical properties of tissue-engineered cartilage. J Orthop Res [Internet] 17(1):130–138. Available from: http://doi.wiley.com/10.1002/jor.1100170119

48. Goodwin TJ, Prewett TL, Wolf DA, Spaulding GF (1993) Reduced shear stress: a major component in the ability of mammalian tissues to form three-dimensional assemblies in simulated microgravity. J Cell Biochem [Internet] 51(3):301–311. Available from: https://doi.org/10.1002/jcb.240510309

49. Marlovits S, Tichy B, Truppe M, Gruber D, Vecsei V (2003) Chondrogenesis of aged human articular cartilage in a scaffold-free bioreactor. Tissue Eng 9(6):1215–1226

50. Saini S, Wick TM (2003) Concentric cylinder bioreactor for production of tissue engineered cartilage: effect of seeding density and hydrodynamic loading on construct development. Biotechnol Prog [Internet] 19(2):510–521. Available from: http://doi.wiley.com/10.1021/bp0256519

51. Williams KA, Saini S, Wick TM (2002) Computational fluid dynamics modeling of steady-state momentum and mass transport in a bioreactor for cartilage tissue engineering. Biotechnol Prog [Internet] 18(5):951–963. Available from: https://doi.org/10.1021/bp020087n

52. Barrila J, Radtke AL, Crabbé A, Sarker SF, Herbst-Kralovetz MM, Ott CM et al (2010) Organotypic 3D cell culture models: using the rotating wall vessel to study host–pathogen interactions. Nat Rev Micro [Internet] 8(11):791–801. Available from: https://doi.org/10.1038/nrmicro2423

53. Wendt D, Marsano A, Jakob M, Heberer M, Martin I (2003) Oscillating perfusion of cell suspensions through three-dimensional scaffolds enhances cell seeding efficiency and uniformity. Biotechnol Bioeng [Internet] 84(2):205–214. Available from: https://doi.org/10.1002/bit.10759

54. Sikavitsas VI, Bancroft GN, Holtorf HL, Jansen JA, Mikos AG (2003) Mineralized matrix deposition by marrow stromal osteoblasts in 3D perfusion culture increases with increasing fluid shear forces. Proc Natl Acad Sci U S A [Internet] 100(25):14683–14688. Available from: http://www.ncbi.nlm.nih.gov/pmc/articles/PMC299759/

55. Gomes ME, Sikavitsas VI, Behravesh E, Reis RL, Mikos AG (2003) Effect of flow perfusion on the osteogenic differentiation of bone marrow stromal cells cultured on starch-based three-dimensional scaffolds. J Biomed Mater Res Part A [Internet] 67A(1):87–95. Available from: https://doi.org/10.1002/jbm.a.10075

56. Shahin K, Doran PM (2015) Shear and compression bioreactor for cartilage synthesis BT – cartilage tissue engineering: methods and protocols. In: Doran PM, editor. New York, NY: Springer New York. Methods Mol Biol 1340:221–233. Available from: https://doi.org/10.1007/978-1-4939-2938-2_16

57. Correia C, Pereira AL, Duarte ARC, Frias AM, Pedro AJ, Oliveira JT et al (2012) Dynamic culturing of cartilage tissue: the significance of hydrostatic pressure. Tissue Eng Part A [Internet] 18(19–20):1979–1991. Available from: http://www.ncbi.nlm.nih.gov/pmc/articles/PMC3463283/

58. Cochis A, Grad S, Stoddart MJ, Farè S, Altomare L, Azzimonti B et al (2017) Bioreactor mechanically guided 3D mesenchymal stem cell chondrogenesis using a biocompatible novel thermo-reversible methylcellulose-based hydrogel. Sci Rep [Internet] 7:45018. Available from: http://www.ncbi.nlm.nih.gov/pmc/articles/PMC5362895/

59. Holt GE, Halpern JL, Dovan TT, Hamming D, Schwartz HS (2005) Evolution of an in vivo bioreactor. J Orthop Res [Internet] 23(4):916–923. Available from: https://doi.org/10.1016/j.orthres.2004.10.005

60. Stevens MM, Marini RP, Schaefer D, Aronson J, Langer R, Shastri VP (2005) In vivo engineering of organs: the bone bioreactor. Proc Natl Acad Sci U S A [Internet] 102(32):11450–11455. Available from: http://www.pnas.org/content/102/32/11450.abstract

61. Gupta N, Liu JR, Patel B, Solomon DE, Vaidya B, Gupta V (2016) Microfluidics-based 3D cell culture models: utility in novel drug discovery and delivery research. Bioeng Transl Med [Internet] 1(1):63–81. Available from: https://doi.org/10.1002/btm2.10013

62. Inamdar NK, Borenstein JT (2011) Microfluidic cell culture models for tissue engineering. Curr Opin Biotechnol [Internet] 22(5):681–689. Available from: http://www.sciencedirect.com/science/article/pii/S0958166911006161

63. Novotný J, Foret F (2017) Fluid manipulation on the micro-scale: basics of fluid behavior in microfluidics. J Sep Sci [Internet] 40(1):383–394. Available from: https://doi.org/10.1002/jssc.201600905

64. Gu H, Duits MHG, Mugele F (2011) Droplets formation and merging in two-phase flow microfluidics. Int J Mol Sci [Internet] 12(4):2572–2597. Available from: http://www.ncbi.nlm.nih.gov/pmc/articles/PMC3127135/

65. Lin L, Chu Y-S, Thiery JP, Lim CT, Rodriguez I (2013) Microfluidic cell trap array for controlled positioning of single cells on adhesive micropatterns. Lab Chip [Internet] 13(4):714–721. Available from: https://doi.org/10.1039/C2LC41070B

66. Geng T, Bredeweg EL, Szymanski CJ, Liu B, Baker SE, Orr G et al (2015) Compartmentalized microchannel array for high-throughput analysis of single cell polarized growth and dynamics. Sci Rep [Internet] 5:16111. Available from: https://doi.org/10.1038/srep16111

67. Horayama M, Shinha K, Kabayama K, Fujii T, Kimura H (2016) Spatial chemical stimulation control in microenvironment by microfluidic probe integrated device for cell-based assay. PLoS One [Internet] 11(12):e0168158. Available from: https://doi.org/10.1371/journal.pone.0168158

68. Jivani RR, Lakhtaria GJ, Patadiya DD, Patel LD, Jivani NP, Jhala BP (2016) Biomedical microelectromechanical systems (BioMEMS): revolution in drug delivery and analytical techniques. Saudi Pharm J [Internet] 24(1):1–20. Available from: http://www.sciencedirect.com/science/article/pii/S131901641300114X

69. Lin J-L, Wang S-S, Wu M-H, Oh-Yang C-C (2011) Development of an integrated microfluidic perfusion cell culture system for real-time microscopic observation of biological cells. Sensors 11(9):8395–8411

70. Mauleon G, Fall CP, Eddington DT (2012) Precise spatial and temporal control of oxygen within in vitro brain slices via microfluidic gas channels. PLoS One [Internet] 7(8):e43309. Available from: https://doi.org/10.1371/journal.pone.0043309

71. Khan DH, Roberts SA, Cressman JR, Agrawal N (2017) Rapid generation and detection of biomimetic oxygen concentration gradients in vitro. Sci Rep [Internet] 7(1):13487. Available from: https://doi.org/10.1038/s41598-017-13886-z

72. Folch A, Ayon A, Hurtado O, Schmidt MA, Toner M (1999) Molding of deep polydimethylsiloxane microstructures for microfluidics and biological applications. J Biomech Eng [Internet] 121(1):28–34. Available from: https://doi.org/10.1115/1.2798038

73. Folch A, Toner M (1998) Cellular micropatterns on biocompatible materials. Biotechnol Prog 14(3):388–392

74. Ling Y, Rubin J, Deng Y, Huang C, Demirci U, Karp JM et al (2007) A cell-laden microfluidic hydrogel. Lab Chip [Internet] 7(6):756–762. Available from: https://doi.org/10.1039/B615486G

75. Martinez AW, Phillips ST, Whitesides GM (2008) Three-dimensional microfluidic devices fabricated in layered paper and tape. Proc Natl Acad Sci [Internet] 105(50):19606–19611. Available from: http://www.pnas.org/content/105/50/19606.abstract

76. Derda R, Laromaine A, Mammoto A, Tang SKY, Mammoto T, Ingber DE et al (2009) Paper-supported 3D cell culture for tissue-based bioassays. Proc Natl Acad Sci [Internet] 106(44):18457–18462. Available from: http://www.pnas.org/content/106/44/18457.abstract

77. Chen CS, Mrksich M, Huang S, Whitesides GM, Ingber DE (1997) Geometric control of cell life and death. Science [Internet] 276(5317):1425–1428. Available from: http://science.sciencemag.org/content/276/5317/1425.abstract

78. Koh W-G, Pishko MV (2006) Fabrication of cell-containing hydrogel microstructures inside microfluidic devices that can be used as cell-based biosensors. Anal Bioanal Chem [Internet] 385(8):1389–1397. Available from: https://doi.org/10.1007/s00216-006-0571-6

79. Tan W, Desai TA (2005) Microscale multilayer cocultures for biomimetic blood vessels. J Biomed Mater Res Part A [Internet] 72A(2):146–160. Available from: https://doi.org/10.1002/jbm.a.30182

80. Boyden S (1962) The chemotactic effect of mixtures of antibody and antigen on polymorphonuclear leucocytes. J Exp Med [Internet] 115(3):453–466. Available from: http://jem.rupress.org/content/115/3/453.abstract

81. Chung S, Sudo R, Mack PJ, Wan C-R, Vickerman V, Kamm RD (2009) Cell migration into scaffolds under co-culture conditions in a microfluidic platform. Lab Chip [Internet] 9(2):269–275. Available from: https://doi.org/10.1039/B807585A

82. Kuo C-T, Liu H-K, Huang G-S, Chang C-H, Chen C-L, Chen K-C et al (2014) A spatiotemporally defined in vitro microenvironment for controllable signal delivery and drug screening. Analyst [Internet] 139(19):4846–4854. Available from: https://doi.org/10.1039/C4AN00936C

83. Takayama S, Ostuni E, LeDuc P, Naruse K, Ingber DE, Whitesides GM (2001) Laminar flows: subcellular positioning of small molecules. Nature [Internet] 411(6841):1016. Available from: https://doi.org/10.1038/35082637

84. Baker BM, Trappmann B, Stapleton SC, Toro E, Chen CS (2013) Microfluidics embedded within extracellular matrix to define vascular architectures and pattern diffusive gradients. Lab Chip [Internet] 13(16):3246–3252. Available from: https://doi.org/10.1039/C3LC50493J

85. Trietsch SJ, Israels GD, Joore J, Hankemeier T, Vulto P (2013) Microfluidic titer plate for stratified 3D cell culture. Lab Chip [Internet] 13(18):3548–3554. Available from: https://doi.org/10.1039/C3LC50210D

86. Nève N, Kohles SS, Winn SR, Tretheway DC (2010) Manipulation of suspended single cells by microfluidics and optical tweezers. Cell Mol Bioeng [Internet] 3(3):213–228. Available from: http://www.ncbi.nlm.nih.gov/pmc/articles/PMC2932633/

87. Courson DS, Rock RS (2009) Fast benchtop fabrication of laminar flow chambers for advanced microscopy techniques. PLoS One [Internet] 4(8):e6479. Available from: https://doi.org/10.1371/journal.pone.0006479

38. Sart S, Tomasi RF-X, Amselem G, Baroud CN (2017) Multiscale cytometry and regulation of 3D cell cultures on a chip. Nat Commun [Internet] 8(1):469. Available from: https://doi.org/10.1038/s41467-017-00475-x

39. Wang X, Phan DTT, George SC, Hughes CCW, Lee AP (2016) Engineering anastomosis between living capillary networks and endothelial cell-lined microfluidic channels. Lab Chip [Internet] 16(2):282–290. Available from: http://www.ncbi.nlm.nih.gov/pmc/articles/PMC4869859/

90. Responte DJ, Lee JK, Hu JC, Athanasiou KA (2012) Biomechanics-driven chondrogenesis: from embryo to adult. FASEB J [Internet] 26(9):3614–3624. Available from: http://www.ncbi.nlm.nih.gov/pmc/articles/PMC3425829/

91. Alexander PG, Song Y, Taboas JM, Chen FH, Melvin GM, Manner PA et al (2013) Development of a spring-loaded impact device to deliver injurious mechanical impacts to the articular cartilage surface. Cartilage [Internet] 4(1):52–62. Available from: http://www.ncbi.nlm.nih.gov/pmc/articles/PMC4297114/

92. Buckwalter JA (2002) Articular cartilage injuries. Clin Orthop Relat Res (402):21–37

93. Repo RU, Finlay JB (1977) Survival of articular cartilage after controlled impact. J Bone Joint Surg Am 59(8):1068–1076

94. Haut RC, Ide TM, De Camp CE (1995) Mechanical responses of the rabbit patello-femoral joint to blunt impact. J Biomech Eng [Internet] 117(4):402–408. Available from: https://doi.org/10.1115/1.2794199

95. Kuiper NJ, Wang QG, Cartmell SH (2014) A perfusion co-culture bioreactor for osteochondral tissue engineered plugs. J Biomater Tissue Eng [Internet] 4(2):162–171. Available from: https://doi.org/10.1166/jbt.2014.1145

96. Chang C-H, Lin F-H, Lin C-C, Chou C-H, Liu H-C (2004) Cartilage tissue engineering on the surface of a novel gelatin–calcium-phosphate biphasic scaffold in a double-chamber bioreactor. J Biomed Mater Res Part B Appl Biomater [Internet] 71B(2):313–321. Available from: https://doi.org/10.1002/jbm.b.30090

97. Mahmoudifar N, Doran PM (2013) Osteogenic differentiation and osteochondral tissue engineering using human adipose-derived stem cells. Biotechnol Prog [Internet] 29(1):176–185. Available from: https://doi.org/10.1002/btpr.1663

98. Goldman SM, Barabino GA (2017) Cultivation of agarose-based microfluidic hydrogel promotes the development of large, full-thickness, tissue-engineered articular cartilage constructs. J Tissue Eng Regen Med [Internet] 11(2):572–581. Available from: https://doi.org/10.1002/term.1954

99. Shi X, Zhou J, Zhao Y, Li L, Wu H (2013) Gradient-regulated hydrogel for interface tissue engineering: steering simultaneous osteo/chondrogenesis of stem cells on a chip. Adv Healthc Mater [Internet] 2(6):846–853. Available from: https://doi.org/10.1002/adhm.201200333

100. Lin H, Lozito TP, Alexander PG, Gottardi R, Tuan RS (2014) Stem cell-based microphysiological osteochondral system to model tissue response to interleukin-1β. Mol Pharm [Internet] 11(7):2203–2212. Available from: http://www.ncbi.nlm.nih.gov/pmc/articles/PMC4086740/

Part VI
In Vivo Models for Osteochondral Regeneration

Chapter 19
Small Animal Models

Alain da Silva Morais, J. Miguel Oliveira, and Rui L. Reis

Abstract Animal assays represent an important stage between in vitro studies and human clinical applications. These models are crucial for biomedical research and regenerative medicine studies, as these offer precious information for systematically assessing the efficacy and risks of recently created biomaterials, medical devices, drugs, and therapeutic modalities prior to initiation of human clinical trials. Therefore, selecting a suitable experimental model for tissue engineering purposes is essential to establish valid conclusions. However, it remains important to be conscious of the advantages and limitations of the various small and large animal models frequently used for biomedical research as well as the different challenges encountered in extrapolating data obtained from animal studies and the risks of misinterpretation. This chapter discusses the various small animal model strategies used for osteochondral defect repair. Particular emphasis will be placed on analyzing the materials and strategies used in each model.

Keywords Small animal models · Scaffolds · Biomaterials · Stem cells · Growth factors · Osteochondral regeneration strategies

A. da Silva Morais (✉)
3B's Research Group – Biomaterials, Biodegradables and Biomimetics, University of Minho, Headquarters of the European Institute of Excellence on Tissue Engineering and Regenerative, Barco, Guimarães, Portugal

ICVS/3B's – PT Government Associate Laboratory, Braga/Guimarães, Portugal
e-mail: alain.morais@dep.uminho.pt

J. M. Oliveira · R. L. Reis
3B's Research Group – Biomaterials, Biodegradables and Biomimetics, University of Minho, Headquarters of the European Institute of Excellence on Tissue Engineering and Regenerative, Barco, Guimarães, Portugal

ICVS/3B's – PT Government Associate Laboratory, Braga/Guimarães, Portugal

The Discoveries Center for Regenerative and Precision Medicine, Headquarters at University of Minho, Guimarães, Portugal

© Springer International Publishing AG, part of Springer Nature 2018   423
J. M. Oliveira et al. (eds.), *Osteochondral Tissue Engineering*,
Advances in Experimental Medicine and Biology 1059,
https://doi.org/10.1007/978-3-319-76735-2_19

19.1 Introduction

The use of animal models for investigation is both a long-standing practice in biological research and medicine and a common matter of discussion in our societies. Animal models are currently used in biomedical research for the following motives:

(i) Similarities to Human. The notable physiological and anatomical similarities between humans and animals, principally mammals, have encouraged researchers to explore a large range of mechanisms and consider novel therapies in animal models before applying their findings to humans. For example, chimpanzees and mice share about 99% and 98% of DNA with humans, respectively [1, 2]. Then, animals have the trend to be affected by different human worrying problems and represent good models for the study of human diseases.

(ii) Feasibility. The management of animal models is relatively easy since different factors can be controlled from the composition of food intake to temperature and lighting. Therefore, compared to human studies, there are less environmental variations. Moreover, the animal lifespan is shorter than humans. Hence, they represent good models, as they can be studied over their entire life cycle or even through several generations [3, 4].

(iii) Drug Safety. Preclinical toxicity testing, pharmacokinetics, and pharmacodynamic profiles of drugs can be investigated on animal models before use in humans. It remains important to evaluate the effectiveness of a drug as potential treatment on animals prior testing on humans. Drug safety profiles need to be established in order to protect the animals, human, and environment. Nevertheless, not all results acquired on animals can be directly translated to humans.

The use of animal models for biological research is restricted by the presence of confounding variables; limited accessibility of imaging for observation, throughput, and usability; and differences between human and animal biology [5]. These points are emphasized by those who refute any value to animal research. Moreover, the place of the animals in our modern society is frequently debated, namely, the right to use animals to benefit human purposes, with the risk that animals could be harmed. These aspects lead regularly in confusing opinions, which made the citizens and politicians to have an unclear picture of the problems. This has been the case during the evaluation of the European Citizen Initiative "Stop Vivisection" recently presented to the European Commission [6]. Despite that, animal studies remain essential to fill the gap between in vitro experimentation and human clinical trials.

Tissue engineering and regenerative medicine (TERM) are innovative research areas dealing with the potential of natural signaling pathways combined with the components of the organism to induce repair and regeneration of organs and tissues. Basically, the principal constituents of a tissue engineering approach are (i) cells, (ii) bioactive signals as growth factors or bioreactors, and (iii) biomaterial scaffolds which act as template for tissue formation [7]. There is a growing demand for new biomaterials to replace damaged osteoarticular tissue. Therefore, orthopedic applications represent one of the main market of tissue engineering [8]. The International

Cartilage Repair Society (ICRS) developed a five-grade cartilage lesion classification score system based on the macroscopic evaluation and the depth of the cartilage defect [9]. In ICRS Grade 0, the cartilage is normal. Grade 1 is divided into 1a, which includes cartilage lesions with a cartilage softening with or without superficial fissures, and 1b, which includes also superficial lesions, with the presence of fissures and cracks. Grade 2 is when cartilage lesion is deeper, extending to less than 50% of the cartilage thickness and with fraying. For classifying a cartilage lesion as a Grade 3, the cartilage injury has to be deeper than 50% of the cartilage thickness as well as down to the calcified layer. Grade 4 lesion is characterized by a complete loss of cartilage thickness and exposure of the underlying bone. In the ICRS classification, osteochondral defect corresponds to the worst case of cartilage lesion (Grade 4). Osteochondral defect management and repair represent a significant challenge in orthopedic surgery because it simultaneously affects both articular cartilage and the underlying subchondral bone. Then, the cartilage, bone, and the cartilage-bone interface have to be taken into account on the development of new strategies to repair an osteochondral defect. Recently, TERM approach emerged as a potential alternative to the current clinical palliative treatments for osteochondral defect repair, because this approach can be efficiently used to regenerate the cartilage, bone, and the cartilage-bone interface.

Because the choice of the appropriate animal model is fundamental to establish pertinent conclusions, the factors that will allow it must be identified and well understood. Before choosing the ideal animal model, it remains crucial to identify correctly the problem that has to be solved in order to obtain the right answer to the right question. Thus, the animal species to be used as well as the experimental design to be established will clearly depend upon the question asked. Animal models used in preclinical studies for osteoarticular tissue engineering goals cannot accurately reproduce the human biomechanical conditions. A preclinical study for bone and cartilage repair may be conducted in large animal models as sheep, goat, or horse. The time for recovery and the dimension of the defect should be enough and sufficient in order to obtain the evidence and allow a robust analysis. Small animal models are crucial in "proof-of-concept" studies where theories are verified and results acquired in vitro are applied in vivo. Small animal models are frequently used to study the pathophysiology and pathogenesis of the disease process. These smaller models are faster, low-cost, easy to handle and house, and easier to implement and study than the large models. They are currently used as the first screening tool for new drugs and treatment development which then warrants further testing in large animal models before clinical trials. But important limitations in translational studies are identified as (i) the limited volume of bone and cartilage defects, (ii) the less thickness of the cartilage, and (iii) the high degree of flexion of those small animals and consequent partial weight-bearing condition, which are important drawbacks when compared with human conditions [10, 11]. Moreover, the drugs, which demonstrated to be efficient in small animal studies, may not be translatable to humans with the same effectiveness [12]. One of the reasons for this might be the well-known difference of anatomy, histology, and physiology between these animals and humans.

The present chapter will focus on the use of small animal models for the development of new strategies for osteochondral defect (Grade 4 of ICRS classification).

19.2 Small Animal Model Strategies for Osteochondral Repair

19.2.1 Mouse

Before applying new product on tissue engineering purposes, initial studies are required to evaluate important issues such as biocompatibility, degradation, and biofunctionality. This evaluation is firstly achieved through the surgical implantation of the product in ectopic subcutaneous sites. These studies are typically performed on small animal species such as mice. These small animal models have some benefits: (i) expenses are low; (ii) large groups of animals can be used; (iii) homogeneous response of strains reduces individual deviations commonly observed in large animal models; (iv) advanced imaging techniques are available such as microCT and bioluminescence imaging; (v) a variety of genetic modifications are commercially available; and (vi) the use of immune-deficient strains allows studies of human cells or grafts without immune response implication.

Different animal models are currently used in research on restoration of osteochondral lesions including medium- (rabbits and dogs) [13–17] and large-sized (sheep and horses) [18–23] animals. However, the use of rodent models (mice and rats) to study osteochondral (OC) lesions is limited, despite the benefits previously described. The main concern regarding these models is their high rate of spontaneous repair after osteochondral defect induction. Despite that, and in order to better understand the cartilage repair process, an osteochondral defect model in mice has been established. Through a small (~0.5–1 cm) medial parapatellar skin incision, the joint capsule was opened and the patella dislocated laterally to expose the trochlear groove articular surface. The full thickness lesion was made in the cartilage with 21–27 G needles using a circular motion (0.4–0.5 mm diameter) until reaching the subchondral bone. Invasion of the subchondral bone was confirmed by the presence of blood resulting from removal of the needle [24–28]. This surgical protocol has recently been applied to evaluate the potential of an injectable cellularized PEG-based scaffold [29] and a 3D alginate-Gelfoam complexes [30] on cartilage repair. The data obtained in both studies provide proof of principle that the resultant structures possess great capacity for articular cartilage repair using tissue engineering approach.

This OC defect procedure allowed the development of a murine model of spontaneous cartilage regeneration. However, from these studies, it remains obvious that spontaneous healing capacity is clearly dependent on mouse age and strains. The spontaneous cartilage recovery is not the only way in sustaining the importance of the mouse strains used in cartilage recovery applications. Recently, Mak et al. [31] have evaluated the impact of intra-articular injections of synovial mesenchymal stem cells (MSCs), isolated from two different strains (C57BL6 and MRL strains) on cartilage repair using the same mouse injury model. They demonstrated that intra-articular injection of these synovial MSCs, isolated from MRL or C57BL6 mice, protects against the joint deterioration that would normally result after a surgically induced focal cartilage defect, although the mechanism of protection does appear to be different between the two strains of mice [31]. Instead the existence of a spontaneous recovery in mice model.

Then, the strain of the mouse showed to have some importance when studying approaches to improve cartilage and bone repair.

Nowadays, an innovative approach aims to analyze genetic and biomolecular mechanisms underlying cartilage repair. For this reason, the use of genetically modified animals represents a powerful tool to investigate the biological mechanisms involved. Mice offer robust benefits for mechanistic in vivo studies due to the accessibility to athymic, transgenic, and knockout strains. Athymic mice, which have a limited cellular immune response, allow initial in vivo study of allogenic and xenogeneic cartilage repair approaches [32–36]. Genetically modified mice, including transgenic and knockout models, are currently used to study the effects of a particular gene or protein on bone and cartilage repair and regeneration in different musculoskeletal diseases [37–39].

19.2.2 Guinea Pig

Guinea pigs (*Cavia porcellus* or *Cavia cobaya*) had a special place in research. This rodent is considered as a suitable model of human skeletal growth pattern because its epiphyses fuse as growth is completed [40]. However, it presented many disadvantages, namely, the fact that growth plate fusion occurs several months after bone growth stops and that guinea pig presents various alignment of the knees, which results in an increased load on medial compartment [41]. Therefore a reduced number of studies have used this animal model for osteochondral repair strategies. Kaar et al. [40] have evaluated the impact of this model on cartilage full thickness defect and concluded that despite the regeneration occurred in all cases, the level of tissue restoration was variable and the degree of repair was independent of the age. Actually, and mostly due to the increased use of genetically engineered mice and rats for specific disease models, the usage of guinea pig in research declined. Furthermore, guinea pigs demonstrated spontaneous cartilage degeneration [42], which, associated with age-related osteophyte formation, subchondral bone changes, and synovitis, made these animals a popular model for the study of osteoarthritis [41, 43].

19.2.3 Rat

Small animal models have been explored in order to address the challenge for osteochondral repair [44, 45]. The use of rats as osteochondral defect model seemed very attractive in order to provide proof-of-concept data. Rat model display some advantages: (i) economically rats are relatively low cost and easy to care of; and (ii) clinically they are more relevant than the mouse model based on their articular cartilage which presents typically also a zonal structure mimicking the one observed in human joints [46]. And as for mice, immune-deficient models are also available. However, articular cartilage is thinner, and defects are much smaller compared with humans; moreover, most defects cannot be set without penetrating the subchondral bone plate. Therefore, the rat model, as well as mice model, seems suitable only for preliminary in vivo assays and not for preclinical studies, but there is a constant requirement to better understand the biology of osteochondral defects. Different approaches have been applied on rat models to evaluate osteochondral defects restoration. Joint surface of rat knee demonstrated some regenerative ability. The major and growing concerns in osteochondral repair remained to evaluate the normal progression of spontaneous osteochondral healing during time, not only regarding the altered area but also in the cartilage surrounding the defect. Therefore, it remained fundamental to define a critical size osteochondral defect model and to establish the subchondral bone plate advancement toward the join surface [47]. Katagiri et al. defined a critical size osteochondral defect as 1.4 mm in diameter in rat and showed that the subchondral bone plate advancement happened quickly [48]. Moreover, they showed that the articular cartilage close to the osteochondral defect presented expression of Interleukin 1 beta (IL1β), fibroblast growth factor 2 (FGF2), and a disturbed FGF receptor 1/FGF receptor 3 balance, resulting in a catabolic activity which potentially could be responsible of an early osteoarthritic disease process.

Instead it is well described that a large osteochondral defect does not repair itself with original cartilage and leads to osteoarthritis; other approaches have been evaluated in order to induce the repair of osteochondral defect. Scaffold-free cell-based strategies have been tested. The transplantation of autologous chondrocytes organized in sheets has been showed to promote the repair mechanism of osteochondral defect [49] compared to synovium cells, described to have the highest potential for both proliferation and chondrogenesis [50]. Another approach was the use of cartilage-like tissue, generated ectopically by muscle-derived cells or amnion-derived cells using bone morphogenetic protein-2, which showed to be effective in repairing articular cartilage defects in rats [51, 52]. However, problems have been reported such as the dedifferentiation of cells with passaging [53]. Therefore, a strategy that mobilizes the endogenous pool of mesenchymal stem cells (MSCs) would offer a cheaper and less invasive alternative. MSCs are widely used as scaffold-free cell strategy for osteochondral defects regeneration [53–57]. Moreover, Yamaguchi demonstrated that exercise could efficiently promote cartilage repair after an MSC intra-articular injection [58]. As with all cell-based strategies, there are significant logistic and operational challenges associated with proper handling and cell storage required to maintain cell viability and vitality. Therefore, in view of all these issues, cell-free-based approaches have been tested. The administration of the myelostimulant granulocyte-colony stimulating factor (G-CSF) [59], a cytokine that serves as a growth factor for the hematopoietic stem cells, or exosome [60] (Fig. 19.1), a cell-secreted nano-sized vesicles present in the MSC secretome, has demonstrated potential for cartilage repair. Both strategies could overcome the impeding restrictions of current cell-based therapies.

Another therapeutic strategy for osteochondral repair is based on the implantation of scaffolds. Since rat model also offers a cost-effective means for in vivo evaluation of degradation characteristics and safety profile of new biodegradable scaffolds

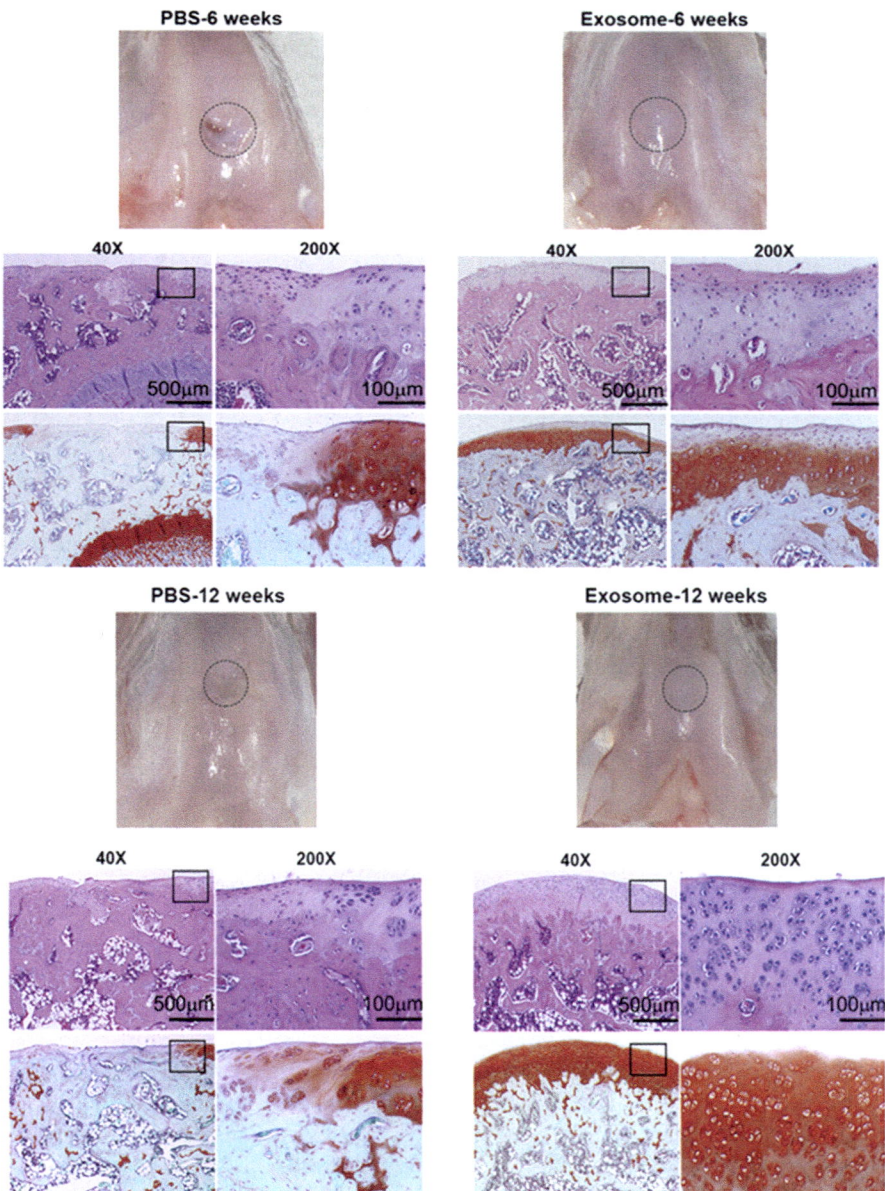

Fig. 19.1 In vivo cartilage repair at 6- and 12-week post-surgery. (Reprinted with permission [60]. Copyright © 2016, Elsevier)

and polymers, cell-free and cell-seeded scaffolds approaches have been investigated. Regarding the acellular scaffold-based strategies, the impact of different biomaterials and structures has been evaluated on cartilage and bone repair. Ferretti et al. used the rat osteochondral defect model to support the use of genipin crosslinked polyethylene glycol hydrogels as an innovative delivery system to control in vivo release of growth factors for improving articular cartilage repair [61]. Nanofiber scaffold, composed of poly(vinyl alcohol) or chondroitin sulfate, has enhanced the endogenous repair process without exogenous cells [62]. The use of cell-free multilayered silk fibroin-based scaffolds, combined or not with TGFb2 and BMP-2 growth factors, has shown to possess an inherent ability to attract endogenous, joint-resident cells capable of differentially differentiating down the osteochondral lineages [63]. Nogami et al. [64] developed a cell-free scaffold composed of human amniotic mesenchymal (HAM) cell-derived extracellular matrix and polylactic-co-glycolic acid. They demonstrated that the implantation of this cell-free scaffold, in rat model of osteochondral defect, promoted in-growth of endogenous cells and resulted in good cartilage repair [64]. More recently the administration of absorbable gelatin sponges, combined with insulin-like growth factor-1 or hyaluronic acid, in rat knee has showed to be efficient in the repair of osteochondral lesions [65]. The potential of cellular scaffolds on osteochondral defect repair in rat model has also been investigated. Within those assays, both cell types and scaffold materials have been studied. Dahlin et al. evaluated the impact of the ability of cocultures of articular chondrocytes and MSCs to repair articular cartilage in osteochondral defects [66]. For that purpose, bovine articular chondrocytes and rat MSCs were seeded separately or in coculture onto electrospun poly(ε-caprolactone) scaffolds and implanted in the defect. The authors demonstrated the potential for the use of cocultures of articular chondrocytes and MSCs for the in vivo repair of cartilage defects [66]. Moreover, the implantation of autologous chondrocyte, cultured in media supplemented with recombinant acid ceramidase and seeded on a biphasic material containing a collagen I top layer and a porous collagen III bottom layer (Bio-Gide), has enhanced cartilage repair in a rat osteochondral defect model [67]. All these studies supported the importance of designing tissue-engineered scaffolds that mimic the physical and biological components of extracellular matrix to produce ideal tissue repair in vivo. Overall, small rodents are attractive models for cartilage research due to the accessibility of immune-deficient and transgenic animals, as well as cheaper to house and purchase. Nevertheless, their translational potential remains limited due to their small joint size and tiny cartilage. In the context of bone and cartilage repair and regeneration, rodent models are most useful for in vivo mechanistic studies, feasibility studies, and preliminary testing of new therapy strategies.

19.2.4 Rabbit

The rabbit model provides a more appropriate small animal model for the assessment of osteochondral defect repair as they have larger joints and are a sufficient size for easy surgical procedures. Moreover they presented a bone plate thickness of 0.4–0.5 mm and a cartilage thickness of 0.25–0.75 mm [68, 69]. As for the previously described models, rabbits are easy to handle and low cost to maintain in-house. However, this model presents some disadvantages, i.e., an increased intrinsic healing due to increased cell density, different load characteristics on the join, and the difficulty to achieve a consistent partial thickness. In all studies that will be cited thereafter, the creation of an osteochondral defect was always based on the same protocol. The rabbits were anesthetized and, through a longitudinal parapatellar incision, the patella was laterally dislocated. All visible bleeding was carefully cauterized. With the knee joint maximally flexed, an osteochondral defect of 3–5 mm in diameter and 2–3 mm deep was created in the load-bearing region of the medial condyle. All debris were removed from the defect with a curette and the edges cleaned with sharp scalpel blade. After, the patella was relocated and the wound sutured in layers [70, 71]. Moreover, the age of the rabbit at the time of the surgery remains important. A histological and radiographic study of the closure of the distal femur, proximal tibia, and proximal fibula demonstrated the New Zealand white rabbits are skeletally mature between 19 and 24 weeks old [72].

Different tissue engineering (TE) strategies have been developed to address osteochondral defect. These approaches are mainly applied for restoration/regeneration of the tissues and based on the use of cells, scaffolds, and growth factors alone or combined. Cell-based approach is one of the current osteochondral repair option. This approach is increasingly explored to deliver biological substitution of the injured tissue, either by injection of chondrocytes or implantation of specific grafts. However, it is limited by the number of cells available for isolation and by the uncontrolled phenotypic alterations in those cells. As such, stem cells have been investigated as cell sources for cartilage and bone engineering due to their well-established ability to generate cartilage-like and bone-like tissues under the appropriate culture conditions. As alternative cell-based approach, the use of platelet-rich plasma (PRP) for the treatment of numerous types of orthopedic disorders, including chondral and osteochondral injuries, has increased recently. PRP is a plasma fraction containing a high concentration of platelets and is rich in many growth factors (GF). These GF take part in the natural process of tissue healing and homeostasis. They present the capability to stimulate cell proliferation, mesenchymal stem cell chemotaxis, and cell differentiation. Nevertheless, the use of PRP in preclinical and clinical studies, in chondral injuries, remains controversial [73].

Recently reported rabbit preclinical studies for the treatment of OC lesions using different scaffold-free strategies are summarized in Table 19.1 (Fig. 19.2).

Current approaches for articular OC repair are centered on the use of hydrogels and scaffolds providing a suitable three-dimensional (3D) environment supporting the growth of cartilaginous and bone repair tissues. These 3D structures are often critical, both in vitro and in vivo, to summarizing the in vivo milieu and allowing cells to modulate their own microenvironment. The ideal scaffolds for OC tissue engineering must to be based on the following basic requirements: porous, biocompatible, biodegradable, and appropriate for cell attachment, proliferation, and differentiation. Therefore the biomaterial is one of the key design

Table 19.1 Recent preclinical studies for OC repair on rabbit model using scaffold-free based approaches

Cells	Growth factors	References
Autologous bone marrow mesenchymal stem cells (BM-MScs)	n.a.	[74]
Autologous BM-MScs aggregated into a spheroid-like structure	n.a.	[75]
Human umbilical cord Wharton's jelly-derived MSCs (hWJMSCs)	n.a.	[76]
Allogenic chondrogenic pre-differentiated MSCs (C-MSCs)	n.a.	[77]
Allogenic magnetically labeled MSC (m-MSC)	n.a.	[78]
Synovial membrane-derived MSC (S-MSC)	n.a.	[79]
Adipose-derived MSCs (Ad-MSCs)	n.a.	[80]
Costal cartilage grafts	n.a.	[81]
Autologous BM-MSCs	Granulocyte-colony stimulating factor (G-CSF)	[82]
Autologous BM-MScs	Platelet-rich fibrin releasate (PRFr)	[83]
Osteochondral autograft transplantation	Platelet-rich plasma (PRP)/platelet-rich fibrin clot	[84]
Osteochondral autograft transplantation	PRP	[85]
Mosaicplasty	PRP	[86]
n.a.	PRP	[87]
n.a.	PRP or PRF + stromal cell-derived factor-1 (SDF-1)	[88]
n.a.	PRP gel	[89]

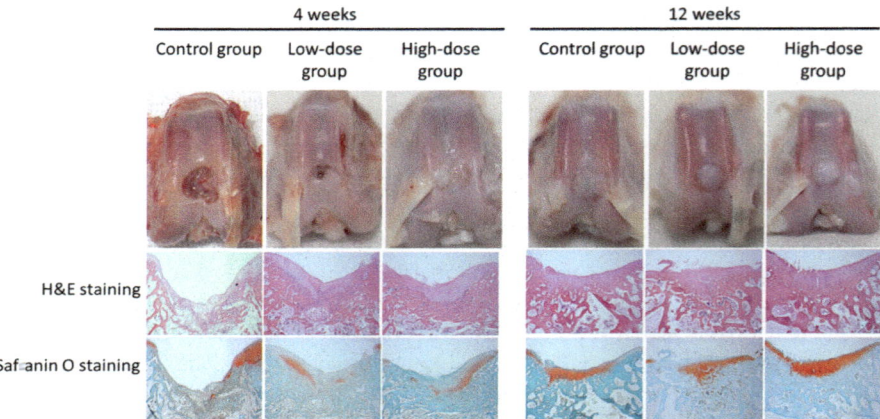

Fig. 19.2 Macroscopic and microscopic findings at 4- and 12-week posttreatment. (Reprinted with permission [82]. Copyright © 2017)

factors to be considered in scaffold- or hydrogel-based OC tissue engineering. Different biomaterials are currently used including naturally or synthetically derived polymers. Some rabbit preclinical studies for the treatment of OC lesions using different hydrogel and scaffold-based strategies are summarized in Table 19.2 (Fig. 19.3).

Table 19.2 Recent preclinical studies for OC repair on rabbit model using hydrogel- and scaffold-based approaches

	Hydrogel/scaffold	Cells/growth factors	References
Natural polymers	Laminin	BM-MSCs	[90]
		IGF-1	
		TGF-β1	
	Gelatin-chitosan	TGF-β1	[91]
	Collagen	SDF-1	[92]
	Gelatin	BM-MSCs	[93, 94]
		Chondrocytes	
Mixture of natural and synthetic polymers	Collagen-SF	PTHrP	[95]
Synthetic polymers	GCH-GCBB	BM-MSCs	[17]
		Chondrocytes	
	PLGA	BM-MSCs	[96, 97]
		BMP-2	
	GelMA/PAM	n.a.	[98]
	OPF	TGF-b3	[13, 99]
		IGF-1	
		BMP-2	
	SF/silk-nanoCaP	n.a.	[100]
	SF/CNF	TGF-β1	[101]
		BMP-2	
	SF/CS – SF/CS/nHA	BM-MSCs	[102]
	HAp	S-MSCs	[103]
	PLDLA/HAp	n.a.	[104]
	HAp/PCL	n.a.	[105]
Synthetic polymers	Hap/DN	n.a.	[106]
	PAMPS/PDMAAm	n.a.	[107]
	βTCP	S-MSCs	[103]
	PLLA-CL-COL I/βTCP	BM-MSCs	[108]

βTCP, Beta-tricalcium phosphate; BMP-2, bone morphogenetic protein-2; CNF, carbon nanofiber; COL I, collagen type I; CS, chitosan; DN, PAMPS/PDMAAm double-network; GCBB, gelatin and ceramic bovine bone; GCH, gelatin, chondroitin sulfate, and sodium hyaluronate; GelMA, methacrylated gelatin; HAp, hydroxyapatite; IGF-1, insulin-like growth factor-1; nanoCaP, nanocalcium phosphate; nHA, nano-hydroxyapatite; OPF, oligo(poly(ethylene glycol) fumarate); PAM, polyacrylamide; PAMPS, poly-(2-acrylamido-2-methylpropane sulfonic acid); PCL, poly(ε-caprolactone); PDMAAm: poly-(N,N0-dimethyl acrylamide); PLDLA, poly-L/D-lactide; PLGA, poly(lactide-co-glycolide); PLLA-CL, poly (L-lactic acid)-co-poly (ε-caprolactone); PTHrP, parathyroid hormone-related protein; SDF-1, stromal cell-derived factor-1; SF, silk fibroin; TGF-β1, transforming growth factor beta 1

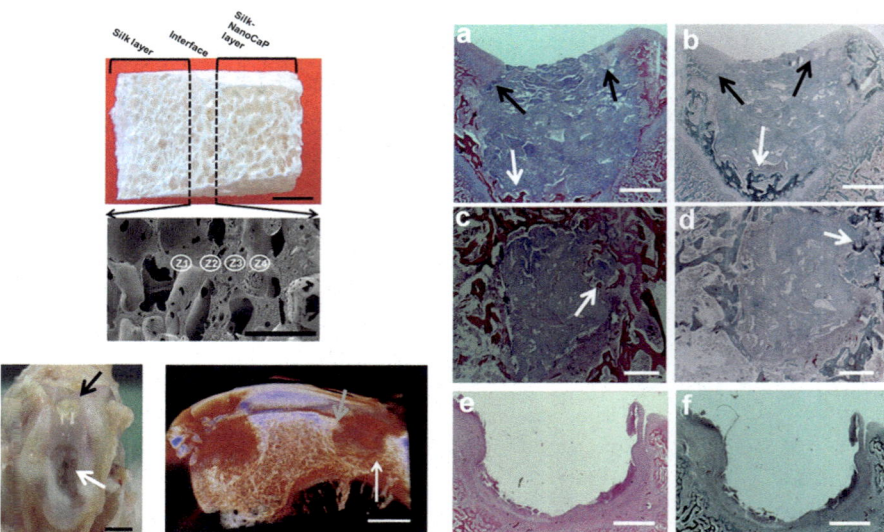

Fig. 19.3 "(**a** - **f**) Histological analysis of the explants from rabbit OCD. (**a, b**) H&E and Masson's trichrome staining's of the longitudinal section of the explants, respectively. (**c, d**) H&E and Masson's trichrome staining's of the cross-section of the explants in the silk-nanoCaP layer, respectively. (**e, f**) H&E staining and Masson's trichrome staining's of the longitudinal section of the defect, respectively. Neocartilage formation in the silk layer of the bilayered scaffolds is indicated by black arrows, and indicates new subchondral bone formation inside the silk-nanoCaP layer of the bilayered scaffolds is indicated by the white arrows. (Reprinted with permission [100]. Copyright © 2015, Elsevier)

19.3 Conclusions

Over the previous few years, progress has been realized to reinforce the use of tissue engineering strategies in preclinical studies and clinical assays aiming the regeneration of OC lesions. In preclinical studies, the main approaches involve the improvement of new biomaterials used for the development of biocompatible scaffolds/hydrogels combined or not with growth factors and/or cells. Grafts, from allogenic or autologous origin, or arthroplasty already proved their possibilities in cartilage repair. Despite this numerous therapeutic proposals for the chondral and osteochondral lesions, it remains difficult to agree on the best treatment to be applied. Before clinical trials, those strategies have to demonstrate their potential during preclinical studies in animal models. Animal studies are essential to establish a proof-of-concept, which will be based on biological responses, degradation time, and dose response of the implanted materials. However, and in order to evaluate the potential of new regenerative strategies on OC defects, small animal models, which include mouse, guinea pig, rat, and rabbit, might not be the most suitable models, since large animal models (e.g., pig, sheep, goat, or horse) more closely resemble to the human tissue.

Acknowledgments Alain da Silva Morais acknowledges ERC-2012-ADG 20120216–321266 (ComplexiTE) for his postdoc scholarship. The research leading to this work has received funding from the Portuguese Foundation for Science and Technology for the funds provided under the program Investigador FCT 2012 and 2015 (IF/00423/2012 and IF/01285/2015).

References

1. Mural RJ et al (2002) A comparison of whole-genome shotgun-derived mouse chromosome 16 and the human genome. Science 296(5573):1661
2. Chimpanzee Sequencing and Analysis Consortium (2005) Initial sequence of the chimpanzee genome and comparison with the human genome. Nature 437(7055):69–87
3. Erickson ZT, Falkenberg EA, Metz GA (2014) Lifespan psychomotor behaviour profiles of multigenerational prenatal stress and artificial food dye effects in rats. PLoS One 9(6):e92132
4. Franklin TB et al (2010) Epigenetic transmission of the impact of early stress across generations. Biol Psychiatry 68(5):408–415
5. Shanks N, Greek R, Greek J (2009) Are animal models predictive for humans? Philos Ethics Humanit Med 4:2
6. Wang X, Ye JD, Wang Y (2007) Influence of a novel radiopacifier on the properties of an injectable calcium phosphate cement. Acta Biomater 3:757
7. O'Brien FJ (2011) Biomaterials & scaffolds for tissue engineering. Mater Today 14(3):88–95
8. PR Newswire (18 Dec 2016) Tissue Engineering – Global Market Outlook (2016–2022).
9. Brittberg M, Winalski CS (2003) Evaluation of cartilage injuries and repair. J Bone Joint Surg Am 85-A(Suppl 2):58–69
10. Vilela CA et al (2015) Cartilage repair using hydrogels: a critical review of in vivo experimental designs. ACS Biomater Sci Eng 1(9):726–739
11. McCoy AM (2015) Animal models of osteoarthritis. Vet Pathol 52(5):803–818
12. Pelletier J-P et al (2010) Experimental models of osteoarthritis usefulness in the development of disease-modifying osteoarthritis drugs/agents. Therapy 7(6):621–634
13. Kim K et al (2013) Osteochondral tissue regeneration using a bilayered composite hydrogel with modulating dual growth factor release kinetics in a rabbit model. J Control Release 168(2):166–178
14. Madry H et al (2013) Cartilage constructs engineered from chondrocytes overexpressing IGF-I improve the repair of osteochondral defects in a rabbit model. Eur Cell Mater 25:229–247
15. Lv YM, Yu QS (2015) Repair of articular osteochondral defects of the knee joint using a composite lamellar scaffold. Bone Joint Res 4(4):56–64
16. Iamaguti LS et al (2012) Reparação de defeitos osteocondrais de cães com implante de cultura de condrócitos homólogos e membrana biossintética de celulose: avaliação clínica, ultrassonográfica e macroscópica. Arq Bras Med Vet Zootec 64:1483–1490
17. Deng T et al (2012) Construction of tissue-engineered osteochondral composites and repair of large joint defects in rabbit. J Tissue Eng Regen Med 8(7):546–556
18. Cokelaere S, Malda J, van Weeren R (2016) Cartilage defect repair in horses: current strategies and recent developments in regenerative medicine of the equine joint with emphasis on the surgical approach. Vet J 214:61–71
19. Schleicher I et al (2013) Biphasic scaffolds for repair of deep osteochondral defects in a sheep model. J Surg Res 183(1):184–192

20. Rautiainen J et al (2013) Osteochondral repair: evaluation with sweep imaging with Fourier transform in an equine model. Radiology 269(1):113–121

21. Pilichi S et al (2014) Treatment with embryonic stem-like cells into osteochondral defects in sheep femoral condyles. BMC Vet Res 10:301

22. Desjardin C et al (2014) Omics technologies provide new insights into the molecular physio-pathology of equine osteochondrosis. BMC Genomics 15(1):947

23. Orth P et al (2013) Improved repair of chondral and osteochondral defects in the ovine troch-lea compared with the medial condyle. J Orthop Res 31(11):1772–1779

24. Eltawil NM et al (2009) A novel in vivo murine model of cartilage regeneration. Age and strain-dependent outcome after joint surface injury. Osteoarthr Cartil 17(6):695–704

25. Fitzgerald J (2017) Enhanced cartilage repair in "healer" mice-new leads in the search for better clinical options for cartilage repair. Semin Cell Dev Biol 62:78–85

26. Fitzgerald J et al (2008) Evidence for articular cartilage regeneration in MRL/MpJ mice. Osteoarthr Cartil 16(11):1319–1326

27. Matsuoka M et al (2015) An articular cartilage repair model in common C57Bl/6 mice. Tissue Eng Part C Methods 21(8):767–772

28. Rai MF et al (2012) Heritability of articular cartilage regeneration and its association with ear wound healing in mice. Arthritis Rheum 64(7):2300–2310

29. Wang J et al (2017) Fabrication of injectable high strength hydrogel based on 4-arm star PEG for cartilage tissue engineering. Biomaterials 120:11–21

30. Wang P et al (2016) Flavonoid compound icariin activates hypoxia inducible factor-1alpha in chondrocytes and promotes articular cartilage repair. PLoS One 11(2):e0148372

31. Mak J et al (2016) Intra-articular injection of synovial mesenchymal stem cells improves cartilage repair in a mouse injury model. Sci Rep 6:23076

32. Jin G-Z et al (2014) Biphasic nanofibrous constructs with seeded cell layers for osteochon-dral repair. Tissue Eng Part C Methods 20(11):895–904

33. Sartori M et al (2017) A new bi-layered scaffold for osteochondral tissue regeneration: in vitro and in vivo preclinical investigations. Mater Sci Eng C Mater Biol Appl 70(Pt 1):101–111

34. Shalumon K et al (2016) Microsphere-based hierarchically juxtapositioned biphasic scaf-folds prepared from poly(lactic-co-glycolic acid) and nanohydroxyapatite for osteochondral tissue engineering. Polymers 8(12):429

35. Sheehy EJ et al (2013) Engineering osteochondral constructs through spatial regulation of endochondral ossification. Acta Biomater 9(3):5484–5492

36. Yan L-P et al (2015) Current concepts and challenges in osteochondral tissue engineering and regenerative medicine. ACS Biomater Sci Eng 1(4):183–200

37. Li S et al (2017) A conditional knockout mouse model reveals a critical role of PKD1 in osteoblast differentiation and bone development. Sci Rep 7:40505

38. Hu K, Olsen BR (2016) Osteoblast-derived VEGF regulates osteoblast differentiation and bone formation during bone repair. J Clin Invest 126(2):509–526

39. Oh H, Chun CH, Chun JS (2012) Dkk-1 expression in chondrocytes inhibits experimental osteoarthritic cartilage destruction in mice. Arthritis Rheum 64(8):2568–2578

40. Kaar TK, Fraher JP, Brady MP (1998) A quantitative study of articular repair in the guinea pig. Clin Orthop Relat Res 346:228–243

41. Kraus VB et al (2010) The OARSI histopathology initiative - recommendations for histologi-cal assessments of osteoarthritis in the guinea pig. Osteoarthr Cartil 18(Suppl 3):S35–S52

42. Bendele AM, Hulma JF (1988) Spontaneous cartilage degeneration in guinea pigs. Arthritis Rheum 31(4):561–565

43. Vázquez-Portalatín N et al (2015) Accuracy of ultrasound-guided intra-articular injections in guinea pig knees. Bone Joint Res 4:1–5

44. Chu CR, Szczodry M, Bruno S (2010) Animal models for cartilage regeneration and repair. Tissue Eng Part B Rev 16(1):105–115

45. Gregory MH et al (2012) A review of translational animal models for knee osteoarthritis. Arthritis 2012:764621

46. Ahern BJ et al (2009) Preclinical animal models in single site cartilage defect testing: a systematic review. Osteoarthr Cartil 17(6):705–713
47. Orth P, Madry H (2015) Advancement of the subchondral bone plate in translational models of osteochondral repair: implications for tissue engineering approaches. Tissue Eng Part B Rev 21(6):504–520
48. Katagiri H, Mendes LF, Luyten FP (2017) Definition of a critical size osteochondral knee defect and its negative effect on the surrounding articular cartilage in the rat. Osteoarthr Cartil 25(9):1531–1540
49. Shimizu R et al (2015) Repair mechanism of osteochondral defect promoted by bioengineered chondrocyte sheet. Tissue Eng Part A 21(5–6):1131–1141
50. Yoshimura H et al (2007) Comparison of rat mesenchymal stem cells derived from bone marrow, synovium, periosteum, adipose tissue, and muscle. Cell Tissue Res 327(3):449–462
51. Nawata M et al (2005) Use of bone morphogenetic protein 2 and diffusion chambers to engineer cartilage tissue for the repair of defects in articular cartilage. Arthritis Rheum 52(1):155–163
52. Wei JP et al (2009) Human amniotic mesenchymal cells differentiate into chondrocytes. Cloning Stem Cells 11(1):19–26
53. Lee JM et al (2012) In vivo tracking of mesenchymal stem cells using fluorescent nanoparticles in an osteochondral repair model. Mol Ther 20(7):1434–1442
54. Chijimatsu R et al (2017) Characterization of mesenchymal stem cell-like cells derived from human iPSCs via neural crest development and their application for osteochondral repair. Stem Cells Int 2017:1960965
55. Itokazu M et al (2016) Transplantation of scaffold-free cartilage-like cell-sheets made from human bone marrow mesenchymal stem cells for cartilage repair: a preclinical study. Cartilage 7(4):361–372
56. Muttigi MS et al (2017) Matrilin-3 co-delivery with adipose-derived mesenchymal stem cells promotes articular cartilage regeneration in a rat osteochondral defect model. J Tissue Eng Regen Med 2017;1–9. https://doi.org/10.1002/term.2485
57. Oshima Y et al (2005) Behavior of transplanted bone marrow-derived GFP mesenchymal cells in osteochondral defect as a simulation of autologous transplantation. J Histochem Cytochem 53(2):207–216
58. Yamaguchi S et al (2016) The effect of exercise on the early stages of mesenchymal stromal cell-induced cartilage repair in a rat osteochondral defect model. PLoS One 11(3):e0151580
59. Okano T et al (2014) Systemic administration of granulocyte colony-stimulating factor for osteochondral defect repair in a rat experimental model. Cartilage 5(2):107–113
60. Zhang S et al (2016) Exosomes derived from human embryonic mesenchymal stem cells promote osteochondral regeneration. Osteoarthr Cartil 24(12):2135–2140
61. Ferretti M et al (2006) Controlled in vivo degradation of genipin crosslinked polyethylene glycol hydrogels within osteochondral defects. Tissue Eng 12(9):2657–2663
62. Coburn JM et al (2012) Bioinspired nanofibers support chondrogenesis for articular cartilage repair. PNAS 109(25):10012–10017
63. Saha S et al (2013) Osteochondral tissue engineering in vivo: a comparative study using layered silk fibroin scaffolds from mulberry and nonmulberry silkworms. PLoS One 8(11):e80004
64. Nogami M et al (2016) A human amnion-derived extracellular matrix-coated cell-free scaffold for cartilage repair: in vitro and in vivo studies. Tissue Eng Part A 22(7–8):680–688
65. Alemdar C et al (2016) Effect of insulin-like growth factor-1 and hyaluronic acid in experimentally produced osteochondral defects in rats. Indian J Orthop 50(4):414–420
66. Dahlin RL et al (2014) Articular chondrocytes and mesenchymal stem cells seeded on biodegradable scaffolds for the repair of cartilage in a rat osteochondral defect model. Biomaterials 35(26):7460–7469
67. Frohbergh ME et al (2016) Acid ceramidase treatment enhances the outcome of autologous chondrocyte implantation in a rat osteochondral defect model. Osteoarthr Cartil 24(4):752–762

68. ASTM F2451-05 (2010) Standard guide for in vivo assessment of implantable devices intended to repair or regenerate articular cartilage. ASTM International, West Conshohocken, PA

69. Chevrier A et al (2015) Interspecies comparison of subchondral bone properties important for cartilage repair. J Orthop Res 33(1):63–70

70. Qiu YS et al (2003) Observations of subchondral plate advancement during osteochondral repair: a histomorphometric and mechanical study in the rabbit femoral condyle. Osteoarthr Cartil 11(11):810–820

71. Gao J et al (2002) Repair of osteochondral defect with tissue-engineered two-phase composite material of injectable calcium phosphate and hyaluronan sponge. Tissue Eng 8(5):827–837

72. Kaweblum M et al (1994) Histological and radiographic determination of the age of physeal closure of the distal femur, proximal tibia, and proximal fibula of the New Zealand white rabbit. J Orthop Res 12(5):747–749

73. Kon E et al (2011) Platelet-rich plasma (PRP) to treat sports injuries: evidence to support its use. Knee Surg Sports Traumatol Arthrosc 19(4):516–527

74. Mahmoud EE et al (2017) Role of mesenchymal stem cells densities when injected as suspension in joints with osteochondral defects. Cartilage https://doi.org/10.1177/1947603517708333

75. Ishihara K et al (2014) Simultaneous regeneration of full-thickness cartilage and subchondral bone defects in vivo using a three-dimensional scaffold-free autologous construct derived from high-density bone marrow-derived mesenchymal stem cells. J Orthop Surg Res 9:98

76. Liu S et al (2017) Repair of osteochondral defects using human umbilical cord Wharton's jelly-derived mesenchymal stem cells in a rabbit model. Biomed Res Int 2017:8760383

77. Dashtdar H et al (2011) A preliminary study comparing the use of allogenic chondrogenic pre-differentiated and undifferentiated mesenchymal stem cells for the repair of full thickness articular cartilage defects in rabbits. J Orthop Res 29(9):1336–1342

78. Mahmoud EE et al (2016) Cell magnetic targeting system for repair of severe chronic osteochondral defect in a rabbit model. Cell Transplant 25(6):1073–1083

79. Li H et al (2016) Osteochondral repair with synovial membrane-derived mesenchymal stem cells. Mol Med Rep 13(3):2071–2077

80. Mehrabani D et al (2015) The healing effect of adipose-derived mesenchymal stem cells in full-thickness femoral articular cartilage defects of rabbit. Int J Organ Transplant Med 6(4):165–175

81. Du D et al (2015) Repairing osteochondral defects of critical size using multiple costal grafts: an experimental study. Cartilage 6(4):241–251

82. Sasaki T et al (2017) The effect of systemic administration of G-CSF on a full-thickness cartilage defect in a rabbit model MSC proliferation as presumed mechanism. Bone Joint Res 6:123–131

83. Wu CC et al (2017) Intra-articular injection of platelet-rich fibrin releasates in combination with bone marrow-derived mesenchymal stem cells in the treatment of articular cartilage defects: an in vivo study in rabbits. J Biomed Mater Res B Appl Biomater 105(6):1536–1543

84. Maruyama M et al (2017) Comparison of the effects of osteochondral autograft transplantation with platelet-rich plasma or platelet-rich fibrin on osteochondral defects in a rabbit model. Am J Sports Med 45(14):3280–3288

85. Boakye LA et al (2015) Platelet-rich plasma increases transforming growth factor-beta1 expression at graft-host interface following autologous osteochondral transplantation in a rabbit model. World J Orthop 6(11):961–969

86. Altan E et al (2014) The effect of platelet-rich plasma on osteochondral defects treated with mosaicplasty. Int Orthop 38(6):1321–1328

87. Smyth NA et al (2016) Platelet-rich plasma may improve osteochondral donor site healing in a rabbit model. Cartilage 7(1):104–111

88. Bahmanpour S et al (2016) Effects of platelet-rich plasma & platelet-rich fibrin with and without stromal cell-derived factor-1 on repairing full-thickness cartilage defects in knees of rabbits. Iran J Med Sci 41(6):507–517

89. Danieli MV et al (2014) Treatment of osteochondral injuries with platelet gel. Clinics 69(10):694–698

90. Gugjoo MB et al (2017) Mesenchymal stem cells with IGF-1 and TGF- beta1 in laminin gel for osteochondral defects in rabbits. Biomed Pharmacother 93:1165–1174

91. Han F et al (2015) Photocrosslinked layered gelatin-chitosan hydrogel with graded compositions for osteochondral defect repair. J Mater Sci Mater Med 26(4):160

92. Chen P et al (2015) Radially oriented collagen scaffold with SDF-1 promotes osteochondral repair by facilitating cell homing. Biomaterials 39:114–123

93. Mazaki T et al (2014) A novel, visible light-induced, rapidly cross-linkable gelatin scaffold for osteochondral tissue engineering. Sci Rep 4:4457

94. Wang CC et al (2016) Expandable scaffold improves integration of tissue-engineered cartilage: an in vivo study in a rabbit model. Tissue Eng Part A 22(11–12):873–884

95. Zhang W et al (2013) The promotion of osteochondral repair by combined intra-articular injection of parathyroid hormone-related protein and implantation of a bi-layer collagen-silk scaffold. Biomaterials 34(25):6046–6057

96. Duan P et al (2014) The effects of pore size in bilayered poly(lactide-co-glycolide) scaffolds on restoring osteochondral defects in rabbits. J Biomed Mater Res A 102(1):180–192

97. Vayas R et al (2017) Evaluation of the effectiveness of a bMSC and BMP-2 polymeric trilayer system in cartilage repair. Biomed Mater 12(4):045001

98. Han L et al (2017) Biohybrid methacrylated gelatin/polyacrylamide hydrogels for cartilage repair. J Mater Chem B 5:731

99. Lu S et al (2014) Dual growth factor delivery from bilayered, biodegradable hydrogel composites for spatially-guided osteochondral tissue repair. Biomaterials 35(31):8829–8839

100. Yan LP et al (2015) Bilayered silk/silk-nanoCaP scaffolds for osteochondral tissue engineering: in vitro and in vivo assessment of biological performance. Acta Biomater 12:227–241

101. Naskar D et al (2017) Dual growth factor loaded nonmulberry silk fibroin/carbon nanofiber composite 3D scaffolds for in vitro and in vivo bone regeneration. Biomaterials 136:67–85

102. Ruan SQ et al (2017) Preparation of a biphase composite scaffold and its application in tissue engineering for femoral osteochondral defects in rabbits. Int Orthop 41(9):1899–1908

103. Shimomura K et al (2017) Comparison of 2 different formulations of artificial bone for a hybrid implant with a tissue-engineered construct derived from synovial mesenchymal stem cells: a study using a rabbit osteochondral defect model. Am J Sports Med 45(3):666–675

104. Żylińska B et al (2017) Osteochondral repair using porous three-dimensional nanocomposite scaffolds in a rabbit model. In Vivo 31(5):895–903

105. Du Y et al (2017) Selective laser sintering scaffold with hierarchical architecture and gradient composition for osteochondral repair in rabbits. Biomaterials 137:37–48

106. Wada S et al (2016) Hydroxyapatite-coated double network hydrogel directly bondable to the bone: biological and biomechanical evaluations of the bonding property in an osteochondral defect. Acta Biomater 44:125–134

107. Higa K et al (2017) Effects of osteochondral defect size on cartilage regeneration using a double-network hydrogel. BMC Musculoskelet Disord 18(1):210

108. Wang C et al (2017) Cartilage oligomeric matrix protein improves in vivo cartilage regeneration and compression modulus by enhancing matrix assembly and synthesis. Colloids Surf B Biointerfaces 159:518–526

Chapter 20
Large Animal Models for Osteochondral Regeneration

Isabel R. Dias, Carlos A. Viegas, and Pedro P. Carvalho

Abstract Namely, in the last two decades, large animal models – small ruminants (sheep and goats), pigs, dogs and horses – have been used to study the physiopathology and to develop new therapeutic procedures to treat human clinical osteoarthritis. For that purpose, cartilage and/or osteochondral defects are generally performed in the stifle joint of selected large animal models at the condylar and trochlear femoral areas where spontaneous regeneration should be excluded. Experimental animal care and protection legislation and guideline documents of the US Food and Drug Administration, the American Society for Testing and Materials and the International Cartilage Repair Society should be followed, and also the specificities of the animal species used for these studies must be taken into account, such as the cartilage thickness of the selected defect localization, the defined cartilage critical size defect and the joint anatomy in view of the post-operative techniques to be performed to evaluate the chondral/osteochondral repair. In particular, in the articular cartilage regeneration and repair studies with animal models, the subchondral bone plate should always be taken into consideration. Pilot studies for chondral and osteochondral bone tissue engineering could apply short observational periods for evaluation of

I. R. Dias (✉) · C. A. Viegas
Department of Veterinary Sciences, Agricultural and Veterinary Sciences School, University of Trás-os-Montes e Alto Douro (UTAD), Vila Real, Portugal

3B's Research Group – Biomaterials, Biodegradables and Biomimetics, Department of Polymer Engineering, University of Minho, Headquarters of the European Institute of Excellence on Tissue Engineering and Regenerative Medicine, AvePark – Parque da Ciência e Tecnologia, Zona Industrial da Gandra, Barco - Guimarães 4805-017, Portugal

Department of Veterinary Medicine, ICVS/3B's – PT Government Associate Laboratory, Braga/Guimarães, Portugal
e-mail: idias@utad.pt

P. P. Carvalho
Department of Veterinary Medicine, University School Vasco da Gama, Av. José R. Sousa Fernandes 197, Lordemão, Coimbra 3020-210, Portugal

CIVG - Vasco da Gama Research Center, University School Vasco da Gama, Coimbra, Portugal

© Springer International Publishing AG, part of Springer Nature 2018   441
J. M. Oliveira et al. (eds.), *Osteochondral Tissue Engineering*,
Advances in Experimental Medicine and Biology 1059,
https://doi.org/10.1007/978-3-319-76735-2_20

the cartilage regeneration up to 12 weeks post-operatively, but generally a 6- to 12-month follow-up period is used for these types of studies.

Keywords Large animal models · Osteochondral tissue · Tissue engineering · Translational studies

Highlights

The defects in large animal models – small ruminants (sheep and goats), pigs, dogs and horses – are generally performed in the stifle joint at the condylar and trochlear femoral areas where spontaneous regeneration should be excluded.

In articular cartilage regeneration and repair studies with animal models, the subchondral bone plate must be taken into consideration.

Pilot studies for chondral and osteochondral bone tissue engineering could apply short observational periods for evaluation of the cartilage regeneration up to 12 weeks post-operatively, but generally a 6- to 12-month follow-up period is used for these types of studies.

20.1 Introduction

Due to the lack of a blood supply and the low cellular density of articular cartilage, this hard tissue has a limited healing capacity. Therefore, focal traumatic events causing cartilage defects rarely resolve spontaneously or are usually repaired by tissue with inferior properties when compared to the original chondral tissue, which may subsequently develop clinical osteoarthritis (OA). Cartilage lesions are typically treated by chondroplasty and palliative debridement techniques, drilling and microfracture repair or restoration techniques that employ autologous chondrocyte implantation, mosaic arthroplasty and osteochondral allograft transplantation [1–5]. Nevertheless, recent advances in osteochondral tissue engineering have prompted research on techniques to repair articular cartilage damage using a variety of transplanted cells or novel replacement devices [6–19].

In this way, animal models of experimental articular cartilage defect are largely used in preclinical and translational research studies to assess novel concepts for chondral and osteochondral tissue engineering treatments aiming for regenerative joint resurfacing [20]. Relatively to laboratory animal models (mouse, rat and rabbit), generally used for pilot and proof of concept in orthopaedic research studies, the large animal models (small ruminants – sheep and goat – pig, dog and horse) have the advantages of having joint size and cartilage thickness and also the clinical lesions most similar to those present in humans [21–23]. These large animal models also exhibit secondary Haversian bone tissue remodelling in the skeletally mature stages of their lifespan, bone tissue macro- and microstructure and composition, biochemical bone properties and bone mineral density more closely to humans. Nevertheless, we can also find variations in bone and cartilage tissue composition and mechanical properties that are species-dependent [24–31]. The most remarkable differences of these animal models in orthopaedic research are their rapid growth, with a predominance of plexiform or lamellar bone in the areas adjacent to the periosteum and endosteum of long bone cortices during the first years of their lifespan (namely, in the dog and the small ruminants), and the quadruped locomotion relatively to the biped locomotion of the human species that use exclusively the hindlimbs for this purpose [30].

Some of these animal species also exhibit significant differences in their metabolism, namely, concerning the gastrointestinal system with a completely different diet and reproductive and endocrine systems relatively to the human species. This marked physiological specificity could also influence hard tissues' composition and microstructural properties. Sheep and goats are small ruminant (polygastric stomach) herbivores, seasonally polyestrous beginning a 10-month period of oestrous cycling when daylight hours diminish in the fall [32]. Pigs, like humans, are true omnivores with a monogastric stomach and an identical digestive physiology [33]. Pigs are polyestrous and cycle approximately every 19.5 days [34]. Dogs are monogastric stomach carnivores that have evolved to an omnivorous characteristic due to domestication, nevertheless maintaining an anatomy, namely, of dentition and digestive system, physiology and diet behaviour of carnivore species [35]. Domestic dogs are nonseasonal monoestrous polytocous [36]. The horse is a domesticated modified monogastric herbivore with a simple stomach and a unique digestive tract that digests portions of its feeds enzymatically first in the foregut and then ferments in the hindgut [37]. The horse is predominantly monovulatory with seasonal breeders with ovulatory oestrous cycle activity being related to long days, between May and October [38].

The ideal animal model should mimic as close as possible the clinical setting, being biologically analogous and recognizable as an appropriate model for cartilage physiology [20, 39]. Nonetheless, this objective of performing valid cartilage defect testing studies has been very difficult to achieve, [40] since 95% of the human cartilage defects do not affect the subchondral bone [41]. On the contrary, in experimental studies using animal models, the subchondral bone is frequently involved in various of these studies [20] with the subsequent advancement of the subchondral bone plate during spontaneous healing of osteochondral defects and following articular cartilage treatment for chondral lesions using, for instance, marrow stimulation or tissue engineering strategies [42]. Another difference between human cartilage lesions and those experimentally induced in animal models is the volume of the cartilage defect which is larger in humans than in animal models, having approximately a mean total volume of 552.25 mm^3, with 10 mm or more in diameter, and usually involving only chondral tissue [20, 22, 41, 43–45]. In the most common animal models used for osteochondral bone repair studies, the total volume of the cartilage defects is reduced relatively to human and also involving the subchondral bone (mean total volume: sheep, 359.54 mm^3; goat,

251.65 mm³; pig, 107.43 mm³; dog, 82.39 mm³; horse, 334.73 mm³) [20]. This fact could be partially justified by the large variation of cartilage thickness between human and the above referred animal species (medial femoral condyle cartilage thickness: human 2.35 mm; sheep, 0.45 mm; goat, 1.1 mm; pig, 1.5 mm; dog, 0.95 mm; horse, 1.75 mm) where humans present the thickest articular cartilage at the stifle joint level [46]. Martino et al. [47] had already referred a thickness of articular cartilage at the femoral condyles ranging from 2.2 to 2.4 mm in humans, also confirmed by the value presented by Frisbie et al. [46] for man. This anatomical difference between species, and also their different body size [48] with a very thin articular cartilage layer present in most of the animal models used in articular cartilage regeneration and repair studies, could justify in part the involvement of the subchondral bone in these studies.

In view of the aforementioned, we have not found yet an ideal preclinical animal model for translational orthopaedic studies concerning the treatment of articular cartilage lesions in humans. Consequently, the choice of an appropriate animal model for chondral or osteochondral tissue engineering studies by researchers should be based on published scientific literature [49–56] and guideline documents of the American Society for Testing and Materials, [57] the International Cartilage Repair Society [58, 59] and the US Food and Drug Administration [60], performing a multifactorial analysis of each animal model.

This review tries to introduce and discuss the advantages and limitations of the most common large animal models used for chondral and osteochondral defects repaired by tissue engineering approaches. These defects are generally surgically induced focal defects; however, other types of direct injuries are also described in scientific literature, which aim to promote cartilage loss and osteoarthritis, such as joint destabilization (medial meniscal release, medial femoral condylar groove creation or anterior cruciate ligament transection in the canine stifle joint) [56, 61–64].

20.2 Large Animal Models

20.2.1 Sheep

Sheep are frequently used as an animal model in orthopaedic research since they are easily available, easy to handle and relatively inexpensive to purchase and maintain (feeding, housing). They have a strong flocking instinct, resulting from a protective group behaviour; they are docile and without any major behavioural adverse traits. The skeletal maturity in sheep is achieved between 2 and 3 years of age. Specifically referring to the articular cartilage defect repair studies, sheep models allow arthroscopic evaluation; nevertheless the species has a large infrapatellar fat pad and demands a marked flexion of the stifle joint to allow the visualization of the femoral condyles [20].

Concerning the thickness of the articular cartilage, there are a few different reports in the scientific literature. Simon [65] refers a cartilage thickness of femoral condyles of 1.68 mm, Lu et al. [66] report a cartilage thickness ranging from 0.4 to 1 mm at stifle joint level, and Frisbie et al. [46] report a 0.45 mm at the medial femoral condyle, among other authors' references to this parameter [67]. Archibald et al. [68] also reported an average articular cartilage thickness of the stifle joint of 1.2 mm for sheep. This variation indicates a large interindividual variation of the volume of osteochondral bone tissue involved among animals of different experimental groups, with consequent variability within the results and difficulty in its analysis and comparison of different studies using ovine models [20]. It is common for large defects in sheep to involve large subchondral bone tissue, which is very dense and hard, making it difficult to create identical bone defects when trephine and fracture techniques are used instead of drilling [20].

The large majority of the cartilage defect studies in sheep are performed in the stifle joint involving the weight-bearing area of one or both femoral condyles and/ or in the femoral trochlea, with a follow-up post-operative period which generally varies between 2 weeks and 18 months (Table 20.1). Typically, defects are either surgically created, ranging from 4 mm to 15 mm in diameter chondral/osteochondral, or impact injury whose primary use is in pivotal studies for which post-operative management variables are not critical [56]. The critical size defect (CSD – it should be considered the smallest wound established at the cartilage tissue, which cannot heal spontaneously during the lifetime of the animal) considered for cartilage defects in sheep is reported as 7 mm [56, 67].

The largest number of preclinical and translational studies for osteochondral tissue engineering that resort to large animal models has been performed in the ovine species, with a marked increase over the last 5 years – from 2013 to the present moment. Nonetheless, the large majority of the studies performed in sheep are in vivo studies, whereas only a few are in vitro studies performed with cells of ovine origin, such as the study of Vahedi et al. [126] which aims to evaluate adipose tissue-derived stem cells (ASCs) from infrapatellar fat pad and to characterize the cell surface markers using antihuman antibodies of these cells, based on the potential of ASCs for cellular therapies to repair injured tissues.

Referring to the in vivo studies in sheep, the large majority of them focus on acute chondral/osteochondral surgical lesions that are treated at the surgical moment by tissue engineering techniques. A minor part of these studies report posterior treatment by means of a second surgical intervention or arthroscopic technique for the injury treatment (Table 20.1). There are also a few studies in the ovine species published in the scientific literature which describe the induction of OA by promotion of articular instability of the stifle joint achieved by unilateral meniscectomy (generally the medial meniscus) and/or anterior cruciate ligament (ACL) resection associated with a post-operative exercise period after a recovery period, for later cell therapy by the intra-articular injection of living cells for repair of the chondral lesions. Example of that is the study of Al Faqeh et al. [127] where all animals were made to run on a hard surface for a distance of 100 m, once daily for 3 weeks after a recovery period of 3 weeks, with posterior intra-articular injection of a single dose

Table 20.1 Chondral/osteochondral tissue engineering studies performed in ovine model for cartilage repair

Authors	Population	Type of defect and localization	Material tested and follow-up period
Homminga et al. [69]	10 adult female Texel sheep	Full-thickness cartilage 5 mm × 10 mm defect in the weight-bearing area of the medial femoral condyle of the stifle joint	Xenograft of costal rabbit perichondrium fixed with human fibrin glue 12 weeks
Hurtig et al. [70]	24 adult female Suffolk-Romanoff crossbred sheep	Cartilage 11.5 mm Ø defects perpendicular to the weight-bearing portion of the femoral condyle of stifle joint	Fresh autograft and frozen allograft osteochondral dowel grafts 3, 6 and 12 months
Pearce et al. [71]	13 mature sheep	Full-thickness cartilage 14 mm deep defect on the weight-bearing surface of the medial femoral condyle of the stifle joint	4.5 mm × 10 mm cylindrical osteochondral autografting held in place by sidewall friction alone; grafts were delivered flush with the surrounding cartilage or left 2 mm proud of the joint surface
Siebert et al. [72]	20 sheep (10 animals per group)	Osteochondral defects in stifle joint	Autologous osteochondral transplants under the influence of bFGF 3 and 6 months
Von Rechenberg et al. [73]	40 adult Swiss, female alpine sheep, 3–4 years of age (4 to 14 animals per group)	Cartilage 6 mm deep defects in both stifle joints using the medial and the lateral femoral condyle of each joint	4.4 mm fresh, untreated auto- and 5.3 mm Ø × 6 mm deep xeno- and photooxidized osteochondral allo- and xenografts 2, 6, 12 and 18 months
Guo et al. [74]	Sheep	Full-thickness cartilage 8 mm Ø × 4 mm in depth defects in stifle joint	Autologous bone marrow MSCs and chondrocytes seeded onto β-TCP scaffold + autogenous periosteal grafts 12 and 24 weeks
Tibesku et al. [75]	4 male Holstein sheep, 87–102 kg	Cartilage 10-mm-lenght hole defect in the load-bearing areas of medial and lateral femoral condyle from one stifle joint	8.0 mm Ø × 10-mm-long osteochondral autografts obtained from one of the stifle joints 3 months
Dorotka et al. [76]	11 adult Austrian sheep (11 female, 1 male), ±80 kg	Subchondral 4.5 mm defects in the right medial femoral condyle of stifle joint (1 and 2.5 cm distal from the intercondylar notch)	Microfracture perforations 2 mm deep and a three-dimensional collagen matrix seeded with autologous chondrocytes 16 weeks

(continued)

Table 20.1 (continued)

Authors	Population	Type of defect and localization	Material tested and follow-up period
Hoemann et al. [77]	18 sheep (4 to 8 animals per group)	Full-thickness chondral 1 cm² defects in the stifle joint (one defect in the medial femoral condyle and the other defect in the trochlea)	Microfracture perforations + chitosan-glycerol phosphate/autologous whole blood 6 months
Russlies et al. [78]	15 adult German sheep, 1–2 years of age (5 animals per group)	Full-thickness osteochondral 7 mm to 9 mm Ø defects – one in the trochlea groove and the other in the medial femoral condyle of one of the stifle joints	Autologous chondrocyte transplantation + periosteum or in combination with a collagen I/III membrane cover 1 year
Tytherleigh-strong et al. [79]	12 adult sheep	Osteochondral defect on the medial femoral condyle	Intra-articular hyaluronan viscosupplementation after osteochondral grafting (mosaic arthroplasty) 12 weeks
Burks et al. [80]	12 sheep	Circular cartilage 10 mm defect in the medial left and right femoral condyle of stifle joint	6 mm circular osteochondral autograft plug + bone graft 6 and 12 months
Frosch et al. [81]	9 sheep	Osteochondral 7.25 mm Ø × 5-mm-deep defects in the medial trochlea of both femora in stifle joints	Round titanium implants with a 2 mm × 7.3 mm Ø were seeded with autologous bone marrow MSCs 6 months
Kandel et al. [82]	12 sheep (3 to 9 animals per group)	Full-thickness osteochondral 4 mm Ø × 9 mm deep defects in the trochlear groove of stifle joint	Biphasic cartilage-calcium polyphosphate constructs ("osteochondral"-type plug) 3, 4 and 9 months
Siebert et al. [83]	16 adolescent merino sheep, ≥50 kg (8 animals per group)	Osteochondral 6.3 mm Ø × 8–10 mm in depth in the weight-bearing surface of each femoral condyle in the stifle joints	Osteochondral autologous transplantations + PBS containing 50 µg bFGF or BMP-2 directly prior to implantation 6 months
Jones et al. [84]	21 sheep (7 animals per group)	Partial-thickness 6 mm Ø × 1.5 mm in depth defects on the trochlea and medial femoral condyle of the stifle joint	Autologous chondrocyte transplantation or type I/III collagen membrane 8, 10 and 12 week

(continued)

Table 20.1 (continued)

Authors	Population	Type of defect and localization	Material tested and follow-up period
Jubel et al. [85]	48 sheep (12 to 20 animals per group)	Chondral defects 4 mm Ø (1 per sheep) in the centre of one medial femoral condyle of stifle joint	Autologous de novo cartilage graft + periosteal flap 26 and 52 weeks
Schlichting et al. [86]	24 sheep	Critical osteochondral defects in the femoral condyles of the stifle joint	Subchondral bone reconstructed with a stiff scaffold or a modified softer one 3 and 6 months
Schagemann et al. [87]	12 sheep (4 animals per group)	Osteochondral femoral defects in the stifle joint (4 defects per animal)	Alginate-gelatin biopolymer hydrogel + autologous chondrocytes or periosteal cells 16 weeks
Streitparth et al. [88]	12 sheep (6 animals per group)	Osteochondral defects in condylar facets of femur in stifle joint	PGLA scaffold or a modified softer one (87% and 55% of the elastic modulus of ovine subchondral bone, respectively) 6 months
Wegener et al. [89]	12 skeletally mature M erino sheep, 18 months of age, ≥60 kg (6 animals per group)	Chondral 8 mm Ø defects in the weight-bearing area of femoral condyle of the stifle joint (2 defects per animal)	Microfracture perforations + PGLA implant 12 weeks
Gille et al. [90]	30 female common German sheep, 12–18 months of age, 46 ± 9.6 kg (6 animals per group)	Full-thickness 7 mm Ø chondral defects in the trochlea and medial condyle of both stifle joints	Two scaffolds in comparison (collagen I/III, Chondro-Gide®; collagen II, Chondrocell®) for covering microfractured defects, both scaffolds colonized in vitro with autologous chondrocytes or scaffold-free microfracture perforation technique 1 year
Kon et al. [91]	12 skeletally mature female Bergamasca-Massese sheep, 70 ± 5 kg, 3–4 years of age (4 animals per group)	Osteochondral 7 mm Ø × 9 mm in depth defects in the weight-bearing areas of the right medial and lateral femoral condyle of the stifle joint (2 defects per animal)	New multilayer gradient nanocomposite scaffold obtained by nucleating collagen fibrils with HA nanoparticles + in vitro cultured autologous chondrocytes 6 months
Kon et al. [92]	12 skeletally mature female Bergamasca-Massese sheep, 70 ± 5 kg, 3–4 years of age (4 animals per group)	Osteochondral 7 mm Ø × 9 mm in depth defects in the weight-bearing areas of the right medial and lateral femoral condyle of the stifle joint (2 defects per animal)	PRP + multilayer gradient, nanocomposite scaffold obtained by nucleating collagen fibrils with HA nanoparticles 6 months

(continued)

Table 20.1 (continued)

Authors	Population	Type of defect and localization	Material tested and follow-up period
Getgood et al. [93]	24 skeletally mature Welsh Mountain sheep, ±4.3 years of age (6 animals per group)	Osteochondral 6 mm Ø × 6 mm in depth defects created in the medial femoral condyle and the lateral trochlea sulcus of the stifle joint	Combination of PRP or concentrated bone marrow aspirate with a biphasic collagen/glycosaminoglycan or left untreated 6 months
Milano et al. [94]	30 sheep	Full-thickness chondral lesion on the medial femoral condyle of the stifle joints	Microfracture alone or associated with 5 weekly injections of autologous conditioned plasma 3, 6 and 12 months
Bell et al. [95]	5 skeletally mature Arcott cross female sheep from specific pathogen-free herds, 2–4 years of age, 65–93 kg	2.5 mm Ø Jamshidi biopsy holes created bilaterally in the medial femoral condyle of the stifle joints (6 defects/joint)	Three distinct molecular weight presolidified chitosan formulation-blood implant with fluorescent chitosan tracer (10 kDa, 40 kDa or 150 kDa chitosan) or untreated (left to bleed) (control left joint) 1 day, 3 weeks and 3 months
Bernstein et al. [96]	28 female merino sheep, 2–4 years of age, 74.54 ± 16.55 kg (7 animals per group)	Osteochondral 7-mm-wide × 25-mm-long hole drilled in the Centre of the load-bearing area of the medial femoral condyle of the right stifle joint (left joint served as control)	Cylindrical plugs of microporous β-TCP (7 mm Ø long × 25 mm long; porosity 43.5 ± 2.4%; pore Ø ~5 μm) with interconnecting pores + seeded and cultured in vitro with autologous chondrocytes for 4 weeks 6, 12, 26 and 52 weeks
Carneiro et al. [97]	10 healthy female sheep, 1–2 years of age, 30–50 kg	Osteochondral 8 mm Ø defects produced in the middle of the trochlear groove of both stifle joints without affecting the subchondral bone but only until the bleeding became evident	Autologous PRP of control defects in the left stifle joint left unfilled 12 weeks
Fan et al. [98]	Sheep	Cylindrical cartilage-subchondral bone defects created in the right femoral of the stifle joint	HA-β-TCP, PLGA-HA-β-TCP dual-layered composite scaffolds or autologous bone chips + autologous cartilage layer on top of the subchondral materials 3 months

(continued)

Table 20.1 (continued)

Authors	Population	Type of defect and localization	Material tested and follow-up period
Kunz et al. [99]	15 mature sheep (5 animals per group)	Traumatically inducted chondral 4.5 mm to 7 mm Ø defect anterior central weight-bearing region of the medial femoral condyles of both stifle joints performed through a 2 cm infrapatellar arthrotomy [302]	After 4 months, autologous osteochondral cartilage repaired using conventional approach, optically guided method or template-guided method 3 months after repair surgery
Martinez-Carranza et al. [100]	12 healthy female SwedishL andrace sheep, 5–7 years of age, 70–99 kg (6 animals per group)	For primary fixation, the implants have an HA-coated peg (10 mm long × 2 mm Ø) that was press-fitted into an undersized (1.8 mm Ø) drill hole in the bone adapted to the contour of the posterior weight-bearing of the medial femoral condylar surface of stifle joint	Monobloc implants (Episurf AB, Stockholm, Sweden): - 1st batch (outer 10 mm Ø, hemispherical radius of 17 mm) (spark plasma sintering technique with a gradient powder blend of 50% HA at the non-articulating surface and pure stainless steel -316 L towards the joint cavity) - 2nd batch (7.5 mm Ø), double-curved (radii 19 and 12 mm) computer-aided design/ computer-aided manufacturing modelled articulating surface, according to CT scans of a standard sheep knee (implant-grade Cr-co with a coating of plasma-sprayed HA – Plasma Biotal ltd. Buxton, GBR) 6 and 12 weeks
Mayr et al. [101]	28 sheep, 2–4 years of age, 74.54 ± 16.55 kg (4 animals per group)	Osteochondral 7 mm Ø × 25 mm in depth defects created in the medial femoral condyle of the left stifle joint; right stifle joint served as control	Microporous β-TCP cylinders seeded with autologous chondrocytes obtained from right patella and cultured for 4 weeks in vitro or left untreated 6, 12, 26 and 52 weeks
Schinhan et al. [102]	18 mature female Austrian Mountain sheep, 71 ± 11 kg	Unicompartmental OA induced in a stable joint by creating a critical size defect 7 mm Ø associated with a weight-bearing regime of 12 weeks (Schinhan et al. [131])	Spongialization alone, spongialization followed by implantation of a hyaluronan matrix with autologous chondrocytes or without chondrocytes 4 months after repair surgery

(continued)

Table 20.1 (continued)

Authors	Population	Type of defect and localization	Material tested and follow-up period
Schleicher et al. [103]	12 female merino sheep, 2 years of age, 41.8 ± 3.6 kg (6 animals per group)	Osteochondral 9.4 mm Ø × 1.1 cm in deep defects in the weight-bearing area of the medial femoral condyle of the right stifle joint	Scaffold A – 8- to 10-mm-deep layer of HA + collagen (ratio of 1:3); the 2 mm upper layer consisted of porcine collagen I/III (combination of the 2 layers by dripping 20 µL collagen onto the plane surface of the cylinder five times, each time followed by a drying period) Scaffold B – Allogenous bone derived from sheep coxae + an upper layer of porcine collagen I/III (equal combination protocol of the 2 layers + lyophilization + γ-ray sterilization) 6 weeks
Schleicher et al. [104]	12 female merino sheep, 3 years of age, 58.4 ± 9.3 kg (6 animals per group)	Osteochondral 9.4 mm Ø × 1.1 cm in deep defects in the weight-bearing area of the medial femoral condyle of the right stifle joint	Allogenous bone cylinders (bone tissue from sheep coxae + upper layer of porcine collagen by dripping 20 µL of collagen five times onto the surface of the bone cylinder, each time followed by a drying period + lyophilization + γ-ray sterilization) + cultured autologous chondrocytes dripped onto the surface 6 weeks, 3 and 6 months
Caminal et al. [105]	9 Ripollesa-Lacona breed sheep, 3 years of age	Chondral 7 mm Ø defect with complete removal of articular cartilage (without affecting the subchondral bone) in the medial femoral condyle of the stifle joint	Cylindrical PLGA scaffolds (7 mm Ø × 2 mm thick) either alone or seeded with $3.3 \times 10^6 \pm 0.4 \times 10^6$ autologous bone marrow MSCs fixed with fibrin glue (Tissucol duo®) 12-month follow-up study
Eldracher et al. [106]	13 adult sheep	Rectangular full-thickness chondral defect in the trochlea of the stifle joints	Microfracture treatment with six subchondral drillings of either 1.0 mm (reflective of the trabecular distance) or 1.8 mm Ø 6 months

(continued)

Table 20.1 (continued)

Authors	Population	Type of defect and localization	Material tested and follow-up period
Fonseca et al. [107]	6 healthy Ripollesa-Lacona breed ewes, 2 year of age	Cylindrical osteochondral 3.5 mm Ø × 5 mm in depth lesions in the caudal aspect of the medial femoral condyle of both stifle joints	To evaluate the feasibility of an arthroscopic approach for the implantation of biocompatible cell-free and cell-loaded (3 × 10^6 autologous chondrocytes) PLGA scaffolds (cylindrical shape, 4 mm Ø × 7 mm high) 11 and 19 weeks
Guillén-García et al. [108]	15 female Manchega sheep, 2–3 year of age	Full-thickness 1 cm^2 incision in the articular cartilage of the medial femoral condyle of the stifle joint using a scalpel	Different doses of 1 and 5 million of chondrocytes or 5 million MSCs deposited on top of the rough face of the membrane used as carrier of the cells sealed on top of the injury with Tissucol® and sutured to the adjacent cartilage (microfracture at the trochlea and normal cartilage from a nearby region used as references) 12 weeks
Martinez-Carranza et al. [109]	10 healthy female Swedish landrace sheep, ±4 years of age, ±74.4 kg (3 and 6 animals per group)	Implants (7.5 mm Ø with a double-curved – Radii 19 and 12 mm – Articulating surface) unilateral medial femoral condyle insertion in the weight-bearing surface of the stifle joint	Monobloc unipolar cobalt-chrome implant with a double coating (1st layer of commercially pure titanium 60 μm and on top of which a layer of HA 60 μm) 6 and 12 months
Filichi et al. [110]	22 adult Sarda ewes, ±5.5 years of age, ±45 kg	Osteochondral 6 mm Ø × 2-mm-deep defects in the medial femoral condyles of both stifle joints (right joint as control)	Male sheep embryonic stemlike cells (500,000 cells were embedded in 60 μL of fibrin glue engrafted into the cartilage defect) 1, 2, 6, 12 and 24 months
Power et al. [111]	80 mature female Welsh Mountain sheep, 3.9 years of age (16 animals per group)	Chondral 8 mm Ø defects in the medial femoral condyle of the right stifle joint associated with 7 evenly spaced 1.5 mm Ø × 3-mm-deep microfracture holes in each defect	Microfracture alone or microfracture + intra-articular rhFGF-18 (100 ng/ml) (administered either as 1 or 2 cycles of 3× weekly injections) 13 and 26 weeks

(continued)

Table 20.1 (continued)

Authors	Population	Type of defect and localization	Material tested and follow-up period
Sanz-Ramos et al. [112]	12 sheep	Focal 6 mm Ø lesions in the medial femoral condyle, without affecting the subchondral bone, on both stifle joints	Expanded ovine chondrocytes (3×10^6 of the cells in 100 μl PBS) on plastic or on 3 mg/ml collagen hydrogels under hypoxic conditions + periosteal flap obtained from the tibial epiphysis sutured to the surrounding cartilage + Tissucol® 3 months
Garcia et al. [113]	12 sheep (3 animals per group)	Full-thickness cartilage defect in the bearing region of the lateral femoral condyle of the stifle joint	To compare the potential for cartilage repair of fresh or cryopreserved amniotic membrane, cryopreserved amniotic membrane previously cultured with bone marrow MSCs and untreated defect as control 2 months
Gelse et al. [114]	10 female adult merino sheep, 3.4 years of age, 70–80 kg	Chondral 5 mm Ø defects in the femoral condyle of the stifle joints with preservation of the calcified cartilage (2 defects/medial condyle separated by 5 mm of intact cartilage)	Native cartilage autografts obtained from the cranial zone of the femoral trochlea with or without microfracture perforation technique 6 and 26 weeks
Hopper et al. [115]	24 skeletally mature female Welsh Mountain sheep, 3–5 year of age (6 animals per group)	Full-thickness osteochondral 6 mm Ø × 8 mm deep defects in the medial femoral condyle of the stifle joint	Biphasic collagen-glycosaminoglycan scaffold (ChondroMimetic®) with MSCs (1×10^6 cells) alone, MSCs + human peripheral blood mononuclear cells (2.0×10^5 cells) (ration 20:1), MSCs + human peripheral blood mononuclear cells (ratio 2:1) and human peripheral blood mononuclear cells alone 26 weeks

(continued)

Table 20.1 (continued)

Authors	Population	Type of defect and localization	Material tested and follow-up period
Howard et al. [116]	40 skeletally mature female Welsh Mountain sheep, ±3.9 years of age, 40–42 kg (5 animals per group)	Full-thickness chondral 8 mm Ø defects in the medial femoral condyle (10 mm distal to the condyle groove junction and aligned with the medial crest of the trochlear groove) of the right stifle joint	Microfracture holes (1.5 mm Ø × 3 mm deep) alone, microfracture + intra-articular rhFGF-18 (30 µg once a week for 3 weeks at 4th, 5th and 6th post-operative week and 16th, 17th and 18th post-operative week) or microfracture + rhFGF-18 (0.064, 0.64, 6.4 or 32 mg/total dose) delivered on a bilayer collagen membrane 13 and 26 weeks
Mohan et al. [117]	18 skeletally mature female Rambouillet × Columbia cross sheep, >3.5 years of age	Osteochondral 6 mm Ø × 6 mm in depth defects in both the medial femoral condyles and lateral femoral condyles of the right stifle joints	To compare between microfracture and an osteochondral approach with microsphere-based gradient plugs of PLGA-based scaffolds with opposing gradients of chondroitin sulphate and β-TCP – Chondrogenic microspheres (PLGA-CS-NaHCO$_3$; 77.5:20:2.5) and osteogenic microspheres (PLGA-β-TCP; 90:10) infiltrated with TGF-β3 or IGF-1 12 months
Zorzi et al. [118]	15 Dorper sheep, 3–5 years of age, 58.02 ± 11.38 kg	Partial-thickness chondral 10 mm Ø defect in the medial femoral condyle of both stifle joints (avoiding any bleeding of the base of the lesion)	Chitosan and type I collagen 1:1 (w/w) scaffold seeded with human ASCs (1 × 10^6 cells), scaffold without cells fixed with 1 ml of fibrin glue (Tissucol®) or untreated lesions as control 6 months
Caminal et al. [119]	8 healthy ewes Ripollesa-Lacona breed sheep, 3 years of age	Cylindrical osteochondral 3.5 mm Ø × 5 mm in depth defects in the caudal aspect of the medial and lateral femoral condyles of both stifle joints	Biocompatible porous PLGA scaffolds non-seeded and seeded with fresh ex vivo expanded autologous progenitor cells derived from different cell sources (cartilage, fat and bone marrow) 6 and 12 months

(continued)

Table 20.1 (continued)

Authors	Population	Type of defect and localization	Material tested and follow-up period
De Barros et al. [120]	8 healthy Texel mix-breed sheep, 18–36 months of age, ±30 kg	Circular chondral 3 mm Ø on the articular surface of the medial femoral trochlea of the stifle joint	Heterologous fibrin sealant developed from the venom of *Crotalus durissus terrificus* (thrombin-like compound derived from the venom of the snake cryoprecipitated in animal fibrinogen and a calcium chloride solution) 7 and 15 days
Hindle et al. [121]	Sheep	Articular cartilage defects on the medial femoral condyle of the stifle joint	To develop an autologous large animal model for perivascular stem cell transplantation and to specifically determine if these implanted cells are retained in articular cartilage defects using a hydrogel as vehicle and collagen membranes 4 weeks
Kitamura et al. [122]	17 skeletally mature female Suffolk sheep, 4–6 years of age, 65–85 kg (5 animals per group)	Osteochondral 6.0 mm defect in the femoral trochlea and the medial femoral condyle of both stifle joints	Poly-(2-acrylamido-2-methylpropanesulfonic acid)/poly-(N,N′-dimethyl acrylamide) DN gel; cylindrical DN gel plug was implanted into the defect of the right joint so that a vacant space of the planned depths of 2.0 mm, 3.0 mm and 4.0 mm was left (left joint with defect with the same depth as the right joint as control) 12 weeks
Mrosek et al. [123]	24 skeletally mature, castrated male sheep (8 animals per group)	Osteochondral 8 mm Ø × 13-mm-deep defect on the medial aspect of the medial femoral condyle in the main weight-bearing area of left stifle joint	Trabecular metal (porous tantalum biocomposites) + periosteal graft, trabecular metal alone or empty defect 16 weeks
Schagemann et al. [124]	12 female sheep, 3–6 years of age, 62–84 kg (6 animals per group)	Osteochondral 6 mm width × 20 mm long × 5 mm in depth defects on the weight-bearing area of the medial femoral condyles of the left stifle joint	To compare the regenerative capacity of two distinct bilayer implants, a novel biomimetic PCL implant or a combination (synthetic implant) of Chondro-Gide and Orthoss (biologic implant) 19 months

(continued)

Table 20.1 (continued)

Authors	Population	Type of defect and localization	Material tested and follow-up period
Yucekul et al. [125]	13 merino sheep (3 control and 5 animals per group)	Osteochondral 8 mm Ø × 10 mm in depth defects on the lateral femoral condyles of the right stifle joint	Three layered cylinder scaffold is composed of a nonwoven PGA (top layer), PLA mixed with a colourant (middle layer) and porous structure composed of a blend of PLA and PCL (bottom layer), coated with type I collagen and β-TCP, hyaluronic acid gel or defects left untreated 6 months

ASCs adipose stem cells, *β-TCP* beta-tricalcium phosphate, *BMP* bone morphogenetic protein, *DBM* demineralized bone matrix, *ECM* extracellular matrix, *EGF* epidermal growth factor, *FGF* fibroblastic growth factor, *GH* growth hormone, *HA* hydroxyapatite, *IGF-I* insulin-like growth factor-I, *MRI magnetic resonance imaging*, *MSCs* mesenchymal stem cells, *OA* osteoarthritis, *PCL* poly(epsilon-caprolactone), *PGA* polyglycolic acid, *PLA* polylactic acid, *PLGA* poly(D,L)lactide-co-glycolide acid, *PRP* platelet-rich plasma, *PMMA* polymethyl methacrylate, *rh* recombinant human, *TGFβ* transforming growth factor beta

of autologous chondrogenic-induced bone marrow mesenchymal stem cells (MSCs) 6 weeks after OA induction (5 mL standard culture medium FD at density of 2 million cells/ml) and euthanasia of the animals 6 weeks after the treatment implementation. Ude et al. [128] developed a similar animal model with the intra-articular administration of a single dose of 2 × 107 autologous PKH26-labelled, chondrogenically induced ASCs and bone marrow MSCs (5 mL injection), while control animals received 5 ml culture medium, also with animal euthanasia 6 weeks after injection. Song et al. [129] applied a concentrated of uncultured autologous bone marrow mononuclear cells or a single dose of ten million autologous bone MSCs suspended in phosphate-buffered saline delivered via direct intra-articular injection to the injured joint, in the 12th week after OA induction or vehicle alone in the control group allowing a period of 8 weeks after injection. In the study of Desando et al. [130], also an autologous bone marrow concentrate and bone marrow MSCs seeded onto hydroxyapatite (HA) (Hyaff®-11) and a 12 weeks period was allowed after the intra-articular injection. In 2012, Schinhan and co-authors [131] induced chondral 7 mm or 14 mm diameter defects, created in the weight-bearing area of the medial femoral condyle of the stifle joint in sheep, to define the CSD of a chondral lesion to induce unicompartmental OA in a stable joint, assessed at the 6th and 12th postoperative week.

Taking into consideration the in vivo chondral/osteochondral tissue engineering studies performed in sheep with induction of surgical defects, it's possible to conclude by the observation of Table 20.1 that in the first studies performed in this scientific field, xeno-, allo- and fresh autografts were applied alone for mosaic arthroplasty [69–71, 73, 75]. In 2003, Siebert and co-authors [72] associated one type of growth factor (GF) to an autologous osteochondral transplant, and in 2006

[83], this research team performed another study associating two different GFs, and Guo et al. [74] presented the first study involving cell therapy by using autologous bone marrow MSCs and chondrocytes seeded onto a scaffold also associated with an autogenous periosteal graft. Dorotka et al. [76] and Hoemann et al. [77] refer for the first time in sheep the use of the microfracture perforation technique associated with a scaffold and autologous cell therapy, being this technique also applied later in association with scaffolds [89], autologous plasma [94] and GFs [111]. In 2005, Russlies and co-authors [78] introduced the use of a collagen membrane cover after autologous chondrocyte transplantation. Frosch et al. [81] refer the use of titanium implants associated with cell therapy and Kandel et al. [82] the use of an "osteochondral"-type plug biphasic cartilage-calcium polyphosphate construct. In 2009, Schagemann and co-authors [87] applied a biopolymer hydrogel associated with cell therapy for osteochondral repair in sheep. Also in 2009, the studies of Streitparth et al. [88] and Wegener et al. [89] used poly(D,L)lactide-co-glycolide acid (PLGA) scaffold alone or associated with the microfracture perforation technique, respectively. Since 2010 until the present moment, numerous studies have been performed testing new tissue engineering approaches to repair osteochondral defects in sheep, using different scaffolds [type I/III collagen, nanocomposite scaffold obtained by nucleating collagen fibrils, HA nanoparticles, hyaluronan matrix, β-tricalcium phosphate (β-TCP), collagen hydrogels, PLGA] colonized with autologous chondrocyte [74, 90–92, 96, 101, 102, 107, 112, 119]. Different cell sources have been tested alone or seeded onto different scaffolds – autologous bone marrow MSCs [74, 105, 108, 119], auto- or allogenic ASCs [118, 119], embryonic stemlike cells [110], allogenic peripheral blood mononuclear cells [115] or perivascular stem cells [121] and also platelet-rich plasma (PRP) that has been used in some studies [22, 93, 97]. More recently, bilayer scaffolds to better mimic the dual chondral and bone tissue components have been used [98, 103, 104, 115, 124], and even a third layer cylinder scaffold has been proposed for osteochondral tissue engineering [125]. Likewise, metal implants such as titanium [81], pure stainless steel [100], cobalt-chrome [109] and tantalum biocomposites [123] have been tested in sheep.

There are also some published studies in sheep concerning the potential occurrence of subchondral bone cyst formation and subchondral bone pathology after the implementation of microfracture perforation technique for repair of articular cartilage defects in stifle joints, namely, the study of Beck et al. [132] where this technique was associated with type I/III collagen scaffold with a post-operative period of 13 and 26 weeks for assessment and control of these complications and also the study of Vikingsson et al. [133] where biodegradable polycaprolactone (PCL) scaffold with double porosity or PCL scaffold attached to a poly-L-lactide acid (PLA) pin anchored to the subchondral bone was used to reduce nodule formation and other tissue alterations after a 4.5 month post-operative period.

20.2.2   Goat

Goat has housing and feeding requirements and also a social behaviour similar to sheep being a species relatively robust and easy to handle. In spite of this, goats are more aggressive and curious than sheep, and they tend to demonstrate dominance within their social group more than sheep. Herds are led by a dominant female and a dominant male. The skeletal maturity is similar to sheep and is achieved around 2–3 years of age [20].

The goat model is very frequently used for cartilage articular defect regeneration and repair research since the stifle joint has an anatomy similar to human. It allows the arthroscopic examination; it has a more thick cartilage than sheep and also limited capacity to heal without iatrogenic intervention like in humans [22, 134]. The thickness of the medial condyle cartilage in goat has been reported by Brehm et al. [134] ranging between 0.8 mm and 2 mm and by Frisbie et al. [46] with a mean value of 1.1 mm resulting in wide variations of the chondral and subchondral bone volume involved in different studies. Goat cartilage articular thickness tolerates the creation of chondral defects more easily, without affecting the osteochondral bone plate, similar to the majority of the clinical cartilage lesions present in human [20, 41]. Also, the goat subchondral bone is softer than in sheep, which facilitates the creation of osteochondral bone defects [20]. Hence, caprine cartilage defects with 150 mm² could be created through 12 mm diameter defects (lower range of common human cartilage lesions), consequently closer to human clinical trials [20, 22].

The large majority of the cartilage defect studies in the goat are performed in the stifle joint, involving lateral or medial femoral condyles and/or femoral trochlea, generally with creation of osteochondral defects and a follow-up post-operative period that varies between 2 weeks and 1 year (Table 20.2). As in sheep, the type of defects commonly performed in goats is also the surgically created 4 mm to 15 mm in diameter chondral/osteochondral and impact injury, and its primary use is in pivotal studies for which post-operative management variables are not critical since protection of load bearing of the operated limbs and exercise protocols are almost impossible to perform in this animal model [22, 56]. The CSD for cartilage defects in goats is reported as being 6 mm, which does not heal after 6 months [162, 163]. Also Simon and Aberman [164] described two untreated lesion models useful for testing articular cartilage repair strategies during 1 year, one 6 mm deep (through the subchondral plate) and one shallow (to the level of the subchondral bone plate) in the middle one-third of the medial femoral condyle of the stifle joint.

Similar to sheep, the goat model has also been used to study cell therapy treatment after induced OA by meniscectomy associated or not with ACL resection. Namely, in the study of Murphy et al. [165], a single dose of ten million of bone marrow MSCs suspended in a dilute solution of sodium hyaluronan was delivered to the injured stifle joint by direct intra-articular injection in the 6th week after OA induction and a 6- and 20-week post-operative period before result evaluation. In another recent study, a single intra-articular dose of 7×10^6 human ACSs in the 9th week after the meniscectomy associated with daily injections of cyclosporin A (10 mg/kg for 7 days followed by a reduced dose of 5 mg/kg for another 7 days) was applied and evaluated 8 weeks after injection of human ASCs [166].

Table 20.2 Cartilage and osteochondral tissue engineering studies performed in caprine model for cartilage repair

Authors	Population	Type of defect and localization	Material tested and follow-up period
Shahgaldi et al. [135]	19 mature goats (4 to 6 animals per group)	5.0 mm Ø × 4–6 mm long holes in the femoral condyle, patella of left stifle joint and right stifle joint intact as controls	Glutaraldehyde-fixed bovine meniscal xenograft, glutaraldehyde-fixed bovine costal cartilage xenograft and viable osteochondral allografts 12 months
Butnariu-Ephrat et al. [136]	Goats	Round 4.0 mm Ø hole in the femoral condyle of the stifle joint	Autogenic and allogeneic bone marrow MSCs implantation
Jackson et al. [137]	24 mature female Spanish goats (6 animals per group)	Medial tibial plateau Wedged defect	Fresh articular cartilage allografts 3, 17, 26 and 52 weeks
Van Susante et al. [138]	32 skeletally mature Dutch milk goats (*Capra hircus sana*), 6 years of age, 48.8–79.3 kg	Full-thickness 10 mm Ø × 4 mm deep cartilage defect in the weight-bearing area of the medial femoral condyle of the stifle joints	Xenografted with isolated rabbit chondrocytes + fibrin glue 3, 8, 13, 26 and 52 weeks
Louwerse et al. [139]	17 skeletal mature female Dutch milk goats, 2 years of age, ±50 kg	Full-thickness subchondral defect ˜0.9 cm Ø in the anterior weight-bearing part of the medial condyle	rhBMP-7 combined with collagen (400 µg/mL, with small particles of autologous ear perichondrium) or not (200 µg/mL BMP-7 peptide) 1, 2 and 4 months
Niederauer et al. [140]	16 Spanish goats	Osteochondral 3 mm Ø × 4 mm in depth defects bilaterally in the medial femoral condyle (high weight bearing) and the distal medial portion of the patellar groove (low weight bearing)	PLGA scaffolds + 45S5 bioglass® and medical grade calcium sulphate as additives to vary stiffness and chemical properties 16 weeks
Lane et al. [141]	8 adult goats	Osteochondral defects in the weight-bearing portion of the medial femoral condyle of the stifle joint	Autogenous plugs transferred from the femoral trochlea to defects in the weight-bearing portion of the medial femoral condyle 6 months

(continued)

Table 20.2 (continued)

Authors	Population	Type of defect and localization	Material tested and follow-up period
Quintavalla et al. [142]	Female Nubian cross goats, >2 years of age	Cylindrical full-thickness osteochondral 4.5 mm Ø × 4.0 mm deep defects on the proximal 1/3 of the medial facet trochlear groove, bordering the axial groove, and on the anterior 1/3 of the central medial condyle	Bone marrow MSCs + gelatin constructs 1, 2, 7 and 14 days
Welch et al. [143]	23 skeletally mature Spanish cross-bred goats, 35–42 kg (3 to 5 animals per group)	Cartilage 12 mm Ø × 15 mm in depth subchondral defects bilaterally in the medial femoral condyles of the stifle joint	Bioresorbable HA (BoneSource®), PMMA (PMMA simplex®) or autogenous bone graft 24 h, 6 weeks, 6 months, 1 and 2 years
Dell'Accio et al. [144]	6 female Saanen goats	Cartilage 6 mm Ø × 0.8 mm deep joint surface defect in the lateral femoral condyle (preserving the mineralized cartilage and subchondral bone)	Autologous chondrocyte implantation + periosteal flap 3, 10 and 14 weeks
Lane et al. [145]	8 adult goats	Full-thickness articular cartilage defects in the weight-bearing portion of the medial femoral condyle of the stifle joint	Autologous osteochondral plugs from the femoral trochlea 6 months
Vasara et al. [146]	14 goats (7 males, 7 females), 2–4 years of age (4 to 5 animals per group)	Circular cartilage 6 mm Ø defects in the upper part of the medial femoral condyle of the stifle joint	Autologous chondrocyte transplantation with chondrocytes implanted under a periosteal flap or a bioabsorbable film and mesh scaffold (copolymer of PLA and PCL combined with a mesh of poly-L/D-lactide 96/4) 3 and 6 months
Brehm et al. [134]	9 female Saanen goats, 3 years of age, 60 kg (2 to 3 animals per group)	Superficial osteochondral 6 mm Ø × 0.8 mm in depth defects in the trochlea femoral of the left stifle joint	Autologous chondrocytes, scaffold-free – Semipermeable membranes, engineered cartilage constructs secured in the defect using a covering periosteal flap alone or in combination with adhesives – PRP or fibrin glue (Tissucol®) or using PRP alone 8 weeks

(continued)

Table 20.2 (continued)

Authors	Population	Type of defect and localization	Material tested and follow-up period
Kirker-head et al. [147]	6 healthy castrated male Nubian goats, 3 years of age, 55–73 kg	28 mm long bone channel and introduction of HemiCAP™ seated subchondrally with its head positioned 5 mm below the osteochondral junction	20 mm long recessed cannulated titanium alloy anchoring screw and a co-Cr alloy resurfacing prosthesis (HemiCAP™) 26 and 52 weeks
Hunziker and Stähli [148]	16 adult Saanen goats	One purely chondral 5 mm width × 10 mm long (in the sagittal plane) defect within the lateral facet of the femoral patellar groove in the stifle joint	Bovine fibrinogen to support a devitalized flap of autologous synovial tissue, sutured to the surrounding articular cartilage with single, interrupted stitches 2–3 h and 3 weeks
Lind and Larsen [149]	8 adult goats	Full-thickness osteochondral 6 mm Ø defects in the femoral condyles of the stifle joint (2 defects animal)	Collagen membrane scaffold + autologous chondrocytes or minced cartilage placed under collagen membrane scaffold 4 months
Lind et al. [150]	20 adult goats	Full-thickness circular osteochondral 6 mm Ø defects in the medial femoral condyles of the stifle joint (2 defects per animal)	Autologous chondrocytes + fibrin hydrogel or autologous chondrocytes + MPEG-PLGA scaffold + fibrin hydrogel 4 months
Custers et al. [151]	9 Dutch milk goats	Cartilage defects in the medial femoral condyle in both stifle joints	Microfracture or placement of an oxidized zirconium implant treatment 10 weeks after osteochondral defect creation 26 weeks
Lane et al. [152]	8 adult goats	Full-thickness 8 mm Ø chondral defect in the medial femoral condyle of stifle joint	Osteochondral autograft plug transfers 4.5 mm Ø × 10–12 mm long (mosaic arthroplasty obtained from the lateral aspect of the femoral trochlea) + perforations treatment 6 months
Miot et al. [153]	27 adult female Saanen goats, >18 months age (10 and 17 animals per group)	Osteochondral 6 mm Ø × 5 mm deep defects in each of the three biopsy sites prior to performing the autologous grafting – proximomedial, distal and lateral locations of the trochlea groove in the left stifle joint	Autologous chondrocytes cultured in hyaluronic acid scaffolds and implanted above HA/hyaluronic acid sponges 8 weeks and 8 months

(continued)

Table 20.2 (continued)

Authors	Population	Type of defect and localization	Material tested and follow-up period
Bekkers et al. [154]	9 adult Dutch milk goats, 3–5 years of age, 75 ± 10 kg	Full-thickness chondral 5 mm cylindrical defects in the central weight-bearing region of both medial femoral condyles	Chondrocytes + bone marrow mononuclear fraction cells embedded in fibrin glue (Beriplast®, Nycomed) or microfracture treatment (4 holes through the subchondral bone with a 1.5 mm K-wire) 6 months
Kon et al. [155]	14 skeletally mature Saanen goats, 55 ± 11 kg	Osteochondral 6 mm Ø × 8 mm in depth defects in the weight-bearing area of the medial and lateral femoral condyles of stifle joint	Biphasic osteochondral coralline aragonite scaffolds with different modifications 6 months
Pei et al. [156]	12 goats (4 animals per group)	Osteochondral 6 mm Ø × 12 mm in depth defect created in the femoral medial condyle weight-bearing areas of both stifle joints	Tissue-engineered osteochondral graft obtained in a double-chamber stirring bioreactor to construct bone and cartilage composites simultaneously of autologous bone marrow MSCs-β-TCP scaffold with or without mechanical stimulation of stir and untreated defect as control 12 and 24 weeks
Geraghty et al. [157]	9 skeletally mature female Spanish goats, 2–4 years of age (3 and 6 animals per group)	Cartilage 6 mm Ø defects created in the medial femoral condyle of the right anterior stifle joint (2 defects/joint)	Marrow stimulation alone or treatment with marrow stimulation augmented with cryopreserved, viable osteochondral allograft which retains viable chondrocytes, chondrogenic growth factors and ECM proteins within the intact architecture of native hyaline cartilage 3 and 12 months
Kon et al. [158]	20 skeletally mature female Saanen goats, >2 years of age, 48 ± 6 kg (6 and 14 animals per group)	Osteochondral 6 mm Ø × 10 mm in depth defects created in the load-bearing medial femoral condyle of the right stifle joint	Aragonite-hyaluronate biphasic scaffold implant or untreated blood clot-filled defects as control 6 and 12 months
Mumme et al. [159]	Goats	Chondral 6 mm Ø defects in the stifle joint	Autologous nasal chondrocytes and articular chondrocytes (control) seeded on a type I/III collagen membrane 3 and 6 months

(continued)

Table 20.2 (continued)

Authors	Population	Type of defect and localization	Material tested and follow-up period
Sun et al. [160]	18 masculine goats, ±22.5 kg	Cylindrical 9 mm Ø × 3 mm in depth defects created in the weight-bearing area of the medial femoral condyle in both stifle joints	Different proportion of human TGF-β1 gene-transduced autologous bone marrow MSCs in sodium alginate (density of 5×10^7cells mL^{-1}) and 102 mmol/L $CaCl_2$ to form calcium alginate gels + mosaic arthroplasty (2 to 4 cylindrical osteochondral 5.5 mm Ø × 3 mm Ø grafts obtained from the medial femoral condyle periphery of the patellofemoral joint) 24 weeks
Wang et al. [161]	6 healthy adult goats (1 goat, sham control group; 5 goats, experimental group)	Defects were created bilaterally on the condylar cartilage and bone to induce temporomandibular joint osteoarthritis accordingly to the method of Ishimaru and Goss [303]	Middle fibrin-rich gel with aggregated platelets and concentrated growth factor applied as a membrane covering the right condylar surface and sutured to the joint capsule (physiologic saline was applied to the left joints as control) 1 month

ASCs adipose stem cells, *β-TCP* beta-tricalcium phosphate, *BMP* bone morphogenetic protein, *DBM* demineralized bone matrix, *ECM* extracellular matrix, *EGF* epidermal growth factor, *FGF* fibroblastic growth factor, *GH* growth hormone, *HA* hydroxyapatite, *IGF-I* insulin-like growth factor-I, *MRI magnetic resonance imaging, MSCs* mesenchymal stem cells, *OA* osteoarthritis, *PCL* poly(epsilon-caprolactone), *PGA*, polyglycolic acid, *PLA* polylactic acid, *PLGA* poly(D,L)lactide-co-glycolide acid, *PRP* platelet-rich plasma, *PMMA* polymethyl methacrylate, *rh* recombinant human, *TGFβ* transforming growth factor beta

The majority of the first in vivo chondral/osteochondral tissue engineering studies performed in goat with induction of surgical defects (Table 20.2) used xeno-, allo- and fresh autografts for mosaic arthroplasty [135, 137, 138, 141, 145] or associated with microfracture perforation technique [152]. After this phase, autologous chondrocytes started to be applied covered by periosteal flaps [144] and also associated with bioabsorbable membranes and/or seeded in scaffolds [134, 146, 149, 153, 159]. Bone marrow MSCs were used alone, associated with gelatin constructs or seeded in β-TCP scaffolds [136, 142, 156]. Growth factors were also applied associated with collagen [139] and recently to fibrin-rich gel with aggregated platelets as a covering defect membrane [161]. Osteoconductive scaffolds alone were also used [140, 143] and biphasic scaffolds to simulate the chondral and the bone tissue components [155, 158]. In a recent study of Sun et al. [160], gene therapy was used by the application of human TGF-β1 gene-transduced autologous bone marrow MSCs, in sodium alginate and $CaCl_2$ to form calcium alginate gels, associated with mosaic arthroplasty. Kangarlu and Gahunia [167] evaluated the capability of MRI (1.5 T and 8 T) to visualize, characterize and qualitatively compare osteochondral defect repair in the stifle joint of goats in the 8th and 16th post-operative weeks. Also, Watanabe et al. [168] used delayed gadolinium-enhanced MRI of cartilage and T2 mapping to evaluate the quality of osteochondral repair tissue after microfracture treatment in the 24th and 48th post-operative weeks.

20.2.3 Pig

The pig is more rarely used as an animal model in articular cartilage repair and regeneration studies due to its large size and high weight, handling difficulties and particular housing and facility requirements [20]. Due to these facts, mini-pig breeds (Göttingen and Yucatan mini-pig, Lee-Sung miniature pig) are usually preferred, being more docile, with an adult weight generally equivalent to the adult human, but also expensive to acquire and to maintain [22]. Pigs are curious, cautiously friendly and social animals that under free-ranging conditions live in social groups but when in confinement or stress demonstrate behavioural problems like aggression towards other pigs or tail biting [169]. The skeletal maturity of pigs is generally reached around 18 months of age [170], but the US Food and Drug Administration admits the age of 42–52 weeks for skeletal maturity in mini-pigs, although mini-pigs, such as the Göttingen breed, reach skeletal maturity between 18 and 22 months of age [22]. This breed demonstrates similarities to humans in physiological and serum biochemical parameters [171], peripheral bone apposition rate and trabecular thickness [22], and the collagen fibre arrangement in chondral tissue is close to those in humans [172]. The swine species also allows arthroscopic examination [173, 174] yet not very suitable to studies that require weight-bearing and exercise protection [22].

There are other advantages in using pig model for partial- or full-thickness chondral defect [175] or osteochondral defect [171] studies, since their cartilage thickness is larger than in other common animal models with approximately the same size and weight [20], and skeletally mature pigs also have intrinsic limited capacity for repair of chondral and osteochondral lesions [22]. Hembry et al. [176] report a cartilage thickness of 2.0 mm, and Chiang et al. [177] and Frisbie et al. [46] report a value of 1.5 mm at the medial femoral condyle level. Archibald et al. [68] reported an average articular cartilage thickness in the stifle joint of 3.2 mm. In spite of this large cartilage thickness in pigs relatively to other common animal models (sheep and goat), the total volume of cartilage defects has been smaller, despite generally including a large percentage of chondral tissue, and consequently closer to that of human clinical lesion [20]. Shimomura et al. [178], considering cartilage thickness, affirm that the pig is the species that closer mimics the human joint and thus should be used to evaluate implantation of tissue-engineered constructs.

The large majority of the cartilage defect studies in the mini-pig are performed in the stifle joint, involving only the medial or both femoral condyles, or in the femoral trochlea, with creation of chondral or osteochondral defects, with 6 mm to 8 mm of diameter or larger dimensions [171] and with a follow-up post-operative period which varies between 2 weeks and 1 year (Table 20.3). The CSD reported for cartilage defects in adult pigs is of 6 mm [185]. Gotterbarm et al. [171] showed that

Table 20.3 Cartilage and osteochondral tissue engineering studies performed in porcine model for cartilage repair

Authors	Population	Type of defect and localization	Material tested and follow-up period
Hunziker and Rosenberg [179]	Yucatan adult mini-pigs	Partial-thickness defects evolving in mature articular cartilage	Chondroitinase ABC or trypsin, fibrin clots and mitogenic growth factors (bFGF, TGF-β1, EGF, IGF-1 or GH) 1 to 48 weeks
Mainil-Varlet et al. [180]	22 skeletally mature Yucatan mini-pigs, 26–60 kg (8 animals per group)	Chondral 4.0 mm Ø, superficial osteochondral and full-thickness defects in the stifle joint cartilage	Tissue-engineered cartilage-like implant, derived from chondrocytes cultured in a novel patented, scaffold-free bioreactor system 4 and 24 weeks
Gal et al. [181]	10 piglets, 3 months of age	4 train drilled through the distal femoral physis in the area of lateral condyle Identical control defect in the medial condyle	Autogenous chondrocytes graft 10 weeks
Liu et al. [182]	Mini-pigs	Large full-thickness articular cartilage defects of stifle joints	Isolated autologous chondrocytes + PGA + Pluronic 24 weeks
Chiang et al. [177]	12 Lee-Sung mini-pigs (5 males, 7 females), 4 ± 1 months of age, 11 ± 3 kg	Chondral full-thickness 7 mm Ø defects in lateral and femoral condyles of both stifle joints	Autologous chondrocyte transplantation + periosteal patch 6 months
Jung et al. [183]	18 adult Göttingen mini-pigs, ±30 months of age, ±38 kg	Osteochondral 6.3 mm Ø × 10 mm in depth defects in the medial patellar groove of both stifle joints	Transferring periosteum – Bone cylinders (obtained from the medial aspect of tibia) + growth factor mixture (BMP-2, 3, 4, 6, 7, 12 and 13, TGF-β3 and acid-FGF-I) 6, 12 and 52 weeks
Chang et al. [184]	15 sexually mature Lee-Sung miniature pigs (8 male and 7 female)	Full-thickness 8 mm Ø × 2 mm deep articular defect and 5 mm deep osteochondral defect in the medial or lateral femoral condyles	Allogenous chondrocytes and gelatin-chondroitin-hyaluronan tri-copolymer scaffold covered with periosteum 18 and 36 weeks
Harman et al. [185]	18 pigs (6 animals per group)	Osteochondral defect created in the weight-bearing region of the medial femoral condyle Control in the contralateral medial femoral condyle	Osteochondral autografts 6 weeks, 3, 6 and 12 months

(continued)

Table 20.3 (continued)

Authors	Population	Type of defect and localization	Material tested and follow-up period
Vasara et al. [175]	57 skeletally immature pigs, 90–100 kg, 8–9 months of age (6 to 14 animals per group)	Full-thickness 6 mm Ø chondral defect in the upper part of the lateral facet of the patellar groove of the femur	Autologous chondrocyte transplantation covered with periosteum or muscle fascia 3 and 12 months
Ando et al. [186]	9 female pigs, 4 months of age	Chondral 8.5 mm Ø × 2.0 mm in depth defects on the medial femoral condyles of the both stifle joints	Porcine synovial tissue MSCs in 6 cm Ø culture dishes (21.3 cm²) for 7 days and prepared as an allograft without any chondrogenic stimulation 6 months
Filová et al. [187]	9 miniature pigs	Two defects located in the non-weight-bearing zone of the femoral trochlea of the stifle joint	Novel scaffold containing sodium hyaluronate, type I collagen and fibrin + autologous chondrocyte 12 weeks and 6 months
Jiang et al. [188]	10 Lee-Sung mini-pigs (4 male, 6 female), 6–7 months of age	Circular osteochondral 8 mm Ø × 8 mm in depth defects at the weight-bearing surface of the two femoral condyles of the right stifle joint	Autologous chondrocytes implantation at low cell seeding density on a biphasic scaffold cylindrical porous plug of DL-PLGA, with its lower body impregnated with β-TCP as the osseous phase 6 months
Baumbach et al. [189]	12 adult Göttingen mini-pigs (3 males, 9 females), ±30 kg	Osteochondral 4.5 mm Ø × 8 mm in depth defects in the medial femoral condyles of both knees	Autologous osteochondral transplantation (press-fit placement of osteochondral grafts 4.6 mm Ø × 10 mm in depth harvested from the patellar groove) 2, 8, 26 and 52 weeks
Gotterbarm et al. [171]	90 Göttingen mini-pigs (62 male, 28 female), ±24 months of age, ±38 kg	Circular full-thickness 46 chondral defects (6.3 mm Ø) and 134 osteochondral defects (5.4 mm or 6.3 mm Ø × 8 mm or 10 mm deep) in the medial facet of the trochlear groove of the stifle joints during various experiments	To evaluate the healing of osteochondral or full-thickness cartilage defects with different treatment options 12 months

(continued)

Table 20.3 (continued)

Authors	Population	Type of defect and localization	Material tested and follow-up period
Petersen et al. [190]	8 Göttingen mini-pigs, ±27 months of age, ±40 kg	Osteochondral full-thickness 4.5 mm Ø × 3.0 ± 0.5 mm deep defects in the weight-bearing zone of both medial femoral condyles of the stifle joints	Autologous chondrocytes + calcium phosphate cylinders carriers (Calcibon®) 26 and 52 weeks
Blanke et al. [191]	5 female adult mini-pigs, 18 months of age, 35–40 kg	Round chondral 5 mm Ø defects in the central part of the femoral trochlea of stifle joint; lesions were clearly separated from each other by at least 3 mm of intact cartilage; subchondral bone plate provisionally intact	Microfracturing perforation + collagen membrane or treated by matrix-associated autologous chondrocyte transplantation (in control lesions, the subchondral bone plate was left intact – Partial-thickness lesion) 12 weeks
Jung et al. [192]	8 mature female Göttingen mini-pigs, ±19 months of age, ±32.5 kg	Chondral 5.4 mm Ø defects in both medial patellar trochlear groove of the stifle joints without penetrating the subchondral bone plate	Autologous bone marrow MSCs + covered with sutured collagen type I/III membrane (Chondro-Gide®) 8 weeks
Li et al. [193]	4 Lee-Sung mini-pigs (2 male, 2 female), 10 months of age, 21–23 kg	Circular full-thickness chondral 7 mm Ø defects in the centre of distal weight-bearing surface of each femoral condyle of both stifle joints	Biodegradable PCL nanofibrous scaffold + allogeneic chondrocytes or xenogenic human MSCs or acellular PCL scaffolds 6 months
Steck et al. [194]	Mini-pigs	Full-thickness cartilage defects in stifle joints	Collagen type I/III membrane + transplantation of expanded autologous MSCs 1, 3 and 8 weeks
Ho et al. [195]	Pigs	Critically sized osteochondral defects located at the medial condyle and patellar groove of the stifle joints	Dual-phase construct of PCL cartilage scaffold and a PCL-TCP osseous matrix + autologous MSCs 6 months
Schneider et al. [196]	18 male adult Göttingen mini-pigs, ±3 years, ±36 kg (6 animals per group)	Full-thickness chondral 6.3 mm Ø defects in the left stifle joint (3 defects/joint)	Cell-free collagen gel, a collagen gel seeded with 2×10^5/ml chondrocytes or left untreated 6, 12 and 52 weeks
Ebihara et al. [197]	12 mini-pigs, 7–8 months of age, 21.3–25 kg	Full-thickness chondral 6 mm Ø × 5 mm deep in the area of the host animal's medial femoral condyle in the left and right (control) stifle joints	Layered chondrocyte sheets prepared on a temperature-responsive culture dish 3 weeks

(continued)

Table 20.3 (continued)

Authors	Population	Type of defect and localization	Material tested and follow-up period
Nakamura et al. [198]	15 Mexican hairless pigs (9 male, 6 female), 13 months of age, ±33.5 kg	Full-thickness osteochondral 8 mm Ø × 2 mm deep defects (approximately 1.5 mm cartilaginous and 0.5 mm bony part) created in the weight-bearing area of the medial femoral condyles in both stifle joints, 10 mm below the terminal ridge; left stifle joint served as a vehicle internal control	Transplantation of allogeneic synovial MSCs injected into the defect 3 months
Betsch et al. [199]	14 adult female Göttingen mini-pigs, 18–30 months of age, 25–35 kg	Full-thickness osteochondral 10 mm Ø × 6 mm in depth defects created in both medial femoral condyles in stifle joints	Biphasic scaffold alone (TRUFIT BGS, Smith & Nephew, USA), scaffold + PRP, scaffold + bone marrow aspiration concentrate and scaffold in combination with bone marrow aspiration concentrate and PRP or scaffold alone (control) 26 weeks
Filová et al. [200]	8 female miniature pigs, 7 month of age, 29 ± 10.5 kg	Circular osteochondral 8 mm Ø defect into the load-bearing part of the right femoral condyle of the stifle joint; 2 defects in the left stifle joints were created as controls	Cell-free hyaluronate/type I collagen/fibrin composite scaffold containing polyvinyl alcohol nanofibres enriched with liposomes, basic FGF and insulin 12 weeks
Moriguchi et al. [201]	6 skeletally mature female miniature pigs	Full-thickness cylindrical 4 mm Ø defect created bilaterally in the medial meniscus in stifle joints	Scaffold-free tissue-engineered construct derived from allogeneic porcine synovial MSCs cultured in monolayers at high density in the presence of ascorbic acid followed by the suspension culture to develop a 3D cell/matrix construct 6 months
Jagodzinski et al. [202]	10 miniature pigs	Full-thickness osteochondral 7 mm Ø × 10 mm in depth defects created in the femoral condyles of the stifle joints (2 defects per animal)	Grafted cylinder upside down or with a combined scaffold containing a spongious bone cylinder (Tutobone®) covered with a collagen membrane (Chondro-Gide®), another group with the same scaffolds but seeded with a stem cell concentrate and empty defects (control) 3 months

(continued)

Table 20.3 (continued)

Authors	Population	Type of defect and localization	Material tested and follow-up period
Wang et al. [203]	12 adult Diannan small-ear pigs, 8 months of age, ±12 kg	Full-thickness cartilage 7 mm Ø × 3–4 mm in depth created in the central weight-bearing surface of the femur condyles of the stifle joints (two defects in each joint)	Combination of DBM and bone marrow MSCs infected with adenovirus-mediated BMP-2 and adenovirus-mediated TGF-β3 1, 2 and 3 months
Ha et al. [204]	6 healthy male mini-pigs, 40–45 kg, ±1.5 years of age	Full-thickness chondral injury created in the trochlear groove of each stifle joint; 3 weeks later, an osteochondral 5 mm Ø × 10 mm deep defect was created followed by an 8 mm Ø × 5 mm deep reaming	Mixture (1.5 ml) of human umbilical cord blood-derived MSCs (0.5 × 10^7 cells/ml), and 4% HA hydrogel composite was transplanted into the defect on the right knee; left stifle joint was untreated to act as the control 12 weeks
Matsuo et al. [205]	12 skeletally mature miniature pigs, 33–50 kg	Full-thickness osteochondral 7 mm Ø × 10 mm in depth defects created in the femoral groove of both stifle joints (4 defects per joint)	To identify the optimal material and implantation method for subchondral bone repair: β-TCP with 75% and 67% porosity and HA with 75% porosity implanted at 0, 2 or 4 mm below the subchondral bone plate 3 months
Peck et al. [206]	Skeletally mature mini-pigs	Full-thickness 6 mm Ø created on the femoral condyles of stifle joints	Autologous synthetic scaffold-free construct with hyaline quality of defects left untreated 2 and 6 months
Sosio et al. [207]	6 pigs	Osteochondral defects created in the trochlea in the stifle joint (6 defects/animal)	Novel interconnecting collagen-HA substitute (to mimic the biological tissue between the chondral and the osseous phase) + chondrocytes, unseeded scaffold or defects left untreated 3 months
Ding et al. [208]	Pig model	Autotransplantation to subcutaneous pockets after 8 weeks of in vitro cartilage construction of bone marrow MSCs, auricular chondrocytes or both and PGA/PLA scaffold control groups	Bone marrow MSCs, auricular chondrocytes or both and PGA/PLA scaffold alone 2, 4 and 8 weeks

(continued)

Table 20.3 (continued)

Authors	Population	Type of defect and localization	Material tested and follow-up period
Fisher et al. [209]	8 male Yucatan mini-pigs, 6 months of age, ±25–35 kg	Full-thickness trochlear groove chondral 4 mm Ø defects without macroscopic removal of the subchondral bone (4 defects/joint)	Acellular HA hydrogel and HA hydrogels containing allogeneic MSCs, TGF-β3 or both 2 and 6 weeks
Muhonen et al. [210]	20 female domestic pigs (*Sus scrofa domesticus*), 4 months of age	Circular full-thickness chondral 8 mm Ø defect created in the right medial femoral condyle of the stifle joint	After 3 weeks, the defects were repaired with rh type II collagen-PLA scaffold or commercial porcine type I/III collagen membrane + autologous chondrocytes, or the defects were only debrided and left untreated 4 months
Zuo et al. [211]	18 miniature pigs, 7–8 months of age, 20–25 kg	Full-thickness 6 mm Ø defect created in the weight-bearing area of each medial femoral condyle in both stifle joints (9 defects/group)	Defects were initially repaired by autologous osteochondral mosaic arthroplasty, and the empty spaces between the multiple plugs were filled with cell-free PLGA scaffolds, tissue-engineered cartilage or bone marrow mononuclear cell-PLGA composites, and empty spaces were left untreated as control 6 months
Lin et al. [212]	28 mini-type pigs, 3 months of age, 15 ± 2.5 kg, gender not considered	Full-thickness 10 mm Ø × 3 mm in depth chondral defects on the load-bearing region of both left and right lateral femoral condyles	Porous chitosan scaffold + chondrocytes (control group); avidinised porous chitosan scaffold + biotinylated chondrocytes; avidinised porous chitosan scaffold + chondrocytes treated with biotin-conjugated anti-CD44; autologous cartilage plug 12 and 24 weeks

ASCs adipose stem cells, *β-TCP* beta-tricalcium phosphate, *BMP* bone morphogenetic protein, *DBM* demineralized bone matrix, *ECM* extracellular matrix, *EGF* epidermal growth factor, *FGF* fibroblastic growth factor, *GH* growth hormone, *HA* hydroxyapatite, *IGF-I* insulin-like growth factor-I, *MRI magnetic resonance imaging, MSCs* mesenchymal stem cells, *OA* osteoarthritis, *PCL* poly(epsilon-caprolactone), *PGA* polyglycolic acid, *PLA* polylactic acid, *PLGA* poly(D,L)lactide-co-glycolide acid, *PRP* platelet-rich plasma, *PMMA* polymethyl methacrylate, *rh* recombinant human, *TGFβ* transforming growth factor beta

chondral and osteochondral defects with 6.3 mm in diameter did not heal in the Göttingen mini-pig, confirming the suitability of this swine breed for articular cartilage research studies.

In pig, the in vivo chondral/osteochondral tissue engineering studies with induction of surgical defects resorted to auto- or allogenic chondrocytes associated with a periosteal flap [175, 177], collagen gel or membrane [191, 196, 197, 210] or an osteoconductive scaffold [180, 182, 184, 187, 190, 193, 212] (Table 20.3). Porcine synovial tissue allogenic MSCs were also used in various studies either alone [186, 198] or seeded into a scaffold [201]. Other types of cells were also used – auto- or allogenic bone marrow MSCs [194, 203, 208, 209], xenogenic umbilical cord blood-derived MSCs [204] and bone marrow mononuclear cells [211] alone or associated with different types of scaffolds. Biphasic scaffolds to mimic the chondral and bone tissue under the articular cartilage are also reported associated with GFs [171], autologous chondrocytes [188, 207], PRP and/or aspirated bone marrow [199].

Duda et al. [213] described, in Yucatan female mini-pigs, the influence of mechanical conditions on delay or complete failure of cartilage healing on osteochondral defects on the lateral surface of the trochlear groove of the stifle joint at the 4th, 6th and 12th post-operative week.

20.2.4 Dog

The dog is considered worldwide as a very friendly and loved companion and domestic animal. The social and ethical issues related to the use of dog as an animal model for preclinical and translational studies are one of the main reasons why this usage is limited. Dogs are docile and cooperative animals, and their joint anatomy and body weight allow arthroscopy examination [20, 22], and, like humans, they have significant difficult-to-heal cartilage lesions [22]. Dogs also manifest cartilage-related pathologies such as osteochondritis dissecans and osteoarthritis [214, 215]. The skeletal maturity in dogs is reached between 12 and 24 months of age [67].

The articular cartilage thickness at the medial femoral condyle has been reported to be 0.95 mm [46] and 1.3 mm [65, 67]. Archibald et al. [68] also reported an average articular cartilage thickness in the stifle joint of 0.7 mm for dogs. Similar to the pig, the cartilage thickness in dogs allows the creation of full-thickness chondral defects, involving only the chondral tissue. Nevertheless, the majority of the studies in dogs report to osteochondral defects [20], and the mean volume of the cartilage defects is generally inferior to the human lesions [20, 43, 44].

The large majority of cartilage defect studies in the dog model are performed in the stifle joint, involving only the medial or both femoral condyles, or in the femoral trochlea, generally with the creation of osteochondral defects and with a follow-up post-operative period that varies between 2 weeks and 18 months (Table 20.4). The typical defects generally studied in dogs are the surgically created 3 mm to 12 mm in diameter chondral/osteochondral defects in the stifle joint, shoulder, elbow or coxofemoral joints, osteochondrosis lesions of the femoral condyle or medial humeral condyle, grades 3 and 4 lesions of the femoral condyle or medial humeral condyle [59], or OA femoral condylar lesions occurring secondary to impact injury or meniscal release and elbow dysplasia [61–64]. Its primary use is in pivotal studies for post-operative assessments and management that closely mimic human con-

Table 20.4 Cartilage and osteochondral tissue engineering studies performed in canine model for cartilage repair

Authors	Population	Type of defect and localization	Material tested and follow-up period
Campbell et al. [216]	42 adult mongrel dogs	Condyle osteotomy at the distal femur level Osteotomy at the distal radius level	Autogenous and homogenous transplants of large articular fragments and half joints with a thin adjoining osseous border 5 to 500 days
Engkvist [217]	13 dogs	Completely chondral excision of patella	Free autologous rib perichondral grafts 2 to 17 months
Stevenson et al. [218]	22 Beagle dogs (5 to 6 animals per group)	Osteotomy at the distal radius level	Orthotopically implanted canine leukocyte antigen-matched and mismatched proximal osteochondral allografts of the radius, fresh and cryopreserved 11 months
Klompmaker et al. [219]	Dogs	Full-thickness round or oval defect performed in the distal femur level	Different porous and chemical polymer composition implants
Oates et al. [220]	14 adult mongrel dogs	Standardized defect on the weight-bearing surface of the medial femoral condyles	Autograft, a fresh allograft or a stored allograft at 4°C in tissue culture medium for 14 days prior to implantation 12 weeks
Shortkroff et al. [214]	6 adult mongrel dogs, ±30 kg	Round or oval chondral and osteochondral defects of 2 mm and 3 mm Ø in the weight-bearing, articulating portions of the medial and lateral femoral condyles	Cultured autologous chondrocytes with periosteal covering fixed with fibrin glue 2 and 4 weeks
Breinan et al. [221]	14 adult dogs	Round or oval holes on the trochlear groove of the distal part of the femur	Autologous chondrocyte implantation + periosteal flap 12 and 18 months
Nehrer et al. [222]	21 mixed-breed adult dogs, 2–4 years of age, ±30 kg	Chondral 4 mm Ø defect, 1 cm and 2 cm proximal from the intercondylar notch, on the median and lateral aspect of the trochlea groove of the left stifle joint	Cultured autologous articular chondrocytes seeded in type I and II collagen GAG matrices + fascia lata flap 15 weeks
Van Dyk et al. [223]	20 adult mongrel dogs (4 animals per group)	Cylindrical osteochondral 10 mm Ø × 10 mm deep defect in the femoral trochlea	Autogenous cancellous bone grafts 2, 4, 8, 12 and 24 weeks

(continued)

Table 20.4 (continued)

Authors	Population	Type of defect and localization	Material tested and follow-up period
Breinan et al. [224]	12 adult mongrel dogs, 1.5–3 years of age, ±25 kg (4 animals per group)	Chondral 4 mm Ø defects were to the depth of the tidemark in the trochlear groove in each stifle joint (2 defects/joint)	Microfracture perforation alone or associated with a type II collagen matrix placed in the defect and type II matrix seeded with cultured autologous chondrocytes 15 weeks
Breinan et al. [225]	34 adult mongrel dogs, ±30 kg	Chondral 4 mm Ø defects in the canine trochlear groove (1.25 cm and 3.25 cm proximal to the intercondylar notch, each slightly lateral or medial to the midline)	Autologous chondrocyte implantation + periosteal flap 1.5, 3 and 6 months
Cook et al. [226]	65 adult dogs	Osteochondral 5 mm Ø × 9 mm deep bilateral defects in the medial femoral condyles	Osteogenic protein-1 (BMP-7) + bovine bone-derived type I collagen device 6, 12, 16, 20 and 52 weeks
Feczkó et al. [227]	50 German shepherd breed dogs	Mosaicplasty osteochondral cylinder defect (3.5 mm, 4.5 mm and 6.5 mm) in trochlear groove of bilateral stifle joint	"Donor site plugs" made from HA (4.5 mm), carbon fibre rods (3.5 mm), polyglyconate-B (4.5 and 6.5 mm), compressed collagen and 2 PCL versions Between 8 and 30 weeks
Lee et al. [228]	6 skeletally mature outbreed hound dogs, 1–3 years of age, ±25 kg	Chondral 4 mm Ø defects in the canine trochlear groove (1.25 mm and 3.25 cm proximal to the intercondylar notch, each slightly lateral or medial to the midline) of the right stifle joint	Autologous articular chondrocyte-seeded type II collagen scaffold 15 weeks
Chen et al. [229]	1 Beagle dog, 1 year old	Osteochondral defects 4.5 mm Ø were drilled in the femoral condyles into the subchondral bone	Biphasic scaffold composed of stratified collagen/PLGA-collagen sponge seeded with autologous bone marrow MSCs 4 months
Glenn et al. [230]	18 adult dogs	Bilateral creation of an outerbridge grade 4 cartilage defect on the stifle joints	Bilateral osteochondral graft implantation to compare fresh osteochondral autografts and allografts 3 and 6 months
Yamazoe et al. [231]	9 beagle dogs, 1–3 years of age, 8.3–10.5 kg	Full-thickness 5 mm Ø × 4.5 mm in depth cylindrical defects on both weight-bearing areas of the femoral condyles in stifle joint	To study the effects of atelocollagen gel containing bone marrow MSCs on repair of osteochondral defect

(continued)

Table 20.4 (continued)

Authors	Population	Type of defect and localization	Material tested and follow-up period
Choi et al. [232]	Beagle dogs	Osteochondral defects created in the trochlear groove of stifle joint	Implantation of canine ASCs or umbilical cord blood-derived MSCs 13 weeks
Sagliyan et al. [233]	10 mature dogs, different breeds, weights and sexes (5 animals per group)	Osteochondral 8 mm Ø × 10 mm in depth defects created in the femoral trochlear groove in both stifle joints	Autologous cancellous grafts obtained from the metaphyseal region of the tibia + intra-articular hyaluronic acid (2 mg/kg, twice administered, immediately post-operatively and 1 month afterwards) (contralateral joint as control) 3 and 6 months
Ng et al. [234]	Adult mongrel dogs, 2–4 years of age, 22.7–27.2 kg	Full-thickness chondral 4.0 mm Ø (completely through subchondral bone plate) defects created in the Centre of the femoral trochlear groove of the stifle joint	Allogeneic adult canine chondrocytes encapsulated in agarose and cultured in serum-free media with TGF-β3 by 28 days 12 weeks
Mokbel et al. [235]	16 domestic mongrel dogs, males, 2–3 years of age, 15–20 kg	Partial-thickness chondral 3 mm Ø × 1 mm in depth defects created weight-bearing area of the lateral femoral condyle of the stifle joint without damaging the subchondral bone	Intra-articular injection of bone marrow MSCs and its involvement in the healing process of experimentally induced, acute and chronic, partial chondral defects 2 and 8 weeks
Sun et al. [236]	Dogs	Osteochondral defects using mosaic arthroplasty	Chondrocyte-β-TCP scaffold composites on top of the defect and osteoblast-β-TCP scaffold composites below the defect (chondrocyte and osteoblast composites were cocultured using a bioreactor)
Igarashi et al. [237]	12 adult female beagle dogs, ±20 kg	Two osteochondral 5 mm Ø × 5 mm in depth defects in the patella groove of each the stifle joint	Acellular novel material based on purified sodium alginate (sea matrix®) implantation with or without bone marrow MSCs 16 weeks

(continued)

Table 20.4 (continued)

Authors	Population	Type of defect and localization	Material tested and follow-up period
Zhang et al. [238]	16 mature dogs, 11.2–14.5 kg	Full-thickness chondral 7 mm Ø defect in the weight-bearing area of the medial femoral condyle + screw hole on the Centre of the defect with 2.5 mm Ø × 17 mm in depth for insertion of the screw and spherical cap connection	Cartilage resurfacing defects with two kinds of titanium alloy plug with different elastic modulus: Titanium 2448 plug and titanium-6 aluminium-4 vanadium plug on osteointegration 12 weeks
Kasemi et al. [239]	12 adult mixed-breed dogs	Full-thickness osteochondral 6 mm Ø × 5 mm in depth defects created in the weight-bearing area of femoral condyles of both stifle joints	Autologous platelet-rich fibrin in one joint (contralateral limb as control) 4, 16 and 24 weeks
Lv and Lu [240]	10 hybrid dogs, 12 months of age, both sexes, 12–15 kg	Osteochondral 6 mm Ø × 4 mm in depth defects created in the subchondral bone at the femoral trochlea of the right stifle joint	HA nanoparticles/collagen I/ copolymer of PLA-hydroxyacetic acid (bony scaffold) and sodium hyaluronate/PLGA (cartilaginous scaffold) + bone marrow MSCs 12 and 24 weeks
Cook et al. [241]	16 adult female dogs	Osteochondral 8 mm Ø defects created in the lateral and medial femoral condyles of one stifle joint	Osteochondral allograft preserved for 28 or 60 days after procurement, and chondrocyte viability was quantified before implantation 6 months
McCarty et al. [242]	8 adult mongrel dogs	Bilateral hindlimb osteochondral graft implantation in the stifle joint after creation of an acute outerbridge grade 4 cartilage defect	Osteochondral autograft and fresh osteochondral allograft 12 months
Kasemi et al. [243]	12 adult mixed-breed dogs, 18–40 kg	Osteochondral 6 mm Ø × 5 mm in depth defects created on the weight-bearing area of the medial femoral condyles of the stifle joint (two defects per dog, one on each limb)	Autologous bone marrow MSCs seeded on autologous platelet-rich fibrin or left empty 4, 16 and 24 weeks

ASCs adipose stem cells, *β-TCP* beta-tricalcium phosphate, *BMP* bone morphogenetic protein, *DBM* demineralized bone matrix, *ECM* extracellular matrix, *EGF* epidermal growth factor, *FGF* fibroblastic growth factor, *GH* growth hormone, *HA* hydroxyapatite, *IGF-I* insulin-like growth factor-I, *MRI* magnetic resonance imaging, *MSCs* mesenchymal stem cells, *OA* osteoarthritis, *PCL* poly(epsilon-caprolactone), *PGA* polyglycolic acid, *PLA* polylactic acid, *PLGA* poly(D,L)lactide-co-glycolide acid, *PRP* platelet-rich plasma, *PMMA* polymethyl methacrylate, *rh* recombinant human, *TGFβ* transforming growth factor beta citions since dogs allow implementation of specific post-operative exercise and rehabilitation protocols [22, 56]. The CSD for cartilage defects in dogs is reported to be of 4 mm [56].

Calandruccio and Gilmer [244] performed full-thickness and superficial round or oval defects at the weight-bearing surface of femur condyles and in the patellar ridge of immature dogs that were evaluated at 2, 4 and 6 weeks to study proliferation, regeneration and repair of articular cartilage. Hale et al. [245] induced full-thickness round or oval osteochondral defects in adult dogs of 1, 3 or 5 mm in diameter bilaterally in femoral condyles of stifle joint assessed 10 months after lesion induction for indentation assessment of biphasic mechanical property deficits in size-dependent osteochondral defect repair.

In dog model, the first in vivo chondral/osteochondral tissue engineering studies with induction of surgical defects report the use of auto- and allogenic osteochondral grafts for their repair [216–218, 220] (Table 20.4). Also, autogenous cancellous bone grafts have been used [223, 233], and recently other studies report again the use of osteochondral autograft or preserved or fresh allografts [230, 241, 242]. Klompmaker et al. [219] applied for the first time an osteoconductive scaffold for repair of an osteochondral defect and with Shortkroff et al. [214] cultured autologous chondrocytes covered by a periosteal flap fixed by fibrin glue. Autologous chondrocyte implantation covered by periosteal flaps has also been used by many other research teams [221, 225] or seeded in collagen matrices [222, 224, 228]. Biphasic scaffolds that mimic the chondral tissue and the underlying bone tissue with autologous bone marrow MSCs [229, 240] or just composed by a β-TCP scaffold seeded with chondrocytes on top of the defect and with osteoblasts below the osteochondral defect [236] were also performed. Finally, and more recently, autologous platelet-rich fibrin, a second-generation platelet-derived product, has also been applied alone [239] or seeded with autologous bone marrow MSCs [243].

20.2.5 Horse

The horse is the largest of animal models recommended for articular cartilage repair and regeneration studies with a mean weight of 400 kg to 500 kg [20, 22]. Horses are very well appreciated as a companion animal, and therefore its utilization as an animal model for scientific research studies implies social and ethical concerns [20]. Proper housing, anaesthetic and surgical facilities and specialized veterinary staff and equipment are needed for feasibility of this large animal model. As horses are robust and long life expectancy animals, they are suitable models to assess resurfacing technologies in chronic lesions and for post-operative evaluation of cartilage repair under very specific and demanding weight-bearing and weight-loading conditions [20]. Like humans, horses manifest cartilage disorders such as osteochondritis dissecans, several types of cartilage injuries and osteoarthritis, with established techniques described in the scientific literature for their treatment that frequently involve arthroscopic techniques for primary intervention and second-look or longitudinal observation for post-operative evaluation [22]. Similar to humans, the horse also demonstrates low intrinsic capacity for repair of articular cartilage lesions like humans [246, 247]. In 1972, Convery and co-authors [246] studied in Shetland ponies the relationship between the full-thickness osteochondral defect of graduated diameter – up to 21 mm Ø, in the weight-bearing areas of femoral condyles – and the completeness of repair, the effect of defects on the opposing articular surface and in the joint as a functional unit. Skeletal maturity in horses is generally achieved around 26 months of age (between 26 and 32 months for the closure of the distal radial physis) [248–251]. Archibald et al. [68] reported an average articular cartilage thickness in the stifle joint in the horse of 3.2 mm, and Frisbie et al. [46] refer a thickness of 1.75 mm to 2 mm at the medial femoral condyle level, very similar to the thickness of the human species. This cartilage thickness allows the realization of partial- or full-thickness cartilage defects with the creation of defects with similar size and volume to those of human lesions. It also allows to create chondral defects alone with a volume of 350 mm^2 without the involvement of the osteochondral bone [20]. However, the living weight of these specimens and the immediate post-operative load bearing of the surgically operated limb are very challenging and do not properly simulate the human post-operative conditions. These aspects should be taken into careful consideration when choosing the anatomical area to perform the cartilage defect [20, 22]. The large majority of the cartilage defect studies in horse models involve the lateral femoral trochlea at the stifle joint, since the patellofemoral compartment is relatively unloaded during housing confinement and controlled walking [20, 22]. The lateral condyle of the metacarpophalangeal joint and the middle carpal bones are also utilized [20], generally with the creation of osteochondral defects, but also in the ankle joints; the follow-up post-operative period varies between 3 weeks and 2 years (Table 20.5). The type of defects generally studied in horses is surgically created 6 mm to 20 mm in diameter chondral/osteochondral, chip fracture or osteochondrosis, and its primary use is for pivotal studies using surgically created or spontaneous defects with both cartilage thickness and morphology and also biomechanic assessment that closely mimic the human species [56]. The CSD for cartilage defects in horses is reported to be 9 m [56, 67], although most of the studies conducted report cartilage defects with larger dimensions relatively to established CSD for the species, ranging from 8 mm to 15 mm in diameter [20] (Table 20.5).

Overall, the horse is considered an important animal model for preclinical and translational studies for the development and evaluation of new cartilage treatments before human clinical trials due to its joint size, chondral tissue morphology and thickness and fully extended upright stifle joint during gait which makes this animal model easier to translate to human when compared to other large animal models used for chondral/osteochondral tissue engineering [22, 285]. Another advantage in using horse as a large animal model for cartilage defect repair research studies relates to the high incidence of articular cartilage lesions in high-performance equines that, consequently, need well-developed cartilage repair techniques [286–288].

Table 20.5 Cartilage and osteochondral tissue engineering studies performed in equine model for cartilage repair

Authors	Population	Type of defect and localization	Material tested and follow-up period
Shamis et al. [252]	13 horses (6 and 7 animals per group)	Partial-thickness 1 cm defect, round defect in the 3rd carpal bone	In one joint, 1 mm Ø × 1 cm deep holes were drilled within the defect, and one joint was used as a control 1 and 3 weeks
Vachon et al. [253]	10 horses, 2–3 years of age	Defects 1.0 × 1.0 cm² were induced bilaterally on the distal articular surface of each radial carpal bone	Periosteal autografts harvested from the proximal portion of the tibia + fibrin adhesive 16 weeks
Vachon et al. [254]	10 horses, 2–3 years of age	Circular osteochondral 1 cm Ø × 4 mm deep defects on both radial carpal bones	Sternal cartilage autografts fixed by biodegradable pins (contralateral carpi were not grafted in the radial carpal bone) 16 weeks
Hendrickson et al. [255]	8 horses (4 animals per group)	Circular full-thickness 12 mm defect on the lateral trochlea of both stifle joints	Chondrocyte implantation in a fibrin vehicle or empty defect (control) 4 and 8 months
Howard et al. [256]	10 horses, 2–3 years of age	Circular 10 mm Ø × 3 mm deep bilateral defect in the radial carpal bone	Sternal cartilage autograft or defects left untreated (control) 12 months
Sams and Nixon [257]	12 horses (6 animals per group)	Circular full-thickness 15 mm defect in both stifle joint	Chondrocyte-collagen composites or defects left untreated (control) 4 and 8 months
Frisbie et al. [258]	10 horses, 2–5 years of age (5 animals per group)	Full-thickness chondral 1 cm² defect created in both radial carpal bones and both medial femoral condyles in stifle joint	Microfracture perforation subchondral of the bone plate under the defect in one carpus and one femoral condyle 4 and 12 months
Nixon et al. [259]	6 young mature horses, 2–6 years of age	Full-thickness cartilage 15 mm lesions bilaterally in the lateral trochlear ridge of the femur at stifle joint	Autogenous fibrin + rhIGF-I (25 μg) 6 months
Fortier et al. [260]	8 adult horses, 3–5 years of age	Cartilage 15 mm Ø × 3 mm deep to the level of subchondral bone, single cartilage defect in the mid-lateral trochlear ridge of each femur at stifle joint	Composites of chondrocytes and polymerized fibrin supplemented with IGF-I 8 months

(continued)

Table 20.5 (continued)

Authors	Population	Type of defect and localization	Material tested and follow-up period
Hidaka et al. [261]	10 horses (5 males, 5 females), 1–6 years of age	Cartilage 15 mm Ø defects in the lateral trochlear ridge (not in the subchondral plate) of each femur in stifle joint	Genetically modified chondrocytes infected with adenovirus vector encoding BMP-7 or as control, an adenovirus vector encoding an irrelevant gene (*Escherichia coli* cytosine deaminase) 4 weeks and 8 months
Litzke et al. [262]	8 skeletally mature horses, both sexes, 9 ± 6 years of age, ±400 kg	Full-thickness cartilage 10 mm Ø defect in the minor load-bearing area on lateral talus of tibiotarsal joints	Autologous chondrocyte transplantation injected beneath the periosteum in one joint (contralateral joint as control) 2 years
Strauss et al. [263]	Horses	Full-thickness cartilage defect	Genetically modified chondrocytes expressing IGF-I or unmodified, control chondrocytes to evaluate articular cartilage adjacent to chondral defect repair 8 months
Barnewitz et al. [264]	8 healthy Haflinger horses, both sexes, 2 years of age, ±400 kg	Full-thickness cartilage 8 mm Ø defects in the tubular bone condyle of the fetlock joint	Autologous cartilage tissue engineering transplants based on resorbable polyglactin/ polydioxanone scaffolds 6 and 12 months
Gratz et al. [265]	8 adult horses, 3–5 years of age	Test samples were obtained from the previous study of Fortier et al. [260]	To develop a quantitative biomechanical method to assess the tensile modulus of repair tissue and its integration in vivo, as well as determine whether supplementation of transplanted chondrocytes with IGF-I affected these mechanical properties
Goodrich et al. [266]	16 adult female horses, 2–6 years of age (4 to 8 animals per group)	Single chondral 15 mm Ø defect in each femoropatellar joint	Arthroscopically grafted chondrocytes genetically modified by an adenovirus vector encoding equine IGF-I 4 and 9 weeks and 8 months

(continued)

Table 20.5 (continued)

Authors	Population	Type of defect and localization	Material tested and follow-up period
Morisset et al. [267]	12 horses	Full-thickness chondral defects created in carpus and stifle joints	Microfracture associated with injection of gene therapy treatment to supply adenoviral vectors carrying the interleukin-1 receptor antagonist protein (IL-1ra) and IGF-1 equine genes 16 weeks
Wilke et al. [268]	6 young mature horses	Circular full-thickness 15 mm cartilage lesions on the lateral trochlear ridge of the femur (no more than 1 mm of the subchondral bone plate was removed)	Injection of a self-polymerizing autogenous fibrin vehicle containing MSCs (previously aspirated from horse sternums) 30 days and 8 months
Frisbie et al. [269]	15 horses, 3 years of age	Two chondral 15 mm Ø defects on the medial trochlear ridge of the femur in the opposite stifle joint (relatively to the autologous cartilage harvested arthroscopically)	Autologous chondrocytes cultured expanded and seeded on a collagen membrane – Porcine small intestine submucosa or collagen membrane alone 4, 8, 12 and 18 months
Frisbie et al. [270]	10 skeletally mature horses	Clinically relevant 15 mm Ø defect within equine femoral trochlea in stifle joint	One-step surgical procedure using autologous cartilage fragments on a polydioxanone scaffold, compared with a two-step autologous chondrocyte implantation technique as well as with empty defects and defects with polydioxanone foam scaffolds alone 4, 8 and 12 months
Fortier et al. [271]	12 horses	Extensive full-thickness 15 mm Ø cartilage defects created on the lateral trochlear ridge of the femur	Autologous bone marrow aspirate concentrate + microfracture or with microfracture alone 3 and 8 months

(continued)

Table 20.5 (continued)

Authors	Population	Type of defect and localization	Material tested and follow-up period
Kon et al. [272]	2 standard-bred trotter horses, 6–7 years of age, 410 and 430 kg	Chondral 10 mm Ø defects (lateral condyle) with superficial debridement of the subchondral bone and deep osteochondral 10 mm Ø × 8–10 mm in depth defects (medial condyle) made in the distal epiphysis of the weight-bearing area of the 3rd metacarpal bone of both forelimbs	3D biomimetic scaffold (fin-Ceramica Faenza S.p.A., Faenza-Italy) obtained by nucleating collagen fibrils with HA nanoparticles, in two configurations, bi- and tri-layered, to reproduce, respectively, chondral and osteochondral anatomy 6 months
McIlwraith et al. [273]	10 horses, 2.5–5 years of age	Cartilage 1 cm² defects arthroscopically created on both medial femoral condyles of the stifle joint debrided to subchondral bone	Intra-articular injection of bone marrow MSCs to augment healing with microfracture compared with microfracture alone + strenuous treadmill exercise simulating race training until completion of the study 6 and 12 months
Nixon et al. [274]	16 horses (8 animals per group)	Paired partial (calcified cartilage intact) or full-thickness (penetrating to but not into the subchondral bone plate) cartilage defects 15 mm Ø in the patellofemoral joint	Autologous chondrocyte implantation or periosteal flap alone 8 weeks
Seo et al. [275]	6 healthy thoroughbred female horses, 5.0 ± 3.4 years of age, 433.2 ± 92.3 kg	Full-thickness osteochondral 4.5 mm Ø × 10 mm in depth defect created on the lateral trochlear ridge of the talus	Bilayer gelatin/β-TCP sponges loaded with bone marrow MSCs, chondrocytes, BMP-2 and PRP 0, 1, 2, 3 and 4 months
Miller et al. [276]	16 skeletally mature horses, 2–5 years of age	Defects 15 mm Ø created on the medial trochlear ridge and debrided down to the subchondral bone in stifle joint	Injectable self-assembling peptide associated or not to microfracture and empty defects as control Every 4 weeks until 12 months
Tsuzuki et al. [277]	6 thoroughbred horses, (2 males, 4 females), 3.3–2.4 years of age, 429.5 ± 118.2 kg	Cartilage 4.5 mm Ø × 3.0 mm in depth created in both 3rd carpal bones; a subchondral hole 2 mm Ø × 35 mm in depth was drilled in the Centre of this cartilage defect but not in the control group	PRP incorporated in gelatin hydrogel microsphere + subchondral drilling 16 weeks

(continued)

Table 20.5 (continued)

Authors	Population	Type of defect and localization	Material tested and follow-up period
Frisbie et al. [278]	12 skeletally mature horses, 2–6 years of age	Cartilage 15 mm Ø defects, extending down to the level of the subchondral bone plate, on the medial aspect of the trochlea at the distal aspect of each femur	Empty defect controls only with fibrin, autologous chondroprogenitor cells + fibrin and allogenic chondroprogenitor cells + fibrin associated with strenuous exercise throughout the 12-month study 12 months
Nixon et al. [279]	6 young mature horses, 2–6 years of age	Two full-thickness cartilage 15 mm Ø defects were formed in the lateral trochlear ridge of the femur in the patellofemoral joint of one randomly selected limb, without involving the subchondral bone plate	Autologous chondrocytes seeded on collagen I/III membrane (ACI-Maix™) and control defect remain empty 12 weeks and 6 months
Ortved et al. [280]	24 horses (8 animals per group)	Full-thickness chondral 15 mm Ø defect created in the lateral trochlear ridge of femur in the stifle joint, involving the subchondral bone plate; contralateral joint used as control	Fibrin-containing autologous chondrocytes transduced with r adeno-associated virus 5-IGF-I or r adeno-associated virus GFP (positive control) or naive, untransduced chondrocytes 8 weeks and 8 months
Seo et al. [281]	5 healthy thoroughbred horses (3 males, 2 females), 3.6 ± 2.3 years of age, 445.8 ± 71.1 kg	Osteochondral 10 mm Ø × 5 mm in depth defects in the medial femoral condyle in one of the stifle joints; the other one was used as control	Synovial flap + fibrin glue and gelatin/β-TCP sponge loaded with MSCs, BMP-2 and PRP; in the control group, the synovial flap was not used 4 months
Fortier et al. [282]	5 horses	Two 10 mm Ø full-thickness cartilage defects created in the trochlear ridge of both stifle joints: One proximal (high load) and another distal (low load)	Microfracture with allograft articular cartilage (BioCartilage) + PRP (microfracture alone as control) 13 months
Rocha junior et al. [283]	6 horses, 7.2 ± 1,3 years of age, 342 ± 1.58 kg	Chondral 1 cm^2 lesions in the lateral femoral trochlea in both stifle joints	Microperforation of the subchondral bone + four weekly injections of kartogenin (20 μM) (ringer lactate solution in the contralateral joint as control) 60 days

(continued)

Table 20.5 (continued)

Authors	Population	Type of defect and localization	Material tested and follow-up period
Yamada et al. [284]	12 mestiços Crioulos bred horses, 2–5 years of age, neutered female and male, ±350 kg (6 animals per group)	Chondral 15 mm Ø defect created on the medial femoral trochlea of stifle joint, involving the subchondral bone plate	Implantation of ASCs + PRP gel or defects left untreated 5 months

ASCs adipose stem cells, *β-TCP* beta-tricalcium phosphate, *BMP* bone morphogenetic protein, *DBM* demineralized bone matrix, *ECM* extracellular matrix, *EGF* epidermal growth factor, *FGF* fibroblastic growth factor, *GH* growth hormone, *HA* hydroxyapatite, *IGF-I* insulin-like growth factor-I, *MRI* magnetic resonance imaging, *MSCs* mesenchymal stem cells, *OA* osteoarthritis, *PCL* poly(epsilon-caprolactone), *PGA* polyglycolic acid, *PLA* polylactic acid, *PLGA* poly(D,L)lactide-co-glycolide acid, *PRP* platelet-rich plasma, *PMMA* polymethyl methacrylate, *rh* recombinant human, *TGFβ* transforming growth factor beta

The horse has also been the focus of some in vitro studies, namely, the study of Vidal et al. [289], which intended to compare the chondrogenic potential of adult equine cryopreserved MSCs and ASCs used for pellet cultures in stromal medium or induced into chondrogenesis with transforming growth factor (TGF)-β3 and bone morphogenetic protein (BMP)-6 during 3, 7, 14 and 21 days. In the study of Henson et al. [290] after induction of full-thickness cartilage disks harvested from the 3rd metacarpal bone, the obtained chondral tissue was single impact loaded with 0.175 J at 0.7 m/s and cultured in DMEM with 1% (vol/vol) hyaluronic acid-chondroitin sulphate-N-acetyl glucosamine (HCNAG) or fibroblast growth factor (FGF)-2 (50 ng/ml) to study the effect of a solution of HCNAG on the repair response of cartilage to single-impact load damage.

There are other interesting studies in the horse such as the work of Shoemaker et al. [291] that performed intra-articular administration of methylprednisolone acetate (100 mg), once a week for four applications, to treat full-thickness 1 cm in diameter bilateral defects in the articular cartilage on the dorsal distal surface of the radial carpal bone evaluated at the 16th post-operative week. Furthermore, Todhunter et al. [292] promoted mild to moderate post-operative exercise associated with intra-articular polysulphated glycosaminoglycan administration to stimulate the repair of rectangular induced osteochondral defects in a weight-bearing surface of the radial carpal bone evaluated at 17th post-operative week, and Bertone et al. [293] compared the articular cartilage from horses with naturally developing osteochondrosis, with normal articular cartilage and healing cartilage obtained from horses with experimentally induced osteochondral fractures in femur and tibia in stifle joint, 40 days post-surgery.

The first in vivo chondral/osteochondral tissue engineering studies in horse, with induction of surgical defects, reported the use of microfracture perforation technique [252, 258], periosteal autograft [253], sternal cartilage autograft alone [256] or fixed by biodegradable pins [254], chondrocyte implantation in a fibrin vehicle

[255] or chondrocyte-collagen composites [257] (Table 20.5). In 1999, Nixon and co-authors [259] introduced recombinant human GFs for the treatment of full-thickness cartilage lesions in horse, followed by a series of studies involving gene therapy, more specifically, using genetically modified chondrocytes infected by adenovirus encoding different GFs [261, 263, 266, 267, 280]. Autologous chondrocyte transplantation alone [262, 274] or associated with GFs [260, 265] or collagen membrane [269, 279] has also been used, as well as other cell types like bone marrow MSCs and ASCs [273, 275, 281, 284]. Autologous bone marrow aspirate concentrate [271] and PRP have also been described as treatment options in the horse, generally associated with allograft, GFs, other cell types, hydrogels or microfracture perforation technique [271, 277, 281, 282, 284].

20.3   Concluding Remarks

Research studies using animal models for orthopaedic research should be designed and developed in accordance to the prevailing European Union legislation for the protection of animals used for scientific purposes – Directive 2010/63/EU [294] – and the Federation of European Laboratory Animal Science Associations (FELASA) guidelines [295–297] which follow the 3R's recommendations of replacement, reduction and refinement of animal experimentation use. Also, animal maintenance and all experimental work should be reviewed and approved by the National Animal Use and Care Committee/National Institutes of Health of each country and by the specific ethical committees of the research institutes or medical centres. Accordingly, researchers are advised to minimize the use of animals through (i) replacement, e.g. with computational models and in vitro or ex vivo studies; (ii) reduction, e.g. with adequate study's experimental design, pilot studies, ex vivo studies and the use of non-invasive or minimally invasive techniques of diagnosis, like ultrasonography, fluoroscopy, X-ray exams and fluorochromes and other advanced imaging techniques such as computed tomography (CT) and MRI and arthroscopy for observation and/or obtaining biopsy samples, which allow the collection of data from the same individual specimen at different time points; and (iii) refinement in such a way that post-operative pain, suffer and distress inflicted to the animals are reduced to minimum levels [298–301]. Also the guidelines from the US Food and Drug Administration [60], the American Society for Testing and Materials and the International Cartilage Repair Society have to be followed for a correct choice of the animal model and the principal published literature in this scientific field [49–56].

By the observation of Tables 20.1, 20.2, 20.3, 20.4 and 20.5 concerning the research in repair of chondral and osteochondral tissues performed in large animal models in the last 25 years, it's possible to verify that these studies focus on implantation of autologous chondrocytes and a wide range of scaffolds and cell-scaffold tissue engineering constructs colonized with cells from different origins, of which some of them have already been translated to human clinical application and even, more recently, to gene therapy [22]. These large animal models have the advantages to allow accurate clinical post-operative evaluations of the induced chondral/osteochondral defects and their treatment, including lameness (pain), subjective function, arthroscopic, radiographic, CT, MRI, micro-CT techniques, histologic scores, histomorphometry and immunohistochemical analysis after euthanasia or biopsy samples obtainment by arthroscopic technique and biomechanical analysis, validating the results obtained with these approaches [56]. Depending on the particular treatment under research, the joint size, anatomy and arthroscopic access usually favour the horse and the small ruminants – sheep and goat – for articular cartilage repair studies [54].

References

1. Boushell MK, Hung CT, Hunziker EB, Strauss EJ, Lu HH (2016) Current strategies for integrative cartilage repair. Connect Tissue Res 6:1–14
2. Chahla J, LaPrade RF, Mardones R, Huard J, Philippon MJ, Nho S, Mei-Dan O, Pascual-Garrido C (2016) Biological therapies for cartilage lesions in the hip: a new horizon. Orthopedics 39(4):e715–e723
3. Richter DL, Schenck RC Jr, Wascher DC, Treme G (2016) Knee articular cartilage repair and restoration techniques: a review of the literature. Sports Health 8(2):153–160
4. Frehner F, Benthien JP (2017) Microfracture: state of the art in cartilage surgery? Cartilage. https://doi.org/10.1177/1947603517700956
5. Tan AR, Hung CT (2017) Concise review: mesenchymal stem cells for functional cartilage tissue engineering: taking cues from chondrocyte-based constructs. Stem Cells Transl Med 6(4):1295–1303
6. Grässel S, Lorenz J (2014) Tissue-engineering strategies to repair chondral and osteochondral tissue in osteoarthritis: use of mesenchymal stem cells. Curr Rheumatol Rep 16(10):452
7. Shen Y, Fu Y, Wang J, Li G, Zhang X, Xu Y, Lin Y (2014) Biomaterial and mesenchymal stem cell for articular cartilage reconstruction. Curr Stem Cell Res Ther 9(3):254–267
8. Kon E, Roffi A, Filardo G, Tesei G, Marcacci M (2015) Scaffold-based cartilage treatments: with or without cells? A systematic review of preclinical and clinical evidence. Arthroscopy 31(4):767–775
9. Kumar R, Griffin M, Butler PE (2016) A review of current regenerative medicine strategies that utilize nanotechnology to treat cartilage damage. Open Orthop J 10:862–876
10. López-Ruiz E, Jiménez G, García MÁ, Antich C, Boulaiz H, Marchal JA, Perán M (2016) Polymers, scaffolds and bioactive molecules with therapeutic properties in osteochondral pathologies: What's new? Expert Opin Ther Pat 26(8):877–890
11. Driessen BJ, Logie C, Vonk LA (2017) Cellular reprogramming for clinical cartilage repair. Cell Biol Toxicol 33(4):329–349. https://doi.org/10.1007/s10565-017-9382-0
12. Hinckel BB, Gomoll AH (2017) Autologous chondrocytes and next-generation matrix-based autologous chondrocyte implantation. Clin Sports Med 36(3):525–548
13. Goldberg A, Mitchell K, Soans J, Kim L, Zaidi R (2017) The use of mesenchymal stem cells for cartilage repair and regeneration: a systematic review. J Orthop Surg Res 12(1):39
14. Huang K, Li Q, Li Y, Yao Z, Luo D, Rao P, Xiao J (2017) Cartilage tissue regeneration: the roles of cells, stimulating factors and scaffolds. Curr Stem Cell Res Ther. https://doi.org/10.2174/1574888X12666170608080722
15. Lolli A, Penolazzi L, Narcisi R, van Osch GJVM, Piva R (2017) Emerging potential of gene silencing approaches targeting anti-chondrogenic factors for cell-based cartilage repair. Cell Mol Life Sci 74(19):3451–3465. https://doi.org/10.1007/s00018-017-2531-z

16. Monibi FA, Cook JL (2017) Tissue-derived extracellular matrix bioscaffolds: emerging applications in cartilage and meniscus repair. Tissue Eng Part B Rev 23(4):386–398. https://doi.org/10.1089/ten.TEB.2016.0431

17. Rai V, Dilisio MF, Dietz NE, Agrawal DK (2017) Recent strategies in cartilage repair: a systemic review of the scaffold development and tissue engineering. J Biomed Mater Res A 105(8):2343–2354. https://doi.org/10.1002/jbm.a.36087

18. Vega SL, Kwon MY, Burdick JA (2017) Recent advances in hydrogels for cartilage tissue engineering. Eur Cell Mater 33:59–75

19. Yang J, Zhang YS, Yue K, Khademhosseini A (2017) Cell-laden hydrogels for osteochondral and cartilage tissue engineering. Acta Biomater 57:1–25. https://doi.org/10.1016/j.actbio.2017.01.036

20. Ahern BJ, Parvizi J, Boston R, Schaer TP (2009) Preclinical animal models in single site cartilage defect testing: a systematic review. Osteoarthr Cartil 17(6):705–713

21. An YA, Friedman RJ (1999) Animal models of articular cartilage defect. In: An YA, Friedman RJ (eds) Animal models in orthopaedic research. CRC Press, Boca Raton, pp 309–325

22. Chu CR, Szczodry M, Bruno S (2010) Animal models for cartilage regeneration and repair. Tissue Eng Part B 16(1):105–115

23. Schneider-Wald B, von Thaden AK, Schwarz MLR (2013) Defect models for the regeneration of articular cartilage in large animals. Orthopade 42(4):242–253

24. Eitel F, Klapp F, Jacobson W, Schweiberer L (1981) Bone regeneration in animals and man. Arch Orthop Traumat Surg 99(1):59–64

25. Aerssens J, Boonen S, Lowet G, Dequeker J (1998) Interspecies differences in bone composition, density, and quality: potential implications for in vivo bone research. Endocrinology 139(2):663–670

26. An YA, Draughn RA (1999) Mechanical properties and testing methods of bone. In: An YA, Friedman RJ (eds) Animal models in orthopaedic research. CRC Press, Boca Raton, pp 139–163

27. An YA, Friedman RJ (1999) Animal selection in orthopaedic research. In: An YA, Friedman RJ (eds) Animal models in orthopaedic research. CRC Press, Boca Raton, pp 39–57

28. Hillier ML, Bell LS (2006) Differentiating human bone from animal bone: a review of histological methods. J Forensic Sci 52(2):249–263

29. Martiniaková M, Grosskopf B, Omelka R, Vondráková M, Bauerová M (2006) Differences among species in compact bone tissue microstructure of mammalian skeleton: use of a discriminant function analysis for species identification. J Forensic Sci 51(6):1235–1239

30. Pearce AI, Richards RG, Milz S, Schneider E, Pearce SG (2007) Animal models for implant biomaterial research in bone: a review. Eur Cell Mater 13:1–10

31. Chevrier A, Kouao AS, Picard G, Hurtig MB, Buschmann MD (2015) Interspecies comparison of subchondral bone properties important for cartilage repair. J Orthop Res 33(1):63–70

32. Jainudeen MR, Wahid H, Hafez ESE (2000) Sheep and goats. In: Hafez B, Hafez ESE (eds) Reproduction in farm animals, 7th edn. Lippincott Williams & Wilkins, Baltimore, pp 172–181

33. Smith AC, Swindle MM (2006) Preparation of swine for the laboratory. ILAR J 47(4):358–363

34. Laber KE, Whary MT, Bingel SA, Goodrich JA, Smith AC, Swindle MM (2002) Biology and disease in swine. In: Fox JG, Anderson LC, Loew FM, Quimby FW (eds) Laboratory animals medicine, 2nd edn. Academic, San Diego, pp 615–673

35. Bradshaw JWS (2006) The evolutionary basis for the feeding behavior of domestic dogs (Canis familiaris) and cats (Felis catus). J Nutr 136(Suppl 7):1927S–1931S

36. Concannon PW (2011) Reproductive cycles of the domestic bitch. Anim Reprod Sci 124(3–4):200–210

37. Simon VA (2010) Nutritional adaptations. In: Simon VA (ed) Adaptations in the animal kingdom. Xlibris corporation, Bloomington, pp 19–28

38. Aurich C (2011) Reproductive cycles of horses. Anim Reprod Sci 124(3–4):220–228

39. Li Y, Chen SK, Li L, Qin L, Wang XL, Lai YX (2015) Bone defect animal models for testing efficacy of bone substitute biomaterials. J Orthop Translat 3(3):95–104
40. Hunziker EB (1999) Biologic repair of articular cartilage. Defect models in experimental animals and matrix requirements. Clin Orthop Relat Res (367 Suppl):S135– S146
41. Hjelle K, Solheim E, Strand T, Muri R, Brittberg M (2002) Articular cartilage defects in 1,000 knee arthroscopies. Arthroscopy 18(7):730–734
42. Orth P, Madry H (2015) Advancement of the subchondral bone plate in translational models of osteochondral repair: implications for tissue engineering approaches. Tissue Eng Part B Rev 21(6):504–520
43. Bouwmeester PS, Kuijer R, Homminga GN, Bulstra SK, Geesink RG (2002) A retrospective analysis of two independent prospective cartilage repair studies: autogenous perichondrial grafting versus subchondral drilling 10 years post-surgery. J Orthop Res 20(2):267–273
44. Hunziker EB (2002) Articular cartilage repair: basic science and clinical progress. A review of the current status and prospects. Osteoarthr Cartil 10(6):432–463
45. Clar C, Cummins E, McIntyre L, Thomas S, Lamb J, Bain L, Jobanputra P, Waugh N. (2005) Clinical and cost-effectiveness of autologous chondrocyte implantation for cartilage defects in knee joints: systematic review and economic evaluation. Health Technol Assess 9(47):iii–iv, ix–x, 1–82
46. Frisbie DD, Cross MW, McIlwraith CWA (2006) Comparative study of articular cartilage thickness in the stifle of animal species used in human pre-clinical studies compared to articular cartilage thickness in the human knee. Vet Comp Orthop Traumatol 19(3):142–146
47. Martino F, Ettorre GC, Patella V, Macarini L, Moretti B, Pesce V, Resta L (1993) Articular cartilage echography as a criterion of the evolution of osteoarthritis of the knee. Int J Clin Pharmacol Res 13(Suppl):35–42
48. Malda J, de Grauw JC, Benders KE, Kik MJ, van de Lest CH, Creemers LB, Dhert WJ, van Weeren PR (2013) Of mice, men and elephants: the relation between articular cartilage thickness and body mass. PLoS One 8(2):e57683
49. Pineda S, Pollack A, Stevenson S, Goldberg V, Caplan A (1992) A semiquantitative scale for histologic grading of articular cartilage repair. Acta Anat (Basel) 143(4):335–340
50. Aigner T, Cook JL, Gerwin N, Glasson SS, Laverty S, Little CB, McIlwraith W, Kraus VB (2010) Histopathology atlas of animal model systems – overview of guiding principles. Osteoarthr Cartil 18(Suppl 3):S2–S6
51. Mainil-Varlet P, Van Damme B, Nesic D, Knutsen G, Kandel R, Roberts S (2010) A new histology scoring system for the assessment of the quality of human cartilage repair: ICRS II. Am J Sports Med 38(5):880–890
52. Heir S, Årøen A, Løken S, Sulheim S, Engebretsen L, Reinholt FP (2010) Intraarticular location predicts cartilage filling and subchondral bone changes in a chondral defect. Acta Orthop 81(5):619–627
53. Hoemann C, Kandel R, Roberts S, Saris DB, Creemers L, Mainil-Varlet P, Méthot S, Hollander AP, Buschmann MD (2011) International cartilage repair society (ICRS) recommended guidelines for histological endpoints for cartilage repair studies in animal models and clinical trials. Cartilage 2(2):153–172
54. Hurtig MB, Buschmann MD, Fortier LA, Hoemann CD, Hunziker EB, Jurvelin JS, Mainil-Varlet P, McIlwraith CW, Sah RL, Whiteside RA (2011) Preclinical studies for cartilage repair: recommendations from the international cartilage repair society. Cartilage 2(2):137–152
55. Fortier LA, Cole BJ, Science MICW (2012) Animal models of marrow stimulation for cartilage repair. J Knee Surg 25(1):3–8
56. Cook JL, Hung CT, Kuroki K, Stoker AM, Cook CR, Pfeiffer FM, Sherman SL, Stannard JP (2014) Animal models of cartilage repair. Bone Joint Res 3(4):89–94
57. American Society for Testing and Materials (2010) ASTM F2451–05 Standard Guide for in vivo Assessment of Implantable Devices Intended to Repair or Regenerate Articular Cartilage. http://www.astm.org/Standards/F2451.htm (date last Accessed 20 June 2017)

58. International Cartilage Repair Society (ICRS) (2011) Preclinical studies for cartilage repair: recommendation from the international cartilage repair society; ICRS recommendation papers. Cartilage 2(2):137–152

59. Hoemann C, Kandel R, Roberts S, Saris DB, Creemers L, Mainil-Varlet P, Méthot S, Hollander AP, Buschmann MD (2011) International cartilage repair society (ICRS) recommended guidelines for histological endpoints for cartilage repair studies in animal models and clinical trials. Cartilage 2(2):153–172

60. Food and Drug Administration: Guidance for industry: Preparation of IDEs and INDs for products intended to repair or replace knee cartilage. http://www.regulations.gov/#!home (date last Accessed 20 June 2017)

61. Luther JK, Cook CR, Cook JL (2009) Meniscal release in cruciate ligament intact stifles causes lameness and medial compartment cartilage pathology in dogs 12 weeks post operatively. Vet Surg 38(4):520–529

62. Garner BC, Stoker AM, Kuroki K, Evans R, Cook CR, Cook JL (2011) Using animal models in osteoarthritis biomarker research. J Knee Surg 24(4):251–264

63. Kuroki K, Cook CR, Cook JL (2011) Subchondral bone changes in three different canine models of osteoarthritis. Osteoarthr Cartil 19(9):1142–1149

64. O'Connell GD, Lima EG, Liming B, Chahine NO, Albro MB, Cook JL, Ateshian GA, Hung CT (2012) Toward engineering a biological joint replacement. J Knee Surg 25(3):187–196

65. Simon WH (1970) Scale effects in animal joints. I. Articular cartilage thickness and compressive stress. Arthritis Rheum 13(3):244–256

66. Lu Y, Hayashi K, Hecht P, Fanton GS, Thabit G 3rd, Cooley AJ, Edwards RB, Markel MD (2000) The effect of monopolar radiofrequency energy on partial-thickness defects of articular cartilage. Arthroscopy 16(5):527–536

67. Cellular, Tissue and Gene Therapies Advisory Committee. Cellular Products for Joint Surface Repair – Briefing Document. Meeting 38, US Food and Drug Administration, 3–4 March 2005

68. Archibald M, Runciman J, Dickey J, Hurtig M Do animal models approximate the subchondral bone and cartilage characteristics of humans? In 4th International Cartilage Repair Society, June 2002, Toronto, Canada

69. Homminga GN, Bulstra SK, Kuijer R, van der Linden AJ (1991) Repair of sheep articular cartilage defects with a rabbit costal perichondrial graft. Acta Orthop Scand 62(5):415–418

70. Hurtig MB, Novak K, McPherson R, McFadden S, McGann LE, Mul drew K, Schachar NS (1998) Osteochondral dowel transplantation for repair of focal defects in the knee: an outcome study using an ovine model. Vet Surg 27(1):5–16

71. Pearce SG, Hurtig MB, Clarnette R, Kalra M, Cowan B, Miniaci A (2001) An investigation of 2 techniques for optimizing joint surface congruency using multiple cylindrical osteochondral autografts. Arthroscopy 17(1):50–55

72. Siebert CH, Miltner O, Weber M, Sopka S, Koch S, Niedhart C (2003) Healing of osteochondral grafts in an ovine model under the influence of bFGF. Arthroscopy 19(2):182–187

73. von Rechenberg B, Akens MK, Nadler D, Bittmann P, Zlinszky K, Kutter A, Poole AR, Auer JA (2003) Changes in subchondral bone in cartilage resurfacing – an experimental study in sheep using different types of osteochondral grafts. Osteoarthr Cartil 11(4):265–277

74. Guo X, Wang C, Duan C, Descamps M, Zhao Q, Dong L, Lü S, Anselme K, Lu J, Song YQ (2004) Repair of osteochondral defects with autologous chondrocytes seeded onto bioceramic scaffold in sheep. Tissue Eng 10(11–12):1830–1840

75. Tibesku CO, Szuwart T, Kleffner TO, Schlegel PM, Jahn UR, Van Aken H, Fuchs S (2004) Hyaline cartilage degenerates after autologous osteochondral transplantation. J Orthop Res 22(6):1210–1214

76. Dorotka R, Windberger U, Macfelda K, Bindreiter U, Toma C, Nehrer S (2005) Repair of articular cartilage defects treated by microfracture and a three-dimensional collagen matrix. Biomaterials 26(17):3617–3629

77. Hoemann CD, Hurtig M, Rossomacha E, Sun J, Chevrier A, Shive MS, Buschmann MD (2005) Chitosan-glycerol phosphate/blood implants improve hyaline cartilage repair in ovine microfracture defects. J Bone Joint Surg Am 87(12):2671–2686

78. Russlies M, Behrens P, Ehlers EM, Bröhl C, Vindigni C, Spector M, Kurz B (2005) Periosteum stimulates subchondral bone densification in autologous chondrocyte transplantation in a sheep model. Cell Tissue Res 319(1):133–142

79. Tytherleigh-Strong G, Hurtig M, Miniaci A (2005) Intra-articular hyaluronan following autogenous osteochondral grafting of the knee. Arthroscopy 21(8):999–1005

80. Burks RT, Greis PE, Arnoczky SP, Scher C (2006) The use of a single osteochondral autograft plug in the treatment of a large osteochondral lesion in the femoral condyle: an experimental study in sheep. Am J Sports Med 34(2):247–255

81. Frosch KH, Drengk A, Krause P, Viereck V, Miosge N, Werner C, Schild D, Stürmer EK, Stürmer KM (2006) Stem cell-coated titanium implants for the partial joint resurfacing of the knee. Biomaterials 27(12):2542–2549

82. Kandel RA, Grynpas M, Pilliar R, Lee J, Wang J, Waldman S, Zalzal P, Hurtig M (2006) CIHR-bioengineering of skeletal tissues team. Repair of osteochondral defects with biphasic cartilage-calcium polyphosphate constructs in a sheep model. Biomaterials 27(22):4120–4131

83. Siebert CH, Schneider U, Sopka S, Wahner T, Miltner O, Niedhart C (2006) Ingrowth of osteochondral grafts under the influence of growth factors: 6-month results of an animal study. Arch Orthop Trauma Surg 126(4):247–252

84. Jones CW, Willers C, Keogh A, Smolinski D, Fick D, Yates PJ, Kirk TB, Zheng MH (2008) Matrix-induced autologous chondrocyte implantation in sheep: objective assessments including confocal arthroscopy. J Orthop Res 26(3):292–303

85. Jubel A, Andermahr J, Schiffer G, Fischer J, Rehm KE, Stoddart MJ, Häuselmann HJ (2008) Transplantation of de novo scaffold-free cartilage implants into sheep knee chondral defects. Am J Sports Med 36(8):1555–1564

86. Schlichting K, Schell H, Kleemann RU, Schill A, Weiler A, Duda GN, Epari DR (2008) Influence of scaffold stiffness on subchondral bone and subsequent cartilage regeneration in an ovine model of osteochondral defect healing. Am J Sports Med 36(12):2379–2391

87. Schagemann JC, Erggelet C, Chung HW, Lahm A, Kurz H, Mrosek EH (2009) Cell-laden and cell-free biopolymer hydrogel for the treatment of osteochondral defects in a sheep model. Tissue Eng Part A 15(1):75–82

88. Streitparth F, Schöttle P, Schlichting K, Schell H, Fischbach F, Denecke T, Duda GN, Schröder RJ (2009) Osteochondral defect repair after implantation of biodegradable scaffolds: indirect magnetic resonance arthrography and histopathologic correlation. Acta Radiol 50(7):765–774

89. Wegener B, Schrimpf FM, Pietschmann MF, Milz S, Berger-Lohr M, Bergschmidt P, Jansson V, Müller PE (2009) Matrix-guided cartilage regeneration in chondral defects. Biotechnol Appl Biochem 53(Pt 1):63–70

90. Gille J, Kunow J, Boisch L, Behrens P, Bos I, Hoffmann C, Köller W, Russlies M, Kurz B (2010) Cell-laden and cell-free matrix-induced chondrogenesis versus microfracture for the treatment of articular cartilage defects: a histological and biomechanical study in sheep. Cartilage 1(1):29–42

91. Kon E, Delcogliano M, Filardo G, Fini M, Giavaresi G, Francioli S, Martin I, Pressato D, Arcangeli E, Quarto R, Sandri M, Marcacci M (2010) Orderly osteochondral regeneration in a sheep model using a novel nano-composite multilayered biomaterial. J Orthop Res 28(1):116–124

92. Kon E, Filardo G, Delcogliano M, Fini M, Salamanna F, Giavaresi G, Martin I, Marcacci M (2010) Platelet autologous growth factors decrease the osteochondral regeneration capability of a collagen-hydroxyapatite scaffold in a sheep model. BMC Musculoskelet Disord 11:220

93. Getgood A, Henson F, Skelton C, Herrera E, Brooks R, Fortier LA, Rushton N (2012) The augmentation of a collagen/glycosaminoglycan biphasic osteochondral scaffold with platelet-rich plasma and concentrated bone marrow aspirate for osteochondral defect repair in sheep: a pilot study. Cartilage 3(4):351–363

94. Milano G, Deriu L, Sanna Passino E, Masala G, Manunta A, Postacchini R, Saccomanno MF, Fabbriciani C (2012) Repeated platelet concentrate injections enhance reparative response of microfractures in the treatment of chondral defects of the knee: an experimental study in an animal model. Arthroscopy 28(5):688–701

95. Bell AD, Lascau-Coman V, Sun J, Chen G, Lowerison MW, Hurtig MB, Hoemann CD (2013) Bone-induced chondroinduction in sheep jamshidi biopsy defects with and without treatment by subchondral chitosan-blood implant: 1-day, 3-week, and 3-month repair. Cartilage 4(2):131–143

96. Bernstein A, Niemeyer P, Salzmann G, Südkamp NP, Hube R, Klehm J, Menzel M, von Eisenhart-Rothe R, Bohner M, Görz L, Mayr HO (2013) Microporous calcium phosphate ceramics as tissue engineering scaffolds for the repair of osteochondral defects: histological results. Acta Biomater 9(7):7490–7505

97. Carneiro MO, Barbieri CH, Neto JB (2013) Platelet-rich plasma gel promotes regeneration of articular cartilage in knees of sheep. Acta Ortop Bras 21(2):80–86

98. Fan W, Wu C, Miao X, Liu G, Saifzadeh S, Sugiyama S, Afara I, Crawford R, Xiao Y (2013) Biomaterial scaffolds in cartilage-subchondral bone defects influencing the repair of autologous articular cartilage transplants. J Biomater Appl 27(8):979–989

99. Kunz M, Devlin SM, Hurtig MB, Waldman SD, Rudan JF, Bardana DD, Stewart AJ (2013) Image-guided techniques improve the short-term outcome of autologous osteochondral cartilage repair surgeries: an animal trial. Cartilage 4(2):153–164

100. Martinez-Carranza N, Berg HE, Hultenby K, Nurmi-Sandh H, Ryd L, Lagerstedt AS (2013) Focal knee resurfacing and effects of surgical precision on opposing cartilage. A pilot study on 12 sheep. Osteoarthr Cartil 21(5):739–745

101. Mayr HO, Klehm J, Schwan S, Hube R, Südkamp NP, Niemeyer P, Salzmann G, von Eisenhardt-Rothe R, Heilmann A, Bohner M, Bernstein A (2013) Microporous calcium phosphate ceramics as tissue engineering scaffolds for the repair of osteochondral defects: biomechanical results. Acta Biomater 9(1):4845–4855

102. Schinhan M, Gruber M, Dorotka R, Pilz M, Stelzeneder D, Chiari C, Rössler N, Windhager R, Nehrer S (2013) Matrix-associated autologous chondrocyte transplantation in a compartmentalized early stage of osteoarthritis. Osteoarthr Cartil 21(1):217–225

103. Schleicher I, Lips KS, Sommer U, Schappat I, Martin AP, Szalay G, Hartmann S, Schnettler R (2013) Biphasic scaffolds for repair of deep osteochondral defects in a sheep model. J Surg Res 183(1):184–192

104. Schleicher I, Lips KS, Sommer U, Schappat I, Martin AP, Szalay G, Schnettler R (2013) Allogenous bone with collagen for repair of deep osteochondral defects. J Surg Res 185(2):667–675

105. Caminal M, Moll X, Codina D, Rabanal RM, Morist A, Barrachina J, Garcia F, Pla A, Vives J (2014) Transitory improvement of articular cartilage characteristics after implantation of polylactide:polyglycolic acid (PLGA) scaffolds seeded with autologous mesenchymal stromal cells in a sheep model of critical-sized chondral defect. Biotechnol Lett 36(10):2143–2153

106. Eldracher M, Orth P, Cucchiarini M, Pape D, Madry H (2014) Small subchondral drill holes improve marrow stimulation of articular cartilage defects. Am J Sports Med 42(11):2741–2750

107. Fonseca C, Caminal M, Peris D, Barrachina J, Fàbregas PJ, Garcia F, Cairó JJ, Gòdia F, Pla A, Vives J (2014) An arthroscopic approach for the treatment of osteochondral focal defects with cell-free and cell-loaded PLGA scaffolds in sheep. Cytotechnology 66(2):345–354

108. Guillén-García P, Rodríguez-Iñigo E, Guillén-Vicente I, Caballero-Santos R, Guillén-Vicente M, Abelow S, Giménez-Gallego G, López-Alcorocho JM (2014) Increasing the dose of autologous chondrocytes improves articular cartilage repair: histological and molecular study in the sheep animal model. Cartilage 5(2):114–122

109. Martinez-Carranza N, Berg HE, Lagerstedt AS, Nurmi-Sandh H, Schupbach P, Ryd L (2014) Fixation of a double-coated titanium-hydroxyapatite focal knee resurfacing implant: a 12-month study in sheep. Osteoarthr Cartil 22(6):836–844

110. Pilichi S, Rocca S, Pool RR, Dattena M, Masala G, Mara L, Sanna D, Casu S, Manunta ML, Manunta A, Passino ES (2014) Treatment with embryonic stem-like cells into osteochondral defects in sheep femoral condyles. BMC Vet Res 10:301

111. Power J, Hernandez P, Guehring H, Getgood A, Henson F (2014) Intra-articular injection of rhFGF-18 improves the healing in microfracture treated chondral defects in an ovine model. J Orthop Res 32(5):669–676

112. Sanz-Ramos P, Duart J, Rodríguez-Goñi MV, Vicente-Pascual M, Dotor J, Mora G, Izal-Azcárate I (2014) Improved chondrogenic capacity of collagen hydrogel-expanded chondrocytes: in vitro and in vivo analyses. J Bone Joint Surg Am 96(13):1109–1117

113. Garcia D, Longo UG, Vaquero J, Forriol F, Loppini M, Khan WS, Denaro V (2015) Amniotic membrane transplant for articular cartilage repair: an experimental study in sheep. Curr Stem Cell Res Ther 10(1):77–83

114. Gelse K, Riedel D, Pachowsky M, Hennig FF, Trattnig S, Welsch GH (2015) Limited integrative repair capacity of native cartilage autografts within cartilage defects in a sheep model. J Orthop Res 33(3):390–397

115. Hopper N, Wardale J, Brooks R, Power J, Rushton N, Henson F (2015) Peripheral blood mononuclear cells enhance cartilage repair in in vivo osteochondral defect model. PLoS One 10(8):e0133937

116. Howard D, Wardale J, Guehring H, Henson F (2015) Delivering rhFGF-18 via a bilayer collagen membrane to enhance microfracture treatment of chondral defects in a large animal model. J Orthop Res 33(8):1120–1127

117. Mohan N, Gupta V, Sridharan BP, Mellott AJ, Easley JT, Palmer RH, Galbraith RA, Key VH, Berkland CJ, Detamore MS (2015) Microsphere-based gradient implants for osteochondral regeneration: a long-term study in sheep. Regen Med 10(6):709–728

118. Zorzi AR, Amstalden EM, Plepis AM, Martins VC, Ferretti M, Antonioli E, Duarte AS, Luzo AC, Miranda JB (2015) Effect of human adipose tissue mesenchymal stem cells on the regeneration of ovine articular cartilage. Int J Mol Sci 16(11):26813–26831

119. Caminal M, Peris D, Fonseca C, Barrachina J, Codina D, Rabanal RM, Moll X, Morist A, García F, Cairó JJ, Gòdia F, Pla A, Vives J (2016) Cartilage resurfacing potential of PLGA scaffolds loaded with autologous cells from cartilage, fat, and bone marrow in an ovine model of osteochondral focal defect. Cytotechnology 68(4):907–919

120. de Barros CN, Miluzzi Yamada AL, Junior RS, Barraviera B, Hussni CA, de Souza JB, Watanabe MJ, Rodrigues CA, Garcia Alves AL (2016) A new heterologous fibrin sealant as a scaffold to cartilage repair-experimental study and preliminary results. Exp Biol Med (Maywood) 241(13):1410–1415

121. Hindle P, Baily J, Khan N, Biant LC, Simpson AH, Péault B (2016) Perivascular mesenchymal stem cells in sheep: characterization and autologous transplantation in a model of articular cartilage repair. Stem Cells Dev 25(21):1659–1669

122. Kitamura N, Yokota M, Kurokawa T, Gong JP, Yasuda K (2016) In vivo cartilage regeneration induced by a double-network hydrogel: evaluation of a novel therapeutic strategy for femoral articular cartilage defects in a sheep model. J Biomed Mater Res A 104(9):2159–2165

123. Mrosek EH, Chung HW, Fitzsimmons JS, O'Driscoll SW, Reinholz GG, Schagemann JC (2016) Porous tantalum biocomposites for osteochondral defect repair: a follow-up study in a sheep model. Bone Joint Res 5(9):403–411

124. Schagemann JC, Rudert N, Taylor ME, Sim S, Quenneville E, Garon M, Klinger M, Buschmann MD, Mittelstaedt H (2016) Bilayer implants: electromechanical assessment of regenerated articular cartilage in a sheep model. Cartilage 7(4):346–360

125. Yucekul A, Ozdil D, Kutlu NH, Erdemli E, Aydin HM, Doral MN (2017) Tri-layered composite plug for the repair of osteochondral defects: in vivo study in sheep. J Tissue Eng. https://doi.org/10.1177/2041731417697500

126. Vahedi P, Soleimanirad J, Roshangar L, Shafaei H, Jarolmasjed S, Nozad Charoudeh H (2016) Advantages of sheep infrapatellar fat pad adipose tissue derived stem cells in tissue engineering. Adv Pharm Bull 6(1):105–110

127. Al Faqeh H, Nor Hamdan BM, Chen HC, Aminuddin BS, Ruszymah BH (2012) The potential of intra-articular injection of chondrogenic-induced bone marrow stem cells to retard the progression of osteoarthritis in a sheep model. Exp Gerontol 47(6):458–464

128. Ude CC, Sulaiman SB, Min-Hwei N, Hui-Cheng C, Ahmad J, Yahaya NM, Saim AB, Idrus RB (2014) Cartilage regeneration by chondrogenic induced adult stem cells in osteoarthritic sheep model. PLoS One 9(6):e98770

29. Song F, Tang J, Geng R, Hu H, Zhu C, Cui W, Fan W (2014) Comparison of the efficacy of bone marrow mononuclear cells and bone mesenchymal stem cells in the treatment of osteoarthritis in a sheep model. Int J Clin Exp Pathol 7(4):1415–1426

30. Desando G, Giavaresi G, Cavallo C, Bartolotti I, Sartoni F, Nicoli Aldini N, Martini L, Parrilli A, Mariani E, Fini M, Grigolo B (2016) Autologous bone marrow concentrate in a sheep model of osteoarthritis: new perspectives for cartilage and meniscus repair. Tissue Eng Part C Methods 22(6):608–619

31. Schinhan M, Gruber M, Vavken P, Dorotka R, Samouh L, Chiari C, Gruebl-Barabas R, Nehrer S (2012) Critical-size defect induces unicompartmental osteoarthritis in a stable ovine knee. J Orthop Res 30(2):214–220

32. Beck A, Murphy DJ, Carey-Smith R, Wood DJ, Zheng MH (2016) Treatment of articular cartilage defects with microfracture and autologous matrix-induced chondrogenesis leads to extensive subchondral bone cyst formation in a sheep model. Am J Sports Med 44(10):2629–2643

133. Vikingsson L, Sancho-Tello M, Ruiz-Saurí A, Martínez Díaz S, Gómez-Tejedor JA, Gallego Ferrer G, Carda C, Monllau JC, Gómez Ribelles JL (2015) Implantation of a polycaprolactone scaffold with subchondral bone anchoring ameliorates nodules formation and other tissue alterations. Int J Artif Organs 38(12):659–666

134. Brehm W, Aklin B, Yamashita T, Rieser F, Trüb T, Jakob RP, Mainil-Varlet P (2006) Repair of superficial osteochondral defects with an autologous scaffold-free cartilage construct in a caprine model: implantation method and short-term results. Osteoarthr Cartil 14(12):1214–1226

135. Shahgaldi BF, Amis AA, Heatley FW, McDowell J, Bentley G (1991) Repair of cartilage lesions using biological implants. A comparative histological and biomechanical study in goats. J Bone Joint Surg Br 73(1):57–64

136. Butnariu-Ephrat M, Robinson D, Mendes DG, Halperin N, Nevo Z (1996) Resurfacing of goat articular cartilage by chondrocytes derived from bone marrow. Clin Orthop Relat Res 330:234–243

137. Jackson DW, Halbrecht J, Proctor C, Van Sickle D, Simon TM (1996) Assessment of donor cell and matrix survival in fresh articular cartilage allografts in a goat model. J Orthop Res 14(2):255–264

138. van Susante JL, Buma P, Schuman L, Homminga GN, van den Berg WB, Veth RP (1999) Resurfacing potential of heterologous chondrocytes suspended in fibrin glue in large full-thickness defects of femoral articular cartilage: an experimental study in the goat. Biomaterials 20(13):1167–1175

139. Louwerse RT, Heyligers IC, Klein-Nulend J, Sugihara S, van Kampen GP, Semeins CM, Goei SW, de Koning MH, Wuisman PI, Burger EH (2000) Use of recombinant human osteogenic protein-1 for the repair of subchondral defects in articular cartilage in goats. J Biomed Mater Res 49(4):506–516

140. Niederauer GG, Slivka MA, Leatherbury NC, Korvick DL, Harroff HH, Ehler WC, Dunn CJ, Kieswetter K (2000) Evaluation of multiphase implants for repair of focal osteochondral defects in goats. Biomaterials 21(24):2561–2574

141. Lane JG, Tontz WL Jr, Ball ST, Massie JB, Chen AC, Bae WC, Amiel ME, Sah RL, Amiel D (2001) A morphologic, biochemical, and biomechanical assessment of short-term effects of osteochondral autograft plug transfer in an animal model. Arthroscopy 17(8):856–863

142. Quintavalla J, Uziel-Fusi S, Yin J, Boehnlein E, Pastor G, Blancuzzi V, Singh HN, Kraus KH, O'Byrne E, Pellas TC (2002) Fluorescently labeled mesenchymal stem cells (MSCs) maintain multilineage potential and can be detected following implantation into articular cartilage defects. Biomaterials 23(1):109–119

143. Welch RD, Berry BH, Crawford K, Zhang H, Zobitz M, Bronson D, Krishnan S (2002) Subchondral defects in caprine femora augmented with in situ setting hydroxyapatite cement, polymethylmethacrylate, or autogenous bone graft: biomechanical and histomorphological analysis after two-years. J Orthop Res 20(3):464–472

144. Dell'Accio F, Vanlauwe J, Bellemans J, Neys J, De Bari C, Luyten FP (2003) Expanded phe-
     notypically stable chondrocytes persist in the repair tissue and contribute to cartilage matrix
     formation and structural integration in a goat model of autologous chondrocyte implantation.
     J Orthop Res 21(1):123–131
145. Lane JG, Massie JB, Ball ST, Amiel ME, Chen AC, Bae WC, Sah RL, Amiel D (2004)
     Follow-up of osteochondral plug transfers in a goat model: a 6-month study. Am J Sports
     Med 32(6):1440–1450
146. Vasara AI, Hyttinen MM, Lammi MJ, Lammi PE, Långsjö TK, Lindahl A, Peterson L,
     Kellomäki M, Konttinen YT, Helminen HJ, Kiviranta I (2004) Subchondral bone reaction
     associated with chondral defect and attempted cartilage repair in goats. Calcif Tissue Int
     74(1):107–114
147. Kirker-Head CA, Van Sickle DC, Ek SW, McCool JC (2006) Safety of, and biological and
     functional response to, a novel metallic implant for the management of focal full-thickness
     cartilage defects: preliminary assessment in an animal model out to 1 year. J Orthop Res
     24(5):1095–1108
148. Hunziker EB, Stähli A (2008) Surgical suturing of articular cartilage induces osteoarthritis-
     like changes. Osteoarthr Cartil 16(9):1067–1073
149. Lind M, Larsen A (2008) Equal cartilage repair response between autologous chondrocytes
     in a collagen scaffold and minced cartilage under a collagen scaffold: an in vivo study in
     goats. Connect Tissue Res 49(6):437–442
150. Lind M, Larsen A, Clausen C, Osther K, Everland H (2008) Cartilage repair with chondro-
     cytes in fibrin hydrogel and MPEG polylactide scaffold: an in vivo study in goats. Knee Surg
     Sports Traumatol Arthrosc 16(7):690–698
151. Custers RJ, Saris DB, Dhert WJ, Verbout AJ, van Rijen MH, Mastbergen SC, Lafeber FP,
     Creemers LB (2009) Articular cartilage degeneration following the treatment of focal carti-
     lage defects with ceramic metal implants and compared with microfracture. J Bone Joint Surg
     Am 91(4):900–910
152. Lane JG, Healey RM, Chen AC, Sah RL, Amiel D (2010) Can osteochondral grafting be
     augmented with microfracture in an extended-size lesion of articular cartilage? Am J Sports
     Med 38(7):1316–1323
153. Miot S, Brehm W, Dickinson S, Sims T, Wixmerten A, Longinotti C, Hollander AP, Mainil-
     Varlet P, Martin I (2012) Influence of in vitro maturation of engineered cartilage on the out-
     come of osteochondral repair in a goat model. Eur Cell Mater 23:222–236
154. Bekkers JE, Creemers LB, Tsuchida AI, van Rijen MH, Custers RJ, Dhert WJ, Saris DB
     (2013) One-stage focal cartilage defect treatment with bone marrow mononuclear cells and
     chondrocytes leads to better macroscopic cartilage regeneration compared to microfracture
     in goats. Osteoarthr Cartil 21(7):950–956
155. Kon E, Filardo G, Robinson D, Eisman JA, Levy A, Zaslav K, Shani J, Altschuler N (2014)
     Osteochondral regeneration using a novel aragonite-hyaluronate bi-phasic scaffold in a goat
     model. Knee Surg Sports Traumatol Arthrosc 22(6):1452–1464
156. Pei Y, Fan JJ, Zhang XQ, Zhang ZY, Repairing YM (2014) The osteochondral defect in goat
     with the tissue-engineered osteochondral graft preconstructed in a double-chamber stirring
     bioreactor. Biomed Res Int 2014:21e9203
157. Geraghty S, Kuang JQ, Yoo D, LeRoux-Williams M, Vangsness CT Jr, Danilkovitch A (2015)
     A novel, cryopreserved, viable osteochondral allograft designed to augment marrow stimula-
     tion for articular cartilage repair. J Orthop Surg Res 10:66
158. Kon E, Filardo G, Shani J, Altschuler N, Levy A, Zaslav K, Eisman JE, Robinson D (2015)
     Osteochondral regeneration with a novel aragonite-hyaluronate biphasic scaffold: up to
     12-month follow-up study in a goat model. J Orthop Surg Res 10:81
159. Mumme M, Steinitz A, Nuss KM, Klein K, Feliciano S, Kronen P, Jakob M, von Rechenberg
     B, Martin I, Barbero A, Pelttari K. Regenerative potential of tissue-engineered nasal chondro-
     cytes in goat articular cartilage defects. Tissue Eng Part A 2016;22(21–22):1286–1295

160. Sun J, Hou XK, Zheng YX (2016) Restore a 9 mm diameter osteochondral defect with gene enhanced tissue engineering followed mosaicplasty in a goat model. Acta Orthop Traumatol Turc 50(4):464–469

161. Wang F, Sun Y, He D, Wang L (2017) Effect of concentrated growth factors on the repair of the goat temporomandibular joint. J Oral Maxillofac Surg 75(3):498–507

162. Shahgaldi BF (1998) Repair of large osteochondral defects: load bearing and structural properties of osteochondral repair tissue. Knee 5:111–117

163. Jackson DW, Lalor PA, Aberman HM, Simon TM (2001) Spontaneous repair of full-thickness defects of articular cartilage in a goat model. A preliminary study. J Bone Joint Surg Am 83-A(1):53–64

164. Simon TM, Aberman HM (2010) Cartilage regeneration and repair testing in a surrogate large animal model. Tissue Eng Part B Rev 16(1):65–79

165. Murphy JM, Fink DJ, Hunziker EB, Barry FP (2003) Stem cell therapy in a caprine model of osteoarthritis. Arthritis Rheum 48(12):3464–3474

166. Ko JY, Lee J, Lee J, Im GI (2017) Intra-articular xenotransplantation of adipose-derived stromal cells to treat osteoarthritis in goat model. Tissue Eng Regen Med 14(1):65–71

167. Kangarlu A, Gahunia HK (2006) Magnetic resonance imaging characterization of osteochondral defect repair in a goat model at 8 T. Osteoarthr Cartil 14(1):52–62

168. Watanabe A, Boesch C, Anderson SE, Brehm W, Mainil Varlet P (2009) Ability of dGEMRIC and T2 mapping to evaluate cartilage repair after microfracture: a goat study. Osteoarthr Cartil 17(10):1341–1349

169. Søndergaard LV, Dagnæs-Hansen F, Herskin MS (2011) Welfare assessment in porcine biomedical research – suggestion for an operational tool. Res Vet Sci 91(3):e1–e9

170. Reiland S (1978) Growth and skeletal development of the pig. Acta Radiol Suppl 358:15–22

171. Gotterbarm T, Breusch SJ, Schneider U, Jung M (2008) The minipig model for experimental chondral and osteochondral defect repair in tissue engineering: retrospective analysis of 180 defects. Lab Anim 42(1):71–82

172. Kaab MJ, Gwynn JA, Notzli HP (1998) Collagen fibre arrangement in the tibial plateau articular cartilage of man and other mammalian species. J Anat 193(Pt 1):23–34

173. Pan Y, Li Z, Xie T, Chu CR (2003) Hand-held arthroscopic optical coherence tomography for in vivo high-resolution imaging of articular cartilage. J Biomed Opt 8(4):648–654

174. Zelle S, Zantop T, Schanz S, Petersen W (2007) Arthroscopic techniques for the fixation of a three-dimensional scaffold for autologous chondrocyte transplantation: structural properties in an in vitro model. Arthroscopy 23(10):1073–1078

175. Vasara AI, Hyttinen MM, Pulliainen O, Lammi MJ, Jurvelin JS, Peterson L, Lindahl A, Helminen HJ, Kiviranta I (2006) Immature porcine knee cartilage lesions show good healing with or without autologous chondrocyte transplantation. Osteoarthr Cartil 14(10):1066–1074

176. Hembry RM, Dyce J, Driesang I, Hunziker EB, Fosang AJ, Tyler JA, Murphy G (2001) Immunolocalization of matrix metalloproteinases in partial-thickness defects in pig articular cartilage. A preliminary report. J Bone Joint Surg Am 83-A(6):826–838

177. Chiang H, Kuo TF, Tsai CC, Lin MC, She BR, Huang YY, Lee HS, Shieh CS, Chen MH, Ramshaw JA, Werkmeister JA, Tuan RS, Jiang CC (2005) Repair of porcine articular cartilage defect with autologous chondrocyte transplantation. J Orthop Res 23(3):584–593

178. Shimomura K, Ando W, Tateishi K, Nansai R, Fujie H, Hart DA, Kohda H, Kita K, Kanamoto T, Mae T, Nakata K, Shino K, Yoshikawa H, Nakamura N (2010) The influence of skeletal maturity on allogenic synovial mesenchymal stem cell-based repair of cartilage in a large animal model. Biomaterials 31(31):8004–8011

179. Hunziker EB, Rosenberg LC (1996) Repair of partial-thickness defects in articular cartilage: cell recruitment from the synovial membrane. J Bone Joint Surg Am 78(5):721–733

180. Mainil-Varlet P, Rieser F, Grogan S, Mueller W, Saager C, Jakob RP (2001) Articular cartilage repair using a tissue-engineered cartilage-like implant: an animal study. Osteoarthr Cartil 9(Suppl A):S6–15

181. Gal P, Necas A, Adler J, Teyschl O, Fabian P, Bibrova S (2002) Transplantation of the autogenous chondrocyte graft to physeal defects: an experimental study in pigs. Acta Vet Brno 71(3):327–332

182. Liu Y, Chen F, Liu W, Cui L, Shang Q, Xia W, Wang J, Cui Y, Yang G, Liu D, Wu J, Xu R, Buonocore SD, Cao Y (2002) Repairing large porcine full-thickness defects of articular cartilage using autologous chondrocyte-engineered cartilage. Tissue Eng 8(4):709–721

183. Jung M, Gotterbarm T, Gruettgen A, Vilei SB, Breusch S, Richter W (2005) Molecular characterization of spontaneous and growth-factor-augmented chondrogenesis in periosteum-bone tissue transferred into a joint. Histochem Cell Biol 123(4–5):447–456

184. Chang CH, Kuo TF, Lin CC, Chou CH, Chen KH, Lin FH, Liu HC (2006) Tissue engineering-based cartilage repair with allogenous chondrocytes and gelatin-chondroitin-hyaluronan tri-copolymer scaffold: a porcine model assessed at 18, 24, and 36 weeks. Biomaterials 27(9):1876–1888

185. Harman BD, Weeden SH, Lichota DK, Brindley GW (2006) Osteochondral autograft transplantation in the porcine knee. Am J Sports Med 34(6):913–918

186. Ando W, Tateishi K, Hart DA, Katakai D, Tanaka Y, Nakata K, Hashimoto J, Fujie H, Shino K, Yoshikawa H, Nakamura N (2007) Cartilage repair using an in vitro generated scaffold-free tissue-engineered construct derived from porcine synovial mesenchymal stem cells. Biomaterials 28(36):5462–5470

187. Filová E, Rampichová M, Handl M, Lytvynets A, Halouzka R, Usvald D, Hlucilová J, Procházka R, Dezortová M, Rolencová E, Kostáková E, Trc T, Stastný E, Kolácná L, Hájek M, Motlík J, Amler E (2007) Composite hyaluronate-type I collagen-fibrin scaffold in the therapy of osteochondral defects in miniature pigs. Physiol Res 56(Suppl 1):S5–16

188. Jiang CC, Chiang H, Liao CJ, Lin YJ, Kuo TF, Shieh CS, Huang YY, Tuan RS (2007) Repair of porcine articular cartilage defect with a biphasic osteochondral composite. J Orthop Res 25(10):1277–1290

189. Baumbach K, Petersen JP, Ueblacker P, Schröder J, Göpfert C, Stork A, Rueger JM, Amling M, Meenen NM (2008) The fate of osteochondral grafts after autologous osteochondral transplantation: a one-year follow-up study in a minipig model. Arch Orthop Trauma Surg 128(11):1255–1263

190. Petersen JP, Ueblacker P, Goepfert C, Adamietz P, Baumbach K, Stork A, Rueger JM, Poertner R, Amling M, Meenen NM (2008) Long term results after implantation of tissue engineered cartilage for the treatment of osteochondral lesions in a minipig model. J Mater Sci Mater Med 19(5):2029–2038

191. Blanke M, Carl HD, Klinger P, Swoboda B, Hennig F, Gelse K (2009) Transplanted chondrocytes inhibit endochondral ossification within cartilage repair tissue. Calcif Tissue Int 85(5):421–433

192. Jung M, Kaszap B, Redöhl A, Steck E, Breusch S, Richter W, Gotterbarm T (2009) Enhanced early tissue regeneration after matrix-assisted autologous mesenchymal stem cell transplantation in full thickness chondral defects in a minipig model. Cell Transplant 18(8):923–932

193. Li WJ, Chiang H, Kuo TF, Lee HS, Jiang CC, Tuan RS (2009) Evaluation of articular cartilage repair using biodegradable nanofibrous scaffolds in a swine model: a pilot study. J Tissue Eng Regen Med 3(1):1–10

194. Steck E, Fischer J, Lorenz H, Gotterbarm T, Jung M, Richter W (2009) Mesenchymal stem cell differentiation in an experimental cartilage defect: restriction of hypertrophy to bone-close neocartilage. Stem Cells Dev 18(7):969–978

195. Ho ST, Hutmacher DW, Ekaputra AK, Hitendra D, Hui JH (2010) The evaluation of a biphasic osteochondral implant coupled with an electrospun membrane in a large animal model. Tissue Eng Part A 16(4):1123–1141

196. Schneider U, Schmidt-Rohlfing B, Gavenis K, Maus U, Mueller-Rath R, Andereya S (2011) A comparative study of 3 different cartilage repair techniques. Knee Surg Sports Traumatol Arthrosc 19(12):2145–2152

197. Ebihara G, Sato M, Yamato M, Mitani G, Kutsuna T, Nagai T, Ito S, Ukai T, Kobayashi M, Kokubo M, Okano T, Mochida J (2012) Cartilage repair in transplanted scaffold-free chondrocyte sheets using a minipig model. Biomaterials 33(15):3846–3851

198. Nakamura T, Sekiya I, Muneta T, Hatsushika D, Horie M, Tsuji K, Kawarasaki T, Watanabe A, Hishikawa S, Fujimoto Y, Tanaka H, Kobayashi E (2012) Arthroscopic, histological and MRI analyses of cartilage repair after a minimally invasive method of transplantation of allogeneic synovial mesenchymal stromal cells into cartilage defects in pigs. Cytotherapy 14(3):327–338

199. Betsch M, Schneppendahl J, Thuns S, Herten M, Sager M, Jungbluth P, Hakimi M, Wild M (2013) Bone marrow aspiration concentrate and platelet rich plasma for osteochondral repair in a porcine osteochondral defect model. PLoS One 8(8):e71602

200. Filová E, Rampichová M, Litvinec A, Držík M, Míčková A, Buzgo M, Košťáková E, Martinová L, Usvald D, Prosecká E, Uhlík J, Motlík J, Vajner L, Amler E (2013) A cell-free nanofiber composite scaffold regenerated osteochondral defects in miniature pigs. Int J Pharm 447(1–2):139–149

201. Moriguchi Y, Tateishi K, Ando W, Shimomura K, Yonetani Y, Tanaka Y, Kita K, Hart DA, Gobbi A, Shino K, Yoshikawa H, Nakamura N (2013) Repair of meniscal lesions using a scaffold-free tissue-engineered construct derived from allogenic synovial MSCs in a miniature swine model. Biomaterials 34(9):2185–2193

202. Jagodzinski M, Liu C, Guenther D, Burssens A, Petri M, Abedian R, Willbold E, Krettek C, Haasper C, Witte F (2014) Bone marrow-derived cell concentrates have limited effects on osteochondral reconstructions in the mini pig. Tissue Eng Part C Methods 20(3):215–226

203. Wang X, Li Y, Han R, He C, Wang G, Wang J, Zheng J, Pei M, Wei L (2014) Demineralized bone matrix combined bone marrow mesenchymal stem cells, bone morphogenetic protein-2 and transforming growth factor-β3 gene promoted pig cartilage defect repair. PLoS One 9(12):e116061

204. Ha CW, Park YB, Chung JY, Park YG (2015) Cartilage repair using composites of human umbilical cord blood-derived mesenchymal stem cells and hyaluronic acid hydrogel in a minipig model. Stem Cells Transl Med 4(9):1044–1051

205. Matsuo T, Kita K, Mae T, Yonetani Y, Miyamoto S, Yoshikawa H, Nakata K (2015) Bone substitutes and implantation depths for subchondral bone repair in osteochondral defects of porcine knee joints. Knee Surg Sports Traumatol Arthrosc 23(5):1401–1409

206. Peck Y, He P, Chilla GS, Poh CL, Wang DA (2015) A preclinical evaluation of an autologous living hyaline-like cartilaginous graft for articular cartilage repair: a pilot study. Sci Rep 5:16225

207. Sosio C, Di Giancamillo A, Deponti D, Gervaso F, Scalera F, Melato M, Campagnol M, Boschetti F, Nonis A, Domeneghini C, Sannino A, Peretti GM (2015) Osteochondral repair by a novel interconnecting collagen-hydroxyapatite substitute: a large-animal study. Tissue Eng Part A 21(3–4):704–715

208. Ding J, Chen B, Lv T, Liu X, Fu X, Wang Q, Yan L, Kang N, Cao Y, Xiao R (2016) Bone marrow mesenchymal stem cell-based engineered cartilage ameliorates polyglycolic acid/polylactic acid scaffold-induced inflammation through M2 polarization of macrophages in a pig model. Stem Cells Transl Med 5(8):1079–1089

209. Fisher MB, Belkin NS, Milby AH, Henning EA, Söegaard N, Kim M, Pfeifer C, Saxena V, Dodge GR, Burdick JA, Schaer TP, Steinberg DR, Mauck RL (2016) Effects of mesenchymal stem cell and growth factor delivery on cartilage repair in a mini-pig model. Cartilage 7(2):174–184

210. Muhonen V, Salonius E, Haaparanta AM, Järvinen E, Paatela T, Meller A, Hannula M, Björkman M, Pyhältö T, Ellä V, Vasara A, Töyräs J, Kellomäki M, Kiviranta I (2016) Articular cartilage repair with recombinant human type II collagen/polylactide scaffold in a preliminary porcine study. J Orthop Res 34(5):745–753

211. Zuo Q, Cui W, Liu F, Wang Q, Chen Z, Fan W (2016) Utilizing tissue-engineered cartilage or BMNC-PLGA composites to fill empty spaces during autologous osteochondral mosaicplasty in porcine knees. J Tissue Eng Regen Med 10(11):916–926

212. Lin H, Zhou J, Cao L, Wang HR, Dong J, Chen ZR (2017) Tissue-engineered cartilage constructed by a biotin-conjugated anti-CD44 avidin binding technique for the repairing of cartilage defects in the weight-bearing area of knee joints in pigs. Bone Joint Res 6(5):284–295

213. Duda GN, Maldonado ZM, Klein P, Heller MO, Burns J, Bail H (2005) On the influence of mechanical conditions in osteochondral defect healing. J Biomech 38(4):843–851

214. Shortkroff S, Barone L, Hsu HP, Wrenn C, Gagne T, Chi T, Breinan H, Minas T, Sledge CB, Tubo R, Spector M (1996) Healing of chondral and osteochondral defects in a canine model: the role of cultured chondrocytes in regeneration of articular cartilage. Biomaterials 17(2):147–154

215. Vainio O (2012) Translational animal models using veterinary patients – an example of canine osteoarthritis (OA). Scand J Pain 3(2):84–89

216. Campbell CJ, Ishida H, Takahashi H, Kelly F (1963) The transplantation of articular cartilage. An experimental study in dogs. J Bone Joint Surg Am 45A:1579–1592

217. Engkvist O (1979) Reconstruction of patellar articular cartilage with free autologous perichondrial grafts. An experimental study in dogs. Scand J Plast Reconstr Surg 13(3):361–369

218. Stevenson S, Dannucci GA, Sharkey NA, Pool RR (1989) The fate of articular cartilage after transplantation of fresh and cryopreserved tissue-antigen-matched and mismatched osteochondral allografts in dogs. J Bone Joint Surg Am 71A(9):1297–1307

219. Klompmaker J, Jansen HW, Veth RP, Nielsen HK, de Groot JH, Pennings AJ (1992) Porous polymer implants for repair of full-thickness defects of articular cartilage: an experimental study in rabbit and dog. Biomaterials 13(9):625–634

220. Oates KM, Chen AC, Young EP, Kwan MK, Amiel D, Convery FR (1995) Effect of tissue culture storage on the in vivo survival of canine osteochondral allografts. J Orthop Res 13(4):562–569

221. Breinan HA, Minas T, Hsu HP, Nehrer S, Sledge CB, Spector M (1997) Effect of cultured autologous chondrocytes on repair of chondral defects in a canine model. J Bone Joint Surg Am 79(10):1439–1451

222. Nehrer S, Breinan HA, Ramappa A, Hsu HP, Minas T, Shortkroff S, Sledge CB, Yannas IV, Spector M (1998) Chondrocyte-seeded collagen matrices implanted in a chondral defect in a canine model. Biomaterials 19(24):2313–2328

223. van Dyk GE, Dejardin LM, Flo G, Johnson LL (1998) Cancellous bone grafting of large osteochondral defects: an experimental study in dogs. Arthroscopy 14(3):311–320

224. Breinan HA, Martin SD, Hsu HP, Spector M (2000) Healing of canine articular cartilage defects treated with microfracture, a type-II collagen matrix, or cultured autologous chondrocytes. J Orthop Res 18(5):781–789

225. Breinan HA, Minas T, Hsu HP, Nehrer S, Shortkroff S, Spector M (2001) Autologous chondrocyte implantation in a canine model: change in composition of reparative tissue with time. J Orthop Res 19(3):482–492

226. Cook SD, Patron LP, Salkeld SL, Rueger DC (2003) Repair of articular cartilage defects with osteogenic protein-1 (BMP-7) in dogs. J Bone Joint Surg Am 85-A(Suppl 3):116–123

227. Feczkó P, Hangody L, Varga J, Bartha L, Diószegi Z, Bodó G, Kendik Z, Módis L (2003) Experimental results of donor site filling for autologous osteochondral mosaicplasty. Arthroscopy 19(7):755–761

228. Lee CR, Grodzinsky AJ, Hsu HP, Spector M (2003) Effects of a cultured autologous chondrocyte-seeded type II collagen scaffold on the healing of a chondral defect in a canine model. J Orthop Res 21(2):272–281

229. Chen G, Sato T, Tanaka J, Tateishi T (2006) Preparation of a biphasic scaffold for osteochondral tissue engineering. Mater Sci Eng 26(1):118–123

230. Glenn RE Jr, McCarty EC, Potter HG, Juliao SF, Gordon JD, Spindler KP (2006) Comparison of fresh osteochondral autografts and allografts: a canine model. Am J Sports Med 34(7):1084–1093

231. Yamazoe K, Mishima H, Torigoe K, Iijima H, Watanabe K, Sakai H, Kudo T (2007) Effects of atelocollagen gel containing bone marrow-derived stromal cells on repair of osteochondral defect in a dog. J Vet Med Sci 69(8):835–839

232. Choi HJ, Kwon E, Lee JI (2009) Safety and efficacy assessment of mesenchymal stem cells from canine adipose tissue or umbilical cord blood in a canine osteochondral defect model. Tissue Eng. Regen Med 6(14):1381–1390

233. Sagliyan A, Karabulut E, Unsaldi E, Yaman Y (2009) Evaluation of the activity of intraarticular hyaluronic acid in the repair of experimentally induced osteochondral defects of the stifle joint in dogs. Vet Med-Czech 54(1):33–40

234. Ng KW, Lima EG, Bian L, O'Conor CJ, Jayabalan PS, Stoker AM, Kuroki K, Cook CR, Ateshian GA, Cook JL, Hung CT (2010) Passaged adult chondrocytes can form engineered cartilage with functional mechanical properties: a canine model. Tissue Eng Part A 16(3):1041–1051

235. Mokbel A, El-Tookhy O, Shamaa AA, Sabry D, Rashed L, Mostafa A (2011) Homing and efficacy of intra-articular injection of autologous mesenchymal stem cells in experimental chondral defects in dogs. Clin Exp Rheumatol 29(2):275–284

236. Sun S, Ren Q, Wang D, Zhang L, Wu S, Sun XT (2011) Repairing cartilage defects using chondrocyte and osteoblast composites developed using a bioreactor. Chin Med J 124(5):758–763

237. Igarashi T, Iwasaki N, Kawamura D, Kasahara Y, Tsukuda Y, Ohzawa N, Ito M, Izumisawa Y, Minami A (2012) Repair of articular cartilage defects with a novel injectable in situ forming material in a canine model. J Biomed Mater Res A 100(1):180–187

238. Zhang Y, Wang J, Wang P, Fan X, Li X, Fu J, Li S, Fan H, Guo Z (2013) Low elastic modulus contributes to the osteointegration of titanium alloy plug. J Biomed Mater Res B Appl Biomater 101((4):584–590

239. Kazemi D, Fakhrjou A, Dizaji VM, Alishahi MK (2014) Effect of autologous platelet rich fibrin on the healing of experimental articular cartilage defects of the knee in an animal model. Biomed Res Int 2014:486436

240. Lv YM, Yu QS (2015) Repair of articular osteochondral defects of the knee joint using a composite lamellar scaffold. Bone Joint Res 4(4):56–64

241. Cook JL, Stannard JP, Stoker AM, Bozynski CC, Kuroki K, Cook CR, Pfeiffer FM (2016) Importance of donor chondrocyte viability for osteochondral allografts. Am J Sports Med 44(5):1260–1268

242. McCarty EC, Fader RR, Mitchell JJ, Glenn RE Jr, Potter HG, Spindler KP (2016) Fresh osteochondral allograft versus autograft: twelve-month results in isolated canine knee defects. Am J Sports Med 44(9):2354–2365

243. Kazemi D, Shams Asenjan K, Dehdilani N, Parsa H (2017) Canine articular cartilage regeneration using mesenchymal stem cells seeded on platelet rich fibrin: macroscopic and histological assessments. Bone Joint Res 6(2):98–107

244. Calandruccio RA, Gilmer WS (1962) Proliferation regeneration, and repair of articular cartilage of immature animals. J Bone Joint Surg 44A:431–455

245. Hale JE, Rudert MJ, Brown TD (1993) Indentation assessment of biphasic mechanical property deficits in size-dependent osteochondral defect repair. J Biomech 26(11):1319–1325

246. Convery FR, Akeson WH, Keown GH (1972) The repair of large osteochondral defects. An experimental study in horses. Clin Orthop Relat Res 82:253–262

247. Koch TG, Betts DH (2007) Stem cell therapy for joint problems using the horse as a clinically relevant animal model. Expert Opin Biol Ther 7(11):1621–1626

248. Dzierzecka M, Wasowski A, Kobryn H (2005) Time-span of the ossification of the distal epiphysis in thoroughbred horses as a criterion of skeleton maturity. Med Weter 61(10):1190–1192

249. Pasolini MP, Meomartino L, Testa A, Fatone G, Potena A, Di Rosa G, Lamagna E (2007) Radiographic assessment of skeletal maturity in the racehorse: statistical validation and correlation with orthopaedic injuries in the standardbred. Ippologia 18(3):15–19

250. Pasolini MP, Santoro P, Lamagna F, Greco M, Labriola S, Orefice R, Meomartino L (2007) Use of two biochemical markers of bone metabolism, bone alkaline phosphatase (B-ALP) and cross linked c-telopeptide of type I collagen (ICTP), in the assessment of skeletal maturity in the standardbred horse. Ippologia 18(3):11–14

251. Luszczynski J, Pieszka M, Kosiniak-Kamysz K (2011) Effect of horse breed and sex in growth rate and radiographic closure time of distal radial metaphyseal growth plate. Livest Sci 141(2–3):252–258

252. Shamis LD, Bramlage LR, Gabel AA, Weisbrode S (1989) Effect of subchondral drilling on repair of partial-thickness cartilage defects of third carpal bones in horses. Am J Vet Res 50(2):290–295

253. Vachon AM, McIlwraith CW, Keeley FW (1991) Biochemical study of repair of induced osteochondral defects of the distal portion of the radial carpal bone in horses by use of periosteal autografts. Am J Vet Res 52(2):328–332

254. Vachon AM, McIlwraith CW, Powers BE, McFadden PR, Amiel D (1992) Morphologic and biochemical study of sternal cartilage autografts for resurfacing induced osteochondral defects in horses. Am J Vet Res 53(6):1038–1047

255. Hendrickson DA, Nixon AJ, Grande DA, Todhunter RJ, Minor RM, Erb H, Lust G (1994) Chondrocyte-fibrin matrix transplants for resurfacing extensive articular cartilage defects. J Orthop Res 12(4):485–497

256. Howard RD, McIlwraith CW, Trotter GW, Powers BE, McFadden PR, Harwood FL, Amiel D (1994) Long-term fate and effects of exercise on sternal cartilage autografts used for repair of large osteochondral defects in horses. Am J Vet Res 55(8):1158–1167

257. Sams AE, Nixon AJ (1995) Chondrocyte-laden collagen scaffolds for resurfacing extensive articular cartilage defects. Osteoarthr Cartil 3(1):47–59

258. Frisbie DD, Trotter GW, Powers BE, Rodkey WG, Steadman JR, Howard RD, Park RD, McIlwraith CW (1999) Arthroscopic subchondral bone plate microfracture technique augments healing of large chondral defects in the radial carpal bone and medial femoral condyle of horses. Vet Surg 28(4):242–255

259. Nixon AJ, Fortier LA, Williams J, Mohammed H (1999) Enhanced repair of extensive articular defects by insulin-like growth factor-I-laden fibrin composites. J Orthop Res 17(4):475–487

260. Fortier LA, Mohammed HO, Lust G, Nixon AJ (2002) Insulin-like growth factor-I enhances cell-based repair of articular cartilage. J Bone Joint Surg Br 84(2):276–288

261. Hidaka C, Goodrich LR, Chen CT, Warren RF, Crystal RG, Nixon AJ (2003) Acceleration of cartilage repair by genetically modified chondrocytes over expressing bone morphogenetic protein-7. J Orthop Res 21(4):573–583

262. Litzke LF, Wagner E, Baumgaertner W, Hetzel U, Josimovic-Alasevic O, Libera J (2004) Repair of extensive articular cartilage defects in horses by autologous chondrocyte transplantation. Ann Biomed Eng 32(1):57–69

263. Strauss EJ, Goodrich LR, Chen CT, Hidaka C, Nixon AJ (2005) Biochemical and biomechanical properties of lesion and adjacent articular cartilage after chondral defect repair in an equine model. Am J Sports Med 33(11):1647–1653

264. Barnewitz D, Endres M, Krüger I, Becker A, Zimmermann J, Wilke I, Ringe J, Sittinger M, Kaps C (2006) Treatment of articular cartilage defects in horses with polymer-based cartilage tissue engineering grafts. Biomaterials 27(14):2882–2889

265. Gratz KR, Wong VW, Chen AC, Fortier LA, Nixon AJ, Sah RL (2006) Biomechanical assessment of tissue retrieved after in vivo cartilage defect repair: tensile modulus of repair tissue and integration with host cartilage. J Biomech 39(1):138–146

266. Goodrich LR, Hidaka C, Robbins PD, Evans CH, Nixon AJ (2007) Genetic modification of chondrocytes with insulin-like growth factor-1 enhances cartilage healing in an equine model. J Bone Joint Surg Br 89(5):672–685

267. Morisset S, Frisbie DD, Robbins PD, Nixon AJ, McIlwraith CW (2007) IL-1ra/IGF-1 gene therapy modulates repair of microfractured chondral defects. Clin Orthop Rel Res 462:221–228

268. Wilke MM, Nydam DV, Nixon AJ (2007) Enhanced early chondrogenesis in articular defects following arthroscopic mesenchymal stem cell implantation in an equine model. J Orthop Res 25(7):913–925

269. Frisbie DD, Bowman SM, Colhoun HA, DiCarlo EF, Kawcak CE, McIlwraith CW (2008) Evaluation of autologous chondrocyte transplantation via a collagen membrane in equine articular defects: results at 12 and 18 months. Osteoarthr Cartil 16(6):667–679

270. Frisbie DD, Lu Y, Kawcak CE, DiCarlo EF, Binette F, McIlwraith CW (2009) In vivo evaluation of autologous cartilage fragment-loaded scaffolds implanted into equine articular defects and compared with autologous chondrocyte implantation. Am J Sports Med 37(Suppl 1):S71–S80

271. Fortier LA, Potter HG, Rickey EJ, Schnabel LV, Foo LF, Chong LR, Stokol T, Cheetham J, Nixon AJ (2010) Concentrated bone marrow aspirate improves full-thickness cartilage repair compared with microfracture in the equine model. J Bone Joint Surg Am 92(10):1927–1937

272. Kon E, Mutini A, Arcangeli E, Delcogliano M, Filardo G, Nicoli Aldini N, Pressato D, Quarto R, Zaffagnini S, Marcacci M (2010) Novel nanostructured scaffold for osteochondral regeneration: pilot study in horses. J Tissue Eng Regen Med 4(4):300–308

273. McIlwraith CW, Frisbie DD, Rodkey WG, Kisiday JD, Werpy NM, Kawcak CE, Steadman JR (2011) Evaluation of intra-articular mesenchymal stem cells to augment healing of micro-fractured chondral defects. Arthroscopy 27(11):1552–1561

274. Nixon AJ, Begum L, Mohammed HO, Huibregtse B, O'Callaghan MM, Matthews GL (2011) Autologous chondrocyte implantation drives early chondrogenesis and organized repair in extensive full- and partial-thickness cartilage defects in an equine model. J Orthop Res 29(7):1121–1130

275. Seo JP, Tanabe T, Tsuzuki N, Haneda S, Yamada K, Furuoka H, Tabata Y, Sasaki N (2013) Effects of bilayer gelatin/β-tricalcium phosphate sponges loaded with mesenchymal stem cells, chondrocytes, bone morphogenetic protein-2, and platelet rich plasma on osteochondral defects of the talus in horses. Res Vet Sci 95(3):1210–1216

276. Miller RE, Grodzinsky AJ, Barrett MF, Hung HH, Frank EH, Werpy NM, McIlwraith CW, Frisbie DD (2014) Effects of the combination of microfracture and self-assembling peptide filling on the repair of a clinically relevant trochlear defect in an equine model. J Bone Joint Surg Am 96(19):1601–1609

277. Tsuzuki N, Oshita N, Seo JP, Yamada K, Haneda S, Furuoka H, Tabata Y, Sasaki N (2014) Effect of platelet-rich plasma-incorporated gelatin hydrogel microspheres and subchondral drilling on equine cartilage defects. J Equine Vet Sci 34(6):820–824

278. Frisbie DD, McCarthy HE, Archer CW, Barrett MF, McIlwraith CW (2015) Evaluation of articular cartilage progenitor cells for the repair of articular defects in an equine model. J Bone Joint Surg Am 97A(6):484–493

279. Nixon AJ, Rickey E, Butler TJ, Scimeca MS, Moran N, Matthews GL (2015) A chondrocyte infiltrated collagen type I/III membrane (MACI® implant) improves cartilage healing in the equine patellofemoral joint model. Osteoarthr Cartil 23(4):648–660

280. Ortved KF, Begum L, Mohammed HO, Nixon AJ (2015) Implantation of rAAV5-IGF-I transduced autologous chondrocytes improves cartilage repair in full-thickness defects in the equine model. Mol Ther 23(2):363–373

281. Seo JP, Kambayashi Y, Itho M, Haneda S, Yamada K, Furuoka H, Tabata Y, Sasaki N (2015) Effects of a synovial flap and gelatin/β-tricalcium phosphate sponges loaded with mesenchymal stem cells, bone morphogenetic protein-2, and platelet rich plasma on equine osteochondral defects. Res Vet Sci 101:140–143

282. Fortier LA, Chapman HS, Pownder SL, Roller BL, Cross JA, Cook JL, Cole BJ (2016) Biocartilage improves cartilage repair compared with microfracture alone in an equine model of full-thickness cartilage loss. Am J Sports Med 44(9):2366–2374

283. Rocha Júnior SS, Mendes HMF, Beier SL, Paz CFR, Azevedo DSD, Lacerda IGO, Correa MG, Faleiros RR (2016) Macroscopic and histological evaluations of equine joint cartilage repair treated with microperforation of the subchondral bone associated or not with intra-articular kartogenin. Pesq Vet Bras 36(4):272–278

284. Yamada ALM, Alvarenga ML, Brandão JS, Watanabe MJ, Rodrigues CA, Hussni CA, Alves ALG (2016) PRP gel scaffold associated with mesenchymal stem cells. Use in experimental chondral defect of equine models. Pesq Vet Bras 36(6):461–467

285. McIlwraith CW, Fortier LA, Frisbie DD, Nixon AJ (2011) Equine models of articular cartilage repair. Cartilage 2(4):317–326

286. Cokelaere S, Malda J, van Weeren R (2016) Cartilage defect repair in horses: current strategies and recent developments in regenerative medicine of the equine joint with emphasis on the surgical approach. Vet J 214:61–71

287. Johnson SA, Frisbie DD (2016) Cartilage therapy and repair in equine athletes. Oper Tech Orthop 26(3):155–165
288. Ortved KF, Nixon AJ (2016) Cell-based cartilage repair strategies in the horse. Vet J 208:1–12
289. Vidal MA, Robinson SO, Lopez MJ, Paulsen DB, Borkhsenious O, Johnson JR, Moore RM, Gimble JM (2008) Comparison of chondrogenic potential in equine mesenchymal stromal cells derived from adipose tissue and bone marrow. Vet Surg 37(8):713–724
290. Henson FMD, Getgood AMJ, Caborn DM, McIlwraith CW, Rushton N (2012) Effect of a solution of hyaluronic acid-chondroitin sulfate-H-acetyl glucosamine on the repair response of cartilage to single impact load damage. Am J Vet Res 73(2):306–312
291. Shoemaker RS, Bertone AL, Martin GS, McIlwraith CW, Roberts ED, Pechman R, Kearney MT (1992) Effects of intra-articular administration of methylprednisolone acetate on normal articular cartilage and on healing of experimentally induced osteochondral defects in horses. Am J Vet Res 53(8):1446–1453
292. Todhunter RJ, Minor RR, Wootton JA, Krook L, Burton-Wurster N, Lust G (1993) Effects of exercise and polysulfated glycosaminoglycan on repair of articular cartilage defects in the equine carpus. J Orthop Res 11(6):782–795
293. Bertone AL, Bramlage LR, McIlwraith CW, Malemud CJ (2005) Comparison of proteoglycan and collagen in articular cartilage of horses with naturally developing osteochondrosis and healing osteochondral fragments of experimentally induced fractures. Am J Vet Res 66(11):1881–1890
294. Directive 2010/63/EU of the European Parliament and of the Council of 22 September 2010 on the Protection of Animals Used for Scientific Purposes. Official Journal of the European Union. http://www.3rs-reduction.co.uk/AR_Dir_2010_63_EU__1_.pdf
295. Rehbinder C, Baneux P, Forbes D, van Herck H, Nicklas W, Rugaya Z, Winkler G (1998) FELASA recommendations for the health monitoring of breeding colonies and experimental units of cats, dogs and pigs. Report of the Federation of European Laboratory Animal Science Associations (FELASA) working group on animal health. Lab Anim 32(1):1–17
296. Rehbinder C, Alenius S, Bures J, de las Heras ML, Greko C, Kroon PS, Gutzwiller A (2000) FELASA recommendations for the health monitoring of experimental units of calves, sheep and goats report of the federation of European Laboratory Animal Science Associations (FELASA) Working Group on Animal Health. Lab Anim 34(4):329–350
297. Mähler Convenor M, Berard M, Feinstein R, Gallagher A, Illgen-Wilcke B, Pritchett-Corning K, Raspa M (2014) FELASA recommendations for the health monitoring of mouse, rat, hamster, Guinea pig and rabbit colonies in breeding and experimental units. Lab Anim 48(3):178–192
298. Balls M (1994) Replacement of animal procedures: alternatives in research, education and testing. Lab Anim 28(3):193–211
299. Auer JA, Goodship A, Arnoczky S, Pearce S, Price J, Claes L, von Rechenberg B, Hofmann-Amtenbrinck M, Schneider E, Müller-Terpitz R, Thiele F, Rippe KP, Grainger DW (2007) Refining animal models in fracture research: seeking consensus in optimising both animal welfare and scientific validity for appropriate biomedical use. BMC Musculoskelet Disord 8:72
300. van Gaalen SM, Kruyt MC, Geuze RE, de Bruijn JD, Alblas J, Dhert WJ (2010) Use of fluorochrome labels in in vivo bone tissue engineering research. Tissue Eng Part B Rev 16(2):209–217
301. Franco NH, Olsson IA (2014) Scientists and the 3Rs: attitudes to animal use in biomedical research and the effect of mandatory training in laboratory animal science. Lab Anim 48(1):50–60
302. Hurtig M1, Chubinskaya S, Dickey J, Rueger D (2009) BMP-7 protects against progression of cartilage degeneration after impact injury. J Orthop Res 27(5):602–11
303. Ishimaru J, Goss AN (1992) A model for osteoarthritis of the temporomandibular joint. J Oral Maxillofac Surg 50(11):1191–5

Index

A

Adipose tissue-derived stem cells (ASCs), 223, 224, 445, 456
Agarose, 196
Alginate (Alg), 194
  chemically cross-linked, 162, 164
  Guardix SG™, 161
  hydrogels, 361
  IPNs and Semi-IPNs, 161–162
  linear polysaccharide, 160
  photopolymerization, 164
  physically cross-linked, 162
  physicochemical properties, 161
  pNIPAAm hydrogels, 161
American Academy of Orthopaedic Surgeons (AAOS), 122
American Association of Tissue Bank, 9
AMIC™, 36
Amnion-derived cells, 428
Amniox®, 47
Anatomic open repair, 91
Anatomic repair techniques, 92
Angiogenesis, 244
  gradient scaffold, 324
  growth factors, 320
  modulation, 317
  neovascularization, 319
  osteochondral environment, 317–321
  osteochondral regeneration, 321
  osteochondral tissue, 315
Anhydro-α-L-galactopyranose, 196
Ankle impingement syndromes, 99–101
Ankle lateral instability
  ATFL and CFL, 89
  medial lesion, 94

microtraumatisms, 90
surgical treatment, 91–92
Ankle osteochondral defects
  etiology, 38
  shearing forces, 39
  treatment, 40
Ankle sprain, 39, 90–94
Ankle trauma, 39
Anterior cruciate ligament (ACL), 116, 341, 445
Anterior drawer test, 90
Anterior talofibular ligament (ATFL), 89
Anterior tibial tendon (ATT), 145
Anti-angiogenic drugs, 318–319
Apoptosis, 258
Arthroscopic lavage, 9
Arthroscopy, 8, 93
Articular cartilage, 6, 66, 116, 210, 215, 303
Articular chondrocyte transplantation (ACI), 220
Articular fracture, 69
Autologous chondrocytes, 11
Autologous osteochondral grafts, 354
Autologous osteochondral transplantation, 33, 48

B

Behaviour of bone marrow mesenchymal stem cells (BMMSCs), 262
Beta-glucan compounds, 195
Bevacizumab, 318
Bilayered scaffolds, 15, 383
Bio/synthetic interpenetrating network (BioSINx), 178

© Springer International Publishing AG, part of Springer Nature 2018
J. M. Oliveira et al. (eds.), *Osteochondral Tissue Engineering*,
Advances in Experimental Medicine and Biology 1059,
https://doi.org/10.1007/978-3-319-76735-2

Printed by Printforce, the Netherlands